FUNDAMENTALS OF WEARABLE COMPUTERS AND AUGMENTED REALITY

FUNDAMENTALS OF WEARABLE COMPUTERS AND AUGMENTED REALITY

Edited by

Woodrow Barfield
Virginia Polytechnic Institute

Thomas Caudell
University of New Mexico

IEA LAWRENCE ERLBAUM ASSOCIATES, PUBLISHERS
2001 Mahwah, New Jersey   London

The book cover depicts a mid-1980s embodiment of Dr. Steve Mann's "Wearable Computer" (WearComp) invention he created for his early experiments in computer mediated reality. The apparatus was used to generate a series of cybernetic photographs, using a collaborative process called "dusting", exhibited in his solo show at Night Gallery (185 Richmond St., Toronto) during the summer of 1985. Cybernetic photography was an early application of collaborative mediated reality spaces, as described in Chapter 19 of this book.

Lawrence Erlbaum Associates, Inc., Publishers
10 Industrial Avenue
Mahwah, NJ 07430

Cover design by Kathryn Houghtaling-Lacey

Library of Congress Cataloging-in-Publication Data

    Fundamentals of wearable computers and augmented reality /
    editors, Woodrow Barfield, Thomas Caudell.
        p. cm.
    Includes bibliographical references and index.
    ISBN 0-8058-2901-6 (cloth : alk. paper)
    ISBN 0-8058-2902-4 (pbk. : alk. paper)
    1. Wearable computers. 2. Human-computer interaction.
    I. Barfield, Woodrow.   II. Caudell, Thomas.
    QA76.592 .F86 2000
    004.16—dc21   00-039311

Books published by Lawrence Erlbaum Associates are printed on acid-free paper, and their bindings are chosen for strength and durability.

Printed in the United States of America
10 9 8 7 6 5 4 3 2 1

This book is dedicated to Jessica YiQi and to my parents Mary and Woodrow Barfield.

—Woodrow Barfield Jr.

I dedicate this work with all my love to my brown-eyed girl, Ann.

—Thomas Caudell

Contents

Preface

The power of the machine imposes itself upon us and we can scarcely conceive of living bodies anymore without it; we are strangely moved by the rapid friction of beings and things and we accustom ourselves, without knowing it, to perceive the forces of the former in terms of the forces dominating the latter.

—circa 1913, Raymond Duchamps-Villon 1876–1918

The concept of blending humans with machines has been in the dreams and nightmares of people since long before the industrial revolution, finding its way into many fables and stories over the centuries. In recent times, this topic weaves in and out of the entire science fiction and fantasy genres, culminating in the quintessential human/machine merger, the Borg, in the Star Trek science fiction series. As futuristic as this seems, the blending of humans with machines is now becoming fact through advances in computer, communications, and human–computer interface technologies. This book introduces the reader to the basic concepts, challenges, and the underlying technologies that are making this happen, and through discussion of applications, answers the question "Why do this?"

This book presents a broad coverage of the technologies and interface design issues associated with wearable computers and augmented reality displays, both rapidly developing fields in computer science, engineering, and human interface design. As general descriptions, wearable computers are fully functional, self-powered, self-contained computers that are worn on the body to provide access to and interaction with information anywhere and at anytime. Closely related is the topic of "augmented reality," an advanced human–computer interface technology that attempts to blend or fuse computer-generated information with our sensations of the natural world.

Because of the close association between wearable computers and augmented reality, we refer to both types of computing technology generically as wearware. Throughout this book the various chapter authors describe the integration of head-mounted displays, digital technology, auditory displays, and body tracking technologies to create wearable computer and augmented reality systems. This technology has the potential of improving the efficiency and quality of human labors, particularly in their performance of engineering, manufacturing, construction, diagnostic, maintenance, monitoring, health care delivery, and transactional activities.

One of the main applications for the technologies described in this book is in the fields of medicine and health care. Using immersive head-mounted displays, physicians are able to examine and interact with virtual representations of patients for the purposes of enhanced training for general medical education, and to enhance specialized surgical, anesthesia, or other crisis and procedural training skills. In addition, augmented reality technologies are enabling researchers and physicians to project medical information directly on or into the patient as a supporting informational aid to enhance case-specific decision making. One clear example is the ability to visualize a tumor's location relative to the patient's surrounding anatomy for better surgical outcomes. In the area of mobile computing, wearable computer technology is just beginning to be used in medical settings, allowing physicians the capability to access medical information whenever and wherever they are. On the patient side of wearable computer technology, nascent wearable computers are already being utilized by some of the diabetic population in the form of Insulin Pump Therapy, with other wearable computer technologies on the immediate horizon to treat a host of other conditions.

Much of the current research directions on the topics of wearable computers and augmented reality can be traced to Ivan Sutherland's seminal dissertation work at MIT, and to more recent ideas associated with the topic of ubiquitous computing. Sutherland, often thought to be the "father of virtual reality," developed a see-through display in which computer-generated graphics were superimposed over the real world. This aspect of interface design, superimposing graphics or text over the real world, is an important feature of wearable computer and augmented reality systems. The ability to project or merge information on objects in the real world has led to what Steve Feiner of Columbia University terms "knowledge enhancement" of the world, that is, projecting information of interest (i.e., knowledge) on objects in the world. Extensions of this idea have led to the development of "smart spaces, rooms or environment."

The ideas associated with ubiquitous computing have also contributed to developments in wearable computers and augmented reality. What better way to access computing resources anywhere and at any time than to be wearing them on your body? In this regard, advances in microelectronics and wireless networking are making truly ubiquitous computing a reality. However, before the general public will be seen wearing computers, components of the wearable computer (the CPU housing unit, input and output devices) will have to look far more like clothing or clothing accessories than the commercial wearable computer systems available now. Thus, a new field is rapidly emerging, that of computational clothing (a chapter in this book). Furthermore, based on developments in microelectronics, sensor technology, and medicine, there is an emerging trend to apply computing resources under the surface of the skin and in some cases to integrate digital technology with the user's physiological systems. Such capabilities will allow computing technology to monitor and control various physiological processes, or to act as a sensory and motor prosthesis.

Since this topic has important implications for human use of technology and the potential for further integration of human physiological systems with digital technology, we close the book with a chapter on computing under the skin, briefly discussing some of the ideas associated with the concept of cyberevolution.

This book is a collection of twenty-five chapters that each address an important aspect of wearable computers and augmented reality, either from the conceptual or from an application framework. Given the wide coverage of topics on issues related to the display of computer-generated images in the environment, this book can be used as a text for computer science, computer engineering, and interface design courses. The book is organized around four topic areas. The first main topic covered in the book contains introductory material and consists of two chapters, "Basic Concepts in Wearable Computers and Augmented Reality" and "Augmented Reality: Approaches and Technical Challenges." The next part of the book, containing five chapters, focuses on the technologies associated with the design of augmented reality and wearable computer displays. One of the issues discussed within several of the chapters presented here is the issue of image registration. In many applications it is necessary to accurately register computer-generated images with objects or locations within the real environment. This important topic receives broad coverage from the chapters in Part II of the book: "A Survey of Tracking Technologies for Virtual Environments," "Optical versus Video See-Through Head-Mounted Displays," "Augmenting Reality Using Affine Object Representations," "Registration Error Analysis for Augmented Reality Systems," and "Mathematical Theory for Mediated Reality and WearCam-Based Augmented Reality."

The third part of the book is on the topic of "augmented reality," which include chapters on the technology as well as on applications. The chapters in this section include: "Studies of the Localization of Virtual Objects in the Near Visual Field," "Fundamental Issues in Mediated Reality, WearComp, and Camera-Based Augmented Reality," "STAR: Tracking for Object-Centric Augmented Reality," "NaviCam: A Palmtop Device Approach to Augmented Reality," "Augmented Reality for Exterior Construction Applications," "GPS-Based Navigation System for the Visually Impaired," and "Boeing's Wire Bundle Assembly Project."

Finally, the last part in the book is on "wearable computers," again with the technology discussed and with numerous applications provided. The chapters in Part IV include: "Computational Clothing and Accessories," "Situation Aware Computing with Wearable Computers," "Collaboration with Wearable Computers," "Tactual Displays for Sensory Substitution and Wearable Computers," "From 'Painting with Lightvectors' to 'Painting with Looks': Photographic/Videographic Applications of WearComp-Based Augmented/Mediated Reality," "Military Applications of Wearable Computers and Augmented Reality," "Medical Applications for Wearable Computing," "Constructing Wearable Computers for Maintenance Applications," "Applications of Wearable Computers and Augmented Reality to Manufacturing," "Computer Networks for Wearable Computers," and "Computing Under the Skin."

In summary, this book presents concepts related to the use and underlying technologies of augmented reality and wearable computer systems. As shown in this book, there are many application areas for this technology such as medicine, manufacturing, training, clothing, and recreation. Wearable computers will allow a much closer association of information with the user than is possible with traditional desktop computers. Future extensions of wearable computers will contain sensors that will allow the wearable device to see what the user sees, hear what the user hears, sense the user's physical state, and analyze what the user is typing. Combining sensors with an intelligent agent will result in a system that will be able to analyze what the user is doing and thus predict the resources he or she will need next or in the near future. We expect to see significant advances in wearware, providing humans tools that we have dreamed about having for centuries. We hope this book helps to stimulate further advances in this field.

We would like to thank those individuals that have either directly contributed to the book or have provided motivation for the project. At Lawrence Erlbaum Associates, Ray O'Connell and Lane Akers were supportive in the initial phases of discussions associated with the theme of the book and Anne Duffy from Lawrence Erlbaum Associates was instrumental in serving as Senior Editor for the project. Corin Huff, also from Lawrence Erlbaum Associates, provided help in obtaining needed material from chapter authors. Most importantly, we would like to thank the chapter authors for providing interesting and stimulating material.

Woodrow Barfield
Thomas Caudell

I

Introduction

1

Basic Concepts in Wearable Computers and Augmented Reality

Woodrow Barfield
Virginia Tech

Thomas Caudell
University of New Mexico

1. INTRODUCTION

This chapter introduces the joint fields of wearable computers and augmented reality, both of which represent the theme of this book—a discussion of information technologies that allow user's to access information anywhere and at any time. In many ways, the design of wearable computers and augmented reality systems has been motivated by two primary goals. The first is driven by the need for people to access information, especially as they move around the environment; the second is motivated by the need for people to better manage information. Until just recently, if a user needed to access computational resources, the user had to go to where the computer resources were located, typically a desktop PC or a mainframe computer. Once the user left the terminal, the flow of information stopped. Now networked wearable computers along with other digital devices allow the user to access information at any time, and at any location. However, the ability to access large amounts of information may not always be beneficial—too much information presented too fast may result in information overload. For this reason, the issue of information management is also important. In this regard, wearable computers and augmented reality systems along with software acting as an intelligent agent can act as a filter between the user and the information. Intelligent agents will allow only the relevant information for a given situation to be projected on a head-mounted display, a hand-held computer, or an auditory display.

Wearable computer systems will likely be a component of other advanced information technology initiatives as well. For example, in the area of "smart spaces," by embedding sensors and microprocessors into everyday things, wearable computer and augmented reality systems will be able to respond to and communicate with objects in the environment (Pentland, 1996). In addition, more and more, wearable computing medical devices will be implanted under the skin to regulate physiological parameters, or to serve as cognitive or sensory prosthesis. One prototype wearable device in this area consists of an electrode that is implanted in the motor cortex of the brain to allow speech-incapable patients to communicate via a computer (Bakay and Kennedy in Siuru, 1999). Gold recording wires pick up electrical signals of the brain and transmit the signals through the skin to a receiver and amplifier outside the scalp. The system is powered by an inductive coil placed over the scalp so that wires for powering the device do not have to pass through the skull. Signal processors are used to separate individual signals from the multiple signals that are recorded from inside the conical electrode tip. In the current implementation of the system, these signals are used to drive a cursor on a computer monitor. Such a device may prove beneficial to the 700,000 Americans who suffer from stroke each year and the tens of thousands more who suffer spinal cord injuries and diseases such as Lou Gehrig's disease. To conclude, there are many design issues that must be addressed before wearable computer systems reach their full potential and gain widespread use from the general public. To this aim, how wearable computer and augmented reality technology is designed, as well as application areas for wearable computers and augmented reality systems, is the topic of this chapter.

2. SENSORY PROSTHESIS

Over the past several thousand years, nature has provided humans the sensory systems that allow them to detect and respond to visual, auditory, olfactory, haptic, and gustatory information. Nature has also provided humans well-developed cognitive abilities that allow decisions to be made under conditions of uncertainty, patterns to be detected that are embedded in noise, and common sense judgments to be made. However, even though we can sense a broad range of stimuli, often over a wide range of energy values, our senses are still limited in many ways (Barfield, Hendrix, Bjorneseth, Kaczmarek, and Lotens, 1995). For example, we see images that represent only part of the electromagnetic spectrum, detect tactile and force feedback sensations across a limited range of energy values, and have marked decrements in spatializing images when sound is the primary cue. Due to these limitations, several types of prosthesis have been developed to extend our sensory, motor, and information processing abilities. For example, to extend the visual modality, we have invented glasses, microscopes, and telescopes; and to extend the auditory modality we have invented microphones, hearing aids, and telephones. Furthermore, to extend the haptic modality we have invented sensors that detect forces, which then transmit the force information back to the human.

We have also invented other types of sensory and motor prosthesis as well. Some of these include artificial hearts and kidneys and artificial arms and legs. However, until only recently, the prostheses that have been developed were designed primarily to enhance the detection capability of our sensory systems or to assist our motor capabilities, and to a far lesser extent, to enhance our cognitive capabilities. With the invention of the computer and developments in digital technology, microelectronics, and wireless networking, this is beginning to change. We are now able to wear digital devices that contain considerable computational resources; these devices clearly enhance our decision-making capabilities. Further, these digital devices may be worn "on the skin" (or body), as is the case with a wearable computer, or "under the skin," as is the case with medical devices. Digital devices worn on the skin have led to exciting developments in the area of computational clothing and digital accessories. In this area of research, computer scientists and interface design specialists are working with clothing designers as well as experts in textiles and fabrics to build wearable computers that look more like clothing and less like computers. Advances in computing under the skin have also led to exciting developments in information technology. In fact, in the near future we may be able to integrate computer chips directly into the nervous system. Along these lines, researchers at Johns Hopkins University and North Carolina State University have developed a computer microchip that is connected to the retina (Liu et al., 1999). The chip is designed to send light impulses to the brain. Thus far the device has allowed fifteen test subjects with blindness resulting from retinal damage to see shapes and detect movement.

3. DESCRIPTION OF SYSTEMS

We describe a wearable computer as a fully functional, self-powered, self-contained computer that is worn on the body. As noted earlier, a wearable computer provides access to information, and interaction with information, anywhere and at anytime. Closely related is the topic of "augmented reality," which can be thought of as an advanced human–computer interface technology that attempts to blend or fuse computer-generated information with our sensations of the natural world. For example, using a see-through head-mounted display (HMD), one may project computer-generated graphics into the environment surrounding the user to enhance the visual aspects of the environment. The main differences between what researcher's term "a wearable computer" versus "augmented reality" is twofold: (1) augmented reality is primarily a technology used to augment our senses and (2) wearable computers are far more mobile than augmented reality systems. With augmented reality, the range of mobility is dependent on the length of the cable connecting the wearable computer system to the computing platform. However, with a wearable computer, the computer and output devices are actually worn on the human's body, allowing a much broader range of mobility. Because of the close association between wearable computers and augmented reality, one can refer to both types of computing technology generically as "wearware" or simply as wearable computer systems. Throughout this book we will describe the integration of HMD, digital technology, auditory displays, and body tracking technologies with augmented reality and wearable computing hardware and software. These technologies have the potential of improving the efficiency and quality of human labors, particularly in their performance of engineering, manufacturing, construction, diagnostic, maintenance, monitoring, and transactional activities.

One of the primary display or output devices for wearable computer systems is a head-mounted display. Currently, there are three main application areas for HMDs, these include virtual reality, augmented reality, and wearable computers. These three application areas are summarized below.

Virtual reality: With virtual reality, a participant uses an HMD to experience an immersive representation of a computer-generated simulation of a virtual world. In this case, the user does not view the real world and is connected to the computer rendering the scene with a cable, typically allowing about 3–4 meters of movement.

Augmented reality: With augmented reality, a participant wears a see-through display (or views video of the real world with an opaque HMD) that allows graphics or text to be projected in the real world

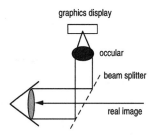

FIG. 1.1. Schematic diagram of an optical-based head-mounted display system for use with augmented reality and wearable computer systems.

(Caudell and Mizell, 1992; Lion, Rosenberg, and Barfield, 1993; Barfield, Rosenberg, and Lotens, 1995) (Figure 1.1). As with the virtual reality experience, the user is connected to the computer rendering the graphics or text with a cable, again allowing about 3–4 meters of movement.

Wearable computers: With wearable computers, the user actually wears the computer and, as in virtual or augmented reality, wears the visual display (hand-held or head-mounted) (Figure 1.2). The wearable computer may be wirelessly connected to a LAN or WAN, thus allowing information to be accessed whenever and wherever the user is in the environment.

As noted earlier, wearable computers allow hands-free manipulation of real objects as does augmented reality displays. However, because virtual reality displays are completely immersive, the user cannot directly see his hands, which makes manipulation of real objects difficult. Of course, with appropriate input devices, manipulation of virtual objects can occur. The see-through display capability that allows text or graphics to be projected within the real world is unique to augmented reality and wearable computer systems. However, all three application areas, virtual reality, augmented reality, and wearable computers, can use non–see-through or opaque displays. With augmented reality and wearable computers, an opaque HMD (monocular or binocular) can show live video with computer-generated text or graphics overlaid over the video.

Wearable computer systems can also be thought of as personal information devices (Starner, Mann, Rhodes, Levine, Healey, Kirsch, Picard, and Pentland, 1997; Rhodes, 1997). With a wearable computer, the user expects his interface to be accessible continually and unchanging, unless specified otherwise. With experience, the user personalizes his system to ensure appropriate responses to everyday tasks. As a result, the user's wearable

FIG. 1.2. User with wearable computer accessing map of a
University campus via wireless networked system.

computer system becomes a mediator for other computers and interfaces,
providing a familiar, dependable interface and set of tools complementing
the abilities the wearable computer infrastructure provides (more processor
power, additional sensing, etc.). With sophisticated user models and corresponding software agents, such an interface can be extended to recognize
and predict the resources needed by the user (Starner et. al., 1997).

There are several dimensions by which wearable computers and augmented reality systems can be evaluated, two of which include the level of
mobility provided by the computing system and the level of scene fidelity
afforded by the rendering platform. The level of scene fidelity refers to the
quality of the image provided by either the virtual reality simulation, real
world scene, or augmented world. One of the primary differences between
virtual reality and augmented reality is in the complexity of the projected
graphical objects. In basic augmented reality systems, only simple wire
frames, template outlines, designators, and text are displayed and animated.
An immediate result of this difference is that augmented reality systems can
be driven by microprocessors either worn on the body or integrated into the

work place. Today's processors have the computational power to transform and plot complex graphics in real time. Unlike full virtual reality systems, augmented reality systems are generally not attempting to generate a complete "world" or realistic scene. Instead, augmented reality systems tend to rely on reality to simulate reality, only superimposing the necessary graphical objects necessary to perform the task at hand. By doing so, many of the human factors issues found in full virtual reality systems such as vertigo and simulation sickness are avoided. The person continues to receive all their orientation cues from the physical visual scene. Display technology, input and output devices, computer architectures, network communication, power supplies, and image registration and calibration techniques are all important aspects of augmented reality systems.

Human's carry their sensors with them as they move around the environment, and they experience the world with the resolution provided by these biological sensors. Wearable computers also allow a high degree of mobility, but not quite that associated with our biological sensors. For example, currently we cannot swim with wearable computers although our biological sensors easily allow this range of mobility. Furthermore, the weight of wearable computers and augmented reality systems further adds to their lack of mobility. Wearable computers allow a high degree of scene fidelity because the real world is viewed either directly with see-through optics or via live video. However, the level of scene fidelity may be less than that associated with augmented reality because wearable computers currently do not have the rendering capability of workstations that are often used (e.g., SGIs) to render graphics for augmented reality environments. Finally, virtual reality in its present form is often low on scene fidelity and very low on mobility within the real world.

4. DESIGN ISSUES

Display technology, input and output devices, power supplies, and image registration techniques are important aspects of wearable computers and augmented reality systems. The following sections briefly discuss these topics.

4.1 Augmented Reality

The ability to project or merge graphics or text with real-world imagery is a characteristic of augmented reality and wearable computer systems (Caudell and Mizell, 1992; Barfield, Rosenberg, and Lotens, 1995). Feiner, MacIntyre, and Seligmann, (1993) refer to this ability as "knowledge enhancement" of the world. As an example, an infrared transmitter can be placed on an object of interest—and once detected by an infrared receiver worn by an individual, information about the object can be accessed through a database and directly projected on the object. Other sensor systems, such as CCD cameras using computer vision techniques, can detect bar codes or other features of an object, allowing the same functionality to occur (Rekimoto, 1997). There are different types of wearable visual display technologies, which can be used to combine real-world objects with computer-generated imagery to form an augmented scene. The two main types of visual display systems supporting wearable computers and augmented reality are shown below.

- *Optical-based systems.* These systems allow the observer to view the real world directly with one or both eyes with computer graphics or text overlaid onto the real world. Optical see-through HMDs are worn like glasses with an optical system attached at a location that does not interfere with visibility.
- *Video-based systems.* These systems can be used to view live video of real-world scenes, combined with overlaid computer graphics or text. Furthermore, monocular (one eye) or binocular (two eyes) displays can be used. Video based see-through displays are opaque displays that use cameras mounted near the users eyes to present live video on the display. Using chroma or luminance keying techniques, the computer then fuses the video with the virtual image(s) to create a video-based augmented reality environment.

4.2   Image Registration

When creating a wearable computer system that can be used to augment the environment with text of graphics, an important visual requirement is that the computer-generated imagery register at some level of accuracy with the surroundings in the real world (Janin, Mizell, and Caudell, 1993a,b; Holloway, 1997. For example, for a medical application where computer-generated images are overlaid onto a patient, accurate registration of the computer-generated imagery with the patient's anatomy is crucial (Bajura, Fuchs, and Ohbuchi, 1992; Lorensen, Cline, Nafis, Kikinis, Altobelli, and Gleason, 1993). In terms of developing scenes for wearable computer systems, the problem of image registration or positioning of the synthetic objects within the scene in relation to the real objects is both a difficult and important technical problem to solve.

Image registration is an important issue regardless of whether one is using a see-through or a video-based HMD to view the augmented reality environment. With applications that require accurate registration, depth information has to be retrieved from the real world in order to carry out the necessary calibration of the real and synthetic environments. Without an accurate knowledge of the geometry of both the real world and computer-generated scene, exact registration is not possible. To properly align video and computer-generated images with respect to each other, several frames of reference must be considered. Janin, Mizell, and Caudell (1993a) (using an HMD) and Lorensen and colleagues (1993) (using a screen-based system) have discussed issues of image calibration in the context of different frames of reference for augmented reality. In Lorensen's medical example, two coordinate systems were necessary, a real-world coordinate system and a virtual-world coordinate system. Alignment of the video and computer-generated imagery was done manually. Lorensen pointed out that this procedure worked well when anatomical features of the patient were easily visible. Janin and colleagues (1993b) presented another technique that can be used to measure the accuracy with which virtual images are registered with real-world images. This method involves the subject aligning a crosshair, viewed through a HMD, with objects of known position and geometry in the real world. Because this crosshair is not head tracked it moves with the user's head; this allows the vector from the position tracker to the real world object to be measured. The alignment procedure is done from multiple viewing positions and orientations, and as reported by Janin and colleagues, it allows a level of accuracy in terms of projection errors of $0.5''$, the approximate resolution of the position sensor. Although these methodologies work reasonably well for systems that have a limited range of mobility, much more research needs to be done before accurate registration of images using wearable computer systems can occur.

4.3 Optics

Another important design issue for wearable computer systems using a video-based head-mounted display relates to the optics of the CCD camera and the optics of the HMD. The focal length of a camera is the distance between the near nodal point of the lens and the focal plane, when the focus of the lens is set to infinity. There are two nodal points in a compound lens system. The front nodal point is where the rays of light entering the lens appear to aim. The rear nodal point is where the rays of light appear to have come from, after passing through the lens. Camera lenses that may be used to design augmented reality scenes using wearable computers may vary in field of view. For example, a typical wide angle lens has a 80 degree field of view, a standard lens a 44 degree field of view, and a typical telephoto lens a 23 degree field of view. The use of a zoom lens that has the capability to change from wide angle to telephoto views of the scene is desirable for video-based wearable computers as this allows the viewer to match the field of view of the real-world scene to the field of view of the computer-generated scene. In addition, depth of field, described as the distance from the nearest to the furthest parts of a scene that are rendered sharp at a given focusing setting, is another important variable to consider for augmented reality displays. Depth of field increases as the lens is stepped down (smaller lens aperture), when it is focused for distant objects, or when the lens has a short focal length. In some cases it may be desirable to have a large depth of field when using a video-based wearable computer in order to maximize the amount of the scene that is in focus. Other important variables to consider when designing a video-based system that displays stereoscopic images include the horizontal disparity (horizontal offset) of the two video cameras and the convergence angle of the two cameras. Some basic information on these variables is provided by Milgram, Drascic, and Grodski (1991) who used video cameras to create a stereo real-world image superimposed with a stereo computer-generated pointer.

If the video of the real world is to appear as if viewed through the user's eyes, the two CCD cameras must act as if they are at the same physical location as the wearer's eyes. Edwards et al. (1992) tried several different camera configurations to find the best combination of these factors for augmented reality displays. They determined that the inter-pupilary nodal distance is an important parameter to consider for applications that require close to medium viewing distances for depth judgments and that off-axisness is hard to get used to when objects are close to the viewer. Placing the cameras in front of the wearer's eyes may result in accurately registered video and computer-generated images, but the image of the outside world seen through the HMD will appear magnified and thus objects will appear closer than they actually are.

4.4   Input Devices

In conjunction with output devices, effective input devices are necessary to allow the user to seamlessly interact with the virtual text or images being presented. The input devices that have evolved for use with wearable computer systems are very diverse and always changing to accommodate user needs. Recent growth in the popularity of wearable computers has sparked an increasing interest in the design and evaluation of input devices (Thomas et. al., 1998).

In the real-world environment, the user is often using one or both hands to perform a task; therefore, the input devices used with wearable computers need to be designed with this requirement in mind. Appropriate input devices need to be utilized to allow the user to efficiently manipulate and interact with objects. For data entry or text input, body-mounted keyboards, speech recognition software, or hand-held keyboards are often used. Devices such as IBM's Intellipoint, track balls, data gloves, and the Twiddler are used to take the place of a mouse to move a cursor to select options or to manipulate data. One of the main advantages of using a wearable computer is that it allows the option of hands-free use. When complete hands-free operation is needed, speech recognition, and posture-based, EMG-based, and EEG-based devices are a few ways for the user to interact with the computer in a completely hands-free manner. Other input devices such as hand-held keyboards, wrist keyboards, or track pads allow the user to input data when complete hands-free use is not necessary.

For some wearable computer applications, another effective tool in providing users with a realistic experience is allowing them to receive haptic feedback to provide a direct physical perception of objects (Kaczmarek and Bach-Y-Rita, 1995; Tan and Pentland, 1997). This directly couples input and output between the computer and user. Tactile and force feedback acts as a powerful addition to augmented reality simulations for problems that involve understanding of 3D structure, shape, or fit, such as in assembly tasks (Barfield, Hendrix, Bjorneseth, Kaczmarek, and Lotens, 1995). The common factors considered in the design of these input devices is that they all must be unobtrusive, accurate, and easy to use on the job.

5.　APPLICATIONS

5.1　Sensor Technology for Wearable Computers and Smart Spaces

One of the interesting extensions of wearable computer systems is the creation of smart spaces. We define smart spaces as environments in which objects have associated with them: (1) a CPU allowing some minimal level of intelligence to exist in the object and (2) sensors that detect the presence of other objects within the environment or the state of the object itself. For example, a wearable computer equipped with a CCD camera can use computer vision techniques to detect the presence of an object in an environment; the detected object may also have a sensor that detects the wearable computer. Once the sensor on the wearable computer detects the presence of another object, the object is recognized and information about that object is accessed through a database; finally, with a see-through display or video system, information about the object is projected onto the object (Rekimoto, 1997; Starner et al., 1997).

In order for any digital system to have an awareness of and be able to react to events in its environment, it must be able to sense the environment. This can be accomplished by incorporating sensors, or arrays of various sensors (sensor fusion), into the system. Sensors are devices that are able to take an analog stimulus from the environment and convert it into electrical signals that can be interpreted by a digital device with a microprocessor. The stimulus can be a wide variety of energy types but most generally it is any quantity, property, or condition that is sensed and converted into an electrical signal (Fraden, 1996).

While sensors are responsible for detecting and converting analog stimuli from the environment, they are only the front end of a complex system that allows intelligence to be distributed within an environment. For wearable computers or other digital systems to be able to react "intelligently" to objects within an environment, the data from the sensors need to be processed and interpreted by microprocessors and software. In creating smart spaces, there are many types of sensors available that can be used to interpret a wide array of input types. They are often classified by the type of stimuli they respond to (Brignell and White, 1994). Table 1.1 shows a few different types of sensors and the types of stimuli they respond to. Some types of sensors used in the creation of smart spaces are mechanical, biological, or electromagnetic (acoustic, electric, optical).

In general, there are two kinds of sensors: *active* and *passive*. The characteristics of these two types of sensors will impact their potential use as components of a wearable computer system. Active sensors require an external power source or *excitation signal* in order to generate their own signal to operate. The excitation signal is then modified by the sensor to produce the output signal. Therefore, active sensors consist of both a transmitting and receiving system. A thermistor is an example of an active sensor. By itself is does not generate a signal, but by passing an electric current through it, temperature can be measured by the variation in the amount of resistance (Fraden, 1996). In contrast, passive sensors directly convert stimulus energy from the environment into an electrical output signal without an external

TABLE 1.1
Sensor Types and Response Stimuli

Sensor Type	Stimulus	Use in Smart Spaces
Mechanical	Position, acceleration, force, shape, mass, displacement, acceleration	Detecting people's/object's position, weight, movements
Biological	Heart rate, body temperature, neural activity, respiration rate	Measuring people's mood, mental state, physical state
Acoustic	Volume, pitch, frequency, phase, changes	Detecting sounds, interpreting speech
Optical	Emissivity, refraction, light wave frequency, brightness, luminance	Computer vision detection, IR motion/presence detection
Environmental	Temperature, humidity	Monitoring the conditions of the environment that people are in

power source. Passive sensors consist only of a receiver. Passive sensors include, for example, infrared motion detectors, which use radiated heat energy from the objects as their source of detection energy.

When building and employing sensors in an intelligent environment (including a wearable computer), there are a number of variables inherent to the particular sensors that need to be addressed. The following is a list of some considerations when dealing with sensors:

- *Accuracy* or *Resolution:* The smallest change in magnitude a sensor can detect.
- *Range* or *Field of View:* The amount of area covered by the sensor.
- *Calibration Error:* The maximum amount of inaccuracy permitted by the sensor.
- *Power Consumption:* The amount of energy required by the sensor to operate.
- *Size:* The physical dimensions of the sensor.
- *Saturation:* The maximum amount of stimulus the sensor can respond to.
- *Repeatability:* Ability to accurately recreate responses under identical stimuli.

- *Sample Rate:* The frequency at which the sensor samples the stimulus.
- *Noise Filtering:* The ability of the sensor to filter out unwanted environmental noise.

By taking into account these variables, designers of smart spaces can select the appropriate sensors to do a particular task. For example, if you are using an infrared beam emitter for object detection in a $10' \times 10'$ room, and the range of the sensor's beam is only $2'$, there is not enough coverage to be useful.

5.2 Industrial Applications

Manufacturing, construction, testing, and maintenance are four classes of complex industrial tasks that currently require a great deal of human involvement. Although automation systems will significantly off-load human workers as advances in robotics technology occur, there will remain a significant subset of these tasks that are either too complex or too costly for automation. Tasks that require dexterous manipulation, fusion of sensory experience and heuristic knowledge, spatial cognitive model building, time-critical decision making, and real-time problem solving are basic human traits that have proven difficult to capture in intelligent machines and will be costly once developed. Said another way, human involvement will continue to be the only economic choice in many industrial environments for the foreseeable future.

Information technology and systems are now an integral part of the modern industrial environment. Computer aided design (CAD) and computer aided manufacturing (CAM) systems are becoming the standard for large manufacturing companies worldwide. Three-dimensional CAD/CAM systems are used to model mechanical designs, visualize stresses or flows calculated from simulations, test for interferences through digital preassembly, and to begin to understand the manufacturability and maintainability of subsystems. Given this, a new question arises: How can manufacturing, construction, testing, and maintenance workers interface efficiently with these engineering data?

Today, the main interface is the physical information artifact paper. Engineering designs are plotted as line drawings, blueprints are produced, test procedures are written, and maintenance manuals are printed. Workers must continually consult documents, diagrams, templates, or tables to get their work done. This is both tedious and distracting and will become more so as the amount of information needed to perform their jobs increases. Even though the engineering design process can improve quality using CAD/CAM systems, it is difficult to maintain this quality through production under the circumstances of a poor worker interface to this information. The printed page responds very slowly to changes or updates in design, leading to out-of-date procedures and manuals and/or to stagnation in product modernization. Commercial aircraft design and manufacture represents an important example of where information technology is making a positive impact.

Modern airliners are complex machines that require a tremendous amount of manual effort to manufacture and assemble. The production of an airliner does not lend itself to complete automation because of the small lot size of parts, where in many cases the average lot size for a part is less than ten. A partial reason for this small size is the custom nature of the product; rarely are two aircraft identical. Automation is not practical in many cases because of the skills required to perform a task. Robots cannot today provide the dexterity and perception of a human required to assemble and test many of the components of this product. Even if a robotic system had the facilities to perform the tasks, the programming costs are exacerbated by the small lot sizes. For these reasons, people will continue to play a major role in the manufacture and assembly of modern aircraft.

Even so, aircraft are becoming more and more complex, as are their manufacturing processes. People in the factories are required to use an increasing amount of information in their jobs. Much of this information is derived from engineering designs for parts and processes stored in CAD systems. In many cases today, this information comes to the factory floor in the form of complicated assembly guides, templates, mylar drawings, wiring lists, and location markings on sheet metal. These information artifacts must be built or printed, cataloged, easily retrievable, and stored for the life of the aircraft. In addition, the worker must continually be trained to used them properly. Engineering change causes a slow, costly, and potentially error-prone propagation of updates through these artifacts and their use.

Attempts have been made to improve the efficiency of the interface between information and the worker in the factory. Occasionally today, a computer screen will be used to augment manufacturing or assembly instructions, indicating the next step in a process or the next location on a diagram to be serviced. Unfortunately, due to the physical displacement of the information screen from the work cell location, this approach is still distracting and does not augment the environment in a way that is natural to human perception and understanding.

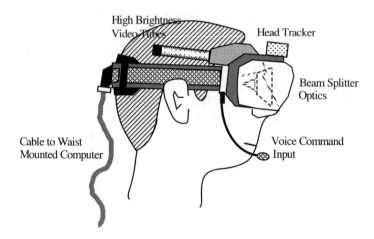

High Brightness Video Tubes

Head Tracker

Beam Splitter Optics

Cable to Waist Mounted Computer

Voice Command Input

FIG. 1.3. A drawing of the components of an augmented reality system. The video sources provide high brightness graphics that are folded into the user's line of sight with beam splitting relay optics. The head is tracked in six degrees of freedom to help create the illusion that the computer information is in the real world. Interaction with the software may occur through voice input if ambient conditions are within tolerance. The user wears the computer that performs the graphics, stores a database of information, and perhaps, communicates with the off-body world.

In 1990, the team at Boeing introduced augmented reality to the worker adding a natural interface to manufacturing, construction, testing, and maintenance information, without the use of physical artifacts (Caudell and Mizell, 1992). The general concept was to provide a "see-thru" virtual reality type head-mounted computer display to the worker, and to use this device to overlay his or her visual field of view with useful and dynamically changing information (Figure 1.3), (hence the name augmented reality). The enabling technologies for this access interface are head-up display headsets (Sutherland, 1968) combined with head position sensing, work place coordinate registration systems, data management systems, and wearable computing. A working hypothesis of the field of augmented reality is that if access to engineering and manufacturing information is appropriately enhanced, then people will be able to perform their tasks with greater ease and accuracy.

Figure 1.4 illustrates the first augmented reality application at Boeing in the area of manufacturing and touch labor: formboards and wire bundle assembly. Wire bundles for aircraft are manufactured by laying individual wires across pegs located in an annotated formboard usually made of a

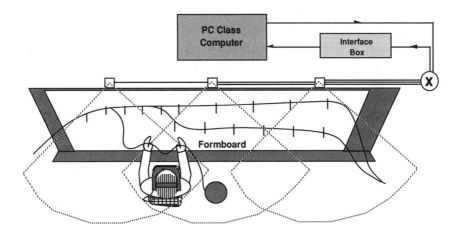

FIG. 1.4. The wiring formboard application where the user has an extended range of operation. The worker is wearing the computer that is performing the graphics and data management functions. The worker sees the path of the current wire as a colored polyline painted on the board.

$3' \times 8'$ piece of plywood. In the traditional method, the board is annotated with a long computer-generated plot of the wire bundle glued to the plywood. The person must be able to see the paths of the wires and read the paper drawing as the bundle grows with each addition of a wire. As an alternative solution, the Boeing team sought to replace the formboard with a general peg board and, at the beginning of a new wire bundle assembly, have the computer indicate the position of the pegs for the new assembly task for the person to install. The user wore a see-thru head-mounted display that had been registered in the coordinate system of the board. After the new pegs were installed, the paths of individual wires are indicated sequentially on the board. When a worker looked through the headset at the augmented reality formboard, the 3D path of the next wire to string was indicated by colored polyline. In addition, the wire gauge and type could have been indicated next to the wire location for quality control. As the person changed his or her perspective on the formboard, the graphical indicators appear to stay in the same 3D location, as if painted on the board. This is accomplished by tracking the six degrees of freedom of the user's headset relative to the workpiece and transforming the graphics to compensate for changes in view location and orientation. With this new approach, workers could now concentrate on the business of quality assembly, unfettered by distracting paper plots and computer screens.

5.3 Medicine

One of the important applications for augmented reality and wearable computers is the visualization of medical information projected onto a patient's body. Currently, MRI and CT images are reviewed by a physician using display technology that is totally detached from the patient. The use of augmented reality displays will allow MRI and CT images to be superimposed over the patients anatomy, which may assist in tasks such as the planning of surgical procedures. Researchers from the Department of Computer Science at the University of North Carolina, Chapel Hill, have pioneered the development of medical applications of augmented reality. For example, Fuchs and Neuman (1993) investigated the use of three-dimensional medical images superimposed over the patient's body for noninvasive visualization of internal human anatomy. Specifically, in their application, a physician wearing an HMD viewed a pregnant woman with an ultrasound scan of the fetus overlaid on the woman's stomach. Walking around the patient allowed the physician to observe the fetus in 3D perspective and to determine its placement relative to other internal organs.

Other researchers, such as Lorensen and colleagues (1993), and Gleason and colleagues at the Surgical Planning Laboratory of Brigham and Woman's Hospital affiliated with Harvard Medical School, have also investigated the use of augmented reality environments for medical visualization. Specifically, Gleason, Kikinis, Black, Alexander, Stieg, Wells, Lorensen, Cline, Altobelli, and Jolesz (1994) used three-dimensional images to assist in preoperative surgical planning and simulation of neurosurgical and craniofacial cases. They built an augmented reality display that allowed the surgeon to superimpose three-dimensional images with the surgeon's operative perspective.

Surgical simulation provides the ability to interact with the reconstructed virtual objects, such as viewing different surgical approaches or displaying only a limited number of anatomic structures or objects. Gleason et al. described the effectiveness of their augmented reality system for intraoperative neurosurgical procedures in the context of sixteen cases. From the same research team, Lorensen et al. (1993) presented a case study that involved the removal of a tumor at the top of a patient's brain. Before making any incisions, the surgical team viewed computer models of the underlying anatomy of the patient's brain surface and tumor, mixed with a live video image of the patient. The combined video and computer image was used to help the surgeon plan a path to the diseased tissue. The video of the patient, enhanced by the computer models, showed the extent of the tumor's intrusion beneath the brain's surface. A future extension of this work could involve a member of the surgical team with a wearable computer located at a remote site also viewing the tumor projected onto the patient and assisting in the diagnosis and planning of the surgical procedure.

6. RESEARCH ISSUES

As noted previously, one of the latest fields of research in the area of output devices is tactual display devices (Burdea, 1996; Tan and Pentland, 1997). These tactual or haptic devices allow the user to receive haptic feedback output from a variety of sources. This allows the user to actually feel virtual objects and manipulate them by touch. This is an emerging technology and will be instrumental in enhancing the realism of wearable augmented environments for certain applications. Tactual displays have previously been used for scientific visualization in virtual environments by chemists and engineers to improve perception and understanding of force fields and of world models populated with impenetrable objects. In addition to tactual displays, the use of wearable audio displays that allow sound to be spatialized are being developed. With wearable computers, designers will soon be able to pair spatialized sound to virtual representations of objects when appropriate to make the wearable computer experience even more realistic to the user.

Furthermore, as the number and complexity of wearable computing applications continue to grow, there will be increasing needs for systems that are faster, are lighter, and have higher resolution displays. Better networking technology will also need to be developed to allow all users of wearable computers to have high bandwidth connections for real-time information gathering and collaboration. In addition to the technology advances that would allow users to wear computers in everyday life, there is also the desire to have users want to wear their computers. To do this, wearable computing needs to be unobtrusive and socially acceptable. By making wearables smaller and lighter, or actually embedding them in clothing, users can conceal them easily and wear them comfortably (Barfield and Baird, 1998).

Several critical research issues must be resolved before the practical application of augmented reality and wearable computer technology is realized. For example, the requirement of an immersive virtual reality display and position sensor is only that the generated scene look realistic enough for the users to do their work and that graphical objects be correctly positioned relative to other graphical objects. The computational load for immersive virtual reality can be arbitrarily large as asymptotically realistic scenes are generated. For augmented reality, the displayed graphics are required to accurately match up against specific points in the user's physical surroundings. The computational load is minute compared to virtual reality since only simple 3D sticklike or firewrame objects and text need be rendered. Evidently, the critical research issues for wearable computer systems are in many cases different from immersive virtual reality. Other hardware and software related research issues are indicated below. The first set of research topics involves hardware and software issues; the second deals with human-centered issues. Although addressing the technological issues are a necessary condition for the successful deployment of augmented reality technology, the many human-centered questions that must also be addressed are equally important. Without an understanding of the human-centered side in addition to the technological side of augmented reality and wearable computers, this technology might be forced-fit into inappropriate environments, leading to a waste of resources, loss of productivity, and a premature abandonment of the approach without fair trails.

6.1 Hardware and Software Issues

1. Extended position tracking: Both high accuracy and long-range position tracking are necessary in many factory and construction environments. The worker will need the freedom to operate untethered over wide areas and in complex structures. Current virtual reality position tracking systems are limited to roughly room-sized volumes and require "clean" environments for accurate operation. Research on improving tracking devices for mobile users outside work environments is also an important topic.

2. Calibration and reregistration: Methods and systems for internally calibrating the display/sensor system and registering the user's coordinate system to the work place are not fully developed. These methods attempt to place the work place, the position/orientation sensor, the virtual screens of the display, and the user's eyes into a common coordinate system. Furthermore, because the user might jostle the display or it might slip out of position, techniques are under development to quickly detect if the user is out of registration, as well as ways to quickly reregister. These registration procedures must be quick and convenient for complete acceptance into the work place.

3. CAD/CAM database interfaces and translators: The data required by the augmented reality and wearable computer system will typically be

found in the traditional CAD/CAM databases, but not in the simplified form needed for optimal augmented reality display. Tools are under development to use knowledge-based methods to transform geometric data into a form easily understandable by the worker when superimposed on the work place. This is a highly domain-dependent approach and more general techniques must be developed.

4.   Smart spaces: As intelligence is imbedded into everyday objects, how should we design information spaces that allow humans to receive the right information at the right time? And, how do we deliver information to mobile users with sufficient security so that privacy is maintained. In addition, we need to develop sensors that are part of a wireless network and that have the capability of detecting and communicating with humans.

5.   Computing under the skin: We need to determine what applications of computing under the skin are important and how to implement such technology. For example, there are many medical benefits associated with integrating microcomputers with the nervous system and other biological systems. Applications could include cognitive and sensory prostheses and the regulation of physiological parameters.

6.   Energy requirements: Power sources for wearable computer systems must be improved so that they are longer lasting and lighter.

6.2   Human-Centered Design Issues

1.   Human 3D perception and reasoning: How do we best match the optics in the augmented reality displays to the characteristics of human vision, such as depth of focus, optical pupil size, and eye strain? Are stereo augmented reality displays necessary for spatial understanding? What role will color play in the display of simplified information? What system characteristics affect the ease of visual fusion or the perception that the virtual graphic is actually part of the physical scene?

2.   Human task performance measurement: How does augmented reality and wearable computer technology really affect a person's ability to perform a task? What types of tasks does it help or hinder? What system characteristics, such as computational time delays or positional errors, affect measures of performance, and how can they be optimized?

3.   Human/sociological issues: How do people perceive augmented reality in their work place? Given a choice, will they use it or revert back to older methods? How does it affect worker satisfaction? Do these factors affect the quality and quantity of their work? Finally, can these effects be quantitatively measured?

4.   Computational clothing: For wearable computer systems to be fully accepted by the general public, we need to design wearables that look more like clothing and clothing accessories than current implementations of wearable computer systems (Post and Orth, 1997). How to do this is a major and current research topic.

In summary, in this chapter we presented concepts related to the use and underlying technologies of augmented reality and wearable computer systems. We discussed how recent advances in augmented reality displays make it possible to enhance the user's local environment with "information." We also briefly touched upon some of the many application areas for this technology such as medicine, manufacturing, training, and recreation. A major point was that wearable computers allow a much closer association of information with the user. By embedding sensors in the wearable to allow it to see what the user sees, hear what the user hears, sense the user's physical state, and analyze what the user is typing, an intelligent agent may be able to analyze what the user is doing and try to predict the resources he or she will need next or in the near future. Using this information, the agent may download files, reserve communications bandwidth, post reminders, or automatically send updates to colleagues to help facilitate the user's daily interactions. This intelligent wearable computer would be able to act as a personal assistant—one that is always around, knows the user's personal preferences and tastes, and tries to streamline interactions with the rest of the world. Finally, our view of the future is that we will be wearing computers that look like clothing but yet contain all the computing capabilities of high-end workstations. These computers will monitor our physiological state, perform the duties of a secretary and butler in managing our everyday life, and protect us from physical harm.

Acknowledgment   Dr. Barfield thanks ONR (N000149710388) for an equipment grant to pursue research in augmented and virtual environments.

REFERENCES

Bajura, M., Fuchs, H., and Ohbuchi, R. (1992). Merging virtual objects with the real world: Seeing ultrasound imagery within the patient, *Computer Graphics, Vol. 26*, No. 2, 203–210.

Bakay, R. E., and Kennedy, P. R. (1999). In B. Siuru, A Brain/Computer Interface, *Electronics Now*, March, 55–56.

Barfield, W., and Baird, K. (1998). Future directions in virtual reality: Augmented environments through wearable computers, *VR '98 Seminar and Workshop on Virtual Reality*, Kuala Lumpur, April 24–25.

Barfield, W., Hendrix, C., Bjorneseth, O., Kaczmarek, K., and Lotens, W. (1995). Comparison of human sensory capabilities with technical specifications for virtual environment equipment, *Presence: Teleoperators and Virtual Environments, Vol. 4*, 329–356.

Barfield, W., Rosenberg, C., and Lotens, W. (1995). Augmented-reality displays, in Barfield, W., and Furness, T. (editors), *Virtual Environments and Advanced Interface Design*, Oxford University Press, 542–575.

Brignell, J., and White, N. (1994). *Intelligent Sensor Systems*, IOP Publishing, Bath, England.

Burdea, G. (1996). *Force and Touch Feedback for Virtual Reality*, John Wiley.

Caudell, T. P., and Mizell, D. W. (1992). Augmented reality: an application of head-up display technology to manual manufacturing processes, *Proc. of Hawiian International Conference on System Sciences*.

Edwards, E. K., Rolland, J. P., and Keller, K. P. (1992). Video see-through design for merging of real and virtual environments, *Proceedings IEEE Virtual Reality Annual International Symposium*, Seattle WA, Sept 18–22, 223–233.

Feiner, S., MacIntyre, B., and Seligmann, D. (1993). Knowledge based augmented reality, *Communications of the ACM, Vol. 36*, No. 7, 53–62.

Fraden, J. (1996). *Handbook of Modern Sensors: Physics, Design, and Application*, Second Edition, AIP Press, New York.

Fuchs, H., and Neuman, U. (1993). A vision telepresence for medical consultation and other applications, *Proceedings of ISRR-93, Sixth International Symposium on Robotics Research*, Hidden Valley PA, October, 1993.

Gleason, P. L., Kikinis, R., Black, P., Alexander, E., Stieg, P. E., Wells, W., Lorensen, W., Cline, H., Altobelli, D., and Jolesz, F. (1994). Intraoperative image guidance for neurosurgical procedures using video registration, Brigham and Womans Hospital, Harvard Medical School.

Holloway, R. L. (1997). Registration error analysis for augmented reality, *Presence: Teleoperators and Virtual Environments, Vol. 6*, 413–432.

Janin, A. L., Mizell, D. W., and Caudell, T. P. (1993a). Calibration of head-mounted displays for augmented reality applications, *Proc. of IEEE VRAIS '93*.

Janin, A. L, Mizell, D. W., and Caudell, T. P. (1993b). Calibration of head-mounted displays for augmented reality applications, *Proceedings IEEE Virtual Reality Annual International Symposium*, Seattle WA, Sept 18–22, 246–255.

Kaczmarek, K. A., and Bach-Y-Rita, P. (1995). Tactile Displays, in Barfield, W., and Furness, T. (editors), *Virtual Environments and Advanced Interface Design*, Oxford University Press, 349–414.

Lion, D., Rosenberg, C., and Barfield, W. (1993). Overlaying three-dimensional computer graphics with stereoscopic live motion video: Applications for virtual environments, *SID Conference*.

Liu, W., McGucken, E., Clements, M., DeMarco, S. C., Vichienchom, K., Hughes, C., et al. (1999). An implantable neuro-stimulator device for a retinal prosthesis, in press, *Intl. Solid-State Circuits Conference*, Feb 1999.

Lorensen, W., Cline, H., Nafis, C., Kikinis, R., Altobelli, D., and Gleason, L. (1993). Enhancing reality in the operating room, *Proceedings Visualization '93*, October 25–29, San Jose, CA, 410–415.

Milgram, P., Drasic, D., and Grodsky, D. (1991). Enhancement of 3D-video displays by means of superimposed stereo-graphics, *Proceedings of the 35th Annual Meeting of the Human Factors Society*, San Francisco CA, Sept 2–6, 1457–1461.

Pentland, A. (1996). Smart rooms, *Scientific American*, April, 68–76.

Post, R., and Orth, M. (1997). Smart fabric, or washable computing. *Proceedings of the First International Symposium on Wearable Computers (ISWC '97)*, Cambridge, MA, 167–8.

Rekimoto, J. (1997). NaviCam: A magnifying glass approach to augmented reality, *Presence: Teleoperators and Virtual Environments, Vol. 6* (No. 4), 399–412.

Rhodes, B. (1997). The wearable remembrance agent: A system for augmented memory. *Proceedings of the First International Symposium on Wearable Computers (ISWC '97)*, Cambridge, MA, 123–8.

Starner, T., Mann, S., Rhodes, B., Levine, J., Healey, J., Kirsch, D., Picard, R., and Pentland, A. (1997). Augmented reality through wearable computing, *Presence: Teleoperators and Virtual Environments, Vol. 6* (No. 4), 384–398.

Sutherland, I. (1968). A head-mounted three dimensional display, *Proc. FJCC*, 757–64, Thompson Books, Washington, DC.

Tan, H. J., and Pentland, S. (1997). Tactual displays for wearable computing, *Proceedings of The First International Symposium on Wearable Computers (ISWC '97)*, Cambridge, MA, 84–89.

Thomas, B., Tyerman, S., and Grimmer, K. (1998). Evaluation of text input mechanisms for wearable computers, *Virtual Reality, Research, Development, and Applications, Vol. 3*, 187–199.

2

Augmented Reality: Approaches and Technical Challenges

Ronald T. Azuma
HRL Laboratories, Malibu, CA

1. INTRODUCTION

Augmented Reality (AR) merges 3-D virtual objects into a 3-D real environment and displays this combination in real time. Unlike Virtual Environments (VEs), AR supplements reality, rather than completely replacing it. This property makes AR particularly well suited as a tool to aid the user's perception of and interaction with the real world. The virtual objects display information that the user cannot directly detect with his own senses. The information conveyed by the virtual objects helps a user perform real-world tasks. AR is a specific example of what Fred Brooks calls *Intelligence Amplification* (IA): using the computer as a tool to make tasks easier for a human to perform (Brooks, 1996). Potential AR application areas include medical visualization, maintenance and repair, annotation, entertainment, and military aircraft navigation and targeting. Chapter 16 discusses AR applications in more detail. AR displays may help user performance in these applications by reducing information search time and errors, while increasing retention and motivation (Neumann and Majoros, 1998).

Building an effective AR system is a difficult task and is an active area of research. This chapter describes, at a high level, the approaches that have been taken so far and the major challenges that system designers face. Section 2 surveys the characteristics of AR systems and the various possible configurations, along with the relative advantages and disadvantages of each. Currently, two of the biggest problems are in registration and sensing: the subjects of Sections 3 and 4. Finally, Section 5 suggests some topics for further work and research.

2. APPROACHES

This section describes the characteristics of AR systems and the available approaches for building an AR system. Section 2.1 explains the basic characteristics of augmentation. There are two ways to accomplish this augmentation: with optical or video technologies. Section 2.2 discusses their characteristics and relative strengths and weaknesses. Blending the real and virtual poses problems with focus and contrast (Section 2.3), and some applications require portable AR systems to be truly effective (Section 2.4). Finally, Section 2.5 summarizes the characteristics by comparing the requirements of AR against those for Virtual Environments.

2.1 Augmentation

Besides *adding* objects to a real environment, Augmented Reality also has the potential to *remove* them. Current work has focused on adding virtual objects to a real environment. However, graphic overlays might also be used to remove or hide parts of the real environment from a user. For example, to remove a desk in the real environment, draw a representation of the real walls and floors behind the desk and "paint" that over the real desk, effectively removing it from the user's sight. This has been done in feature films. Doing this interactively in an AR system will be much harder, but this removal may not need to be photorealistic to be effective.

Augmented Reality might apply to all senses, not just sight. So far, researchers have focused on blending real and virtual images and graphics. However, AR could be extended to include sound. The user would wear headphones equipped with microphones on the outside. The headphones would add synthetic, directional 3-D sound, while the external microphones would detect incoming sounds from the environment. This would give the system a chance to mask or cover up selected real sounds from the environment by generating a masking signal that exactly canceled the incoming real sound (Durlach and Mavor, 1995; Barfield, Rosenberg, and Lotens, 1995). While this would not be easy to do, it might be possible. Another example is haptics. Gloves with devices that provide tactile feedback might augment real forces in the environment. For example, a user might run his hand over the surface of a real desk. Simulating such a hard surface virtually is fairly difficult, but it is easy to do in reality. Then the tactile effectors in the glove can augment the feel of the desk, perhaps making it feel rough in certain spots. This capability might be useful in some applications, such as providing an additional cue that a virtual object is at a particular location on a real desk (Wellner, 1993).

2.2 Optical versus Video

A basic design decision in building an AR system is how to accomplish the combining of real and virtual. Two basic choices are available: optical and video technologies. Each has particular advantages and disadvantages. This section compares the two and notes the trade-offs. For additional discussion, see Rolland, Holloway and Fuchs (1994) or Chapter 5.

A see-through HMD is one device used to combine real and virtual. Standard *closed-view HMDs* do not allow any direct view of the real world. In contrast, a *see-through HMD* lets the user see the real world, with virtual objects superimposed by optical or video technologies.

Optical see-through HMDs work by placing optical combiners in front of the user's eyes. These combiners are partially transmissive, so that the user can look directly through them to see the real world. The combiners are also partially reflective, so that the user sees virtual images bounced off the combiners from head-mounted monitors. This approach is similar in nature to *Head-Up Displays* (HUDs), commonly used in military aircraft, except that the combiners are attached to the head. Thus, optical see-through HMDs have sometimes been described as a "HUD on a head" (Wanstall. 1989). Figure 2.1 shows a conceptual diagram of an optical see-through HMD. Figure 2.2 shows two optical see-through HMDs made by Raytheon.

The optical combiners usually reduce the amount of light that the user sees from the real world. Since the combiners act like half-silvered mirrors, they only let in some of the light from the real world, so that they can reflect some of the light from the monitors into the user's eyes. For example, the HMD described in Holmgren (1992) transmits about 30% of the incoming light from the real world. Choosing the level of blending is a design problem. More sophisticated combiners might vary the level

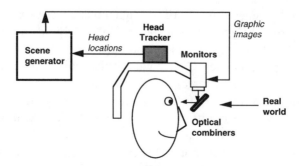

FIG. 2.1. Optical see-through HMD conceptual diagram.

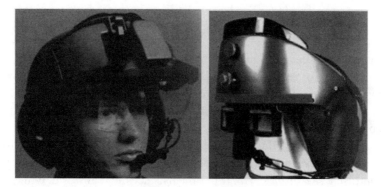

FIG. 2.2. Two optical see-through HMDs, made by Raytheon.

of contributions based upon the wavelength of light. For example, such a combiner might be set to reflect all light of a certain wavelength and none at any other wavelengths. This would be ideal with a monochrome monitor. Virtually all the light from the monitor would be reflected into the user's eyes, while almost all the light from the real world (except at the particular wavelength) would reach the user's eyes. However, most existing optical see-through HMDs do reduce the amount of light from the real world, so they act like a pair of sunglasses when the power is cut off.

In contrast, video see-through HMDs work by combining a closed-view HMD with one or two head-mounted video cameras. The video cameras provide the user's view of the real world. Video from these cameras is combined with the graphic images created by the scene generator, blending the real and virtual. The result is sent to the monitors in front of the user's eyes in the closed-view HMD. Figure 2.3 shows a conceptual diagram of a video see-through HMD. Figure 2.4 shows an actual video see-through HMD, with two video cameras mounted on top of a flight helmet.

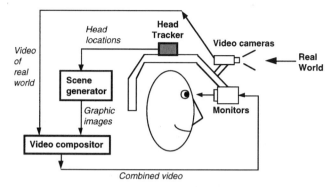

FIG. 2.3. Video see-through HMD conceptual diagram.

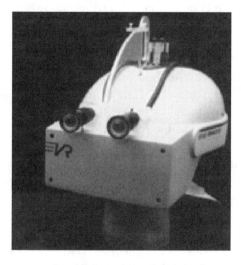

FIG. 2.4. An actual video see-through HMD. (Courtesy Jannick
Rolland, Frank Biocca, and the University of North Carolina Chapel
Hill Department of Computer Science. Photo by Alex Treml.)

Video composition can be done in more than one way. A simple way
is to use chroma-keying: a technique used in many video special effects.
The background of the computer graphic images is set to a specific color,
say green, which none of the virtual objects use. Then the combining
step replaces all green areas with the corresponding parts from the video
of the real world. This has the effect of superimposing the virtual objects over the real world. A more sophisticated composition would use
depth information. If the system had depth information at each pixel for
the real-world images, it could combine the real and virtual images by a

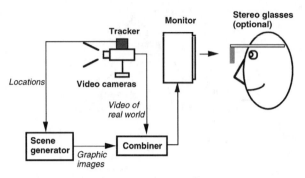

FIG. 2.5.   Monitor-based AR conceptual diagram.

pixel-by-pixel depth comparison. This would allow real objects to cover virtual objects and vice versa.

AR systems can also be built using monitor-based configurations, instead of see-through HMDs. Figure 2.5 shows how a monitor-based system might be built. In this case, one or two video cameras view the environment. The cameras may be static or mobile. In the mobile case, the cameras might move around by being attached to a robot, with their locations tracked. The video of the real world and the graphic images generated by a scene generator are combined, just as in the video see-through HMD case, and displayed in a monitor in front of the user. The user does not wear the display device. Optionally, the images may be displayed in stereo on the monitor, which then requires the user to wear a pair of stereo glasses. The ARGOS system is an example of a monitor-based configuration.

Finally, a monitor-based optical configuration is also possible. This is similar to Figure 2.1 except that the user does not wear the monitors or combiners on her head. Instead, the monitors and combiners are fixed in space, and the user positions her head to look through the combiners. This is typical of Head-Up Displays on military aircraft, and at least one such configuration has been proposed for a medical application (Peuchot, Tanguy, and Eude, 1995).

The rest of this section compares the relative advantages and disadvantages of optical and video approaches, starting with optical. An optical approach has the following advantages over a video approach:

1. *Simplicity:* Optical blending is simpler and cheaper than video blending. Optical approaches have only one "stream" of video to worry about: the graphic images. The real world is seen directly through the

combiners, and that time delay is generally a few nanoseconds. Video blending, on the other hand, must deal with separate video streams for the real and virtual images. Both streams have inherent delays in the tens of milliseconds. Digitizing video images usually adds at least one *frame time* of delay to the video stream, where a frame time is how long it takes to completely update an image. A monitor that completely refreshes the screen at 60 Hz has a frame time of 16.67 ms. The two streams of real and virtual images must be properly synchronized or temporal distortion results. Also, optical see-through HMDs with narrow field-of-view combiners offer views of the real world that have little distortion. Video cameras almost always have some amount of distortion that must be compensated for, along with any distortion from the optics in front of the display devices. Since video requires cameras and combiners that optical approaches do not need, video will probably be more expensive and complicated to build than optical-based systems.

2. *Resolution:* Video blending limits the resolution of what the user sees, both real and virtual, to the resolution of the display devices. With current displays, this resolution is far less than the resolving power of the fovea. Optical see-through also shows the graphic images at the resolution of the display device, but the user's view of the real world is not degraded. Thus, video reduces the resolution of the real world, whereas optical see-through does not.

3. *Safety:* Video see-through HMDs are essentially modified closed-view HMDs. If the power is cut off, the user is effectively blind. This is a safety concern in some applications. In contrast, when power is removed from an optical see-through HMD, the user still has a direct view of the real world. The HMD then becomes a pair of heavy sunglasses, but the user can still see.

4. *No eye offset:* With video see-through, the user's view of the real world is provided by the video cameras. In essence, this puts his "eyes" where the video cameras are. In most configurations, the cameras are not located exactly where the user's eyes are, creating an offset between the cameras and the real eyes. The distance separating the cameras may also not be exactly the same as the user's interpupillary distance (IPD). This difference between camera locations and eye locations introduces displacements from what the user sees compared to what he expects to see. For example, if the cameras are above the user's eyes, he will see the world from a vantage point slightly taller than he is used to. Video see-through can avoid the eye-offset problem through the use of mirrors to create another set of optical paths that mimic the paths directly into the user's eyes. Using

those paths, the cameras will see what the user's eyes would normally see without the HMD. However, this adds complexity to the HMD design. Offset is generally not a difficult design problem for optical see-through displays. While the user's eye can rotate with respect to the position of the HMD, the resulting errors are tiny. Using the eye's center of rotation as the viewpoint in the computer-graphics model should eliminate any need for eye tracking in an optical see-through HMD (Holloway, 1997).

Video blending offers the following advantages over optical blending:

1.   *Flexibility in composition strategies:*   A basic problem with optical see-through is that the virtual objects do not completely obscure the real world objects, because the optical combiners allow light from both virtual and real sources. Building an optical see-through HMD that can selectively shut out the light from the real world is difficult. In a normal optical system, the objects are designed to be in focus at only one point in the optical path: the user's eye. Any filter that would selectively block out light must be placed in the optical path at a point where the image is in focus, which obviously cannot be the user's eye. Therefore, the optical system must have *two* places where the image is in focus: at the user's eye and at the point of the hypothetical filter. This makes the optical design much more difficult and complex. No existing optical see-through HMD blocks incoming light in this fashion. Thus, the virtual objects appear ghost like and semitransparent. This damages the illusion of reality because occlusion is one of the strongest depth cues. In contrast, video see-through is far more flexible about how it merges the real and virtual images. Since both the real and virtual are available in digital form, video see-through compositors can, on a pixel-by-pixel basis, take the real, or the virtual, or some blend between the two to simulate transparency. Because of this flexibility, video see-through may ultimately produce more compelling environments than optical see-through approaches.

2.   *Wide field of view:*   Distortions in optical systems are a function of the radial distance away from the optical axis. The further one looks away from the center of the view, the larger the distortions get. A digitized image taken through a distorted optical system can be undistorted by applying image processing techniques to unwarp the image, provided that the optical distortion is well characterized. This requires significant amounts of computation, but this constraint will be less important in the future as computers become faster. It is harder to build wide field-of-view displays with optical see-through techniques. Any distortions of the user's view of the real world must be corrected *optically*, rather than digitally, because

the system has no digitized image of the real world to manipulate. Complex optics are expensive and add weight to the HMD. Wide field-of-view systems are an exception to the general trend of optical approaches since they are simpler and cheaper than video approaches.

3. *Real and virtual view delays can be matched:* Video offers an approach for reducing or avoiding problems caused by temporal mismatches between the real and virtual images. Optical see-through HMDs offer an almost instantaneous view of the real world but a delayed view of the virtual. This temporal mismatch can cause problems. With video approaches, it is possible to delay the video of the real world to match the delay from the virtual image stream. For details, see Section 4.3.

4. *Additional registration strategies:* In optical see-through, the only information the system has about the user's head location comes from the head tracker. Video blending provides another source of information: the digitized image of the real scene. This digitized image means that video approaches can employ additional registration strategies unavailable to optical approaches. Section 4.4 describes these in more detail.

5. *Easier to match the brightness of real and virtual objects:* This is discussed in Section 3.3.

Both optical and video technologies have their roles, and the choice of technology depends on the application requirements. Many of the mechanical assembly and repair prototypes use optical approaches, possibly because of the cost and safety issues. If successful, the equipment would have to be replicated in large numbers to equip workers on a factory floor. In contrast, most of the prototypes for medical applications use video approaches, probably for the flexibility in blending real and virtual and for the additional registration strategies offered.

2.3 Focus and Contrast

Focus can be a problem for both optical and video approaches. Ideally, the virtual should match the real. In a video-based system, the combined virtual and real image will be projected at the same distance by the monitor or HMD optics. However, depending on the video camera's depth-of-field and focus settings, parts of the real world may not be in focus. In typical graphics software, everything is rendered with a pinhole model, so all the graphic objects, regardless of distance, are in focus. To overcome this, the graphics could be rendered to simulate a limited depth of field, and the video camera might have an autofocus lens.

In the optical case, the virtual image is projected at some distance away from the user. This distance may be adjustable, although it is often fixed. Therefore, while the real objects are at varying distances from the user, the virtual objects are all projected to the same distance. If the virtual and real distances are not matched for the particular objects that the user is looking at, it may not be possible to clearly view both simultaneously.

Contrast is another issue because of the large dynamic range in real environments and in what the human eye can detect. Ideally, the brightness of the real and virtual objects should be appropriately matched. Unfortunately, in the worst case scenario, this means the system must match a very large range of brightness levels. The eye is a logarithmic detector, where the brightest light that it can handle is about eleven orders of magnitude greater than the smallest, including both dark-adapted and light-adapted eyes. In any one adaptation state, the eye can cover about six orders of magnitude. Most display devices cannot come close to this level of contrast. This is a particular problem with optical technologies, because the user has a direct view of the real world. If the real environment is too bright, it will wash out the virtual image. If the real environment is too dark, the virtual image will wash out the real world. Contrast problems are not as severe with video, because the video cameras themselves have limited dynamic response, and the view of both the real and virtual is generated by the monitor, so everything must be clipped or compressed into the monitor's dynamic range.

2.4 Portability

In almost all Virtual Environment systems, the user is not encouraged to walk around much. Instead, the user navigates by "flying" through the environment, walking on a treadmill, or driving some mockup of a vehicle. Whatever the technology, the result is that the user stays in one place in the real world.

Some AR applications, however, will need to support a user who will walk around a large environment. AR requires that the user actually be at the place where the task is to take place. "Flying," as performed in a VE system, is no longer an option. If a mechanic needs to go to the other side of a jet engine, she must physically move herself and the display devices she wears. Therefore, AR systems will place a premium on portability, especially the ability to walk around outdoors, away from controlled environments. The scene generator, the HMD, and the tracking system must all be self-contained and capable of surviving exposure to the environment. If this capability is achieved, many more applications that have not been tried will become available. For example, the ability to annotate the surrounding environment could be useful to soldiers, hikers, or tourists in an unfamiliar new location.

2.5 Comparison against Virtual Environments

The overall requirements of AR can be summarized by comparing them against the requirements for Virtual Environments, for the three basic subsystems that they require.

1. *Scene generator:* Rendering is not currently one of the major problems in AR. VE systems have much higher requirements for realistic images because they completely replace the real world with the virtual environment. In AR, the virtual images only supplement the real world. Therefore, fewer virtual objects need to be drawn, and they do not necessarily have to be realistically rendered in order to serve the purposes of the application. For example, in the annotation applications, text and 3-D wireframe drawings might suffice. Ideally, photorealistic graphic objects would be seamlessly merged with the real environment (see Section 7), but more basic problems have to be solved first.

2. *Display device:* The display devices used in AR may have less stringent requirements than VE systems demand, again because AR does not replace the real world. For example, monochrome displays may be adequate for some AR applications, whereas virtually all VE systems today use full color. Optical see-through HMDs with a small field of view may be satisfactory because the user can still see the real world with his peripheral vision; the see-through HMD does not shut off the user's normal field of view. Furthermore, the resolution of the monitor in an optical see-through HMD might be lower than what a user would tolerate in a VE application, since the optical see-through HMD does not reduce the resolution of the real environment.

3. *Tracking and sensing:* While in the previous two cases AR had lower requirements than VE, that is not the case for tracking and sensing. In this area, the requirements for AR are much stricter than those for VE systems. A major reason for this is the registration problem, which is described in the next section. The other factors that make the tracking and sensing requirements higher are described in Section 4.

3. REGISTRATION

3.1   The Registration Problem

One of the most basic problems currently limiting Augmented Reality applications is the registration problem. The objects in the real and virtual worlds must be properly aligned with respect to each other, or the illusion that the two worlds coexist will be compromised. More seriously, many applications *demand* accurate registration. For example, imagine a needle biopsy application. If the virtual object is not where the real tumor is, the surgeon will miss the tumor and the biopsy will fail. Without accurate registration, Augmented Reality will not be accepted in many applications.

Registration problems also exist in Virtual Environments, but they are not nearly as serious because they are harder to detect than in Augmented Reality. Since the user only sees virtual objects in VE applications, registration errors result in visual-kinesthetic and visual-proprioceptive conflicts. Such conflicts between different human senses may be a source of motion sickness (Pausch, Crea, and Conway, 1992). Because the kinesthetic and proprioceptive systems are much less sensitive than the visual system, visual–kinesthetic and visual–proprioceptive conflicts are less noticeable than visual–visual conflicts. For example, a user wearing a closed-view HMD might hold up her real hand and see a virtual hand. This virtual hand should be displayed exactly where she would see her real hand, if she were not wearing an HMD. But if the virtual hand is wrong by five millimeters, she may not detect that unless actively looking for such errors. The same error is much more obvious in a see-through HMD, where the conflict is visual–visual.

Furthermore, a phenomenon known as *visual capture* (Welch, 1978) makes it even more difficult to detect such registration errors. Visual capture is the tendency of the brain to believe what it sees rather than what it feels, hears, etc. That is, visual information tends to override all other senses. When watching a television program, a viewer believes the sounds come from the mouths of the actors on the screen, even though they actually come from a speaker in the TV. Ventriloquism works because of visual capture. Similarly, a user might believe that her hand is where the virtual hand is drawn, rather than where her real hand actually is, because of visual capture. This effect increases the amount of registration error users can tolerate in Virtual Environment systems. If the errors are systematic, users might even be able to adapt to the new environment, given a long exposure time of several hours or days (Welch, 1978).

Augmented Reality demands much more accurate registration than Virtual Environments (Azuma, 1993). Imagine the same scenario of a user holding up her hand, but this time wearing a see-through HMD. Registration errors now result in visual–visual conflicts between the images of the virtual and real hands. Such conflicts are easy to detect because of the resolution of the human eye and the sensitivity of the human visual system to differences. Even tiny offsets in the images of the real and virtual hands are easy to detect.

What angular accuracy is needed for good registration in Augmented Reality? A simple demonstration will show the order of magnitude required. Take out a dime and hold it at arm's length, so that it looks like a circle. The diameter of the dime covers about 1.2 to 2.0 degrees of arc, depending on your arm length. In comparison, the width of a full moon is about 0.5 degrees of arc! Now imagine a virtual object superimposed on a real object, but offset by the diameter of the full moon. Such a difference would be easy to detect. Thus, the angular accuracy required is a small fraction of a degree. The lower limit is bounded by the resolving power of the human eye itself. The central part of the retina is called the *fovea*, which has the highest density of color-detecting cones, about 120 per degree of arc, corresponding to a spacing of half a minute of arc (Jain, 1989). Observers can differentiate between a dark and light bar grating when each bar subtends about one minute of arc, and under special circumstances they can detect even smaller differences (Doenges, 1985). However, existing HMD trackers and displays are not capable of providing one minute of arc in accuracy, so the present achievable accuracy is much worse than that ultimate lower bound. In practice, errors of a few pixels are detectable in modern HMDs.

Registration of real and virtual objects is not limited to AR. Special-effects artists seamlessly integrate computer-generated 3-D objects with live actors in film and video. The difference lies in the amount of control available. With film, a director can carefully plan each shot, and artists can spend hours per frame, adjusting each by hand if necessary, to achieve perfect registration. As an interactive medium, AR is far more difficult to work with. The AR system cannot control the motions of the HMD wearer. The user looks where he wants, and the system must respond within tens of milliseconds.

Registration errors are difficult to adequately control because of the high accuracy requirements and the numerous sources of error. These sources of error can be divided into two types: static and dynamic. *Static* errors are the ones that cause registration errors even when the user's viewpoint and the objects in the environment remain completely still. *Dynamic* errors are the ones that have no effect until either the viewpoint or the objects begin moving.

For current HMD-based systems, dynamic errors are by far the largest contributors to registration errors, but static errors cannot be ignored either. The next two sections discuss static and dynamic errors and what has been done to reduce them. See Holloway (1997) for a thorough analysis of the sources and magnitudes of registration errors.

3.2 Static Errors

The four main sources of static errors are:

- Optical distortion
- Errors in the tracking system
- Mechanical misalignments
- Incorrect viewing parameters (e.g., field of view, tracker-to-eye position and orientation, interpupillary distance)

1. *Distortion in the optics:* Optical distortions exist in most camera and lens systems, both in the cameras that record the real environment and in the optics used for the display. Because distortions are usually a function of the radial distance away from the optical axis, wide field-of-view displays can be especially vulnerable to this error. Near the center of the field of view, images are relatively undistorted, but far away from the center, image distortion can be large. For example, straight lines may appear curved. In a see-through HMD with narrow field-of-view displays, the optical combiners add virtually no distortion, so the user's view of the real world is not warped. However, the optics used to focus and magnify the graphic images from the display monitors can introduce distortion. This mapping of distorted virtual images on top of an undistorted view of the real world causes static registration errors. The cameras and displays may also have nonlinear distortions that cause errors (Deering, 1992).

Optical distortions are usually systematic errors, so they can be mapped and compensated. This mapping may not be trivial, but it is often possible. For example, Robinett and Rolland (1992) describe the distortion of one commonly used set of HMD optics. The distortions might be compensated by additional optics. Edwards, Rolland, and Keller (1993) describe such a design for a video see-through HMD. This can be a difficult design problem, though, and it will add weight, which is not desirable in HMDs. An alternate approach is to do the compensation digitally. This can be done by image warping techniques, both on the digitized video and on the graphic images. Typically, this involves predistorting the images so that they will appear undistorted after being displayed (Watson and Hodges, 1995). Another way to perform digital compensation on the graphics is to apply the predistortion functions on the vertices of the polygons, in screen space, before rendering (Rolland and Hopkins, 1993). This requires subdividing polygons that cover large areas in screen space. Both digital compensation methods can be computationally expensive, often requiring special hardware to accomplish in real time. Holloway (1997) determined that the additional system delay required by the distortion compensation adds more registration error than the distortion compensation removes, for typical head motion.

2. *Errors in the tracking system:* Errors in the reported outputs from the tracking and sensing systems are often the most serious type of static registration errors. These distortions are not easy to measure and eliminate, because that requires another "3-D ruler" that is more accurate than the tracker being tested. These errors are often nonsystematic and difficult to fully characterize. Almost all commercially available tracking systems are not accurate enough to satisfy the requirements of AR systems. Section 5 discusses this important topic further.

3. *Mechanical misalignments:* Mechanical misalignments are discrepancies between the model or specification of the hardware and the actual physical properties of the real system. For example, the combiners, optics, and monitors in an optical see-through HMD may not be at the expected distances or orientations with respect to each other. If the frame is not sufficiently rigid, the various component parts may change their relative positions as the user moves around, causing errors. Mechanical misalignments can cause subtle changes in the position and orientation of the projected virtual images that are difficult to compensate. While some alignment errors can be calibrated, for many others it may be more effective to "build it right" initially.

4. *Incorrect viewing parameters:* Incorrect viewing parameters, the last major source of static registration errors, can be thought of as a special case of alignment errors where calibration techniques can be applied. Viewing parameters specify how to convert the reported head or camera locations into viewing matrices used by the scene generator to draw the graphic images. For an HMD-based system, these parameters include:

- Center of projection and viewport dimensions
- Offset, both in translation and orientation, between the location of the head tracker and the user's eyes
- Field of view

Incorrect viewing parameters cause systematic static errors. Take the example of a head tracker located above a user's eyes. If the vertical translation offsets between the tracker and the eyes are too small, all the virtual objects will appear lower than they should.

In some systems, the viewing parameters are estimated by manual adjustments, in a nonsystematic fashion. Such approaches proceed as follows: Place a real object in the environment and attempt to register a virtual object with that real object. While wearing the HMD or positioning the cameras, move to one viewpoint or a few selected viewpoints and manually adjust the location of the virtual object and the other viewing parameters until the registration "looks right." This may achieve satisfactory results if the environment and the viewpoint remain static. However, such approaches require a skilled user and generally do not achieve robust results for many viewpoints. Achieving good registration from a single viewpoint is much easier than registration from a wide variety of viewpoints using a single set of parameters. Usually what happens is satisfactory registration at one viewpoint, but when the user walks to a significantly different viewpoint, the registration is inaccurate because of incorrect viewing parameters or tracker distortions. This means that many different sets of parameters must be used, which is a less than satisfactory solution.

Another approach is to directly measure the parameters, using various measuring tools and sensors. For example, a commonly used optometrist's tool can measure the interpupillary distance. Rulers might measure the offsets between the tracker and eye positions. Cameras could be placed where the user's eyes would normally be in an optical see-through HMD. By recording what the camera sees, through the see-through HMD, of the real environment, one might be able to determine several viewing parameters. So far, direct measurement techniques have enjoyed limited success (Janin, Mizell and Caudell, 1993).

View-based tasks are another approach to calibration. These ask the user to perform various tasks that set up geometric constraints. By performing several tasks, enough information is gathered to determine the viewing parameters. For example, Azuma and Bishop (1994) asked a user wearing an optical see-through HMD to look straight through a narrow pipe mounted in the real environment. This sets up the constraint that the user's eye must be located along a line through the center of the pipe. Combining this with other tasks created enough constraints to measure all the viewing parameters. Caudell and Mizell (1992) used a different set of tasks, involving lining up two circles that specified a cone in the real environment. Oishi and Tachi (1996) move virtual cursors to appear on top of beacons in the real environment. All view-based tasks rely upon the user accurately performing the specified task and assume the tracker is accurate. If the tracking and sensing equipment is not accurate, then multiple measurements must be taken and optimizers used to find the "best-fit" solution (Janin, Mizell, and Caudell, 1993).

For video-based systems, an extensive body of literature exists in the robotics and photogrammetry communities on camera calibration techniques; see the references in Lenz and Tsai (1988) for a start. Such techniques compute a camera's viewing parameters by taking several pictures of an object of fixed and sometimes unknown geometry. These pictures must be taken from different locations. Matching points in the 2-D images with corresponding 3-D points on the object sets up mathematical constraints. With enough pictures, these constraints determine the viewing parameters and the 3-D location of the calibration object. Alternately, they can serve to drive an optimization routine that will search for the best set of viewing parameters that fits the collected data. Several AR systems have used camera calibration techniques, including Bajura (1993), Drascic and Milgram (1991), Tuceryan et al. (1995), and Whitaker et al. (1995), and many others.

3.3 Dynamic Errors

Dynamic errors occur because of system delays, or lags. The *end-to-end system delay* is defined as the time difference between the moment that the tracking system measures the position and orientation of the viewpoint to the moment when the generated images corresponding to that position and orientation appear in the displays. These delays exist because each component in an Augmented Reality system requires some time to do its job. The delays in the tracking subsystem, the communication delays, the time it takes the scene generator to draw the appropriate images in the frame buffers, and the scanout time from the frame buffer to the displays all contribute to end-to-end lag. End-to-end delays of 100 ms are fairly typical on existing systems. Simpler systems can have less delay, but other systems have more. Delays of 250 ms or more can exist on slow, heavily loaded, or networked systems.

End-to-end system delays cause registration errors only when motion occurs. Assume that the viewpoint and all objects remain still. Then the lag does not cause registration errors. No matter how long the delay is, the images generated are appropriate, since nothing has moved since the time the tracker measurement was taken. Compare this to the case with motion.

FIG. 2.6. Effect of motion and system delays on registration. Picture on the left is a static scene. Picture on the right shows motion. (Courtesy of the University of North Carolina Chapel Hill Department of Computer Science.)

For example, assume a user wears a see-through HMD and moves her head. The tracker measures the head at an initial time t. The images corresponding to time t will not appear until some future time t_2, because of the end-to-end system delays. During this delay, the user's head remains in motion, so when the images computed at time t finally appear, the user sees them at a different location than the one they were computed for. Thus, the images are incorrect for the time they are actually viewed. To the user, the virtual objects appear to "swim around" and "lag behind" the real objects. This was graphically demonstrated in a videotape of UNC's ultrasound experiment shown at SIGGRAPH '92 (Bajura, Fuchs, and Ohbuchi, 1992). In Figure 2.6, the picture on the left shows what the registration looks like when everything stands still. The virtual gray region above the tip of the wand represents what the ultrasound wand is scanning. This virtual region should be attached to the tip of the real ultrasound wand. This is the case in the picture on the left, where the tip of the wand is visible at the bottom of the picture, to the left of the "UNC" letters. But when the head or the wand moves, large dynamic registration errors occur, as shown in the picture on the right. The tip of the wand is now far away from the virtual region. Also note the motion blur in the background, which is caused by the user's head motion.

System delays seriously hurt the illusion that the real and virtual worlds coexist because they cause large registration errors. With a typical end-to-end lag of 100 ms and a moderate head rotation rate of 50 degrees per second, the angular dynamic error is 5 degrees. At a 68 cm arm length, this results in registration errors of almost 60 mm. System delay is the largest single source of registration error in existing AR systems, outweighing all others *combined* (Holloway, 1997).

Methods used to reduce dynamic registration fall under four main categories:

- Reduce system lag
- Reduce apparent lag
- Match temporal streams (with video-based systems)
- Predict future locations

1. *Reduce system lag:* The most direct approach is simply to reduce, or ideally eliminate, the system delays. If there are no delays, there are no dynamic errors. Unfortunately, modern scene generators are usually built for throughput, not minimal latency (Foley et al., 1990). It is sometimes possible to reconfigure the software to sacrifice throughput to minimize latency. For example, the SLATS system completes rendering a pair of interlaced NTSC images in one field time (16.67 ms) on Pixel-Planes 5 (Olano et al., 1995). Being careful about synchronizing pipeline tasks can also reduce the end-to-end lag (Wloka, 1995; Jacobs, Livingston, and State, 1997).

System delays are not likely to completely disappear anytime soon. Some believe that the current course of technological development will automatically solve this problem. Unfortunately, it is difficult to reduce system delays to the point where they are no longer an issue. Recall that registration errors must be kept to a small fraction of a degree. At the moderate head rotation rate of 50 degrees per second, system lag must be 10 ms or less to keep angular errors below 0.5 degrees. Just scanning out a frame buffer to a display at 60 Hz requires 16.67 ms. It might be possible to build an HMD system with less than 10 ms of lag, but the drastic cut in throughput and the expense required to construct the system would make alternate solutions attractive. Minimizing system delay is important, but reducing delay to the point where it is no longer a source of registration error is not currently practical.

2. *Reduce apparent lag:* Image deflection is a clever technique for reducing the amount of apparent system delay for systems that only use head orientation (Burbidge and Murray, 1989; Regan and Pose, 1994; Riner and Browder, 1992; So and Griffin, 1992). It is a way to incorporate more recent orientation measurements into the late stages of the rendering pipeline. Therefore, it is a feed-forward technique. The scene generator renders an image much larger than needed to fill the display. Then just before scanout, the system reads the most recent orientation report. The orientation value is used to select the fraction of the frame buffer to send to the display, since small orientation changes are equivalent to shifting the frame buffer output horizontally and vertically.

Image deflection does not work on translation, but image warping techniques might (Mark, McMillan, and Bishop, 1997). After the scene generator renders the image based upon the head tracker reading, small adjustments in orientation and translation could be done after rendering by warping the image. These techniques assume knowledge of the depth at every pixel, and the warp must be done much more quickly than rerendering the entire image.

3. *Match temporal streams:* In video-based AR systems, the video camera and digitization hardware impose inherent delays on the user's view of the real world. This is potentially a blessing when reducing dynamic errors, because it allows the temporal streams of the real and virtual images to be matched. Additional delay is added to the video from the real world to match the scene generator delays in generating the virtual images. This additional delay to the video stream will probably not remain constant, since the scene generator delay will vary with the complexity of the rendered scene. Therefore, the system must dynamically synchronize the two streams.

Note that while this reduces conflicts between the real and virtual, now *both* the real and virtual objects are delayed in time. While this may not be bothersome for small delays, it is a major problem in the related area of telepresence systems and will not be easy to overcome. For long delays, this can produce negative effects such as pilot-induced oscillation.

4. *Predict:* The last method is to predict the future viewpoint and object locations. If the future locations are known, the scene can be rendered with these future locations, rather than the measured locations. Then when the scene finally appears, the viewpoints and objects have moved to the predicted locations, and the graphic images are correct at the time they are viewed. For short system delays (under ~80 ms), prediction has been shown to reduce dynamic errors by up to an order of magnitude (Azuma and Bishop, 1994). Accurate predictions require a system built for real-time measurements and computation. Using inertial sensors makes predictions more accurate by a factor of 2–3. Predictors have been developed for a few AR systems (Emura and Tachi, 1994; Zikan et al., 1994b), but the majority were implemented and evaluated with VE systems (see the reference list in Azuma and Bishop (1994)). More work needs to be done on ways of comparing the theoretical performance of various predictors (Azuma, 1995; Azuma and Bishop, 1995) and in developing prediction models that better match actual head motion (Akatsuka and Bekey, 1998; Wu and Ouhyoung, 1995).

3.4 Vision-Based Techniques

Mike Bajura and Ulrich Neumann (Bajura and Neumann, 1995) point out that registration based solely on the information from the tracking system is like building an "open-loop" controller. The system has no feedback on how closely the real and virtual actually match. Without feedback, it is difficult to build a system that achieves perfect matches. However, video-based approaches can use image processing or computer vision techniques to aid registration. Since video-based AR systems have a digitized image of the real environment, it may be possible to detect features in the environment and use those to enforce registration. They call this a "closed-loop" approach, since the digitized image provides a mechanism for bringing feedback into the system.

This is not a trivial task. This detection and matching must run in real time and must be robust. This often requires special hardware and sensors. However, it is also not an "AI-complete" problem because this is simpler than the general computer vision problem.

For example, in some AR applications it is acceptable to place fiducials in the environment. These fiducials may be LEDs (Bajura and Neumann, 1995) or special markers (Mellor, 1995; Neumann and Cho, 1996; Neumann and Park, 1998). An ultrasound application at UNC Chapel Hill used colored dots as fiducials (State et al., 1996). The locations or patterns of the fiducials are assumed to be known. Image processing detects the locations of the fiducials; then those are used to make corrections that enforce proper registration.

These routines assume that one or more fiducials are visible at all times; without them, the registration can fall apart. But when the fiducials are visible, the results can be accurate to one pixel, which is as about close as one can get with video techniques. Figure 2.7, taken from Neumann and Cho (1996), shows virtual assembly instructions displayed directly on an aircraft part. Colored dots on the aircraft part serve as the fiducials. Mellor (1995) uses dots with a circular pattern as the fiducials to achieve nearly perfect registration. Hoff and Nguyen (1996) use multiple cameras and multiple fiducials for a PC assembly demonstration. Figure 2.8 demonstrates merging virtual objects with the real environment, using colored dots as the fiducials in a video-based approach. In the picture on the left, the stack of cards in the center are real, but the ones on the right are virtual. Notice that they penetrate one of the blocks. The image on the right shows the accuracy of the registration by displaying virtual edge lines over the real blocks and table top (State et al., 1996). Fiducial-based tracking has become a popular

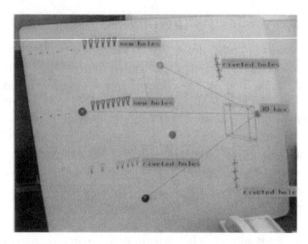

FIG. 2.7. Colored dots guide registration of manufacturing instructions on an aircraft part. (Courtesy Ulrich Neumann, USC.)

FIG. 2.8. Virtual cards and edge lines registered with real blocks
and table. (©1996 Gentaro Hirota and Andrei State, Courtesy of the
University of North Carolina Chapel Hill Department of Computer
Science.)

approach. Many AR systems built after 1996 partially or completely rely
upon tracking fiducials.

Instead of fiducials, Uenohara and Kanade (1995) use *template matching*
to achieve registration. Template images of the real object are taken from a
variety of viewpoints. These are used to search the digitized image for the
real object. Once that is found, a virtual wireframe can be superimposed
on the real object.

Recent approaches in video-based matching avoid the need for any
calibration. Kutukalos and Vallino (1998) represent virtual objects in a

non-Euclidean, affine frame of reference that allows rendering without knowledge of camera parameters. Iu and Rogovin (1996) extract contours from the video of the real world and then use an optimization technique to match the contours of the rendered 3-D virtual object with the contour extracted from the video. Note that calibration-free approaches may not recover all the information required to perform all potential AR tasks. For example, these two approaches do not recover true depth information, which is useful when compositing the real and the virtual.

Techniques that use fiducials as the sole tracking source determine the *relative* projective relationship between the objects in the environment and the video camera. While this is enough to ensure registration, it does not provide all the information one might need in some AR applications, such as the absolute (rather than relative) locations of the objects and the camera. Absolute locations are needed to include virtual and real objects that are not tracked by the video camera, such as a 3-D pointer or other virtual objects not directly tied to real objects in the scene.

Additional sensors besides video cameras can aid registration. Both Mellor (1995) and Grimson et al. (1994, 1995) use a laser rangefinder to acquire an initial depth map of the real object in the environment. Given a matching virtual model, the system can match the depth maps from the real and virtual until they are properly aligned, and that provides the information needed for registration.

Another way to reduce the difficulty of the problem is to accept the fact that the system may not be robust and may not be able to perform all tasks automatically. Then it can ask the user to perform certain tasks. The system in Sharma and Molineros (1997) expects manual intervention when the vision algorithms fail to identify a part because the view is obscured. The calibration techniques in Tuceryan et al. (1995) are heavily based on computer vision techniques, but they ask the user to manually intervene by specifying correspondences when necessary.

3.5 Current Status

The registration requirements for AR are difficult to satisfy, but a few systems have achieved good results. Azuma and Bishop (1994) use an open-loop system that shows registration typically within ±5 millimeters from many viewpoints for an object at about arm's length. Closed-loop systems, however, have demonstrated nearly perfect registration, accurate to within a pixel (Bajura and Neumann, 1995; Mellor, 1995; Neumann and Cho, 1996; State et al., 1996). The Mixed Reality Systems Laboratory demonstrated an AR "air hockey" game that was sufficiently accurate to allow two people to play the game in real time (Ohshima et al., 1998). Yokokohji, Sugawara, Yoshikawa (2000) demonstrated accurate registration even during rapid head movements.

The registration problem is far from solved. Many systems assume a static viewpoint, static objects, or even both. Even if the viewpoint or objects are allowed to move, they are often restricted in how far they can travel. Registration is shown under controlled circumstances, often with only a small number of real-world objects, or where the objects are already well-known to the system. For example, registration may only work on one object marked with fiducials, and not on any other objects in the scene. Much more work needs to be done to increase the domains in which registration is robust. Duplicating registration methods remains a nontrivial task, owing to both the complexity of the methods and the additional hardware required. If simple yet effective solutions could be developed, that would speed the acceptance of AR systems.

4.   SENSING

Accurate registration and positioning of virtual objects in the real environment requires accurate tracking of the user's head and sensing the locations of other objects in the environment. The biggest single obstacle to building effective Augmented Reality systems is the requirement of accurate, long-range sensors and trackers that report the locations of the user and the surrounding objects in the environment. For details of tracking technologies, see the surveys in Ferrin (1991) and Meyer, Applewhite, and Biocca (1992) and Chapter 5 of Durlach and Mavor (1995). Commercial trackers are aimed at the needs of Virtual Environments and motion capture applications. Compared to those two applications, Augmented Reality has much stricter accuracy requirements and demands larger working volumes. No tracker currently provides high accuracy at long ranges in real time. More work needs to be done to develop sensors and trackers that can meet these stringent requirements.

Specifically, AR demands more from trackers and sensors in three areas:

- Greater input variety and bandwidth
- Higher accuracy
- Longer range

4.1   Input Variety and Bandwidth

VE systems are primarily built to handle output bandwidth: the images displayed, sounds generated, etc. The input bandwidth is tiny: the locations of the user's head and hands, the outputs from the buttons and other control devices, etc. AR systems, however, will need a greater variety of input sensors and much more input bandwidth. There are a greater variety of possible input sensors than output displays. Outputs are limited to the five human senses. Inputs can come from anything a sensor can detect. Robinett (1992) speculates that Augmented Reality may be useful in any application that requires displaying information not directly available or detectable by human senses by making that information visible (or audible, touchable, etc.). For example, prototype medical applications use CT, MRI, and ultrasound sensors as inputs. Other future applications might use sensors to extend the user's visual range into infrared or ultraviolet frequencies, and remote sensors would let users view objects hidden by walls or hills. Conceptually, anything not detectable by human senses but detectable by machines might be transduced into something that a user can sense in an AR system.

Range data are a particular input that is vital for many AR applications (Aliaga, 1997; Breen et al., 1996). The AR system knows the distance to the virtual objects, because that model is built into the system. But the AR system may not know where all the real objects are in the environment. The system might assume that the entire environment is measured at the beginning and remains static thereafter. However, some useful applications will require a dynamic environment, in which real objects move, so the objects must be tracked in real time. However, for some applications a depth map of the real environment would be sufficient. That would allow real objects to occlude virtual objects through a pixel-by-pixel depth value comparison. Acquiring this depth map in real time is not trivial. Sensors such as laser rangefinders might be used. Many computer vision techniques for recovering shape through various strategies (e.g., "shape from stereo" or "shape from shading") have been tried. Wloka and Anderson (1995) use intensity-based matching from a pair of stereo images to do depth recovery. Kanbara et al. (2000) uses edge-based matching. Recovering depth through existing vision techniques is difficult to do robustly in real time.

Finally, some annotation applications require access to a detailed database of the environment, which is a type of input to the system. For example, an architectural application of "seeing into the walls" assumes that the system has a database of where all the pipes, wires, and other hidden objects are within the building. Such a database may not be readily available, and even if it is, it may not be in a format that is easily usable. For example, the data may not be grouped to segregate the parts of the model that represent wires from the parts that represent pipes. Thus, a significant modeling effort may be required and should be taken into consideration when building an AR application.

4.2 High Accuracy

The accuracy requirements for the trackers and sensors are driven by the accuracies needed for visual registration, as described in Section 3. For many approaches, the registration is only as accurate as the tracker. Therefore, the AR system needs trackers that are accurate to around a millimeter and a tiny fraction of a degree, across the entire working range of the tracker.

Few trackers can meet this specification, and every technology has weaknesses. Some mechanical trackers are accurate enough, although they tether the user to a limited working volume. Magnetic trackers are vulnerable to distortion by metal in the environment, which exists in many desired AR application environments. Ultrasonic trackers suffer from noise and are difficult to make accurate at long ranges because of variations in the ambient temperature. Optical technologies (Janin et al., 1994; Kim, Richards, and Caudell, 1997) have distortion and calibration problems. Inertial trackers drift with time. Of the individual technologies, optical technologies show the most promise due to trends toward high-resolution digital cameras, real-time photogrammetric techniques, and structured light sources that result in more signal strength at long distances. Future tracking systems that can meet the stringent requirements of AR will probably be hybrid systems (Azuma, 1993; Durlach and Mavor, 1995; Foxlin, 1996; Zikan et al., 1994a), such as a combination of inertial and optical technologies (You, Neumann, Azuma, 1999; Welch, 1995). Using multiple technologies opens the possibility of covering for each technology's weaknesses by combining their strengths.

Attempts have been made to calibrate the distortions in commonly used magnetic tracking systems (Bryson, 1992; Ghazisaedy et al., 1995; Livingston and State, 1997). These have succeeded at removing much of the gross error from the tracker at long ranges, but not to the level required by AR systems (Holloway, 1997). For example, mean errors at long ranges can be reduced from several inches to around one inch.

Welch and Bishop (1997) introduced a new mathematical approach called SCAAT (Single Constraint at a Time) that incorporates partial results as soon as they are measured in the tracking system, rather than waiting for an entire set of measurements to be taken as is normally done. This results in faster update rates, lower latency, and more accurate results. Although this approach was tested and demonstrated on a custom optical-based tracker, it should also apply to almost all other tracking technologies.

The requirements for registering other sensor modes are not nearly as stringent. For example, the human auditory system is not very good at localizing deep bass sounds, so a subwoofer can often be placed many feet away from the intended location of the sound source.

4.3 Long Range

Few trackers are built for accuracy at long ranges, since most VE applications do not require long ranges. *Motion capture* applications track an actor's body parts to control a computer-animated character or for the analysis of an actor's movements. This is fine for position recovery, but not for orientation. Orientation recovery is based upon the computed positions. Even tiny errors in those positions can cause orientation errors of a few degrees, which is too large for AR systems.

Three scalable tracking systems for HMDs have been described in the literature (Ward et al., 1992; Sowizral and Barnes, 1993; Foxlin, Harrington, and Pfeiffer, 1998). A scalable system is one that can be expanded to cover any desired range, simply by adding more modular components to the system. This is done by building a cellular tracking system, where only nearby sources and sensors are used to track a user. As the user walks around, the set of sources and sensors changes, thus achieving large working volumes while avoiding long distances between the current working set of sources and sensors. While scalable trackers can be effective, they are complex and by their very nature have many components, making them relatively expensive to construct.

The Global Positioning System (GPS) is used to track the locations of vehicles almost anywhere on the planet. It might be useful as one part of a long-range tracker for AR systems. However, by itself it will not be sufficient. The best reported accuracy is approximately one centimeter through carrier-phase GPS. That is not sufficiently accurate to recover orientation from a set of positions on a user. Also, carrier-phase GPS systems may not maintain centimeter accuracy if contact with the satellites is occasionally broken.

Tracking an AR system outdoors in real time with the required accuracy has not been demonstrated and remains an open problem, although steps have been taken in that direction (Feiner, MacIntyre, and Höllerer, 1997; Azuma, Hoff, Neely, and Sarfaty, 1999; Behringer, 1999; Piekarski, Gunther, Thomas, 1999).

5. FUTURE DIRECTIONS

This section identifies areas and approaches that require further research to produce improved AR systems.

Hybrid approaches: Future tracking systems may be hybrids, because combining approaches can cover weaknesses. The same may be true for other problems in AR. For example, current registration strategies generally focus on a single strategy. Future systems may be more robust if several techniques are combined. An example is combining vision-based techniques with prediction. If the fiducials are not available, the system switches to open-loop prediction to reduce the registration errors, rather than breaking down completely. The predicted viewpoints in turn produce a more accurate initial location estimate for the vision-based techniques.

Real-time systems and time-critical computing: Many VE systems are not truly run in real time. Instead, it is common to build the system, often on UNIX, and then see how fast it runs. This may be sufficient for some VE applications. Since everything is virtual, all the objects are automatically synchronized with each other. AR is a different story. Now the virtual and real must be synchronized, and the real world "runs" in real time. Therefore, effective AR systems must be built with real-time performance in mind. Accurate timestamps must be available. Operating systems must not arbitrarily swap out the AR software process at any time, for arbitrary durations. Systems must be built to guarantee completion within specified time budgets, rather than just "running as quickly as possible." These are characteristics of flight simulators and a few VE systems (Krueger, 1992). Constructing and debugging real-time systems is often painful and difficult, but the requirements for AR demand real-time performance.

Perceptual and psychophysical studies: Augmented Reality is an area ripe for psychophysical studies. How much lag can a user detect? How much registration error is detectable when the head is moving? Besides questions on perception, psychological experiments that explore *perfor-mance* issues are also needed. How much does head-motion prediction improve user performance on a specific task? How much registration error is tolerable for a specific application before performance on that task degrades substantially? Is the allowable error larger while the user moves her head versus when she stands still? Furthermore, not much is known about potential optical illusions caused by errors or conflicts in the simultaneous display of real and virtual objects (Durlach and Mavor, 1995).

Few experiments in this area have been performed. Jannick Rolland, Frank Biocca, and their students conducted a study of the effect caused by eye displacements in video see-through HMDs (Rolland et al., 1995). They found that users partially adapted to the eye displacement, but they also had negative aftereffects after removing the HMD. Steve Ellis' group at NASA Ames has conducted work on perceived depth in a see-through HMD (Ellis and Bucher, 1994; Ellis and Menges, 1995; Ellis et al., 1997; Ellis and Menges, 1997). ATR has also conducted a study (Utsumi et al., 1994).

Portability: Section 2.4 explained why some potential AR applications require giving the user the ability to walk around large environments, even outdoors. This requires making the equipment self-contained and portable. Therefore, practical AR systems will rely heavily on the development of wearable computers. All the standard issues with wearable computers (weight, power, ergonomics, etc.) affect AR systems as well. AR may provide a natural metaphor for interacting with a wearable computer in some applications (Starner et al., 1997).

Multimodal displays: Almost all work in AR has focused on the visual sense: virtual graphic objects and overlays. But Section 3.1 explained that augmentation might apply to all other senses as well. In particular, adding and removing 3-D sound is a capability that could be useful in some AR applications.

Social and political issues: Technological issues are not the only ones that need to be considered when building a real application. There are also social and political dimensions when getting new technologies into the hands of real users. Sometimes, perception is what counts, even if the technological reality is different. For example, if workers perceive lasers to be a health risk, they may refuse to use a system with lasers in the display or in the trackers, even if those lasers are eye safe. Ergonomics and ease of use are paramount considerations. Whether AR is truly a cost-effective solution in its proposed applications has yet to be determined. Another important factor is whether or not the technology is perceived as a threat to jobs, as a replacement for workers, especially with many corporations undergoing recent layoffs. AR may do well in this regard, because it is intended as a tool to make the user's job easier, rather than something that completely replaces the human worker. Although technology transfer is not normally a subject of academic papers, it is a real problem. Social and political concerns should not be ignored during attempts to move AR out of the research lab and into the hands of real users (Curtis, Mizell, Gruenbaum, and Janin, 1998).

6. CONCLUSION

Augmented Reality is far behind Virtual Environments in maturity. Several commercial vendors sell complete, turnkey Virtual Environment systems. However, no commercial vendor currently sells an HMD-based Augmented Reality system. A few monitor-based "virtual set" systems are available, but today AR systems are primarily found in academic and industrial research laboratories.

The first deployed HMD-based AR systems will probably be in the application of aircraft manufacturing. The merged company of Boeing and McDonnell Douglas has been exploring this application (Nash, 1997; Neumann and Cho, 1996) with both optical and video approaches. Boeing has performed trial runs with workers using a prototype system. Annotation and visualization applications in restricted, limited-range environments are deployable today, although much more work needs to be done to make them cost effective and flexible. Applications in medical visualization will take longer. Prototype visualization aids have been used on an experimental basis, but the stringent registration requirements and ramifications of mistakes will postpone common usage for many years. AR will probably be used for medical training before it is commonly used in surgery.

The next generation of combat aircraft will have helmet-mounted sights with graphics registered to targets in the environment (Wanstall, 1989). These displays, combined with short-range steerable missiles that can shoot at targets off-boresight, give a tremendous combat advantage to pilots in dogfights. Instead of having to be directly behind the target in order to shoot at it, a pilot can now shoot at anything within a 60–90 degree cone of the aircraft's forward centerline. Russia and Israel currently have systems with this capability, and the United States is expected to field the AIM-9X missile with its associated helmet-mounted sight in 2002 (Dornheim and Hughes, 1995; Dornheim, 1995a). Registration errors due to delays are a major problem in this application (Dornheim, 1995b).

Augmented Reality is a relatively new field, where most of the research efforts have occurred in the past six years, as shown by the references listed at the end of this chapter. The SIGGRAPH "Rediscovering Our Fire" report identified Augmented Reality as one of four areas where SIGGRAPH should encourage more submissions (Mair, 1994). Because of the numerous challenges and unexplored avenues in this area, AR will remain a vibrant area of research for at least the next several years.

One area where a breakthrough is required is tracking an HMD outdoors at the accuracy required by AR. If this is accomplished, several interesting applications will become possible. Two examples are described here: navigation maps and visualization of past and future environments.

The first application is a navigation aid to people walking outdoors. These individuals could be soldiers advancing upon their objective, hikers lost in the woods, or tourists seeking directions to their intended destination. Today, these individuals must pull out a physical map and associate what they see in the real environment around them with the markings on the 2-D map. If landmarks are not easily identifiable, this association can be difficult to perform, as anyone lost in the woods can attest. An AR system makes navigation easier by performing the association step automatically. If the user's position and orientation are known, and the AR system has access to a digital map of the area, then the AR system can draw the map in 3-D directly upon the user's view. The user looks at a nearby mountain and sees graphics directly overlaid on the real environment explaining the mountain's name, how tall it is, how far away it is, and where the trail is that leads to the top.

The second application is visualization of locations and events as they were in the past or as they will be after future changes are performed. Tourists that visit historical sites, such as a Civil War battlefield or the Acropolis in Athens, Greece, do not see these locations as they were in the past, due to changes over time. It is often difficult for a modern visitor to imagine what these sites really looked like in the past. To help, some historical sites stage "Living History" events where volunteers wear ancient clothes and reenact historical events. A tourist equipped with an outdoors AR system could see a computer-generated version of living history. The HMD could cover up modern buildings and monuments in the background and show, directly on the grounds at Gettysburg, where the Union and Confederate troops were at the fateful moment of Pickett's charge. The gutted interior of the modern Parthenon would be filled in by computer-generated representations of what it looked like in 430 BC, including the long-vanished gold statue of Athena in the middle. Tourists and students walking around the grounds with such AR displays would gain a much better understanding of these historical sites and the important events that took place there. Similarly, AR displays could show what proposed architectural changes would look like before they are carried out. An urban designer could show clients and politicians what a new stadium would look like as they walked around the adjoining neighborhood, to better understand how the stadium project will affect nearby residents.

After the basic problems with AR are solved, the ultimate goal will be to generate virtual objects that are so realistic that they are virtually indistinguishable from the real environment. Photorealism has been demonstrated in feature films, but accomplishing this in an interactive application will be much harder. Lighting conditions, surface reflections, and other properties must be measured automatically, in real time. More sophisticated lighting, texturing, and shading capabilities must run at interactive rates in future scene generators. Registration must be nearly perfect, without manual intervention or adjustments. While these are difficult problems, they are probably not insurmountable. It took roughly twenty-five years to progress from drawing stick figures on a screen to the photorealistic dinosaurs in "Jurassic Park." Within another twenty-five years, we should be able to wear a pair of AR glasses outdoors to see and interact with photorealistic dinosaurs eating a tree in our backyard.

Acknowledgments   Some of the material in this chapter appeared in the paper "A Survey of Augmented Reality," published in *Presence: Teleoperators and Virtual Environments*, Volume 6, #4 (August 1997) by MIT Press.

I thank the following individuals and organizations for sending pictures to include with this chapter:

Andrei State and Linda Houseman, The University of North Carolina at Chapel Hill Department of Computer Science
Ulrich Neumann, University of Southern California
Jannick Rolland, Center for Research and Engineering in Optics and Lasers (CREOL) at the University of Central Florida.

REFERENCES AND BIBLIOGRAPHY

The best starting point on the World Wide Web is Jim Vallino's AR page at: http://www.cs.rit.edu/~jrv/research/ar/

Akatsuka, Y., and Bekey, G. (1998). Compensation for End to End Delays in a VR System. *Proceedings of IEEE VRAIS '98*, Atlanta, March, 156–159.

Aliaga, D. G. (1997). Virtual Objects in the Real World. *Communications of the ACM, 40*, (3), 49–54.

Azuma, R. (1993). Tracking Requirements for Augmented Reality. *Communications of the ACM, 36* (7), 50–51.

Azuma, R. T. (1995). *Predictive Tracking for Augmented Reality*. Ph.D. dissertation, Department of Computer Science, University of North Carolina at Chapel Hill. Available as UNC-CH CS Dept. technical report TR95-007.

Azuma, R., and Bishop, G. (1994). Improving Static and Dynamic Registration in a See-Through HMD. *Computer Graphics* Annual Conference Series 1994 (*Proceedings of SIGGRAPH '94*), Orlando, July, 197–204.

Azuma, R., and Bishop, G. (1995). A Frequency-Domain Analysis of Head-Motion Prediction. *Computer Graphics* Annual Conference Series 1995 (*Proceedings of SIGGRAPH '95*), Los Angeles, August, 401–408.

Azuma, R., Hoff, B., Neely, H., Sarfaty, R. (1999). A Motion-Stabilized Outdoor Augmented Reality System. *Proceedings of IEEE VR '99*, Houston, March, 252–259.

Bajura, M. (1993). *Camera Calibration for Video See-Through Head-Mounted Display*. Technical Report TR93-048. University of North Carolina Chapel Hill Department of Computer Science.

Bajura, M., Fuchs, H., and Ohbuchi, R. (1992). Merging Virtual Reality with the Real World: Seeing Ultrasound Imagery within the Patient. *Computer Graphics (Proceedings of SIGGRAPH '92), 26* (2), Chicago, July, 203–210.

Bajura, M., and Neumann, U. (1995). Dynamic Registration Correction in Video-Based Augmented Reality Systems. *IEEE Computer Graphics and Applications, 15* (5), 52–60.

Barfield, W., Rosenberg, C., and Lotens, W. A. (1995). Augmented-Reality Displays. In Barfield, W., and Furness, T. A., III (editors). *Virtual Environments and Advanced Interface Design*. Oxford University Press, 542–575.

Behringer, R. (1999). Registration for Outdoor Augmented Reality Applications Using Computer Vision Techniques and Hybrid Sensors. *Proceedings of IEEE VR '99*, Houston, March, 244–251.

Breen, D. E., Whitaker, R. T., Rose, E., and Tuceryan, M. (1996). Interactive Occlusion and Automatic Object Placement for Augmented Reality. *Proceedings of Eurographics '96*, Futuroscope-Poitiers, August, 11–22.

Brooks, F. P., Jr. (1996). The Computer Scientist as Toolsmith II. *Communications of the ACM, 39* (3), 61–68.

Bryson, S. (1992). Measurement and Calibration of Static Distortion of Position Data from 3D Trackers. *Proceedings of SPIE Vol. 1669: Stereoscopic Displays and Applications III*, San Jose, February, 244–255.

Burbidge, D., and Murray, P. M. (1989). Hardware Improvements to the Helmet-Mounted Projector on the Visual Display Research Tool (VDRT) at the Naval Training Systems Center. *SPIE Proceedings Vol. 1116 Head-Mounted Displays*, 52–59.

Caudell, T. P., and Mizell, D. W. (1992). Augmented Reality: An Application of Heads-Up Display Technology to Manual Manufacturing Processes. *Proceedings of Hawaii International Conference on System Sciences*, January, 659–669.

Curtis, D., Mizell, D., Gruenbaum, P., and Janin, A. (1998). Several Devils in the Details: Making an AR App Work in the Airplane Factory. *Proceedings of International Workshop on Augmented Reality '98 (IWAR)*, San Francisco, November, 47–60.

Deering, M. (1992). High Resolution Virtual Reality. *Computer Graphics (Proceedings of SIGGRAPH '92), 26* (2), Chicago, July, 195–202.

Doenges, P. K. (1985). Overview of Computer Image Generation in Visual Simulation. *Course Notes, 14: ACM SIGGRAPH 1985*, San Francisco, July.

Dornheim, M. A. (1995a). U.S. Fighters to Get Helmet Displays After 2000. *Aviation Week and Space Technology, 143* (17), 46–48.

Dornheim, M. A. (1995b). Helmet-Mounted Sights Must Overcome Delays. *Aviation Week and Space Technology, 143* (17), 54.

Dornheim, M. A., and Hughes, D. (1995). U.S. Intensifies Efforts to Meet Missile Threat. *Aviation Week and Space Technology, 143* (16), 36–39.

Drascic, D., and Milgram, P. (1991). Positioning Accuracy of a Virtual Stereographic Pointer in a Real Stereoscopic Video World. *SPIE Proceedings Volume 1457—Stereoscopic Displays and Applications II*, San Jose, February, 302–313.

Durlach, N. I., and Mavor, A. S., editors. (1995). *Virtual Reality: Scientific and Technological Challenges*. Washington, DC: National Academy Press.

Edwards, E., Rolland, J., and Keller, K. (1993). Video See-Through Design for Merging of Real and Virtual Environments. *Proceedings of IEEE VRAIS '93*, Seattle, September, 222–233.

Ellis, S. R., Bréant, F., Menges, B., Jacoby R., and Adelstein, B. D. (1997). *Proceedings of IEEE VRAIS '97*, Albuquerque, March, 138–145.

Ellis, S. R., and Bucher, U. J. (1994). Distance Perception of Stereoscopically Presented Virtual Objects Optically Superimposed on Physical Objects by a Head-Mounted See-Through Display. *Proceedings of 38th Annual Meeting of the Human Factors and Ergonomics Society*, Nashville, October, 1300–1305.

Ellis, S. R., and Menges, B. M. (1995). Judged Distance to Virtual Objects in the near Visual Field. *Proceedings of 39th Annual Meeting of the Human Factors and Ergonomics Society*, San Diego, 1400–1404.

Ellis, S. R., and Menges, B. M. (1997). Judgments of the Distance to Nearby Virtual Objects: Interaction of Viewing Conditions and Accommodative Demand. *Presence: Teleoperators and Virtual Environments, 6*, (4), August, 452–460.

Emura, S., and Tachi, S. (1994). Compensation of Time Lag between Actual and Virtual Spaces by Multi-Sensor Integration. *Proceedings of the 1994 IEEE International Conference on Multisensor Fusion and Integration for Intelligent Systems*, Las Vegas, October, 463–469.

Feiner, S., MacIntyre, B., and Höllerer, T. (1997). A Touring Machine: Prototyping 3D Mobile Augmented Reality Systems for Exploring the Urban Environment. *Proceedings of First International Symposium on Wearable Computers*, Cambridge, October, 74–81.

Ferrin, F. J. (1991). Survey of Helmet Tracking Technologies. *SPIE Proceedings Vol. 1456: Large-Screen Projection, Avionic, and Helmet-Mounted Displays*, 86–94.

Foley, J. D., van Dam, A., Feiner, S. K., and Hughes, J. F. (1990). *Computer Graphics: Principles and Practice*, 2nd edition. Reading, MA: Addison-Wesley.

Foxlin, E. (1996). Inertial Head-Tracker Sensor Fusion by a Complementary Separate-Bias Kalman Filter. *Proceedings of VRAIS '96*, Santa Clara, April, 185–194.

Foxlin, E., Harrington, M., and Pfeiffer, G. (1998). Constellation: A Wide-Range Wireless Motion-Tracking System for Augmented Reality and Virtual Set Applications. *Proceedings of SIGGRAPH '98*, Orlando, July, 371–378.

Ghazisaedy, M., Adamczyk, D., Sandin, D. J., Kenyon, R. V., and DeFanti, T. A. (1995). Ultrasonic Calibration of a Magnetic Tracker in a Virtual Reality Space. *Proceedings of VRAIS '95*, Research Triangle Park, March, 179–188.

Grimson, W., Lozano-pérez, T., Wells, W., Ettinger, G., White, S., and Kikinis, R. (1994). An Automatic Registration Method for Frameless Stereotaxy, Image Guided Surgery, and Enhanced Reality Visualization. *Proceedings of IEEE Conference on Computer Vision and Pattern Recognition*, Los Alamitos, June, 430–436.

Grimson, W. E. L., Ettinger, G. J., White, S. J., Gleason, P. L., Lozano-Pérez, T., Wells, W. M., III, and Kikinis, R. (1995). Evaluating and Validating an Automated Registration System for Enhanced Reality Visualization in Surgery. *Proceedings of Computer Vision, Virtual Reality, and Robotics in Medicine '95*, Nice, April, 3–12.

Hoff, W. A. and Nguyen, K. (1996). Computer vision-based registration techniques for augmented reality. *Proceedings of Intelligent Robots and Computer Vision XV, SPIE vol. 2904*, Boston, November, 538–548.

Holloway, R. (1997). Registration Error Analysis for Augmented Reality. *Presence: Teleoperators and Virtual Environments, 6* (4), August, 413–432.

Holmgren, D. E. (1992). *Design and Construction of a 30-Degree See-Through Head-Mounted Display.* Technical Report TR92-030. University of North Carolina Chapel Hill Department of Computer Science.

Iu, S.-L., and Rogovin, K. W. (1996). Registering Perspective Contours with 3-D Objects without Correspondence Using Orthogonal Polynomials. *Proceedings of VRAIS '96*, Santa Clara, April, 37–44.

Jacobs, M. C., Livingston, M. A., and State, A. (1997). Managing Latency in Complex Augmented Reality Systems. *Proceedings of 1997 Symposium on Interactive 3D Graphics*, Providence, April, 49–54.

Jain, A. K. (1989). *Fundamentals of Digital Image Processing.* Prentice-Hall.

Janin, A. L., Mizell, D. W., and Caudell, T. P. (1993). Calibration of Head-Mounted Displays for Augmented Reality Applications. *Proceedings of IEEE VRAIS '93*, Seattle, September, 246–255.

Janin, A., Zikan, K., Mizell, D., Banner, M., and Sowizral, H. (1994). A Videometric Head Tracker for Augmented Reality. *SPIE Proceedings Volume 2351: Telemanipulator and Telepresence Technologies*, Boston, November, 308–315.

Kanbara, M., Okuma, T., Takemura, H., Yokoya, N. (2000). A Stereoscopic Video See-through Augmented Reality System Based on Real-time Vision-based Registration. *Proceedings of IEEE VR 2000*, New Brunswick, March, 255–262.

Kim, D., Richards, S.W., and Caudell, T. P. (1997). An Optical Tracker for Augmented Reality and Wearable Computers. *Proceedings of IEEE VRAIS '97*, Albuquerque, March, 146–150.

Krueger, M. W. (1992). Simulation versus Artificial Reality. *Proceedings of IMAGE VI Conference*, Scottsdale, July, 147–155.

Kutulakos, K. N., and Vallino, J. R. (1998). Calibration-Free Augmented Reality. *IEEE Transactions on Visualization and Computer Graphics, 4* (1), January–March, 1–20.

Lenz, R. K., and Tsai, R. Y. (1988). Techniques for Calibration of the Scale Factor and Image Center for High Accuracy 3-D Machine Vision Metrology. *IEEE Transactions on Pattern Analysis and Machine Intelligence, 10* (5), 713–720.

Livingston, M. A., and State, A. (1997). Magnetic Tracker Calibration for Improved Augmented Reality Registration. *Presence: Teleoperators and Virtual Environments, 6*, (5), October, 532–546.

Mair, S. G. (1994). Preliminary Report on SIGGRAPH in the 21st Century: Rediscovering Our Fire. *Computer Graphics, 28* (4), 288–296.

Mark, W. R., McMillan, L., and Bishop, G. (1997) Post-Rendering 3D Warping. *Proceedings of 1997 Symposium on Interactive 3D Graphics*, Providence, April, 7–16.

Mellor, J. P. (1995). Realtime Camera Calibration for Enhanced Reality Visualization. *Proceedings of Computer Vision, Virtual Reality, and Robotics in Medicine '95*, Nice, April, 471–475.

Meyer, K., Applewhite, H. L., and Biocca, F. A. (1992). A Survey of Position-Trackers. *Presence: Teleoperators and Virtual Environments, 1* (2), 173–200.

Nash, J. (1997). Wiring the Jet Set. *Wired, 5* (10), October, 128–135.

Neumann, U., and Cho, Y. (1996). A Self-Tracking Augmented Reality System. *Proceedings of VRST '96*, Hong Kong, July, 109–115.

Neumann, U., and Majoros, A. (1998). Cognitive, Performance, and System Issues for Augmented Reality Applications in Manufacturing and Maintenance. *Proceedings of IEEE VRAIS '98*, Atlanta, March, 4–11.

Neumann, U., and Park, J. (1998). Extendible Object-Centric Tracking for Augmented Reality. *Proceedings of IEEE VRAIS '98*, Atlanta, March, 148–155.

Oishi, T., and Tachi, S. (1996). Methods to Calibrate Projection Transformation Parameters for See-Through Head-Mounted Displays. *Presence: Teleoperators and Virtual Environments, 5* (1), 122–135.

Olano, M., Cohen, J., Mine, M., and Bishop, G. (1995). Combating Graphics System Latency. *Proceedings of 1995 Symposium on Interactive 3D Graphics*, Monterey, April, 19–24.

Ohshima, T., Satoh, K., Yamamoto, H., and Tamura, H. (1998). AR^2 Hockey: A Case Study of Collaborative Augmented Reality. *Proceedings of IEEE VRAIS '98*, Atlanta, March, 268–275.

Pausch, R., Crea, T., and Conway, M. (1992). A Literature Survey for Virtual Environments: Military Flight Simulator Visual Systems and Simulator Sickness. *Presence: Teleoperators and Virtual Environments, 1* (3), 344–363.

Peuchot, B., Tanguy, A., and Eude, M. (1995). Virtual Reality as an Operative Tool During Scoliosis Surgery. *Proceedings of Computer Vision, Virtual Reality, and Robotics in Medicine '95*, Nice, April, 549–554.

Piekarski, W., Gunther, B., Thomas, B. (1999). Integrating Virtual and Augmented Realities in an Outdoor Application. *Proceedings of International Workshop on Augmented Reality '99 (IWAR)*, San Francisco, October, 45–54.

Regan, M., and Pose, R. (1994). Priority Rendering with a Virtual Reality Address Recalculation Pipeline. *Computer Graphics* Annual Conference Series 1994 (*Proceedings of SIGGRAPH '94*), Orlando, July, 155–162.

Riner, B., and Browder, B. (1992). Design Guidelines for a Carrier-Based Training System. *Proceedings of IMAGE VI*, Scottsdale, July, 65–73.

Robinett, W. (1992). Synthetic Experience: A Proposed Taxonomy. *Presence: Teleoperators and Virtual Environments, 1* (2), 229–247.

Robinett, W., and Rolland, J. (1992). A Computational Model for the Stereoscopic Optics of a Head-Mounted Display. *Presence: Teleoperators and Virtual Environments, 1* (1), 45–62.

Rolland, J. P., and Hopkins, T. (1993). *A Method of Computational Correction for Optical Distortion in Head-Mounted Displays*. Technical Report TR93-045. University of North Carolina Chapel Hill Department of Computer Science.

Rolland, J., Biocca, F., Barlow, T., and Kancherla, A. (1995). Quantification of Adaptation to Virtual-Eye Location in See-Thru Head-Mounted Displays. *Proceedings of IEEE VRAIS '95*, Research Triangle Park, March, 56–66.

Sharma, R., and Molineros. J. (1997). Computer Vision-Based Augmented Reality for Guiding Manual Assembly. *Presence: Teleoperators and Virtual Environments, 6*, (3), June, 292–317.

So, R. H. Y., and Griffin, M. J. (1992). Compensating Lags in Head-Coupled Displays Using Head Position Prediction and Image Deflection. *Journal of Aircraft, 29* (6), 1064–1068.

Sowizral, H., and Barnes, J. (1993). Tracking Position and Orientation in a Large Volume. *Proceedings of IEEE VRAIS '93*, Seattle, September, 132–139.

Starner, T., Mann, S., Rhodes. B., Levine. J., Healey, J., Kirsch, D., Picard. R.W., and Pentland. A. (1997). Augmented Reality through Wearable Computing. *Presence: Teleoperators and Virtual Environments, 6* (4), August, 386–398.

State, A., Hirota, G., Chen, D. T., Garrett, B., and Livingston, M. (1996). Superior Augmented Reality Registration by Integrating Landmark Tracking and Magnetic Tracking. *Computer Graphics* Annual Conference Series 1996 (*Proceedings of SIGGRAPH '96*), New Orleans, August, 429–438.

Tuceryan, M., Greer, D. S., Whitaker, R. T., Breen, D., Crampton, C., Rose, E., and Ahlers, K. H. (1995). Calibration Requirements and Procedures for Augmented Reality. *IEEE Transactions on Visualization and Computer Graphics, 1* (3), 255–273.

Uenohara, M., and Kanade, T. (1995). Vision-Based Object Registration for Real-Time Image Overlay. *Proceedings of Computer Vision, Virtual Reality, and Robotics in Medicine '95*, Nice, April, 13–22.

Utsumi, A., Milgram, P., Takemura, H., and Kishino, F. (1994). Effects of Fuzziness in Perception of Stereoscopically Presented Virtual Object Location. *SPIE Proceedings Volume 2351: Telemanipulator and Telepresence Technologies*, Boston, November, 337–344.

Wanstall, B. (1989). HUD on the Head for Combat Pilots. *Interavia, 44*, April, 334–338.

Ward, M., Azuma, R., Bennett, R., Gottschalk, S., and Fuchs, H. (1992). A Demonstrated Optical Tracker with Scalable Work Area for Head-Mounted Display Systems. *Proceedings of 1992 Symposium on Interactive 3D Graphics*, Cambridge, March, 43–52.

Watson, B., and Hodges, L. (1995). Using Texture Maps to Correct for Optical Distortion in Head-Mounted Displays. *Proceedings of IEEE VRAIS '95*, Research Triangle Park, March, 172–178.

Welch, R. B. (1978). *Perceptual Modification: Adapting to Altered Sensory Environments*. Academic Press.

Welch, G. (1995). Hybrid Self-Tracker: An Inertial/Optical Hybrid Three-Dimensional Tracking System. UNC Chapel Hill Dept. of Computer Science Technical Report TR95-048, 21 pages.

Welch, G., and Bishop, G. (1997). SCAAT: Incremental Tracking with Incomplete Information. *Computer Graphics* Annual Conference Series 1997 (*Proceedings of SIGGRAPH '97*), Los Angeles, August, 333–344.

Wellner, P. (1993). Interacting with Paper on the DigitalDesk. *Communications of the ACM, 36* (7), 86–96.

Whitaker, R. T., Crampton, C., Breen, D. E., Tuceryan, M., and Rose, E. (1995). Object Calibration for Augmented Reality. *Proceedings of Eurographics '95*, Maastricht, August, 15–27.

Wloka, M. M. (1995). Lag in Multiprocessor Virtual Reality. *Presence: Teleoperators and Virtual Environments, 4* (1), 50–63.

Wloka, M. M., and Anderson, B. G. (1995). Resolving Occlusion in Augmented Reality. *Proceedings of 1995 Symposium on Interactive 3D Graphics*, Monterey, April, 5–12.

Wu, J.-R., and Ouhyoung, M. (1995). A 3D Tracking Experiment on Latency and Its Compensation Methods in Virtual Environments. *Proceedings of UIST '95*, Pittsburgh, November, 41–49.

Yokokohji, Y., Sugawara, Y., and Yoshikawa, Y. (2000). Accurate Image Overlay on Video See-through HMDs Using Vision and Accelerometers. *Proceedings of IEEE VR 2000*, New Brunswick, March, 247–254.

You, S., Neumann, U., Azuma, R. (1999). Hybrid Inertial and Vision Tracking for Augmented Reality Registration. *Proceedings of IEEE VR '99*, Houston, March, 260–267.

Zikan, K., Curtis, W. D., Sowizral, H., and Janin, A. (1994a). Fusion of Absolute and Incremental Position and Orientation Sensors. *SPIE Proceedings Volume 2351: Telemanipulator and Telepresence Technologies*, Boston, November, 316–327.

Zikan, K., Curtis, W. D., Sowizral, H. A., and Janin, A. L. (1994b). A Note on Dynamics of Human Head Motions and on Predictive Filtering of Head-Set Orientations. *SPIE Proceedings Volume 2351: Telemanipulator and Telepresence Technologies*, Boston, November, 328–336.

II

Technology

3

A Survey of Tracking Technology for Virtual Environments

Jannick P. Rolland[1,2,], Larry D. Davis[2], and Yohan Baillot[2]

[1]*School of Optics*

[2]*School of Electrical Engineering and Computer Science, University of Central Florida*

ABSTRACT

Tracking for virtual environments is necessary to record the position and the orientation of real objects in physical space and to allow spatial consistency between real and virtual objects. This chapter presents a top-down classification of tracking technologies aimed more specifically at head tracking, organized in accordance with their physical principles of operation.

Five main principles were identified: time-frequency measurement, spatial scan, inertial sensing, mechanical linkages, and direct-field sensing. We briefly describe each physical principle and present implementations of that principle. Advantages and limitations of these implementations are discussed and summarized in tabular form. A few hybrid technologies are then presented and general considerations of tracking technology are discussed.

1. INTRODUCTION

Human exploration in virtual environments requires technology that can accurately measure the position and the orientation of one or several users as they move and interact in the environment. This is referred to as tracking users in the environment. The position and the orientation of each user are measured with respect to the virtual environment coordinate system. A common approach for tracking a user in a virtual environment is to define a local coordinate system at the head of the user and measure the position and the orientation of this coordinate system with respect to a reference coordinate system. This chapter most generally reviews technologies used to track real-world features at human scale in virtual environments, such as head or limb motion tracking.

The tracking technologies employed span a variety of engineering fields that include optics, electromagnetics, electronics, and mechanics. This multi-disciplinary combination often makes it challenging to understand the working principle of a given tracking system. To facilitate understanding, we propose a top-down taxonomy on the technology that emphasizes the underlying physical principles of operation and the types of measurements involved. We chose such taxonomy because it allows summarizing a large body of work in a manner that we hope will stimulate going beyond the applications and the requirements for tracking and learning more about the various underlying technologies themselves. We also hope it will stimulate the generation of new ideas to the tracking problem.

Previous surveys of tracking technologies and their use in Virtual Reality can be found in Ferrin (1991), Rodgers (1991), Meyer et al. (1992), Bhatnagar (1993), Burdea & Coiffet (1993), Durlach & Mavor (1994), and Fuchs, (1996). The present review brings a top-down perspective on the technology, classifying the various technological implementations by principle of operation. We distinguish between technologies that use only one physical principle and those that use a combination of principles, that later are referred to as hybrid systems. In the literature, hybrid technologies usually refer to the combination of various technological implementations (e.g., optical and mechanical). A system will be specified as hybrid if either various principles of operation or various technological implementations are used.

The proposed classification is inspired in part by Chavel's perspective on range measurement techniques (Chavel & Strand, 1984). We identified five main principles of operation: time-frequency measurement, spatial scan, inertial sensing, mechanical linkages, and direct-field sensing. The classification is presented along with a description of the principles involved and examples. The latter are meant as a representative rather than a comprehensive selection. For each tracking principle, a table summarizes the physical phenomenon involved, the measured variable, the characteristics (e.g., accuracy, resolution, and range of operation), as well as the advantages and limitations of the technique. The tables were assembled from published literature and available patents and while some of the numbers will become obsolete with technological progress, we hope they provide some guidelines for what the technology can provide at a point in time. For the hybrid systems, the tables are omitted because data are mostly unavailable. However, the characteristics, the advantages, and the disadvantages of each system are provided in a tabular format.

Several subclassifications proposed in this chapter are in concordance with various research publications on tracking systems (Wang et al., 1990; Ferrin, 1991; Burdea & Coiffet, 1993; Fuchs, 1996). In Appendix A, we provide some definitions of terms commonly associated with tracking for virtual environments as well as symbols employed in this chapter.

2. FREQUENCY AND TIME MEASUREMENTS

In this section, we discuss tracking systems whose operating principles are based upon the use of time or frequency measurements. Typically, these devices measure the time of propagation of a signal, compare the phase difference of a measured signal to a reference, or use frequency measurement techniques to indirectly measure time differences. By taking advantage of a priori knowledge of the system configuration, these systems can be used to extract relative or absolute position and orientation data.

It is to be pointed out that frequency and time measurements are, by far, the most precise measurement techniques. In fact, the precision that may be obtained in this type of measurement usually ranges between 10^{-8} and 10^{-14} (in up to date frequency and time systems, precision may eventually be higher). This is true, naturally, only if frequency and time measurements are used alone. Under those conditions, many modern measurements use frequency and time systems for various determinations.

2.1 Ultrasonic Time-of-Flight Measurements

Time of Flight (TOF) systems rely on the measure of distance between features attached on one side to a reference and on the other side to a moving target. These distances are determined by measuring the time of propagation of pulsed signals between pairs of points under the assumption that the speed of propagation of the signals is constant. Because the speed of propagation can be measured precisely, given that ultrasonic trackers are simple, rugged and relatively low cost, this method is interesting.

Commonly, ultrasonic trackers utilize three or more ultrasonic emitters on the target and three or more receivers on the reference (e.g., Logitech, 1991). The emitters and the receivers are transducers (e.g., piezo-electric ceramics, electromagnetic and electrostatic transducers, and spark-gap emitters), usually mounted on a triangular structure. Details on the various transducers can be found in Fraden (1997).

The relative spatial positions of the emitters on the target and the receivers on the reference are known. In the scheme presented, each emitter sends an ultrasonic pulse sequentially. It is important to note that all the receivers detect each pulse to ensure that the emitter plane is uniquely defined within some boundary constraints. The spatial position of the emitter with respect to the plane defined by the receivers is measured by triangulation, as shown in Figure 3.1. After determination of the spatial position of at least three emitters, the orientation and the position of the target is known, making the overall system a six-degree-of freedom finder. The emitted frequency is typically around 40 Khz to prevent the user from hearing it. (This frequency is also the most commonly available for piezoelectric transducers and receivers).

The advantage of an ultrasonic system is that the emitting unit held by the user is small and lightweight. Moreover, the system does not suffer from distortion. However, there are several drawbacks to such a system. First, the accuracy of the system depends on the constancy of the velocity of sound. Although, the speed of ultrasonic waves varies primarily with temperature, it also varies with pressure, humidity, turbulence, and therefore position. Other limitations are signal attenuation which tends to limit the range of tracking, ultrasonic ambient noise, and the low update rate. Ultrasonic noise

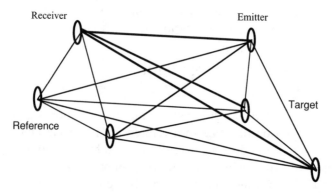

FIG. 3.1.   A simplified Time-of-Flight tracking system.

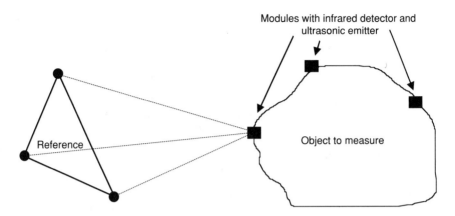

FIG. 3.2. Principle of the wireless US-Control ultrasonic tracking system.

is produced by Cathodic Ray Tube (CRT) sweeping cycles, disk drives, or reflections of the emitted signals. The low update rate results from the sequential triple emission of sound signals and the relatively slow speed of sound. A general approach to improve the update rate is to code the signals in order to send them simultaneously. Several frequencies may be used, as demonstrated in Arranz & Flanigan, (1994).

It has also been suggested that an infrared signal may be used to trigger the ultrasonic emission, thus making the system wireless (Fuchs, 1996). The principle of operation of such a system, the US Control ultrasonic tracking system, is shown in Figure 3.2. An infrared-triggered, ultrasonic system is composed of an infrared emitter, three ultrasonic receivers placed on the reference, and modules placed upon the target features to be sensed. Each module located on the target consists of an infrared receiver and an ultrasonic emitter installed on a small chip. The association of the ultrasonic emitters on the modules and the ultrasonic receiver on the reference constitutes a time-of-flight ultrasonic tracking system. The infrared beam, sent by the reference at the beginning of each acquisition, triggers the firing of the ultrasonic signal emitted by the modules. This setup relies on the fact that the time of flight of infrared waves (speed in air $\sim 3.0 \times 10^8$ m/s) is negligibly small compared to that of ultrasonic waves (speed in air 350 m/s). This system is able to localize the position of several modules simultaneously, making it a three degrees-of-freedom position finder. However, if the geometry between the modules is known, the tracking system becomes a six degrees-of-freedom position and orientation finder. This system has the same advantages and limitations of a conventional ultrasonic system. Table 3.1 summarizes the characteristics of Ultrasonic TOF systems.

TABLE 3.1
Summary Table of the Characteristic of Ultrasonic TOF Systems.

Physical phenomenon	Acoustic pulse propagation
Measured variable	Time of flight
Degrees of freedom (d.o.f.)	Some systems (Honeywell) measure only the orientation (2 or 3 d.o.f.). Others have position and orientation capabilities (6 d.o.f.).
Accuracy (position/orientation)	0.5–6 mm/0.1–0.6 degree
Resolution (position/orientation)	0.1–0.5 mm/0.02–0.5 degree
Update rate	25–200 Hz
Range/Total Orientation Span	250–4500 mm/45 degrees
Advantages	Small, light, no distortion
Limitations	Sensitive to temperature, pressure, humidity, occlusion and ultrasonic noise from CRT sweep frequency or disk drives. Low update rate.
Examples	Honeywell helmet tracking system, 3-D mouse from Alps Electric, RedBaron (Logitech, 1991), Lincoln laboratory Wand (Roberts, 1966), Mattel Power Glove, Sciences Accessories Space Pen. US Control localization senssor, OWL from Kantec, Intersense Inc.

2.2   Pulsed Infrared Laser-Diode

Pulsed infrared laser-diode tracking uses frequency and time techniques with an infrared laser beam. This principle was used in a hybrid system that will be described in Section 7.5.

2.3   GPS

The Global Positioning System (GPS) is a large-scale, time-frequency tracking system. The GPS uses 24 satellites arranged in orbit such that four satellites can be "seen" from any point on the earth at a given time. In addition, there are six monitoring stations, four ground antennas, a master control station, and a backup master control station (Farrell & Barth 1999). There are two levels of service available: a standard-positioning service (SPS) and a precise-positioning service (PPS), which is available only to users authorized by the U.S. government. Each satellite in the GPS has an atomic clock with a predictable accuracy of 340 ns for SPS (Farrell & Barth 1999). The accuracy of the atomic clock is critical because a clock error of 1 ms can produce a horizontal measurement error of 300 km (Farrell & Barth 1999). The master control station controls the orbit of the satellites and corrects the clock for each satellite as needed.

Let us give some practical indications on frequency and time modern techniques. Usually all frequency stable or ultrastable sources exhibit stability better than 10^{-10}/day (usual quartz oscillators are 10^{-11}/day and atomic clocks usually range in the 10^{-12} to 10^{-14} range). What does 10^{-10} practically mean? It is extremely large, because, in the radio navigation system, a 10^{-10} error per day means an error of 3000 m. But it is also extremely small, because it means 0.25 s on a 80 years human life or a 40 mm error on the distance between the earth and the moon.

Theoretically, the system can determine the position of a user with a GPS receiver by receiving a signal from at least three satellites and computing the time of arrival of the respective signals. In practice, however, the GPS receiver clock has an unknown bias. Therefore, four signals from GPS satellites must be received, from which it is possible to determine the position of the receiver and the clock bias. The GPS has an accuracy of approximately 100 meters with SPS but up to date frequency and time systems for localization are much better. PPS systems are in principle at least 10 times better. The main drawback of the GPS system is the inability to locate the receiver without a direct line of sight to the satellites. A precise system for the overall scale of operation, the differential GPS, uses emitting ground stations that refine the resolution to the order of one-tenth of a meter (Farrell & Barth 1999). A GPS type of technology applied at the scale of a typical indoors virtual environment would yield high accuracy and precision for tracking human scale functions. In all cases, recent advances in frequency and time techniques make those techniques extremely attractive for future applications (see for instance Proc. of the 1999 Joint Meeting of EFTF/IEEE IFCS on frequency and time systems and applications – IEEE catalog 99CH36313). Table 3.2 summarizes the characteristics of the GPS.

TABLE 3.2
Summary Table of the Characteristics of the Global Positioning System.

Physical phenomenon	Line-of-sight radio signal
Measured variable	Time of arrival ranging
Degrees of freedom (d.o.f.)	Horizontal and vertical position (2 d.o.f.)
Accuracy (position)	~100 m horizontal, 156 m vertical (SPS), at least 10 times better for PPS, much smaller for differential GPS
Resolution (position)	25.46 m (SPS), 8.51 m (PPS)
Update rate	<1000 Hz (frequency of timing signal)
Range	Global
Advantages	Worldwide availability, uniform accuracy
Limitations	Sensitive to occlusion, currently suited to large scale tracking

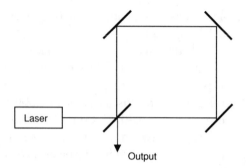

FIG. 3.3. Schematic view of a FOG gyroscope.

2.4 Optical Gyroscopes

Gyroscopes are instruments used to make angular velocity measurements. A sub-class of gyroscopes, the optical gyroscopes, operates on the TOF principle. Fiber Optics Gyroscopes (FOG) and Ring Laser Gyroscopes (RLG) use the time of propagation of light to determine the angular velocity of a target. They are light, durable, and low in power consumption.

A FOG relies on interferometry. Consider a free space interferometer, shown in Figure 3.3. A laser beam is divided in two waves that travel within the interferometer in opposite directions. In the absence of rotation, both waves combine to form an out of phase interference pattern due to consecutive pi phase shifts at each mirror reflection.[1] For a clockwise rotation of the device, the wavefront propagating counterclockwise travels a shorter path than the wavefront propagating clockwise, producing a shifted interference pattern at the output. The number of interference fringes produced is proportional to the angular velocity (Meyer-Arendt, 1995).

A RLG utilizes a ring laser cavity. It resembles the FOG except that it has an amplifying medium within the cavity to stimulate the emission of radiation (i.e., make the light a laser). Upon rotation of the device, two waves of slightly different frequencies propagate in opposite directions. The frequency of the signal observed at the output of the laser is the difference in frequencies of the two waves. The angular velocity of the target can be determined based upon the output signal frequency (Fraden, 1997). Table 3.3 summarizes the characteristics of phase-difference trackers.

---

[1]Note that there is a pi phase shift upon each mirror reflection. For the half-silvered mirror, however,there is a pi phase shift on one side only. This results in a phase shift difference of pi between the two paths.

TABLE 3.3
Summary Table of the Characteristic of Optical Gyroscopes.

Physical phenomenon	Interference of light
Measured variable	Frequency of interference fringes
Degrees of freedom (d.o.f.)	3-axis orientation (3 d.o.f.)
Accuracy (orientation)	0.5 degrees
Resolution (orientation)	0.1 degrees
Update rate	200 Hz
Advantages	Fast, accurate
Limitations	Drift due to successive integrations, sensitive to vibration
Examples	Crossbow DMU-FOG, Honeywell Space Systems

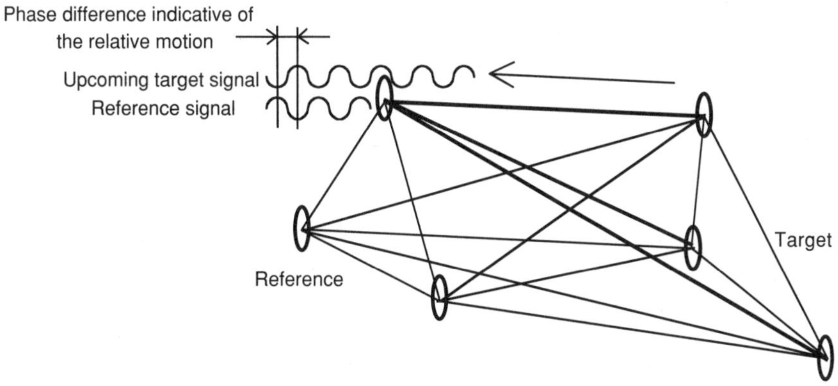

FIG. 3.4. Working principle of phase coherent tracking system.

2.5 Phase-Difference

Phase-difference systems measure the relative phase of an incoming signal from a target and a comparison signal of the same frequency located on the reference. As in the common ultrasonic approach, the system is equipped with three emitters on the target and three receivers on the reference, as shown in Figure 3.4.

Ivan Sutherland explored the use of an ultrasonic phase-difference head tracking system and reported preliminary results (Sutherland, 1968). In Sutherland's system, each emitter sent a continuous sound wave at a specific frequency. All the receivers detected the signal simultaneously. For each receiver, the signal phase was compared to that of the reference signal. A displacement of the target from one measure to another produced a modification of the phases that indicated the relative motion of the emitters with respect to the receivers. After three emitters had been localized, the orientation and position of the target could be calculated. It is important to note that the maximum motion possible between two measurements is limited by the wavelength of the signal. Current systems use solely ultrasonic waves that typically limit the relative range of motion between two measurements to 8 mm. Future systems may include phase-difference measurements of optical waves as a natural extension of the principle that may find best application in hybrid systems. Because it is not possible to measure the phase of light waves directly, interferometric techniques can be employed to this end. The relative range of motion between two measurements will be limited to be less than the wavelength of light unless the ambiguity is eliminated using hybrid technology.

The main disadvantage of phase-difference ultrasonic trackers is their vulnerability to cumulative errors in the measurement process. Other limitations are their sensitivities to environmental conditions (e.g., temperature, pressure, humidity, and ultrasonic noise) and multiple reflections in the environment. Finally, trackers based on phase-difference measurements are limited to relative motion measurements. They will need to be associated with another measuring scheme if absolute measurements are necessary. Such a scheme will also limit cumulative errors obtained from sole relative measures.

An advantage of phase-difference trackers is their ability to generate high data rates because the phase can be measured continuously. It is then possible to use filtering to overcome environmental perturbations. As a result, accuracy and resolution are improved compared to those of TOF ultrasonic trackers. Table 3.4 summarizes the characteristics of phase-difference trackers.

3.   SPATIAL SCAN

A spatial scan tracking system employs optical devices to determine the position and orientation of an object by scanning a working volume. Spatial scan trackers use one of two possible working principles to compute the position and the orientation of a target: the analysis of 2-D projections of image features or the determination of sweep-beam angles. The optical sensors used are typically charge-coupled device (CCD) cameras, lateral-effect photodiodes, or four-quadra (4Q) detectors. A CCD array is a detector receiving an in-focus or out-of-focus image at the focal plane of a camera,

TABLE 3.4
Summary Table of the Characteristics of Phase-Difference Trackers.

Physical phenomenon	Phase difference sensing (e.g. ultrasonic, optical)
Measured variable	Phase difference
Degrees of freedom (d.o.f.)	Orientation and position (6 d.o.f.)
Accuracy	Unknown
Resolution	0.1 mm, 0.1 degree, 1/32 of the maximum range
Update rate	Independent of the range of operation
Range	Unknown
Advantages	Less sensitive to noise than TOF systems, high data rate
Limitations	Error increases in time since relative measurements. Sensitive to occlusion. Possible ambiguity in reported measures
Examples	Sutherland-Seitz-Pezaris head mounted display position tracker (Sutherland, 1968).

depending upon the application. A lateral-effect photodiode is a 1D or 2D array that directly reports the location of the centroid of detected energy (Wang et al., 1990; Chi, 1995). A 4Q detector is a planar component that generates two signals specifying the coordinates of the estimated centroid of the incoming out-of-focus light beam on its surface. The 4Q detector signals are useful for directly controlling two axes of some pointing system gimbals (see Section 8.4). Any device that estimates centroids is designed to work optimally without out-of-focus imagery.

A possible subclassification of spatial scan systems is outside-in versus inside-out. Wang first proposed this terminology for a subclass of optical trackers that use beacons as target features (Wang, 1990). We propose to extend these two classes to pattern recognition and beam sweeping systems, to indicate and emphasize their common physical principles. In the outside- in configuration, the sensors are fastened to the fixed reference. In the inside-out configuration, the sensors are attached to the mobile target.

Optical tracking systems typically have good update rates because they interact with the environment at the speed of light. The measurement accuracy and resolution tend to worsen with increased distance of the target from the sensor (a function of the working volume). This is because the relative distance between two points on the sensor appears smaller as the target gets farther away, making the points harder to resolve spatially. Optical noise, spurious light, and ambiguity of the sensed surface are sources of errors. Most optical tracking systems use infrared light to minimize the effects of optical noise and spurious light. However, in systems that utilize feature detection, the accuracy of the estimation process often depends upon how many target features are detected. Thus, another source of error is inaccurate reporting of the target position and orientation due to missing/occluded features. Fortunately, this source of error can be controlled by correct placement/choice of target features.

3.1   Outside-In

Outside-in systems employ video cameras placed on the reference to record features of the target. This widely used technique has two subclassifications, multiscopy and pattern recognition. We refer to *multiscopy* as an outside-in technique that employs multiple imaging sensors. The simplest multiscopy system uses only two cameras (stereoscopy) as shown in Figure 3.5. A simple example of such a system is the human visual system that perceives 3-D shapes of objects from two viewpoints (i.e., right and left eye position). Multiscopy, therefore, will employ two or more video cameras to compute the spatial position of a target feature by triangulation. The measurement of several features allows determining the orientation of the target. A tracking system may always use additional views either to refine a measure using an appropriate sensor fusion technique or to compensate for potential occlusions. Most of the systems define a plane on the target by detecting several features to measure the orientation and the position of the target (6 DOFs) (Horn, 1987). Some systems, however, could measure a subset of the DOFs to meet the needs of an application.

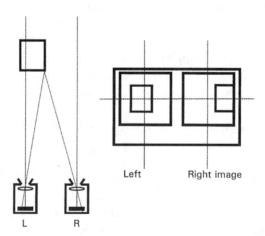

FIG. 3.5.   Principle of the optical stereoscopic tracking system.

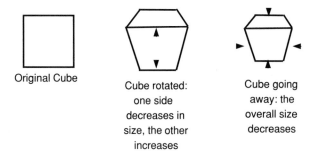

Original Cube

Cube rotated:
one side
decreases in
size, the other
increases

Cube going
away: the
overall size
decreases

FIG. 3.6. Pattern recognition method. The 3-D shape of the features of the cube is known and the image analysis allows reconstructing position and orientation.

Pattern recognition uses one camera and a known geometrical arrangement (pattern) of a set of features on a target (Gennery, 1992; Horn, 1987). The recorded 2-D pattern on the image is a function of the position and orientation of the target. For example, considering a cube structure originally placed perpendicularly to the visual axis, we can detect the slant of the cube by the size of one side compared to the other, as shown in Figure 3.6. If the cube is moved further away from the camera, the overall size is reduced. The combination of these analyses can be used to calculate the orientation and position of the cube. In this example, the tracking system constitutes an orientation and position finder (6 DOFs).

Pattern recognition is also used to reconstruct the motion characteristics of the human body (Simon et al., 1993; Barrett et al., 1994; Regh & Kanade, 1994). To this end, numerous algorithms have been developed, most often without use of landmarks or sensors on the target. If no landmark is used, this method needs complex algorithms to recover the position and orientation from the image of the object. To reduce the processing time, these algorithms can be implemented in electronic circuitry (Okereke & Ipson, 1995) or as artificial neural networks (ANN) rather than in software (Chan et al., 1992; Colla et al., 1995).

Another approach to pattern recognition is the optical formation of a regular pattern in space. The pattern is projected on the tracked objects and imaged by the camera. The shape of the objects can be determined by analysis of the projected pattern. There are two ways of producing regular patterns of light in space. The first method is to produce interference between two or more laser beams (Dewiee, 1989). The advantages of this method are the opportunity to produce fringes with a very small spatial period and equal spacing between the fringes in the region where the beams overlap. The smallest possible period is half the wavelength, which occurs when two interfering beams propagate in opposite directions and form a standing optical wave. The disadvantages to producing a spatial pattern by interference are the small beam overlap region, which makes it difficult to track objects in large environments, and the need to form a pattern with specific geometry.

A second method based on optical imaging is to employ a diffracted beam created by a grating. In this case, the shape of an object can be extracted by analyzing the spacing between the fringes of the diffraction pattern superimposed on the object (Chavel & Strand, 1984). The advantages of this method are the simplicity in forming a spatial pattern of arbitrary geometry using Fourier optics approaches, the lack of restrictions on the size of the environment to be tracked, and the simplicity of the setup. The disadvantage of this scheme is the limited resolution obtained. While these methods are typically used for 3-D shape extraction from a 2-D view, Harding and Harris (1983) demonstrated that it could be used for motion tracking. We therefore postulate that it could be potentially extended to tracking human motion as well. Table 3.5 summarizes the characteristics of outside-in optical trackers.

TABLE 3.5
Summary Table of the Characteristics of Outside-In Optical Trackers.

Physical phenomenon	Projection of an optical pattern
Measured variable	Shape of target features in an image acquired via a camera. Position and orientation for most of the applications.
Degrees of freedom (d.o.f.)	Position finder for each feature (3 d.o.f.). Orientation and position finder if feature geometry is known (6 d.o.f.).
Accuracy	0.1–0.45 mm/1/2800 of cameras field of view/2–15 mrad
Resolution	1/1000 to 1/65536 of cameras field of view/0.01 to 0.1 mm
Update rate	50–400 Hz
Range/Total Orientation Span	Up to 6000 mm/8.8 to 27 degrees
Advantages	Good update rate
Limitations	Sensitive to optical noise, spurious light, ambiguity of surface and occlusion
Examples	Sirah TC242 from Micromaine, CAE system (CAE Electronics, 1991), Optotrak from Northern Digital, LED array or pattern for Helmet tracking from Honeywell, Selspot tracker from SELCOM, Elite from BTS, Multitrak from Simulis, OrthoTRACK and Expert Vision from Motion Analysis Corporation, Vicon 370E from Oxford Metrics, CoSTEL (Cappozzo, 1983), pattern recognition methods.

3.2 Inside-Out

In an inside-out configuration, the sensors are on the target and the emitters are on the reference as shown in Figure 3.7. We distinguish between two techniques: one based on 2-D projection of image features, referred to as videometric, and one based on sweep-beam angles, referred to as beam scanning. The videometric technique uses optical sensors (e.g., CCDs) placed on the target (e.g., head of the user), whereas the beam scanning system uses rotating beams emitted from the reference and detected by the sensors located on the target. While it is not necessarily trivial, it can be noted that both techniques rely on scanning principles. In the videometric technique, the CCD detector of a camera serves as the scanning element, whereas the rotating beam technique uses rotating mirrors to scan the working volume. Both techniques are able to measure position and orientation if the system is equipped with a sufficient number of sensors and features.

The inside-out configuration typically yields higher resolution and accuracy in orientation than the outside-in. The same rotation of a target around a point (e.g., the head of a user rotating around the neck) will produce more displacement on the CCD sensor in an inside-out than in an outside-in configuration. This can be explained by noting that the ratio of the radius of the trajectory of the features being tracked following either rotation of the target or rotation of the camera is smaller in the outside-in than in the inside-out configuration. This is illustrated in Figure 3.7. In spite of their classification as inside-out configurations, systems based on beam scanning techniques do not share this advantage.

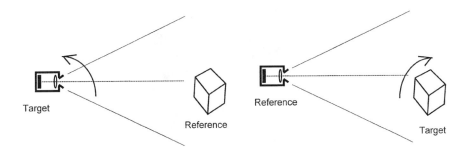

FIG. 3.7. The inside-out and outside-in configurations: (left) Inside-out configuration: The rotation of the camera, held by the target in this case, produces a large motion of the image of the cube on the CCD camera, (right) Outside-in configuration: The rotation of the cube, the target in this case, produces a small motion of the image on the CCD camera.

3.2.1 Videometric

The videometric technique shown in Figure 3.8 employs several cameras placed on a target (e.g., the head of a user). The reference has a pattern of features (e.g., the ceiling panels) whose locations in 3-D space are known. The cameras acquire different views of this pattern. The 2-D projections of the pattern on the sensor can be used to define a vector going from the sensor to a specific feature on the pattern. The position and orientation of the target are computed from at least three vectors constructed from the sensor(s) to the features. The system shown in Figure 3.8 has been built with four cameras located on an helmet-mounted display (target) and a ceiling (reference) covered with sequentially fired infrared LEDs. Theoretically, it is possible to use one camera to track the target. However, redundant measures improve tracking and multiple cameras allow a larger range of

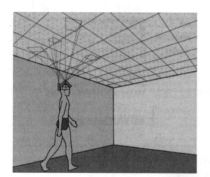
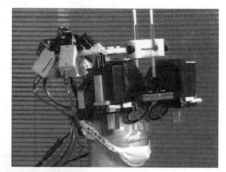

FIG. 3.8. Inside-out configuration: Opto-ceiling tracking system at the Department of Computer Science at the University of North Carolina at Chapel Hill. At the top left, the working principle of the tracker is shown with the fields of view of four cameras. On the right, an early head-mounted display equipped with the tracking device is shown. At the bottom, a more recent version utilizing the SCAAT algorithm (Welch & Bishop, 1997).

motion, constantly keeping the reference (ceiling) in view. A mathematical technique called "space resection by collinearity" is used to recover the position and the orientation of a target (Azuma & Ward, 1991). More recently the capabilities of the tracking system have been expanded by utilizing sensor data immediately when it is obtained (Welch & Bishop, 1997). Furthermore, in principle, this tracking system has the advantage of being scalable by adding ceiling panels. In practice, the cost inferred may outweigh the benefits of a larger tracking volume.

3.2.2 Beam Scanning

This technique uses scanning optical beams emitted from a reference location. Sensors located on the target detect the time of sweep of the beams across their surfaces. The time of sweep is used with the angular velocity to determine the angle from the reference axis to each sensor. The angle from the reference axis to each sensor is then used to determine the position and the orientation of the target.

An example of a beam scanning tracking system is the Honeywell system, used to determine the orientation of the head of a pilot in a cockpit. Given a known location of the helmet, the Honeywell helmet-tracking method computes the angle of the beam on the sensors from the time of sweep (Ferrin, 1991). Figure 3.9 illustrates this principle for two beams and two sensors. In this case only azimuth and elevation of the target can be measured. In more complex configurations where several emitters and sensors are used, the 3-D position and orientation of the target can be computed by triangulation from the angle measurements.

The Minnesota scanner tracking method employs a scanning laser beam to compute the distance between fixed sensors attached to the structure of

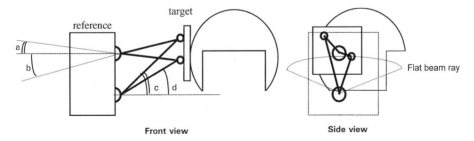

FIG. 3.9. Inside-out configuration: Structure of a beam scanning tracking system used for the determination of pilots' head-orientation in airplane cockpits (Honeywell).

TABLE 3.6
Summary Table of the Characteristics of Inside-Out Trackers.

Physical phenomenon	Spatial scan
Measured variable	Beam position or sweep detection
Degrees of freedom	Position and orientation
Accuracy	2–25 mrad
Resolution	Diminishes with the range of operation
Update rate	up to 1500 Hz
Range	In principle, scalability is unlimited for the UNC tracker
Advantages	Better resolution than outside-in systems, scalability
Limitations	Occlusion sensitive
Examples	OptoCeiling from UNC at Chapel Hill, Honeywell helmet rotating infrared beam system (Ferrin, 1991), LC Technology rotating mirror system (Starks, 1991), Minnesota Scanner (Cappozzo, et al., 1983; Sorensen et al., 1989), CODA (Miller, 1987), Self-tracker project (Bishop, 1984).

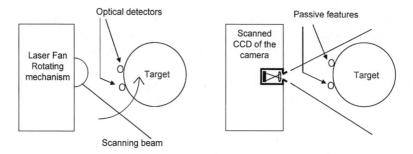

FIG. 3.10. Similarity between the inside-out beam scanning (left) and the outside-in videometric (right) configurations.

the scanner and sensors attached to the user. The distance is computed by counting the elapsed time between the two sensors during a sweeping cycle (Cappozzo et al., 1983; Sorensen et al., 1989).

The scanning-beam technique, while an inside-out configuration, does not share the advantage of higher accuracy and resolution in orientation like videometric tracking systems. Given that the receivers are on the target, which makes it an inside-out configuration, the scanning of the working volume is done from the reference. Paradoxically, we show in Figure 3.10 that such a configuration can be likened to an outside-in configuration where a camera is attached to the reference and scans the scene. Characteristics of inside-out trackers are summarized in Table 3.6.

4. INERTIAL SENSING

Inertia is defined as the property of matter which manifests itself as a resistance to any change in the momentum of a body (Parker, 1984). Therefore, an inertial reference frame is a coordinate system in which a body does not experience a change in momentum. In an inertial coordinate system, the mass of an object is determined according to Newton's Second Law, as opposed to the Newton's Law of Gravitation.

Inertial sensors use physical phenomena to measure acceleration and rotation relative to the inertial reference frame of the earth. The coordinate systems of these sensors are not inertial, due to the changing accelerations (centripetal or linear) of their frame of reference. However, inertial tracking systems are composed of inertial sensors, whose data can be used to determine the absolute position and orientation of an object. In this section, we examine the inertial sensing devices that enable inertial tracking methods.

4.1 Mechanical Gyroscope

A mechanical gyroscope, in its simplest form, is a system based on the principle of conservation of angular momentum, which states that an object rotated at high angular speed in the absence of external moments, conserves its angular velocity. The gyroscope contains a wheel mounted on a frame so that the external moments due to friction are minimized. This allows the target to turn around the wheel without experiencing a change in the direction of its axis, as illustrated in Figure 3.11. The orientation of the

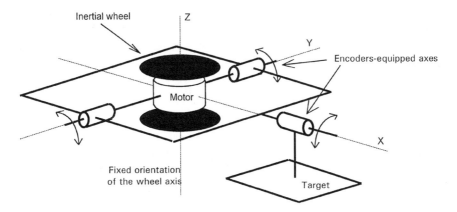

FIG. 3.11. Structure of a mechanical gyroscope.

TABLE 3.7

Summary Table of the Characteristics of Mechanical Gyroscopes.

Physical phenomenon	Inertia
Measured variable	Orientation between the axis of rotation of the wheel and an axis attached to the target
Degrees of freedom (d.o.f.)	Orientation finder (1 to 3 d.o.f.)
Accuracy	0.2 degree, static drift 0.01 deg/s, dynamic drift 0.25 deg/s
Resolution	0.032 degree
Update rate	50 Hz
Total Orientation Span	132 degrees in yaw, 360 degrees in roll.
Advantages	No reference needed
Limitations	Error increases with time since measurements are relative to previous ones leading to drift of the axis with time.
Examples	Gyrotrac2 and Gyrotrac3 from VR systems, Gyropoint from Gyration.

target can be computed from the angles reported by rotational encoders mounted on the frame. The working principle of the encoders is described in Appendix A. Each gyroscope provides one reference axis in space. At least two gyroscopes are needed to find the orientation of an object in space.

A main advantage of this tracking system is that it does not require an external reference to work. The axis of the rotating wheel is the reference. The main problem of gyroscopes, however, comes from the inertial momentum of the wheel that does not remain parallel to the axis of rotation because of the remaining friction between the axis of the wheel and the bearings. This causes a drift in the direction of the wheel axis with time. Taking relative measurements of the orientation rather than absolute measurements can minimize this drift. As a consequence, the system suffers from accumulated numerical errors but a periodic re-calibration of the system will insure greater accuracy over time.

4.2   Accelerometers

An accelerometer measures the linear acceleration of the object to which it is attached. An accelerometer can be specified as a single-degree-of-freedom device, which has some kind of mass, a spring-like supporting system, and a frame structure with damping properties. Accelerometers come in many forms including capacitive, nulling, piezoelectric, and thermal accelerometers. We discuss the forms more commonly used in tracking for virtual environments, the capacitive and the piezoelectric.

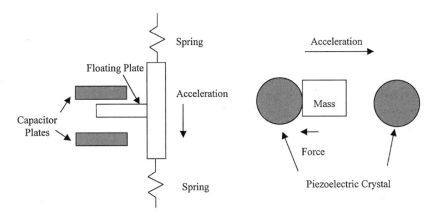

FIG. 3.12. Structure of capacitive and piezoelectric accelerometers.

Capacitive accelerometers utilize a spring-mass system to tune a capacitive element. As shown in Figure 3.12, a capacitor is constructed with a "floating plate" between two fixed plates. The floating plate is attached to a spring-mass system. As the unit accelerates, the mass is pushed in the direction opposite the acceleration. The result lies in the floating plate moving away from one fixed plate and towards the other. The motion of the floating plate changes the capacitance between the fixed plates, providing a measure of the amount of acceleration. The same concept is applied with other capacitor and/or spring mass configurations.

Another type of accelerometer is the piezoelectric. The piezoelectric effect occurs when pressure is applied to a polarized crystal, resulting in a mechanical deformation that produces an electric charge (i.e., a current). Shown in Figure 3.12, a piezoelectric accelerometer contains a mass mounted between two piezo-electric crystals. As the unit accelerates, the mass is pushed in the direction opposite the acceleration, producing a pressure on the crystal. The pressure on the crystal produces a voltage proportional to the amount of force applied, providing a measure of the amount of acceleration. An interesting variation on this method is found in Besson, et al., (1993).

Position information is obtained from an accelerometer by twice integrating the resultant acceleration, assuming that the initial conditions of the target (position and speed) are known. However, this numeric integration produces errors in position that accumulate over time. Many accelerometers are lightweight (micro-machined accelerometers are available), and all accelerometers have an absolute reference. Accelerometer characteristics are summarized in Table 3.8.

TABLE 3.8
Summary Table of the Characteristics of Accelerometers.

Physical phenomenon	Mass inertia
Measured variable	Depends on implementation.
Degree of freedom (d.o.f.)	Position along one axis only (1 d.o.f.)
Accuracy	0.0075 g (0.75% of gravitational acceleration)
Resolution	0.0001 g (0.01% of gravitational acceleration)
Update rate	Up to 1500 Hz
Range	Unlimited
Advantages	No reference needed; light
Limitations	Errors in position due to integration
Examples	Patriot Sensors 3001 series accelerometers, Sekia accelerometers

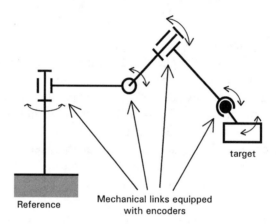

FIG. 3.13. Structure of a typical mechanically linked tracking system.

5. MECHANICAL LINKAGES

This type of tracking system uses mechanical linkages between the reference and the target (Jau, 1991). Two types of linkages are used in mechanical tracking systems. The first type is an assembly of mechanical parts that provides the user with multiple rotation capabilities, as shown in Figure 3.13. The orientation of the linkages is computed from the various linkages angles measured with incremental encoders or potentiometers. The other type of mechanical linkage uses wires that are rolled on coils. A spring system ensures tension is applied to the wires to measure the distance accurately. The degrees of freedom of mechanical linkage trackers depend

TABLE 3.9
Summary Table of the Characteristics of Mechanical Linkages Trackers.

Physical phenomenon	Mechanical linkages
Measured variable	Angle measured by rotating encoder(s)
Degrees of freedom (d.o.f.)	Position and orientation (Up to 6 d.o.f.)
Accuracy	0.1–2.5 mm
Resolution	0.05–1.5 mm/0.15–1 degree
Update rate	Depends on data aqusition capabilities (about 300 Hz)
Range/Total Orientation Span	1.8 m/40 degrees; Limited by the weight and deformation of the mechanical structure with distance from reference.
Advantages	Good accuracy, precision, update rate, and lag. No environmental linked error.
Limitations	Encoder resolution, limitation of motion.
Examples	FaroArm(1), Phantom(1), Spidar(2), Anthropomorphic Remote Manipulator from NASA (Jau, 1991), Argonne Remote Manipulator (ARM), Fake Space Binocular Omni-Oriented Monitor (BOOM), GE Handyman Manipulator, MITI position tracker, Noll Box, Rediffusion, ADL-1 (Shooting Star Technology), Wrightrac (Magellan Marketing), PROBE-IC and PROBE-IX (Immersion Human Interface), Sutherland Head Mounted Display project, University of Tsukuba Master Manipulator, Spidar II (wire tracker) (Hirata & Sato, 1995)

upon the mechanical structure of the tracker. While six degrees of freedom are most often provided, typically only a limited range of motions is possible because of the kinematics of the joints and the length of each link. Also, the weight and the deformation of the structure increase with the distance of the target from the reference and impose a limit on the working volume. Mechanical linkage trackers have been successfully integrated into force-feedback systems used to make the virtual experience more interactive (Brooks et al., 1990; Massie, 1993). Table 3.9 summarizes their characteristics.

6. DIRECT-FIELD SENSING

6.1 Magnetic Field Sensing

By circulating an electric current in a coil, a magnetic field is generated in the coil. The field at some distance r has the following polar components: B_r (along the radial direction) and B_θ (perpendicular to the radial direction)

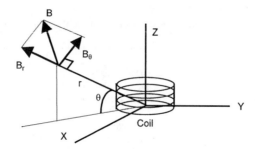

FIG. 3.14.   Rediating electromagnetic field components.

represented in Figure 3.14. If a receiver (some magnetic field sensor) is placed in the vicinity, the magnetic field from the coil induces a magnetic flux in the receiver. This is referred to as magnetic coupling between the emitting coil and the receiver. The sensor output resulting from the induced flux could then be measured. The flux in the vicinity of the receiver is a function of the distance of the receiver from the coil and of its orientation relative to the coil.

To measure the position and orientation of a receiver in space, the emitter must be composed of three coils with orthogonal magnetic fields. This defines a spatial referential from which one magnetic field can exit in any direction, defined by a combination of the three elementary orthogonal directions. On the receiver, three sensors measure the components of the flux received as a consequence of magnetic coupling. Based on these measures, the system determines the position and orientation of the receiver with respect to the emitter attached to the reference (Raab, 1977; Raab et al., 1979). It is typically required that $r \gg R$ and $r \gg L$, where r is the distance of the receiver from the coil, and R and L are the radius and the length of the coil, respectively, as shown in Figure 3.14. A practical solution actually involves the emission of three orthogonal fields: one in the estimated direction of the target and two others in the orthogonal directions (Raab et al., 1979).

Because magnetic trackers are inexpensive, lightweight, and compact, they are widely used in virtual environments. However, they have several issues that must be considered if accurate tracking is desired. First, magnetic trackers introduce lag (see Appendix A) into the desired application. Also, the working volume of the tracker is limited by the attenuation of the coupling signal with distance, additionally affecting the accuracy and resolution achievable by the tracker. Unfortunately, due to the effects of significant electromagnetic fields on human anatomy, the field cannot be increased indefinitely. Lastly, the update rate may be limited if filtering is applied to smooth the received signals, forcing a trade between the working volume, the accuracy and resolution, and the update rate desired.

6.1.1 Sinusoidal Alternating Current (AC)

This type of magnetic tracker is based on alternating the current feeding the emitting coils. This produces a changing magnetic field. The current induced by the changing field in each of three receiving coils is proportional to the product of the amplitude of the magnetic flux and the frequency of the field oscillations. A problem with this system is the generation of eddy currents in the vicinity of metallic objects, which create an opposing field distorting the emitted magnetic field (Bryson, 1992) and possibly leading to tracking errors. The variation in amplitude of the signal is what produces eddy currents by induction in metal sheets. However, if the metallic objects are static, a lookup table, can in principle, be pre-computed to account for the distortions. Characteristics of AC magnetic trackers are summarized in Table 3.10.

TABLE 3.10
Summary Table of the Characteristics of Alternating Current Magnetic Trackers.

Physical phenomenon	Magnetic coupling of two coils, one of which is fed with sinusoidal alternating current
Measured variable	Amplitude at the output of the receiving coil
Degrees of freedom (d.o.f.)	Orientation and position (6 d.o.f.)
Accuracy	0.8 mm to 25 mm (75 mm at 5 m)/0.15 to 3 degrees
Resolution	0.04 mm to 0.8/0.025 to 0.1 degree
Update rate	15–120 Hz divided by the number of emitters
Lag	4–20 ms
Range/Total Orientation Span	up to 5000 mm/360 degrees
Advantages	No occlusion problem, high update rate, low lag, inexpensive, small.
Limitations	Small working volume, distortion of accuracy with distance, sensitive to electromagnetic noise in the 8 Hz–1000 Hz range and metallic objects
Examples	Fastrack, Isotrack, Insidetrack and Ultratrack from Polhemus (Polhemus, 1972), Honeywell, Rediffusion Zeis, Ferranti, Israeli government.

6.1.2   Pulsed Direct Current (DC)

In contrast to the previously described system, direct current magnetic trackers use a pulsed, constant current to excite sensors capable of detecting a constant magnetic flux . To this end, the receiver may utilize magnetrons (described in Appendix A) as sensing devices (Ascension, 1991). During the rising portion of the DC pulse, eddy currents are generated as in AC systems. However, a DC system would wait until these currents die out to take a measurement to eliminate distortion due to eddy currents. Theoretically, one would have to wait an infinitely long time for the eddy currents to fully vanish; therefore, DC systems wait the longest possible time (determined by the acquisition rate of tracking) before making a measurement. For example, to make one measurement of magnetic flux per second, a DC system would produce the following sequence: 1) At time t equals 0 a DC magnetic field is induced; 2) the system waits 999 ms, the longest time allowed by the acquisition rate; 3) At t equals 1 s, the DC system measures the magnetic field. Thus, the pulsed DC method significantly reduces the influence of eddy currents on the accuracy of measurements.

A problem with DC magnetic trackers is the distortion of the magnetic field by ferro-magnetic objects and other sources of electromagnetic fields (such as computer monitors). A calibration procedure at startup of the system should measure the magnetic field bias produced by both the Earth's electromagnetic field and other sources to optimize the system performance. Characteristics of pulsed DC magnetic trackers are summarized in Table 3.11.

6.1.3   Magnetometer/Compass

Magnetometers measure the orientation of an object with respect to the magnetic field of the Earth. Types of magnetic field sensors include fluxgate, Hall effect, magneto-resistive, and magneto-inductive sensors. When used for tracking purposes, magnetometers use magneto-inductive sensors. Such sensors operate, on the change in inductance of a coil in the presence of an external magnetic field component parallel to the axis of coil. If the coil is used in a LCR oscillator, the output frequency of the oscillator changes. This change in frequency indicates the orientation of the coil with respect to the external magnetic field. Using three sensors, the orientation of an object with respect to the magnetic field can be completely determined.

TABLE 3.11
Summary Table of the Characteristics of Pulsed Direct Current Magnetic Trackers.

Physical phenomenon	Magnetic coupling
Measured variable	Amplitude in output of the receiving sensor
Degrees of freedom (d.o.f.)	Orientation and position (6 d.o.f.)
Accuracy	2.5 mm/0.1–0.5 degree
Resolution	0.8 mm/0.1 degree
Update rate	144 Hz
Lag	22 ms
Range	600–2400 mm radius of radiating magnetic sphere
Advantages	No occlusion problem, small, high update rate, low lag, inexpensive
Limitations	Small working volume, accuracy degraded with distance, sensitive to electromagnetic noise in the 8 Hz–1000 Hz range and to ferromagnetic objects
Examples	Ascension Bird, Big Bird and Flock of Birds (Ascension, 1991).

TABLE 3.12
Summary Table of the Characteristics of Magnetometers.

Physical phenomenon	Magnetic field sensing
Measured variable	Depends on implementation
Degrees of freedom (d.o.f.)	Orientation with respect to magnetic north (1 d.o.f.)
Accuracy	Up to 1 degree
Resolution	5 nT (nano-Teslas)
Update rate	300 Hz
Advantages	No reference needed
Limitations	Unknown
Examples	Component of Precision Navigation (TCMVR50), Crossbow 539 Magnetometer

One problem with this technology is that the Earth's electromagnetic field is inhomogeneous and yields angular errors in the orientation measurements. As noted in previous methods, relative measurements can be implemented to compensate for these errors. This technique works well if the field is quasi-constant between measurements. Additionally, this technology is sensitive to disturbances in the ambient magnetic field. Table 3.12 lists magnetometers characteristics.

6.2   Inclinometers: Gravitational Field Sensing

An inclinometer detects changes in orientation with respect to the inertial reference frame of the earth. Because the direction of gravitational acceleration is constant, the inclinometer uses sensing schemes which indicate this direction, indirectly defining the amount of rotation with respect to the direction of gravity. Common implementations use electrolytic or capacitive sensing of fluids.

One method is to measure the relative level of fluids in two branches of a tube to compute inclination. In this case, the inclinometer measures the change in capacitance based upon the level of fluid in the tube. Another, less common implementation is that of an optical inclinometer, shown in Figure 3.15. A viscous, opaque liquid is placed between two optical vertical panels (Fuchs, 1996). One of the panels produces uniform light that is received by a linear array of photosensitive detectors on the other panel. The viscous liquid maintains its perpendicular orientation with respect to the Earth's gravitational field, and a number of photosensitive detectors are exposed (which activates them), while others remain occluded, due to the opacity of the liquid. This number of sensors activated indicates the orientation of the liquid, and, therefore, the orientation of the gravitational field with respect to the target, making it a one-degree-of-freedom orientation finder. We postulate that a problem with this sensor is the slow reaction time imposed by the viscosity of the liquid. The vibration and acceleration of the sensor also will affect the measurements. Inclinometer characteristics are summarized in Table 3.13.

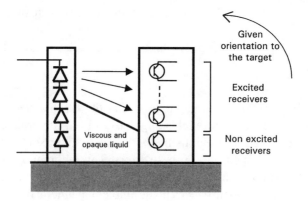

FIG. 3.15.   Principle of operation and structure of an optical inclinometer.

TABLE 3.13
Summary Table of the Characteristics of Inclinometers.

Physical phenomenon	Gravitational field sensing
Measured variable	Relative heights; capacitance
Degrees of freedom (d.o.f.)	Orientation with respect to gravitational direction (1 d.o.f.)
Accuracy	Up to 0.1 degrees
Resolution	Up to 0.03 degrees
Update rate	Up to 40 Hz
Advantages	No reference needed
Limitations	Reaction time degraded by viscosity of liquid
Examples	EX-TILT 2000 from AOSI, Microstrain FAS, Crossbow CXT102E, Sekia NG360

7. HYBRID SYSTEMS

Hybrid technology refers to a combination of various tracking technologies (e.g. optical and mechanical) and/or sensor types. We would like to extend that definition to include also systems based on different principles of operation, such as time-frequency measurement combined with inertial sensing. While hybrid technologies increase the complexity of a tracking system (and likely its cost), they are adopted either to access variables that one technology cannot easily provide (relative and absolute measurements), or to make exhaustive measurements. In the latter case, when associated with filtering and predictive techniques, sensor fusion techniques are used to associate incomplete data sets coming from different sensor types. Five examples of hybrid systems are presented to illustrate how hybrid systems may be built: inertial, inside-out/inertial, magnetic/videometric, and two time-frequency/mechanical linkages/videometric systems. It is beyond the scope of this survey to present a comprehensive review of all hybrid systems that have possibly been conceived.

7.1   Hybrid Inertial Platforms

We shall present two hybrid inertial platforms. The first system is composed of three accelerometers and three gyroscopes mounted on a target. The accelerometers measure the acceleration of the target along three independent perpendicular axes and the gyroscopes measure the orientation of the target about the same axes. Gyrometers could be used instead of

TABLE 3.14
Summary Table of the Hybrid Inertial Platforms Characteristics.

Physical phenomenon	Direct field sensing and inertia
Measured variables	Depends on implementation
Degrees of freedom (d.o.f.)	Orientation and position finder (6 d.o.f.)
Accuracy	Unknown
Resolution	10 ug in acceleration, 0.002 degree in rotation
Update rate	Unknown
Range	Up to 10 g for acceleration, 500 deg/s in rotation
Advantages	Compact and no reference needed.
Limitations	Drift, integration errors
Examples	Motion Pack from Systron-Donner

gyroscopes to access the angular velocity rather than the orientation. By integration of angular velocity, orientation can be estimated. Similarly, the double integration of the measured accelerations leads to the spatial position. The main limitation of gyroscopes is drift. The main limitation of accelerometers is the integration process that leads to additional errors.

The second platform relies on three accelerometers, two inclinometers, and a magnetometer. As seen earlier, a magnetometer can determine the direction of magnetic north, providing the orientation of a target along the Y-axis (Yaw). Inclinometers placed along the X- and Z-axes can provide orientation of a target about these axes as well. These platforms without sources on a reference constitute six degrees of freedom self-trackers that are compact and light weight (Foxlin, 1996).

7.2 Inside Out/Inertial

Azuma proposed such a system to improve inside-out optical tracking systems (Azuma, 1995). The improvement relies upon using Kalman filtering to predict head motion. The filter inputs are the outputs of an inertial platform added to the head-mounted display. A standard Kalman filter is used to predict head position while an Extended Kalman Filter is used for orientation, due to the non-linearity of the quaternions used to represent orientation. The inertial sensor includes three angular rate gyroscopes (i.e. gyrometers) and three linear accelerometers, as discussed in 7.1. Azuma reported that this technique helped greatly in removing what is known as "swimming" of virtual images. Accurate registration still remains a

TABLE 3.15
Summary Table of Advantages and Disadvantages of Inside-Out/Inertial Hybrid Systems.

	Inside-Out	*Inertial*	*Hybrid*
Characteristics/ Advantages	Measures orientation and position, accurate.	Compact, no occlusion problems give stable solution for orientation and position predictions and small lag when output filtered by Kalman predictor.	Compact, accurate, Small lag, stable.
Limitations	Unstable, resolution may degrade with distance (implementation dependent), occlusions sensitive, processing lag.	Long term drift of the orientation.	Occlusion sensitive.

challenge. By use of this technology, Azuma further showed improvements in interactive speed by a factor of 5 to 10 compared to techniques using no prediction tracking and by a factor of 2 to 3 compared to techniques with no inertial sensors. This platform allows the resolution of small fast movements in a shorter time than the original configuration would have allowed.

7.3   MAGNETIC/VIDEOMETRIC

This system, developed by State, is composed of a magnetic and an inside-out videometric tracker (State et al., 1996). The cameras of the videometric tracker, as well as the receivers of the magnetic tracker, are placed on the target (see Figure 3.5). The system is used to measure the position and orientation of the head of a user in a virtual environment with respect to stationary objects in that environment. As a consequence we refer to the head of the user as the target, and the objects in the world as the references. The inside-out videometric system detectes dual color-coded landmarks placed on the objects. Two cameras placed on the target were used because the detection of at least three landmarks is necessary to recover the position and orientation of the target.

TABLE 3.16
Summary Table of the Advantages and Disadvantages of Magnetic/Videometric Hybrid Systems.

	Pulse Direct Current (Magnetic)	Videometric	Hybrid
Characteristics/ Advantages	Robust, no occlusion sensitivity, inexpensive, fast, measure orientation and position.	Accurate, insensitive to electromagnetic noise and ferromagnetic object, measures Orientation and position.	Accurate, robust, insensitive to environment and occlusions, measure position and orientation.
Limitations	Non accurate, sensitive to electromagnetic noise and ferromagnetic objects.	Unstable, sensitive to occlusions, lag due to processing.	Lag due to processing of videometric data.

The magnetic tracking system is used to detect the gross positions and orientations of a target and determine the field of view within which the image processing is performed. A global non-linear equation solver and a local least-squares minimization are used to determine the effective field of view. The magnetic tracker is calibrated according to parameters given by the videometric system. The magnetic tracker is also used to remove any possible ambiguity in the occurrence of multiple solutions and to verify that the solution provided by the videometric system is reasonable, given that instabilities may occur. The system has the robustness of a magnetic tracker and the accuracy of a videometric tracker (State et al., 1996).

7.4   TOF/Mechanical Linkages/ Videometric Position Tracker

This hybrid system, developed by Maykynen et al. (1994), includes a pointing device and a range finder. The pointing device is based on optical technology and is composed of an infrared emitter and a 4Q (four quadrant) detector. Infrared light from the pointing device is reflected off a feature on a target to be localized. The 4Q detector actuates a two-axis motorized gimbal that keeps the feature being localized at the center of the sensor. The use of a 4Q detector allows direct control of the gimbal motors by using the voltage available at the detector output. The optical

TABLE 3.17
Summary Table of the Advantages and Disadvantages of TOF/Mechanical
Linkages/Videometric Hybrid Systems.

	Videometric/ Mechanical Linkages Pointing	TOF Infrared Reflective Range Measurement	Hybrid
Characteristics/ Advantages	Pointing accurate, by 4Q detector, line of sight measurement, determine azimuth and elevation.	Fast range measurements, which allows accuracy through averaging, long range, and wireless.	Position measurements, fast, accurate, long range, and wireless.
Limitations	Mechanical gimbal conditions accuracy and precision of azimuth and elevation determination.	Expensive range measurement detector can only be used for a target at a time.	Expensive, track only one target at a time.

axis of the pointing device is coaxial to the beam axis of the rangefinder. The rangefinder method of measurement is based on the principle of time-of-flight of infrared waves. Once the pointing device is aimed, the range finder measures the distance from the reference to the target. To determine the distance, the range finder sends an infrared beam to the target equipped with a reflective mirror (e.g. sign paint and cat's eye reflector).

Illustration of the principle is shown in Figure 3.16. Incremental encoders attached to the axis of the pointing device determine the elevation and azimuth. Thus the position in space of the target can be determined, as shown in Figure 3.17.

The accuracy and resolution of the system relies upon the characteristics of the motorized parts in the gimbals as well as on the electronics used in determining the distance to the target. Because the speed of the measuring process is high, numerous measurements can be averaged to improve the accuracy of a measure. However, the ability to only track one feature at a time limits the system. The system is mostly used for the measurement of large structures in outdoor environments. However, we see no fundamental limitation that would prohibit its use for applications on a smaller scale. Depending upon the accuracy and resolution requirements, high-speed electronics may be necessary to provide nanoscale resolution.

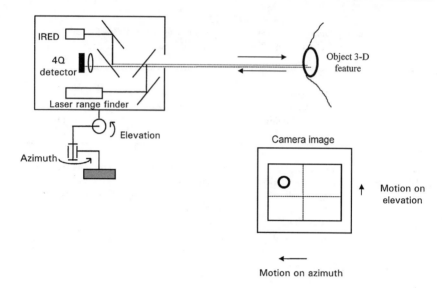

FIG. 3.16. Structure of a TOF/mechanical linkages/videometric tracking system.

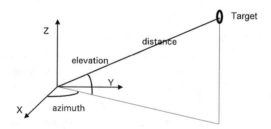

FIG. 3.17. Geometrical view of the measurement method of a TOF/mechanical linkages/videometric tracking system.

7.5 TOF/Mechanical Linkages/ Videometric 5-DOFS Tracker

The orientation of an object can be determined by tracking the position of at least three features with two or more locating systems, such as those introduced in the previous section. This time-frequency/mechanical linkages/videometric 5 DOFs tracker uses an original method to perform this function with a unique device. The tracking system is composed of an infrared range finder and a pointing device similar to the previous system. The system is illustrated in Figure 3.18. This tracking system was originally proposed to teach robot paths (Mäkynen, 1995).

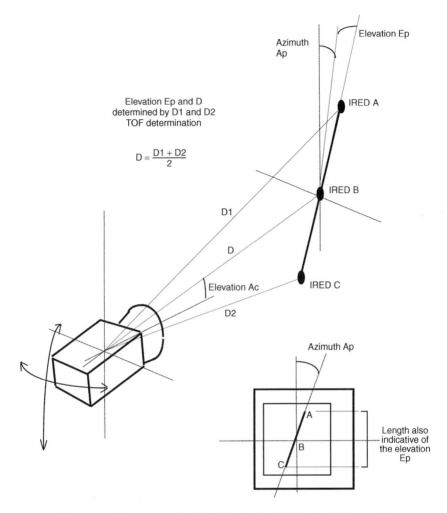

FIG. 3.18. Structure of a TOF/Mechanical Linkages/Videometric tracking system.

The target, a pen in this example, is equipped with three LEDs firing in the infrared. The pointing device aims constantly at the center of the diode, providing its elevation and azimuth to the processing unit via encoders mounted on the axis of the holding gimbal. The processing unit triggers the bottom and top tips of the pen where two LEDs are located. The TOFs of the two beams are measured by the rangefinder. The averaging of the two TOFs yields the distance of the central LED. As a result, the position of the center of the pen can be computed by the use of the elevation and azimuth variables.

TABLE 3.18

Summary Table of the Advantages and Disadvantages of TOF/Mechanical
Linkages/Videometric 5 DOFs Hybrid Systems.

	Videometric/ Mechanical Linkages Pointing	TOF Infrared Range Finding and Orientation Determination	Hybrid
Characteristics/ Advantages	Measures azimuth and elevation.	Determine orientation by TOF difference, fast, long range and accurate.	2 Orientations and 3 position measures, fast, accurate, long range.
Limitations	Videometric system and gimbal mechanics limit the accuracy and resolution in position.	Need wires to bring back to the control unit the received signals, expensive time detector.	Expensive, control wires needed, accuracy and resolution in position limited by mechanics of the gimbal and the quality of the videometric system.

Orientation in pitch of the pen is determined from the times of arrival of the infrared pulses. If the pen is vertical, the times of arrival are the same. If the pen is tilted away from the vertical, one pulse is delayed with respect to the other, and the delay is a function of the amplitude of the pitch. The actual working of the system involves a sequential firing of the two diodes with a delay Td between each firing. Such a delay is necessary in practice to distinguish between the two signals emitted. One of the signals is therefore delayed by a time Td after reception for comparison. The pointing device equipped with a CCD sensing array instead of a 4Q detector as previously adopted measures the orientation in roll. The detection of the positions of the two extreme diodes at the time they fire leads to the identification of the roll of the pen. These two orientations and the position of the central point of the pen yield a 5 DOFs tracking system.

8. DISCUSSION

After reviewing some unique advantages and disadvantages of various tracking technologies, whether in isolation or in hybrid configurations, we shall now discuss common issues to these technologies. First, we shall ask and discuss whether there are fundamental limitations in aiming for finer and finer accuracy and resolution. Next, we shall discuss a critical challenge for virtual environments, the capability for real-time operation. We shall then address the issue of scalability of the technology, an issue especially relevant to large virtual environments where users are physically navigating through the virtual world (e.g. larger indoors or outdoors settings). Finally, general considerations for the choice of tracking technologies are discussed.

8.1   Fundamental Limitations in Accuracy

A review of tracking devices according to their fundamental principle of operation may examine the theoretical limitations in accuracy and resolution of these systems. Accuracy is a measure of the absolute error of either position or orientation of an object in the tracker coordinate system. It is also an estimation of the position or orientation of an object after the system noise has been fully accounted for and averaged. A measure of accuracy, therefore, assumes that a large number of samples have been collected to yield unbiased estimates of the mean values of the underlying distributions in position and orientation.

In this regard, the fundamental limitation of obtaining perfect accuracy in position and orientation is the Heisenberg uncertainty relation. In our macroscopic world, this limitation is negligible. But, for instance, STM microscopy can also be considered as a tracking technology and the uncertain relations play a major role in this case. Other limitations to perfect accuracy are thermal noise in electronic circuits and quantum noise in the sensors.

8.2   Fundamental Limitations
in Resolution

A measure of resolution, on the other hand, quantifies the noise of the tracking system. Resolution is a measure of the spread of an underlying statistical distribution in either position or orientation. Resolution can thus be mathematically defined as the square root of the second central moment of the considered distribution (Frieden, 1983). All trackers are theoretically limited in resolution by a quantification unit. For example, optical trackers are theoretically limited by the size of a quantum of light, while ultrasonic trackers are limited by the wavelength of the signals used. However, at the working scale of the technology, where the application is the tracking of human scale features (e.g. the head), the resolution sought by the tracking technology considered is much larger than the limiting quantification units considered. An exception, perhaps, is the case where secondary parameters are measured with technologies that supersede microscale technology. An example is the measurement of head acceleration with nanoscale technology developed by NASA (Tom Caudell, personal communication, 1998). It has been shown that nanoscale accelerometers are limited in resolution due to interference of gravitational fields that may constitute a fundamental limitation in such measurements. As finer scale technologies are developed and investigated even beyond nanoscale technologies, other fundamental limitations are likely to be discovered. While such resolution requirements are likely beyond most applications, it is important to understand such phenomena and be able to set lower bounds on what can be achieved and what cannot. Nanoscale technology, for example, is now at the cutting edge but may be part of tomorrow's everyday technology.

Today, two common practical limitations in resolution result from the state-of-the-art electronics and the manufacturing of various components of the trackers. Electronics is an essential component in the emission, detection, and processing of the measured variables. The finite speed of the electronic signals produces a lag in the measurements. The limited bandwidth of these signals also limits the data acquisition rates. As an example, one may increase the operating frequency used in an ultrasonic phase coherent system to increase the resolution. A limit in the case of ultrasonic waves, depending on their amplitude, may be simply the viscosity of the air: the higher the frequency, the larger the attenuation. However, the response time of the electronic devices would limit the maximum data acquisition rates, also limiting the resolution of the tracking system. Optical data processing devices present promises for future improvements because of the larger bandwidth of the optical signals and the shorter switching times.

The manufacturing specifications of the emitting and sensing components of the tracking system often limit the resolution of current tracking technology, but not in any fundamental way. As examples, the current resolution of a CCD array, or the architectural layout and the geometry of emitting and sensing sources, limit the resolution. Achievable resolutions rely essentially on the progress of technology, including micro-scale and nano-scales electronics, mechanics, and opto-electronics. However, even if optical switching devices, for example, were to successfully replace electronic devices, and manufacturing errors were negligible, other components or factors of the virtual reality system would limit the achievable resolution of an application (e.g. monitor resolution). For example, the resolution of the display used in the visualization and the natural occurrences of mechanical vibrations are sources of limitation as well. There is no need to seek higher resolution for the tracking system than can be delivered by individual components of the overall virtual reality system.

8.3   Real-Time Capability

An issue of critical importance for trackers is their real-time capability. A virtual reality system qualifies as real-time if the virtual world reacts synchronously to the actions produced by a user. Because this capability is currently not reachable, the preferred term of interactive speed is commonly used. The difficulty in achieving interactive-speed results from the reception of non-synchronous signals coming from the real world (we see the real hand moving) and the virtual world (we see the virtual hand moving on the display). The signals from the real world appear as they are produced if seen directly (see the real hand), whereas the signals from the virtual world appear when the processing that produces them (time taken from end-to-end process of the virtual environment) is completed. Moreover, virtual signals are often generated following the detection of a real signal (the detection of a hand motion), thus aggravating the problem.

The total lag produced by this process comes from the establishment of the measurement conditions, the time to complete the measure before data is available, the filtering, the signal propagation and transmission times, and the synchronization between the tracking system, the computer, and the display. Different implementations may also have different temporal performances (Jacoby et al., 1996).

To minimize the effect of lag, Kalman filtering has been used to predict the position of a target according to the present and past speed and position parameters (Kalman, 1960; Azuma, 1995). The time of prediction can be tuned to equalize the lag produced by the system to produce virtual signals with the impression of an interactive-speed response. However, the predicted position and orientation are solely estimations produced from the last measurements and do not exactly reflect a real position. As a result, the lag problem is often replaced by noticeable registration errors.

In applications requiring registration of real and virtual objects, the lag, low update rates, and the errors in position and orientation are hindrances. Motion sickness can ensue if these variables are incorrect because of visual-proprioceptive conflicts (Kennedy & Stanney, 1997). The severity of the observed errors is a function of other system parameters (e.g. effectiveness of visual cues, use of sound) and the speed of various moving parts among others. Evaluating a tracker in the context of specific applications is a key requirement.

8.4 Scalability

Another important issue of tracker technology development is the potential for scalability of the technology. In certain cases, it becomes the driving factor for adopting an approach to tracking. At the University of North Carolina at Chapel Hill, for example, the opto-ceiling tracker was essentially developed using an inside-out configuration because such an approach was believed to have the potential for natural scalability indoors. Such a system was described in section 3.2.1 and illustrated in Figure 3.7. In this case, the trade-off was between scalability and the need to wear three cameras on the head that appeared highly displeasing. This problem can be, and is being addressed by adopting custom-made, miniature camera configurations (Welch & Bishop, 1997). The first implementation of the tracking system succeeded in demonstrating tracking in a small volume ($\sim 10 \times 10$ feet), and the second implementation demonstrated some scalability of the system (a factor of 2 in one dimension). Practical issues, however (e.g. the requirement for precise calibration of all LED-panels and associated cost), set some limit on scalability. Self-calibration was attempted to remedy this problem but the optimization becomes rapidly untractable for larger and larger volumes (Gottschalk & Hughes, 1993).

Scalability is fairly challenging. Indeed, for most technologies, theoretical as well as practical considerations have to be carefully examined. After reflecting on the scalability issues of the various technologies described in this document, we postulate that, most systems are scalable for indoor settings if the expense is not of primary concern and hybrid technologies can be considered. Most systems are not, however, scalable to handle large navigation settings such as outdoor navigation. Likely, outdoor tracking will not require the high accuracy and precision typical of most indoor settings. Generally speaking, however, scalability may imply that more complex algorithms (e.g. fusion algorithms) will be necessary as systems are scaled and hybrid technologies are implemented.

8.5 General Considerations

Finally, we summarize some general considerations of tracking technology for virtual environments. Ultrasonic systems typically suffer from ultrasonic noise sources in the environment, while other time-frequency-based systems, such the GPS, suffer from occlusion. Phase difference systems based on ultrasonic sources suffer from environment noise sources, while those based on light sources may offer attractive solutions for relative tracking measurements. Direct field sensing trackers, such as magnetic trackers, seem to be most employed in virtual environments because of their robustness and their low price, even though distortions of the magnetic field typically cause large tracking errors. Spatial scan trackers give excellent accuracy and resolution, but typically suffer from occlusion. Moreover, some of these systems are difficult to implement and thus tend to be expensive. Mechanical linkages have the best accuracy, update rate, lag, and resolution, but they impose constraints of motion on certain degrees of freedom. Inertial platform and other reference-less trackers are especially well adapted for fast reaction time and long-range motion of the user, but they suffer from drift and are best used in hybrid configurations. Given an application and an environment (e.g. small scale versus large scale, the potential for environment noise and occlusion), a sole technology or hybrid technology may be selected for optimal performance and trade-offs.

9.   CONCLUSION

This broad technical review examined existing trackers categorized according to their physical principle of operation to explore their similarities and differences. We briefly discussed the physical principle of each technology, as well as the technology advantages and drawbacks. Such taxonomy based on physical principle of operation was proposed to facilitate developing new and improved ways to track features of the real world, as well as assist in the choice of a tracking system to best fit an application. At present, a major limitation of state-of-the-art tracking technologies is the difficulty in achieving interactive-speed performance for complex virtual environments. The limitation is often a system limitation, given that it depends on rendering speed as well as on tracking acquisition and transfer to the rendering engine. The tracker itself is often not the limiting factor.

ACKNOWLEDGMENTS

We are grateful to Raymond Besson from the University of Besancon (France) for his comments and inputs on time-frequency devices. From the University of Central Florida, we thank Jim Parsons for his help on definitions and references, and Boris Zel'dovich for his help with the geometry of the time-frequency trackers. From the University of North Carolina at Chapel Hill, we thank Stefan Gottschalk and Gary Bishop for providing us with information on the opto-ceiling tracking system and Warren Robinett for stimulating discussions on general aspects of this review. Finally, we thank Tom Caudell from the University of New Mexico for his comments and encouragements to complete this work.

REFERENCES

Arranz, A. V., and Flanigan, J. W. (1994). "New tracking system for virtual reality based on ultrasonics and using transputers," *Actes des Journees in L'interface des mondes reels et virtuels*, Montpellier, 359–368.

Ascension Technology Corporation, (1991). "Flock of Birds, Vermont and production description for high-speed tracking of multiples bird receivers and operation of multiple receiver and transmitter configurations," *Trade Literature, April*, 1991.

Ascension Technology Corporation, (1991). "Question and answers about the Bird family of 6D input devices," *Trade Literature*, Vermont.

Azuma, R. (1995). "Predictive tracking for augmented reality," *Ph.D. Dissertation*, UNC-CS at Chapel Hill, TR-95-007.

Azuma, R., and Ward, W. (1991). "Space resection by Colinearity: Mathematics behind the optical ceiling head-tracker," Department of Computer Sciences, University of North Carolina at Chapel Hill, *Technical Report*, TR91-048.

Barrett, S. F., Jerath, M. R., Rylander III, H. G., and Welch, A. J. (1994). "Digital tracking and control of retinal images" *Optical Engineering*, 33(1), 150–159.

Besson, R. J., Bourquin, R. F., Dulmet, B. M., and Maitre, P. C. (1993). "Piezo Electric Resonator Differential Accelerometer," US Patent 5193392.

Bhatnagar, D. K. (1993). "Position trackers for Head Mounted Display systems: A survey," Department of Computer Sciences at the University of North Carolina at Chapel Hill, March 29.

Bishop, T. G. (1984). "Self-Tracker: A Smart Optical Sensor on Silicon," *Ph.D. Dissertation*, UNC-CS at Chapel Hill, TR84-002.

Brooks, F. P., Ouh-Young, M., Batter, J. J., and Kilpatrick, P. J. (1990) "Project GROPE - haptics displays for scientific visualization," *Computer Graphics*, 24(4), 177–185.

Bryson, S. (1992). "Measurement and calibration of static distortion of position data from 3-D trackers," NASA RNR, *Technical Report*, RNR-92-011.

Burdea, G., and Coiffet, P. (1993). *Virtual Reality*, Hermes Editions, France, CAE Electronics, (1991). "The visual display system you wear," CAE-7-3739-BR, *Trade Literature*.

Cappozzo, A., Leo, T., and Macellari, V. (1983). "The COSTEL kinematics monitoring system: Performance and use in human movement measurements," *Biomechanics*, VIII-B.

Chan, K. C., Xie, J., and Shirinzadeh, B. (1992). "Using Neural Network for Motion Analysis," *Control'92*, Perth 2–4.

Chavel, P., and Strand, P. (1984). "Range measurement using Talbot diffraction imaging of gratings," *Applied Optics*, 23(6), 862–870.

Chi, V. L. (1995). "Noise model and performance analysis of outward-looking optical trackers using lateral effect photo diodes," *Technical Report*, TR95-012, University of North Carolina at Chapel Hill.

Colla, A. M., Trogu, L., Zunino, R., and Bailey, E. (1995). "Digital Neuro-Implementation of Visual Motion Tracking Systems," *Proceedings of The Spie 1995 IEEE Workshop*, 0-7803-2739-X/95, IEEE.

Dewiee, T. (1989). "Range finding by the diffraction method," *Laser and Optics*, 8(4), 119–124.

Durlach, N. I., and Mavor, A. S. (1994). *Virtual Reality: Scientific and Technical Challenges*, National Academy Press, Washington DC.

Ferrin, F. J. (1991). "Survey of helmet tracking technologies," *Proceedings of the SPIE*, Large Screen Projection, Avionics and Helmet Mounted Displays, 1456, 86–94.

Farrell, J. A., and Barth, M. (1999). *The Global Positioning System and Inertial Navigation*, McGraw-Hill, New York, NY.

Foxlin, E. (1996). "Inertial head tracker sensor fusion by a complementary separate bias Kalman filter," *Proceedings of the IEEE VRAIS'96*, 185–194.

Fraden, J. (1997). *Handbook of Modern Sensors: Physics, Designs, and Applications*, American Institute of Physics, Woodbury, NY.

Frieden, B. R. (1983) *Probability, Statistical Optics, and Data Testing*, Springer-Verlag, New York.

Fuchs, P. (1996) " Les interfaces de la réalité virtuelle," *Interfaces-Les Journees de Montpellier*, Montpellier, France.

Gennery, D. B. (1993) "Visual tracking of known three dimensional objects," *International Journal of Computer Vision*, 7(3), 243–270.

Gottschalk, S., and Hughes, J. (1993). "Auto-calibration for virtual environments tracking hardware," *Computer Graphics*, Proceedings of SIGGRAPH, 65–72.

Harding, K. G., and Harris, J. S. (1983)."Projection moiré interferometer for vibration analysis," *Applied Optics*, 22(6), 856–861.

Hirata, Y., and Sato, M. (1995). "A measuring method of finger position in virtual work space," *Technical Report of the Precision and Intelligence Laboratory*, 4259, Nagatsuta-Cho, Midori- Ku, Yokohama 227, Japan.

Horn, B. K. P. (1987). "Closed-form solution of absolute orientation using unit quaternions," *Optics. Society, Am.*, 4(4), 629–642.

Jacoby, R. H., Adelstein, B. D., and Ellis, S. R. (1996). "Improved temporal response in virtual environments through system hardware and software reorganization," *Proceedings of SPIE*, 2653, Stereoscopic displays and Virtual Reality Systems III.

Jau, B. (1991). "Technical support Package on Anthropomorphic Remote Manipulator," NASA TECH BRIEF, 15(4), *from JPL Invention Report* (Report No. NPO-17975/7222), JPL Technology Utilization Office, Pasadena, CA.

Kalman, R. E. (1960). "A new approach to linear filtering and prediction problems," *Transactions of the ASME, J. Basic Eng.* 82, D, 35–45.

Kennedy, R. S., and Stanney, K. M. (1997). "Aftereffects in virtual environment exposure: psychometric issues," in *Design of Computing Systems: Social and Ergonomic Considerations*, ed. by M. J. Smith, G. Salvendy, and R. J. Koubek , 897-900. Elsevier, Amsterdam, Logitech (1991) *Press Kit.*

Mäkynen, A. J., Kostamovaara, J. T., and Myllylä, R. A. (1994). "Tracking laser radar for 3D shape measurements of large industrial objects based on the time-of-flight laser range-finding and position-sensitive detection techniques," *IEEE Transactions on Instrumentation and Measurement*, 43(1), 40–49.

Mäkynen, A. J., Kostamovaara, J. T., and Myllylä, R. A. (1995). "Laser-radar-based three dimensional sensor for teaching robot paths," *Optical Engineering*, 34(9).

Massie, T. H. (1993). "Design of a 3 degree of freedom force reflecting haptic interface," *SB Thesis*, Department of EE and CS at MIT.

Meyer, K., Applewhite, H. L., and Biocca, F. A. (1992). "A survey of position trackers," *Presence*, 1(2), 173–200, MIT Press.

Meyer-Arendt, J. R. (1995). *Introduction to Classical and Modern Optics*, Prentice Hall, Englewood Cliffs, New Jersey, 07632.

Miller, J. A. A. (1987). "Motion analysis using the twelve markers CODA-3 measurement system," *Proceedings of The 1987 Applied Mechanics, Bioengineering and Fluids Engineering Conference*, Cincinnati, Ohio, 343–344, ASME'87.

Noe, P., and Zabaneh, K. (1994). "Relative GPS," *IEEE Position Location and Navigation Symposium*, 586–590.

Okereke, O. C., and Ipson, S. S. (1995). "The design and implementation of hardwired 3D video position-tracking transducer," *IEEE Transactions on Instrumentation and Measurements*, 44(5), 958–964.

Parker, S. P. (1984). *McGraw-Hill Dictionary of Scientific & Technical Terms, 3rd Ed.*, McGraw-Hill, New York, NY.

Polhemus, Navigation Science Division, (1972). "Operation and maintenance for the 3D space head tracker," *OPM* 1056-006, 3–1 to 3–15.

Potter, J. H. (1967). *Handbook of the Engineering Sciences*, D. Van Nostrand Company Inc..

Raab, F. (1977). "Remote object position locator," *US Patent* 4054881, October.

Raab, F., Bood, E., Steiner, O., and Jones, H. (1979). "Magnetic position and orientation tracking system", *IEEE Transactions on Aerospace and Electronics Systems*, AES-15(5), 709–717.

Rehg, J. M., and Kanade, T. (1994). "Visual tracking of self-occluding articulated objects," *CMU Technical Report*, CMU-CS-94-224.

Roberts, L. (1996). "The Lincoln Wand," *AFIPS Conference Proceedings*, November 1966 Fall Joint Computer Conference, 29, 223–227.

Rodgers, A. G. (1991). "Advances in head tracker technology- A key contributor to helmet vision system performance and implementation," *Society for Information Display International Symposium, Digest of Technical Papers*, May 6–10, 22,127–130.

Simon, D. A., Hebert, M. and Kanade, T. (1993). "Real-time 3D pose estimation using a high-speed range sensor," *CMU Technical Report*, CMU-RI-TR-93-24.

Sorensen, B. R., Donath, M., Yang, G. B., and Starr, R. C. (1989). "The Minnesota scanner: A prototype sensor for three-dimensional tracking of moving body segments," *IEEE Transactions on Robotics and Automation*, 5(4), 499–509.

Souders, M. (1966). *The Engineer's Companion, A Concise Handbook of Engineering Fundamentals*, Wiley, Piedmont, CA.

State, A., Hirota, G., Chen, D. T., Garrett, W. F., and Livingston, M. A. (1996). "Superior Augmented Reality Registration by Integrating Landmark Tracking and Magnetic Tracking," *Proceedings of SIGGRAPH '96. In Computer Graphics Proceedings, Annual Conference Series*, ACM SIGGRAPH, 429–438.

Sutherland, I. E. (1964). "A head-mounted three-dimensional display," *Fall Joint Computer Conference, AFIPS Conference Proceedings*, 33, 757–764.

Wang, J. F. (1990). "A Real-time Optical 6D Tracker for Head-Mounted Display Systems," UNC-CS at Chapel Hill, *Ph.D. Dissertation*, TR90-011.

Wang, J. F., Azuma, R., Bishop, G., Chi, V., Eyles, J., and Fuchs, H. (1990). "Tracking a head-mounted display in a room-sized environment with head-mounted cameras," *Proceedings of SPIE 1990* Technical Symposium on Optical Engineering and Photonics in Aerospace Sensing, Orlando, Florida.

Welch, G., and Bishop, G. (1997). "SCAAT: Incremental tracking with incomplete information," *Proceedings of ACM SIGGRAPH*, 333–344.

APPENDIX A: DEFINITIONS

Accuracy: Error between a real and a measured position X* for each spatial position. This number is evaluated by taking numerous measures at a given location and orientation, and comparing the computed mean to the real value. In this review, the extreme value of the error will be given for each system. A system with an accuracy A will report a position within ±A of the actual position.

CCD: Charge-Coupled Device. Sensitive photoelectric array measuring the light energy striking each pixel.

Degree-of-freedom: Capability of motion in translation or rotation. There are six degrees of freedom: translation along X, translation along Y, translation along Z, rotation around X (pitch), rotation along Y (Yaw), and rotation along Z (roll).

Interactive Speed: Attribute of a virtual reality system that reacts "in time" according to actions taken by a user. Such a system must be fast enough to allow a user to perform a task at hand satisfactorily.

Lag: Delay between the measurement of a position and orientation by a tracking apparatus and the report to a device (e.g. scene generator, force feedback apparatus) requiring the orientation and position values.

LED: Light Emitting Diode. Photoelectric emitting device used as a light signal.

Magnetron: A semi-conducting device in which the flow of electrons is controlled by an externally applied magnetic field.

Pitch: Rotation in the vertical plane including the line of sight around the X axis shown in Figure 3.A1. Pitch is also called heading.

Real-time: Attribute of a virtual reality system in which the virtual world reacts synchronously to the actions of a user. This capability is practically not reachable since the processing time is not zero, so the preferred term of interactive speed is used.

Reference: Part of a tracking system considered fixed with respect to the motion of a target.

Resolution: Smallest resolvable change in position and orientation. A measure of resolution is the standard deviation of the underlying

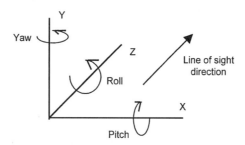

FIG. 3.A1   Referential commonly employed in Virtual Reality.

distribution of measurements around the mean of a measured position or orientation.

Roll: Rotation in the plane perpendicular to the line of sight around the Z axis shown in Figure 3.A1.

Target: Feature (e.g. object, landmark, human feature) to be localized by the tracking process.

Update rate: Maximum frequency of report of position or orientation.

User: Person interacting in a virtual world. Can be a target.

Yaw: Rotation in the horizontal plane including the line of sight around the Y axis shown in Figure 3.A1.

SYMBOLS EMPLOYED IN THIS DOCUMENT

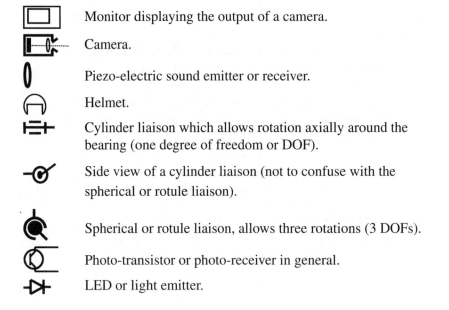

Monitor displaying the output of a camera.

Camera.

Piezo-electric sound emitter or receiver.

Helmet.

Cylinder liaison which allows rotation axially around the bearing (one degree of freedom or DOF).

Side view of a cylinder liaison (not to confuse with the spherical or rotule liaison).

Spherical or rotule liaison, allows three rotations (3 DOFs).

Photo-transistor or photo-receiver in general.

LED or light emitter.

4

Optical versus Video
See-Through
Head-Mounted Displays

Jannick P. Rolland
School of Optics/CREOL
University of Central Florida

Henry Fuchs
Department of Computer Science
University of North Carolina at Chapel Hill

1. INTRODUCTION

One of the most promising and challenging future uses of head-mounted displays (HMDs) is in applications where virtual environments enhance rather than replace real environments. This is referred to as *augmented reality* (Bajura et al., 1992). To obtain an enhanced view of the real environment, users wear see-through HMDs to see 3D computer-generated objects superimposed on their real-world view. This see-through capability can be accomplished using either an optical (shown in Fig. 4.1) or a video see-through HMD (shown in Fig. 4.2). We shall discuss the trade-offs between optical and video see-through HMDs with respect to technological and human factor issues and discuss our experience designing, building, and testing these HMDs.

With optical-see-through HMDs, the real world is seen through half-transparent mirrors placed in front of the user's eyes, as shown in Fig. 4.1. These mirrors are also used to reflect the computer-generated images into the user's eyes, thereby optically combining the real- and virtual world views. With a video see-through HMD, the real-world view is captured

FIG. 4.1. Optical see-through head-mounted display. (Photo courtesy of KaiserElectro-Optics.)

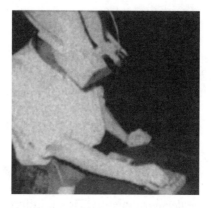

FIG. 4.2. A custom optics video see-through head-mounted display developed at UNC-CH. Edwards et al. (1993) designed the miniature video cameras. The viewer was a large FOV opaque HMD from Virtual Research.

with two miniature video cameras mounted on the head gear as shown in Fig. 4.2, and the computer-generated images are electronically combined with the video representation of the real world (Edwards et al., 1993; State et al., 1994).

See-through HMDs were first developed in the 1960s. Ivan Sutherland's 1965 and 1968 optical see-through and stereo HMDs were the first computer-graphics based HMDs that used miniature CRTs for display devices, a mechanical tracker to provide head position and orientation in real time, and a hand-tracking device (Sutherland, 1965; Sutherland, 1968). Almost all subsequent see-through HMDs have been optical see-through. The VCASS system (Buchroeder et al., 1981, Furness, 1986), the Tilted Cat HMD (Droessler and Rotier, 1990), and the CAE Fiber-Optic HMD (Barrette, 1992) are examples of optical see-through HMDs. Several of these systems have been developed by Kaiser Electronics and McDonnell Douglas (Kandebo, 1988). A hybrid optical/video see-through HMD is the VDC HMD recently developed by SEXTANT Avionique (Desplat, 1997). This HMD superimposes information from three channels: the real scene viewed through a half-silvered mirror, symbolic graphical information, and information captured via infrared cameras looking at the real scene as well. The latter is equivalent to video see-through operating in the infrared instead of in the visible. A primary aim of these various military systems is to train aircraft pilots at reduced cost and risk. Another aim is to effectively display information in air navigation and combat.

While the Air Force engaged in the development of various optical see-through HMDs, research in effective visualization conducted in both academia and other research laboratories started exploring the potential use of such devices as well. Developments in 3D scientific and medical visualization were initiated in the 1980's at the University of North Carolina at Chapel Hill (Brooks, 1992). Optical see-through displays have also been developed for applications such as engineering (Caudell & Mizell, 1992; Feiner et al., 1993) and medical applications (Peuchot et al., 1995; Edwards et al., 1995; Holloway, 1995; Wright et al., 1995; Rolland et al., 1997). A low-cost optical see-through HMD was also developed by former Virtual I/O Corporation to target perhaps less specialized and demanding applications. The systems targeted at specific applications will now be discussed. Specific issues of the technology of see-through HMDs will then be presented.

2. SOME PAST AND CURRENT APPLICATIONS OF OPTICAL AND VIDEO SEE-THROUGH HMDs

Current applications of augmented reality using see-through technologies shall be reviewed to guide the discussion of technology development for optical and video see-through displays.

2.1   Medical Data Visualization

The need for accurate visualization and diagnosis in health care is crucial. One of the main developments of medical care has been imaging.
Since the discovery of x-rays in 1895 by Wilhelm Roentgen, and the first
x-ray clinical application a year later by two Birmingham (UK) doctors,
x-ray imaging and other medical imaging modalities (e.g., CT, Ultrasound,
NMR) have emerged. Medical imaging allows the viewing of aspects of
the interior architecture of living beings that were unseen before. With the
advent of imaging technologies, opportunities for minimally invasive surgical procedures have arisen. Imaging and visualization can be used to guide
needle biopsy, laparoscopic, endoscopic, and catheter procedures. Such
procedures do require additional training because the physicians do not see
the natural structures seen in open surgery. For example, the natural eye and
hand coordination is not available during laparoscopic surgery. Visualization techniques associated with see-through HMD promise to help restore
some of the lost benefits of open surgery, for example by projecting a virtual
image directly on the patient, eliminating the need for remote monitors.

We now briefly discuss examples of recent and current research conducted with: 1) optical see-through HMDs at UNC-CH, at the University
of Central Florida (UCF), and at the United Medical and Dental Schools
of Guy's and Saint Thomas's Hospitals in England; 2) video-see-through
at UNC-CH; and 3) hybrid optical/video see-though at the University of
Blaise Pascal, Clermont Ferrand, France.

A rigorous analysis of errors for an optical see-through HMD targeted
toward the application of optical see-through HMD to craniofacial reconstruction was conducted at UNC-CH (Holloway, 1995). The superimposition of CT skull data onto the head of the real patient would give the
surgeons "x-ray vision." The promise of that system was that viewing the
data *in situ* allows surgeons to make better surgical plans because they
would be able to see the complex relationships between the bone and soft
tissue more clearly. Holloway found that the largest registration error between real and virtual objects in optical see-through HMDs was caused
by delays in presenting updated information associated with tracking. A
detailed analysis of Holloway's work can be found in Chapter 6 of this book.

In the Optical Diagnostics and Applications (ODA) Laboratory at UCF,
Jannick Rolland and colleagues, in a collaboration with Donna Wright
from the Radiology Department at UNC-CH, are currently developing an
augmented reality tool for visualization of human anatomical joints in
motion (Wright et al., 1995; Kancherla et al., 1995; Rolland et al., 1997;

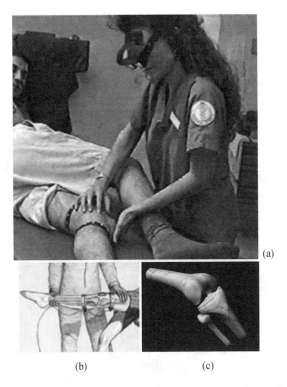

(a)

(b)   (c)

FIG. 4.3. (a) The VRDA tool will allow superimposition of virtual anatomy on a model patient. (b) An illustration of the view of the HMD user (courtesy of Andrei State). (c) A rendered frame of the knee-joint bone structures animated based on a kinematic model of motion developed by Baillot and Rolland (1998) that will be integrated in the tool.

Parsons & Rolland, 1998; Baillot & Rolland, 1998; Baillot et al., 1999; Baillot, 1999). An illustration of the tool using an optical see-through HMD for visualization of anatomy is shown in Fig. 4.3. In the first prototype we have concentrated on the positioning of the leg around the knee joint. The joint is accurately tracked optically by using three infrared video cameras to locate active infrared markers located around the joint. The first obtained results of the optical superimposition of the graphical knee-joint on a leg model seen through one of the lenses of our stereoscopic bench prototype display are shown in Fig. 4.4 (Baillot et al., 2000).

An optical see-through HMD coupled with optical tracking devices positioned along the knee joint of a model patient are used to visualize the 3D computer-rendered anatomy directly superimposed on the real leg in

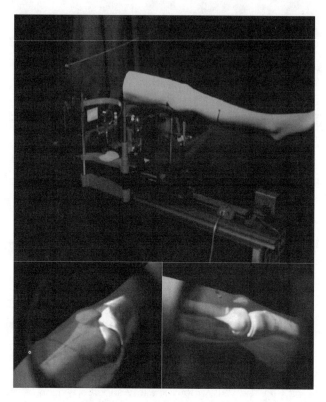

FIG. 4.4. First demonstration of the superimposition of a graphical knee-joint superimposed on a leg model for use in the VRDA
tool: (a) a picture of the benchprototype setup; a snapshot of the
superimposition through one lens of the setup in (b) a diagonal
view and (c) a side view.

motion, as shown in Fig. 4.5. The user may further manipulate the joint
and investigate the joint motions. From a technology point of view, the
field of view (FOV) of the HMD should be sufficient to capture the knee-
joint region, and the tracking devices and image-generation system must
be fast enough to track typical knee-joint motions during manipulation at
interactive speed. The challenge of capturing accurate knee-joint motions
using optical markers located on the external surface of the joint was addressed in Rolland et al. (1997). The application aims at developing a more
advanced tool for teaching dynamic anatomy, advanced in the sense that
the tool allows combination of the senses of touch and vision. We aim
this tool to specifically impart better understanding of bone motions during
radiographic positioning for the radiological science (Wright et al., 1995).

FIG. 4.5. Optical superimposition of internal anatomy using a bench prototype HMD.

To support the need for accurate motions of the knee joint in the VRDA tool, an accurate kinematic model of joint motion based on the geometry of the bones and collision detection algorithms was developed (Baillot et al, 2000). The dynamic registration of the leg with the simulated bones is reported elsewhere (Outters et al., 1999; Argotti et al., 2000). High accuracy optical tracking methods, carefully designed and calibrated HMD technology, and appropriate computer graphics model for stereo pair generation play an important role in achieving accurate registration (Vaissie & Rolland, 2000; Rolland, Quin, et al., 2000).

At the United Medical and Dental Schools of Guy's and Saint Thomas's Hospitals in England, researchers are projecting simple image features derived from preoperative magnetic resonance and computer-tomography images into the light path of a stereo operating microscope, with the aim to allow surgeons to visualize underlying structures during surgery. The first prototype used low contrast color displays (Edwards et al., 1995). The current prototype uses high contrast monochrome displays. The microscope is tracked intraoperatively and the optics are calibrated (including zoom and focus) using a pinhole camera model. The intraoperative coordinate frame is registered using anatomical features and fiducial markers. The image features used in the display are currently segmented by hand. These include the outline of a lesion, the track of key nerves and blood vessels, and bone landmarks. This computer-guided surgery system can be said to be equivalent to an optical see-through system operating on a microscopic scale. In this case the real scene is now viewed through magnifying optics but the eye of the observer is still the direct detecting device as in optical see-through.

At UNC, Henry Fuchs and colleagues are currently developing techniques using merging of video and graphical images for augmented reality (AR). The goal is to develop a system displaying live ultrasound data in real time and properly registered in 3D space within a scanned subject. This would be a powerful and intuitive visualization tool as well. The first application developed was the visualization of a human fetus during ultrasound echography. Figure 4.6 shows the real-time ultrasound images which appear to be pasted in front of the patient's body, rather than fixed within it (Bajura et al., 1992). Real-time imaging and visualization remains a challenge. Figure 4.7 shows a latter non real-time implementation of the

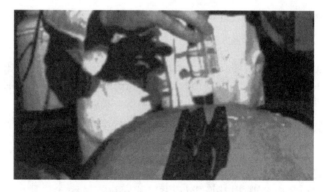

FIG. 4.6.   Real-time acquisition and superimposition of ultrasound slice images on a pregnant woman.

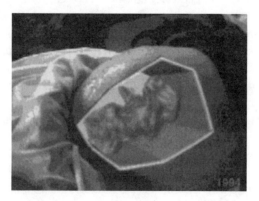

FIG. 4.7.   Improved rendering of fetus inside the abdomen.

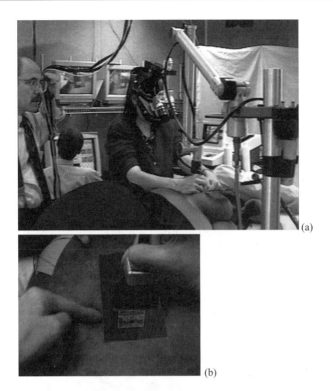

FIG. 4.8. Ultrasound guided biopsy (a) Laboratory setup during evaluation of the technology with Etta Pisano and Henry Fuchs (b) A view through the HMD.

visualization where the fetus is rendered more convincingly within the body (State et al., 1994).

More recently, knowledge from this technology was applied to developing a visualization method for ultrasound–guided biopsies of breast lesions that were detected during mammography screening procedures as shown in Fig. 4.8 (State et al., 1996). This application was motivated from the challenges we observed during a biopsy procedure while collaborating on research with Etta Pisano, head of the Mammography Research Group at UNC-CH. The goal was to be able to locate any tumor within the breast as quickly and accurately as possible. The technology of video see-through developed by Fuchs and colleagues was applied to this problem. The conventional approach to biopsy is to follow-up the insertion of a needle in the breast tissue on a remote monitor displaying real-time 2D ultrasound depth images. Such a procedure typically requires five insertions of the needle to maximize the chances of biopsy of the lesion. In the case where the lesion is located fairly deep in the breast tissue, the procedure is difficult and can be lengthy (e.g., one to two hours is not atypical for deep lesions). Several challenges remain to be overcome before the technology developed can actually be tested in the clinic, including accurate and precise tracking and a technically reliable HMD. The technology may have applications in guided laparoscopy, endoscopy, or catheterization as well.

At the University of Blaise Pascal in Clermont Ferrand, France, Peuchot and colleagues developed several augmented reality visualization tools based on hybrid optical and video see-through to assist surgeons in scoliosis surgery (Peuchot et al., 1994; Peuchot et al., 1995). Scoliosis is a deforming process of the normal spinal alignment. The visualization system, shown in Fig. 4.9 is from an optics point of view the simplest see-through system one may conceive. It is first of all fixed on a stand and it is designed as a viewbox positioned above the patient. The surgeon is positioned above the viewbox to see the patient, and the graphical information is superimposed on the patient as illustrated in Fig. 4.10. The system

FIG. 4.9. Laboratory prototype of the hybrid optical/video see-through AR tool for guided scoliosis surgery developed by Peuchot (1995) at the University of Blaise Pascal, France.

FIG. 4.10. Graphics illustration of current and future use of computer-guided surgery according to Bernard Peuchot.

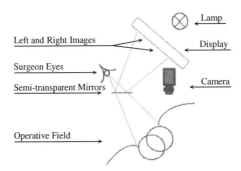

FIG. 4.11. Optical scheme of the hybrid optical/video see-through AR tool shown in Fig. 4.9.

includes a large monitor where a stereo pair of images is displayed and half-silvered mirrors that allow the superimposition of the real and virtual objects. The monitor is optically conjugated to a plane through the half-silvered mirrors and the spine under surgery is located within a small volume around that plane. An optical layout of the system is shown in Fig. 4.11.

It is important to note that the method developed for this application employs a hybrid optical/video technology. In this case, video is essentially used to localize real objects in the surgical field and optical see-through is used as the visualization tool for the surgeon. In the system, vertebrae are located in space by automatic analysis of the perspective view from a single video camera of the pellets located on the vertebrae. Knowing the underlying geometry of the pellet arrangements, a standard algorithm such as the inverse perspective algorithm is used to extract the 3D information from the projections observed in the detector plane (Dhome et al., 1989). The method relies heavily on accurate video tracking of vertebral displacements. High-accuracy algorithms were developed to support the application including development of subpixel detectors and calibration techniques (Peuchot, 1993, 1994). The method has been validated on vertebral specimens and accuracy of submillimeters in depth has been demonstrated.

The success of the method can be attributed to the fine calibration of the system, which, contrary to most systems, does not assume a pinhole camera model for the video camera. Moreover, having a fixed viewer with no optical magnification, contrary to typical HMDs, and a constant average plane of surgical operation, reduces the complexity of problems such as registration and visualization. It can be shown for example that rendered depth errors are minimized when the virtual image plane through the optics (i.e., a simple half-silvered mirror in Peuchot's case) is located in the average plane of the 3D virtual object visualized (Rolland, Ariely, & Gibson, 1995).

Furthermore, Peuchot's system avoids challenging tracking problems, optical distortion compensation, and some issues of accommodation and convergence related to HMDs (Robinett & Rolland, 1992; Rolland & Hopkins, 1993). Some tracking and distortion issues will be further discussed in Sections 3.1 and 3.2, respectively. However, good registration of real and virtual objects in a static framework is a first step to good calibration in a dynamic framework and Peuchot's results are state of the art in this regard.

While the first system developed used one video camera, the methods have been extended to include multiple cameras with demonstrated accuracy and precision of 0.01 mm (Peuchot, personal communication, 1998). Peuchot deliberately chose the hybrid system developed over a video see-through approach because "it allows the operator to work in his real environment with a perception space that is real." Peuchot judged this point to be critical in a medical application like surgery (Peuchot, personal communication, 1998).

2.2 Visualization for Manufacturing and Assembly Tasks

A difficulty with a complex manufacturing assembly task is the need to have sufficient registration of real and virtual information so that workers may perform their jobs without any risk of errors due to limitations in the apparatus. However, accuracy and precision in the order of a millimeter as required in medical data visualization are not necessary. Moreover, only the HMD user moves in this case as opposed to cases where both virtual and real objects move as they are registered against each other. The VRDA tool for visualization of joint motions discussed previously, for example, must account for head and object (i.e., anatomical joint) motions.

Caudell & Mizell (1992) built an optical see-through system for facilitating the electrical wiring of an airplane that requires positioning a large number of wires according to some diagram. In a conventional wiring job, the workers assemble a set of wires to be later incorporated in the airplane on a foam board where drawings of the assembly are provided. In addition, diagrams of the wiring are also provided to guide the assembly. In the case of augmented reality, the assembly board is blank and the wiring diagram is projected on the board to guide the wiring process. This approach has the potential advantage that as the wiring is updated, modifications can be done quickly in the software. Moreover, the successful use of this technology should enable cost reductions and efficiency improvements in the electrical wiring of aircraft manufacturing and potentially other aspects of the overall assembly process. In the summer of 1997, a six-week pilot project was conducted to compare the augmented reality technology to the traditional foam board. The experiment led to the conclusion that the technology is not yet ready to deploy. The technology is being updated with a new generation of hardware and software to prepare for the next experiment planned in the Fall 1998 (Mizell, personal communication, 1998). Details of this research can be found in Chapter 14 of this book.

Another engineering application is that of providing assistance with complex maintenance tasks. A proof of concept of such an application was developed by Steven Feiner at Columbia University (Feiner et al., 1993). The system developed, known as KARMA, uses a knowledge-based graphics component in order to aid the user in an end-user laser printer maintenance task. Feiner et al. used graphics superimposed on the laser writer to provide information on various tasks and used ultrasound tracker sensors on the printer's moving parts to reflect movements of the real-world objects in the virtual scene. Following the development of this proof of concept, Feiner further developed the software to demonstrate the use of augmented reality in aiding architectural construction, inspection, and renovation (Feiner et al., 1997). Current developments of the technology focus on exploring the possibilities of wearable augmented reality systems for use outdoors and multiuser augmented reality systems (MacIntyre & Feiner, 1998).

3. SIMILAR AND UNIQUE MAIN FEATURES OF OPTICAL AND VIDEO SEE-THROUGH TECHNOLOGY

As suggested in the description of the various example applications, the main goal of augmented reality systems is to merge virtual objects into the view of the real scene so that the user's visual system suspends disbelief into perceiving the virtual objects as part of the real environment. Current systems are far from perfect, and system designers typically end up making a number of application-dependent trade-offs. We shall list and discuss these trade-offs in order to guide the choice of technology depending upon the type of application considered.

In both systems, optical or video, there are two image sources: the real world and the computer-generated world; these two image sources are to be merged. Optical see-through HMDs take what might be called a "minimally obtrusive" approach; that is, they leave the view of the real world nearly intact and attempt to augment it by merging a reflected image of the computergenerated scene into the view of the real world. Video see-through HMDs are typically more obtrusive in the sense that they block out the real-world view in exchange for the ability to merge the two views more convincingly. In recent developments, narrow field of view video see-through HMDs have replaced large field of view HMDs, thus reducing the area where the real world captured through video and the computer-generated images are merged to a small part of the visual scene. In any case, a fundamental tradeoff is whether the additional features afforded by the more obtrusive approach justify the loss of the unobstructed real-world view.

The trade-offs between optical and video see-through HMDs with respect to technological and human factors issues from our experience designing, building, using, and assessing these HMDs are discussed. These trade-offs are also discussed with respect to current systems, and those that can be built with today's technology. Improvements for future developments

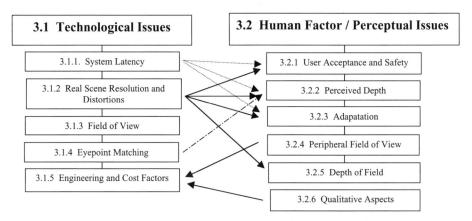

FIG. 4.12. Outline of Sections 3.1 and 3.2.

are suggested. It is important to realize, however, that many of the results may not bring results until perhaps five to ten years from now as many technological and human factors challenges remain. The specific issues now discussed are illustrated in Fig. 4.12. While most issues addressed could be easily discussed under both technological and human-factors/perceptual issues, given that the two are closely interrelated in HMD systems, we have chosen to classify each issue where it is most adequately addressed at this time given the state of the art of the technology. For example, delays in HMD systems are addressed under technology because technological improvements are actively being pursued to minimize delays. Remaining delays certainly have several impacts on various human factors issues (e.g., perceived location of objects in depth; user acceptance). Therefore Fig. 4.12 simply provides a map through this section of the chapter and the multiple arrows indicate some of the interrelationships of each issue to either of the two main categories: technological and human-factors/perceptual issues.

3.1 Technological Issues

The technological issues discussed in this section include latency of the system, resolution and distortion of the real scene, field of view (FOV), eyepoint matching of the see-through device, and engineering and cost factors. While we shall discuss properties of both optical and video see-through HMDs, it must be noted that contrary to optical see-through HMDs, there are no commercially available products for video see-through HMDs. Therefore, discussions with such systems should be considered more carefully as findings may be particular to only a few current systems. Nevertheless, we shall provide as much insight as possible in what we have learned up to date with such systems as well.

3.1.1 System Latency

An essential component of see-through HMDs is the capacity to properly register a user's surrounding and the synthetic space. The geometric calibration between the tracking devices and the HMD optics is assumed to be performed. The major impediment to achieving registration is the gap in time, referred to as lag, between the moment when the HMD position is measured and the moment when the synthetic image for that position is fully rendered and presented to the user.

Lag is the largest source of registration error in most current HMD systems (Holloway, 1995). This lag in typical systems is between 60 and 180 ms. The head of a user can move during such a period of time, and the discrepancy in perceived scene and supposed scene can destroy the illusion of the synthetic objects being fixed in the environment. The synthetic objects can "swim" around significantly in such a way that they may not even seem to be part of the real object to which they belong. For example, in the case of ultrasound-guided biopsy, the computer-generated tumor may appear to be located outside the breast while tracking the head of the user. This swimming effect has been demonstrated and minimized by predicting HMD position instead of simply measured positions (Azuma & Bishop, 1994).

Current HMD systems are lag limited as a consequence of tracker lag, the complexity of rendering, and displaying the images. Tracker lag is often not the limiting factor. If displaying the image is the limiting factor, novel display architectures supporting frameless rendering can help solve the problem (Bishop et al., 1994). Frameless rendering consists in continuously updating an image, as information becomes available instead of updating entire frames at a time. The trade-offs between lag and image quality are currently investigated (Scher-Zagier, 1997). If we assume we are limited by the speed of rendering an image, eye-tracking capability can be useful in the sense that one only needs to quickly update information around the gaze point of the user (Thomas et al., 1989; Rolland, Yoshida, et al., 1998; Vaissie and Rolland, 2000).

Lag Minimization in Video See-Through HMDs One of the major advantages of video see-through HMDs is the potential capability of reducing the relative latencies between the 2D real and synthetic images as a consequence of both types of images being digital. Jacobs et al. (1997) review techniques for managing latency in augmented reality video see-through systems. Manipulation in space and in time of the images is applied to register them. Three-dimensional registration is computationally extensive, if at all robust, and challenging for interactive speed. The spatial approach to forcing registration in video see-through systems is to correct registration errors by imaging landmark points in the real world and registering virtual objects with respect to them (State et al., 1996). One approach to eliminate temporal delays between the real and computer-generated images in such a case is to capture a video image and draw the graphics on top of the video image. Then the buffer is swapped and the combined image is presented to the HMD user. In such a configuration, no delay apparently exists between the real and computer-generated images. If the actual latency of the computer-generated image is large with respect to the video image, however, it may cause sensory conflicts between vision and proprioception because the video images no longer correspond to the real-world scene. Any manual interactions with real objects could suffer as a result.

Another approach to minimizing delays in video see-through HMDs is to delay the video image until the computer-generated image is rendered. This approach is only valid when two streams are available and combined. Bajura & Neumann (1995) applied chroma keying, for example, to dynamically image a pair of red light-emitting diodes (LEDs) placed on two real objects (one stream) and then registered two virtual objects with respect to them (second stream). By tracking more landmarks, better registration of real and virtual objects may be achieved (Tomasi & Kanade, 1991). The limitation of the approach taken is the attempt to register three-dimensional scenes using two-dimensional constraints. If the user rotates his head rapidly or if a real-world object moves, there may be no "correct" transformation for the virtual scene image. In order to align all of the landmarks, one must either allow errors in registration of some of the landmarks or perform a nonlinear warping of the virtual scene that may create undesirable distortions of the virtual objects. The nontrivial solution to this problem is to increase the speed of the system until scene changes between frames are small and can be approximated with simple 2D transformations.

In a similar vein, it is also important to note that the video view of the real scene will normally have some lag due to the time it takes to acquire and display the video images. Thus, video seethrough HMDs will normally be slightly delayed with respect to the real world even without adding delay to match the synthetic images. This delay may increase if an image-processing step is applied to either enforce registration or perform occlusion. The key issue is whether the delay in the system is too great for the user to adapt to it. This subject has been treated at length in the teleoperation literature (Held & Durlach, 1987).

Lag Minimization in Optical See-Through HMDs   Systems using optical see-through HMDs have no means to introduce artificial delays to the real scene. Therefore, the system may need to be optimized for low latency, as suggested, perhaps less than 60 ms where predictive tracking can be effective (Azuma & Bishop, 1994). For any remaining lag, users may have to limit their actions to using slow head motions. Applications where speed of movement can be readily controlled, such as in the VRDA tool described earlier, can highly benefit from optical see-through technology (Rolland and Arthur, 1997). The advantage of not introducing artificial delays is that real objects will always be where they are perceived to be, and this may not only be highly desired but importantly crucial for a broad range of applications.

Lag Minimization and Eye Tracking   We shall note that most current HMDs have the shortcoming to lack integrated effective interaction capabilities combining head and eye tracking (Rolland, Yoshida, et al., 1998). The interaction capability is ordinarily limited to the use of head and hand tracking to measure the position and orientation of the user's head or hand and to generate scenery from the user's perspective (Ferrin, 1991). Thus, for situations that require fast response times or difficult coordination skills, interaction capability supported by manual input devices becomes inadequate. For those cases, eye movement could be used in conjunction with manual input devices to provide effective interaction methods. Various interaction methods can thus be realized through the use of hand, body, and eye movements (Bolt, 1981; Bryson, 1991; Jacoby & Ellis, 1992). Since the eyes respond to stimulus ~150 ms faster than the hand (Colgate, 1968; Oster & Stern, 1980; Girolamo, 1991), they can be used for fast and effective input, selection, and control methods. Vaissie and Rolland (2000) also recently demonstrated that eye tracking is essential in HMDs for accurate rendered depth. A question of investigation is how can eye-tracking capability be best integrated in optical or video see-through systems. To our knowledge eye-tracking capability has been tested, yet not fully integrated, in optical see-through systems (Barrette, 1992; Desplat, 1997). It has not yet been considered in video see-through systems.

Furthermore, it is important to note that image rendering can also take advantage of the physiological limitation of the eyes. It has been well known since Reymond Dodge in the 1900s that when the eyes move, information processing is suppressed. This is known, in the modern literature, as saccadic suppression (Dodge, 1903; Volkman et al., 1978; Volkman, 1986). Therefore, while the gaze point is in rapid motion, the image update does not have to occur at full resolution and the fine detail of the scene can be rendered when the gaze point is considered fixed. The speed of smooth pursuit movements is mostly relevant for discussing speed of rendering. It is typically 100 degrees/second (Goldberg et al., 1991). Furthermore, it is widely accepted in the vision literature that it takes typically 100 ms to process new visual information (numbers from 80 to 150 ms are argued among visual scientists). As a result, a fixation is typically defined as a 100 ms pause in eye movement (ASL, 1997). Finally, tracking of eye movements may help predict motion of the user in the virtual environment. Thus, one of the authors (JR) postulates that tracking eye movements may play a fundamental role not only in providing unique means of interaction in the VE, rendering accurate depth, but also in minimizing system lag.

3.1.2   Real-Scene Resolution and Distortion

The best real-scene resolution a see-through-device can provide is that perceived with the naked eye under unit magnification of the real scene. Certainly under microscopic observation as described by Hill (Edwards et al., 1995), the best scene resolution goes beyond that obtained with a naked eye. It is also assumed that the see-through device has no image-processing capability. A resolution extremely close to that obtained with the naked eye is easily achieved with an optical see-through HMD because the optical interface to the real world is simply a thin parallel plate (e.g., glass plate) positioned between the eyes and the real scene. Such an interface typically introduces only very small amounts of optical aberrations to the real scene: For example, for a real-point object seen through a 2 mm planar parallel plate placed in front of a 4 mm diameter eye pupil, the diffusion spot due to spherical aberration would subtend a 2×10^{-7} arc-minute visual angle for a point object located 500 mm away. Spherical aberration is one of the most common and simple aberrations in optical systems. Such a degradation of image quality is negligible compared to the ability of the human eye to resolve a visual angle of 1 minute of arc. Similarly, planar plates introduce low distortion of the real scene, typically below 1%. There is no distortion only for the chief rays that pass the plate parallel to its normal.[1]

In the case of a video see-through HMD, real-scene images are digitized by miniature cameras (Edwards et al., 1993) and converted to an analog signal, which is fed to the HMD. The images are then viewed through the HMD viewing optic that typically uses eyepiece design. The perceived resolution of the real scene can thus be limited by the resolution of the video cameras or the HMD viewing optics. Currently available miniature video cameras typically have a resolution of 640×480, which is also near the resolution limit of the miniature displays currently used in HMDs.[2] Depending upon the magnification and the field of view of the viewing optics various effective visual resolutions may be reached. While the miniature displays and the video cameras seem to currently limit the resolution of most systems, such performance may improve with higher resolution detectors and displays.

In assessing video see-through systems, one must distinguish between narrow and wide FOV devices. Large FOV eyepiece designs (≥ 50 degree FOV) are known to be extremely limited in optical quality as a consequence of optical aberrations that accompany large FOVs, pixelization that may become more apparent under large magnification, and the exit pupil size that must accommodate the size of the pupils of a person's eyes. Thus, even with higher resolution cameras and displays, video see-through HMDs may remain limited in their ability to provide a real-scene view of high resolution if conventional eyepiece designs continue to be used. In the case of small to moderate FOV video see-through HMDs (10 to 20 degrees) the resolution is still typically a lot less than the resolving power of the human eye.

A new technology, referred to as tiling, may overcome some of the current limitations of conventional eyepiece design for large FOVs (Kaiser Electro-Optics, 1994). The idea is to use multiple narrow FOV eyepieces coupled with miniature displays to completely cover (or tile) the user's FOV. Because the individual eyepieces have a fairly narrow FOV, higher resolution, nevertheless currently less than the human visual system, can be achieved. An additional challenge however is in the assembly process and in rendering seamless views from multiple displays.

Theoretically, distortion is not a problem in video see-through systems since the cameras can be designed to compensate for the distortion of the

---

[1] A chief ray is defined as a ray that emanates from a point in the FOV and passes through the center of the pupils of the system. The exit pupil in an HMD is the entrance pupil of the human eye.

[2] The number of physical elements is typically 640×480. One can use signal processing to interpolate between lines to get higher resolutions.

optical viewer, as demonstrated by Edwards et al. (1993). However, if the goal is to merge real and virtual information, as in ultrasound echography, having a warped real scene increases the complexity of the synthetic image generation significantly (State et al., 1994). Real-time video correction can be used at the expense of an additional delay in the image generation sequence. An alternative is to use low distortion video cameras at the expense of a narrower FOV, merge unprocessed real scenes with virtual scenes, and warp the merged images. Warping can be done using for example real-time texture mapping to compensate for the distortion of the HMD viewing optics as a last step (Rolland & Hopkins, 1993; Watson & Hodges, 1995).

The need for high, real-scene resolution is highly task dependent. Demanding tasks such as surgery or engineering training, for example, may not be able to tolerate much loss in real-scene resolution. Because the large FOV video see-through systems we have experience with are seriously limited in terms of resolution, narrow FOV video see-through HMDs are currently preferred. An additional critical issue in aiming toward narrow FOV video see-through HMDs, independently of resolution, is the need to match the viewpoint of the video cameras with the viewpoint of the user, an unresolved issue with large FOV systems discussed in Section 3.2.3. Also, methods for matching video and real scenes for large FOV tiled displays must be developed. Now, simply considering resolution and given the growing availability of highresolution flat-panel displays, we do not see why the resolution of see-though HMDs cannot gradually increase for both small and large FOV systems. The development and marketing of miniature high-resolution technology must be undertaken to achieve resolutions that match that of the human visual system.

3.1.3 Field of View (FOV)

A generally challenging issue of HMDs is providing the user with an adequate FOV for a given application. For most applications, having a large binocular FOV means that fewer head movements are required to perceive an equivalently large scene. However, in many cases, one would prefer to have a large binocular FOV without trading off the amount of binocular overlap that is necessary for stereo vision (Rash, 1999). In these cases, the monocular FOV itself must be optimized. We believe that a large FOV is especially important for tasks that require grabbing and moving objects and that it provides increased situation awareness when compared to narrow FOV devices (Slater & Wilbur, 1997). The situation with see-through devices is somewhat different from that of fully opaque HMDs in that the aim of using the technology is different from that of immersing the user in a virtual environment.

Overlay and Peripheral FOV   The term overlay FOV is defined as the region of the FOV where graphical information and real information are superimposed. The peripheral FOV is the real-world FOV beyond the overlay FOV. For immersive opaque HMDs no such distinction is made; one refers simply to the FOV. It is important to note that the overlay FOV may only need to be narrow for certain augmented reality applications. For example, in a visualization tool such as the VRDA tool, only the knee-joint region is needed in the overlay FOV. In the case of computer-guided breast biopsy, the overlay FOV could be as narrow as the synthesized tumor. The real scene need not necessarily be synthesized. The available peripheral FOV however is critical for situation awareness and is most often required for various applications whether it is provided as part of the overlay or around the overlay. If provided around the overlay, the transition from real to virtual imagery must be made as seamless as possible, an issue of investigation that has not yet been addressed in video see-through HMDs.

Optical see-through HMDs typically provide from 20 to 60 degrees overlay FOV via the half-transparent mirrors placed in front of the eyes, a characteristic that may appear somewhat limited but promising for a variety of applications whose working visualization distance is within arm reach. Those include various medical visualization and engineering tasks. Larger FOVs have been obtained, up to 82.5 × 67 degrees, at the expense of reduced brightness, increased complexity, and massive, expensive technology (Welch & Shenker, 1984). Such FOVs may have been required for performing various navigation tasks in real and virtual environments, but are likely not required in most augmented reality applications. Those tasks include, for example, air pilot navigational tasks in either simulators and test air flights to assess the technology. While this chapter focuses on binocular HMDs typically operating in the visible, night vision goggles are HMDs and such systems have been extensively used in air combat (Rash, 1999). We would also like to mention that the Apache displays, in fact monocular HMDs, were also extensively used during Desert Storm missions as a precursor perhaps of binocular systems of the future (M. Shenker, personal communication, 1998). Optical see-through HMDs, however, whether or not they have a large overlay FOV have been typically designed open enough that the user can use his/her peripheral vision around the device, thus increasing the total real-world FOV to numbers that match closely one's natural FOV. An annulus of obstruction usually results from the mounts of the thin see-through mirror similar to the way that our vision may be partially occluded by a frame when wearing eyeglasses.

In the design of video see-through HMDs, a difficult engineering task is matching the frustum of the eye with that of the camera, as discussed in Section 3.1.4. While such matching is not so critical for far field viewing, it is certainly important for near field visualization. This difficult matching problem has led the consideration of narrower fields of view systems. A compact, 40 × 30 degrees FOV design, intended for optical see-through HMD but adaptable to video see-through, was proposed by Manhart et al. (1993). Video see-through HMDs, on the other hand, can provide in terms of a see-through FOV, the FOV displayed with the opaque type viewing optic that typically ranges from 20 to 90 degrees. In such systems where the peripheral FOV of the user is occluded, the effective real world FOV is often smaller than in optical see-through systems. When using a video see-through HMD, we found in a recent human-factors study that users needed to perform larger head movements to scan an active field of vision required for a task than with the unaided eye (Biocca & Rolland, 1998). We predict that the need to make larger head movements would not arise as much with see-through HMDs with equivalent overlay FOVs but larger peripheral FOVs because users are provided with increased peripheral vision, and thus additional information, to more naturally perform the task.

Increasing Peripheral FOV in Video See-Through HMDs

An increase in peripheral FOV in video see-through systems can be accomplished in two ways: 1) in a folded optical design, as used for optical see-through HMDs, but with an opaque mirror instead of a half transparent mirror, or 2) in a nonfolded design but with nonenclosed mounts. The latter calls for innovative opto-mechanical design since optics heavier than in either optical or folded video see-through must be supported. Folded systems only require a thin mirror in front of the eyes, and the heavier optical components are placed around the head. The trade-off with folded systems, however, is a significant reduction in the overlay FOV.

Trade-Off Resolution and FOV

While the resolution of a display is defined in the graphics community as the number of pixels, the relevant measure of resolution for HMDs is the number of pixels per angular FOV, also referred to as angular resolution. Indeed, what is of importance for usability is the angular subtends of a pixel at the eye of the HMD user. Most current high-resolution HMDs achieve higher resolution at the expense of a reduced FOV. That is, they use the same miniature, high-resolution CRTs but with optics of less magnification to achieve higher angular resolution. This results in a FOV that is often too narrow for certain applications. The current solutions proposed to improve resolution without trading FOV are either tiling techniques or head-mounted projective displays.

Tiling   One of the few demonstrations of high-resolution, large FOV displays are the tiled displays. They consist in placing a series of miniature displays side by side, thus forming an array of displays in front of the eyes where each element of the array has an associated magnifying lens. Another approach employs large high-resolution displays, or light valves, and transports the high-resolution images to the eyes by imaging optics coupled to a bundle of optical fibers (Thomas et al., 1989). When rendering the images at the gaze point with higher accuracy than the surrounding image, such displays yield high-resolution insets. Displays with high-resolution inset also aim to achieve high resolution and large FOV (Fernie, 1995). The tiled displays certainly bring new practical and computational challenges that need to be confronted. If a see-through capability is desired (e.g., to display virtual furniture in an empty room), it is currently unclear whether the technical problems associated with providing overlay can be solved.

Head-Mounted Projective Displays   HMDs of the projection type have been designed and demonstrated for example by Kojima & Ojika (1997), Parsons & Rolland (1998), Rolland, Parsons, et al. (1998); Hua et al. (2000). Kojima used a conventional projection screen in his prototype. Parsons and colleagues developed a first prototype head-mounted projective display, shown in Fig. 4.13, in order to demonstrate that an undistorted virtual 3D image could be rendered when projecting a stereo pair of images on a bent sheet of microretroreflector cubes. Rolland and colleagues are developing the next generation prototypes of the technology. The system presents various advantages over conventional HMDs including distortion-free images, occluded virtual objects from real objects interposition, no image cross-talks for multiuser participants, and the potential for a wide FOV (i.e., up to 120 degrees).

3.1.4   Viewpoint Matching

In video see-through HMDs, the camera viewpoint (i.e., the entrance pupil) must be matched to the viewpoint of the observer (i.e., the entrance pupil of the eye). The viewpoint of a camera or eye is equivalent to the center of projection used in the computer graphics model employed to

FIG. 4.13. Proof of concept prototype of a head-mounted projective display with microreflector sheeting.

compute the stereo images and is taken here to be the center of the entrance pupil of the eye or camera (Vaissie & Rolland, 2000). In earlier video see-through designs at UNC-CH, Edwards et al. (1993) investigated ways to mount the cameras to minimize errors in viewpoint matching. The error minimization versus exact matching was a consequence of working with wide FOV systems. If the viewpoints of the cameras do not match the viewpoints of the eyes, the user experiences a spatial shift in the perceived scene that may lead to perceptual anomalies, as further discussed under human factors issues (Biocca & Rolland, 1998). Error analysis should then be conducted in such a case to match the need of the application.

For cases when the FOV is small (less than about 20 degrees) exact matching in viewpoints is possible. Because the cameras cannot be physically placed at the actual eyepoints, mirrors can be employed to fold the optical path (much like a periscope) in order to make the cameras' viewpoints correspond to the real eyepoints as shown in Fig. 4.13 (Edwards et al., 1993; Colucci & Chi, 1994). Although such geometry solves the shift in viewpoint problem, it increases the optical path length, which reduces the field of view, for the same reason that optical see-through HMDs tend to have smaller fields of view. Thus, video see-through HMDs must either trade their large FOVs for correct real-world viewpoints or require the user to adapt to the shifted viewpoints as further discussed in Section 3.2.3.

Finally, correctly mounting the video cameras in a video see-through HMD requires that the HMD has interpupillary distance (IPD) adjustment. The video cameras must then be slaved to that adjustment for the views obtained by the video cameras to match those that would have been obtained with naked eyes, given the IPD of a user. Thus the cameras must be separated by the appropriate IPD. If one were to account for eye movements in video see-through HMDs, the level of complexity in slaving the camera viewpoint to the user viewpoint would be highly increased. To our knowledge it has not yet been considered.

3.1.5   Engineering and Cost Factors

Most HMD designs usually suffer from low resolution, limited FOV, poor ergonomic design, and heavy weight. To overcome any of these limitations, one must face new challenges and further trade-offs. A good ergonomic design requires a HMD that is light enough to not weigh much more than a pair of eyeglasses, or folds around the user's head in order for the center of gravity of the device to fall near the center of rotation of the head (Rolland, 1994). This aims toward maximum comfort and usability. Reasonably lightweight HMD designs currently suffer narrow FOVs, on the order of 20 degrees. To our knowledge, there are currently no large FOV stereo see-through HMDs of any type that are comparable in weight to a pair of eyeglasses. Rolland predicts that it could be achieved with some emerging technology of projection HMDs (Rolland, Parsons et al., 1998). However, it must be noted that such technology may not be well suited to all visualization schemes as it requires a projection screen somewhere in front of the user, not necessarily attached to the user's head.

With optical see-through HMDs, the folding can be accomplished with either an on-axis or an off-axis design. Off-axis designs are more elegant and also by far more attractive since they free the user from seeing the ghost images that plague current on-axis designs. The reason off-axis designs are not commercially available is that very few prototypes have been built and those that have been built have been classified (M. Shenker, personal communication). Moreover, off-axis systems are difficult to design and build (Shenker, 1994). A nonclassified, off-axis design has been designed by Rolland (1994) at UNC-CH, and recently further analyzed (Rolland, 2000). Several factors including cost have prohibited building a first prototype as well. There is expectation that new generations of computer-controlled fabrication and testing will change this trend.

Since their beginning, high-resolution HMDs have been CRT based. Early systems were even monochrome, but color CRTs using color wheels or frame sequential color have been fabricated and incorporated into HMDs (Allen, 1993). Five years ago, we may have thought that, today, high-resolution color flat-panel displays would be the first choice for HMDs. While it is slowly happening, miniature CRTs may not be fully obsolete today. It has certainly been predicted for years that CRTs would become obsolete because they will never yield the compact, lightweight designs that could be conceived with flat-panel miniature displays. The current optimism, however, lies in new technologies such as reflective LCDs, micro-electromechanical systems (MEMS)-based displays, and nano-technology-based displays.

3.2   Human-Factor and Perceptual Issues

Assuming that many of the technological challenges described have been addressed and high performance HMDs can be built, a key human factor issue for see-through HMDs is that of user acceptance and safety. This will be discussed first. We shall then discuss technicalities of perception in such displays. An ultimate see-through display is one that provides quantitative and qualitative visual representations of scenes that conform to a predictive model (e.g. conform to that given by the real world if it is what is intended). This includes 1) accuracy and precision of rendered and perceived location of objects in depth; 2) accuracy and precision of rendered and perceived size of real and virtual objects in a scene; and 3) an unobstructed peripheral FOV, which is important for many tasks from situation awareness to simple manipulation of objects and accessories.

3.2.1   User Acceptance and Safety

A fair question for either type of technology is "Will anyone actually wear one of these devices for extended periods?" The answer will doubtlessly be application and technology specific but will be reduced to the issue of whether the advanced capabilities afforded by the technology offset the problems induced by the encumbrance and sensory conflicts associated with it.

In particular, one of us (JR) thinks that video see-through HMDs may meet with resistance in the work place since they take away the direct real-world view in order to augment it. It is an issue of trust that may be difficult to overcome for some users. If wide-angle FOV video see-through HMDs are used, the problem is exacerbated in safety-critical applications.

A key difference in such applications may turn out to be the failure mode of each technology. A technology failure in the case of optical see-through HMDs may leave the subject without any computer-generated images but still with the real-world view. In the case of video see-through, it may leave the user with complete suppression of the real-world view, as well as the computer-generated view. HF however is of the opinion that, because the video view occupies such a small fraction (~10 degree visual angle) of the scene in recent developments of the technology, the issue has became less critical. This is especially true of flip-up and down devices such as that developed at UNC-CH in recent years (Colucci & Chi, 1995).

Certainly image quality and its trade-offs are critical issues related to user acceptance for all types of technology. In a personal communication, Martin Shenker, a senior optical engineer with over twenty years of experience designing HMDs, pointed out that there exists no current standards of image quality and technology specifications for design, calibration, and maintenance of HMDs. This is a current concern at a time where the technology may be adopted in various visualization tasks and various users groups including children.

3.2.2 Rendered and Perceived Location of Objects in Depth

Occlusion The ability to perform occlusion in see-through HMDs is an important issue of comparison between optical and video see-through HMDs. One of the most important differences between these two technologies is how they handle the depth cue known as occlusion (or interposition). In real life, an opaque object can block the view of another object so that part or all of it is not visible. While there is no problem in making computer-generated objects occlude each other in either system, it is considerably more difficult to make real objects occlude virtual objects and vice versa unless the real world for an application is predefined and has been modeled in the computer. Even then, one would need to know the exact location of a user with respect to that real environment. This is however not the case of most augmented reality applications where the real world is constantly changing and on the fly acquisition is all the information one will ever have of the real world. Occlusion is a strong monocular cue to depth perception and may be required in certain applications (Cutting & Vishton, 1995).

In both systems, computing occlusion between the real and virtual scenes requires a depth map of both scenes. A depth map of the virtual scene is usually available (for z-buffered image generators), but a depth map of the real scene is a much more difficult problem. Although one could create a depth map in advance from a static real environment, many applications require on-the-fly image acquisition of the real scene. For example, in the VRDA tool described earlier, each model patient will have a different knee, and a computer model of someone else's real knee may not be useful. While progress in this area is being made (Tomasi & Kanade, 1991; Laveau & Faugeras, 1994), the problem is far from solved. Thus, occlusion cues for either type of display will be limited by the state of the art in this area. We can now move on to a discussion of the trade-offs with respect to occlusion for each type of see-through HMD.

Assuming the system has a depth map of the real environment, video see-through HMDs are perfectly positioned to take advantage of this information. They can, on a pixel-by-pixel basis, selectively block out the view of either scene or even blend them to minimize edge artifacts. One of the chief advantages of video see-through HMDs is that they handle this problem so well.

The situation for optical see-through HMDs is more complex. Existing optical see-through HMDs blend the two images with beam splitters, which blend the real and virtual images uniformly throughout the FOV. Normally, the only control the designer has is the amount of reflectance versus transmittance of the beam splitter, which can be chosen to match the brightness of the displays with the expected light levels in the real-world environment. If the system has a model of the real environment, it is possible to cause real objects to occlude virtual ones simply by not drawing the occluded parts of the virtual objects. The only light will then be from the real objects, giving the illusion that they are occluding the virtual ones. Such an effect requires one to operate in a darkened room with light directed where needed. This technique has been used by CAE Electronics in their flight simulator: When the pilots look out the window, they see computer-generated objects. If they look inside the cockpit, however, the appropriate pixels of the computer-generated image are masked so they can see the real instruments. They keep the room fairly dark so that this technique will work (Barrette, 1992). David Mizell from Boeing Seattle and Tom Caudell now at the University of New Mexico are also using this technique; they refer to it as "fused reality" (Caudell & Mizell, 1992).

Whereas optical see-through HMDs can allow real objects to occlude virtual objects, the reverse is even more challenging since normal beam splitters have no way of selectively blocking out the real environment. There are at least two possible partial solutions to this problem. The first solution is to spatially control the light levels in the real environment and to use displays that are bright enough so that the virtual objects mask the real ones by reason of contrast. This approach is used in flight simulators for creating the virtual instruments. This may be a solution for a few applications. A possible second solution would be to locally attenuate the real world view by using an addressable filter device placed on the see-through mirror. It is possible to generate partial occlusion in this manner because the effective beam of light entering the eye from some point in the scene covers only a small area of the beam splitter, the eye pupil being typically 2 to 4 mm in photopic vision. A problem with this approach is that the user does not focus on the beam splitter, but rather somewhere in the scene. A point in the scene maps to a disk on the beam splitter, and various points in the scene map to overlapping disks on the beam splitter. Thus, any blocking done at the beam splitter may occlude more of the scene than expected, which might lead to odd visual effects. A final possibility is that some applications may work acceptably without properly rendered occlusion cues. That is, in some cases, the user may be able to use other depth cues, such as head-motion parallax, to resolve the ambiguity caused by the lack of occlusion cues.

Rendered Locations of Objects in Depth   We shall distinguish between errors in the rendered and perceived location of objects in depth. The former yields the latter. One can conceive, however, that errors in perceived location of objects in depth can also occur even in the absence of errors in rendered depths as a result of an incorrect computational model for stereo pair generation or a suboptimal presentation of the stereo images. This is true for both optical and video see-through HMDs. Indeed if the technology is adequate to support a computational model, and the model accounts for required technology and corresponding parameters, the rendered locations of objects in depth as well as the resulting perceived locations of objects in depth will follow expectations. Vaissie and Rolland have shown some limitations of the choice of a static eyepoint in computational models for stereo pair generation for virtual environments, and have demonstrated errors in rendered and thus perceived location of objects in depths (Vaissie & Rolland, 2000). The ultimate goal is to derive a computational model and required technology that yield desired perceived location of objects in depth. Errors in rendered depth typically result from inaccurate display calibration and parameter determination such as the FOV, the frame buffer overscan, the eyepoints' location, conflicting or noncompatible cues to depth, and optical aberrations including residual distortions.

FOV and Frame Buffer Overscan   Errors of a few degrees in FOV, which are easily made if no calibration is conducted, can lead to significant errors in rendered depths depending on the imaging geometry. For some medical and computer-guided surgery applications for example, errors of several millimeters are likely unacceptable. For various navigation tasks, they may be considered negligible. The FOV and the overscan of the frame buffer that must be measured and accounted for to yield accurate rendered depths are critical parameters for stereo pair generation in HMDs (Rolland, Ariely, & Gibson, 1995). These parameters must be set correctly regardless of the specifics, optical or video see-through, of the technology.

Specification of Eyepoint Location   The location of the eyepoints of the user used to render the stereo images from two correct viewpoints must be specified for accurate rendered depth. This applies to both optical and video see-through HMDs. In addition for video see-through HMDs, the real-scene video images must be acquired from the correct viewpoint (Biocca & Rolland, 1998).

For the computer graphics generation component, three choices of eyepoint locations within the human eye have been proposed: the nodal point of the eye[3] (Robinett & Rolland, 1992; Deering, 1992); the entrance pupil of the eye (Rolland, 1994; Rolland et al., 1995); and the center of rotation of the eye (Holloway, 1995). Rolland, Ariely, & Gibson, (1995) discuss that the choice of the nodal point would in fact yield errors in rendered depth in all cases whether the eyes are tracked or not. For a device with eye-tracking capability, the entrance pupil of the eye should be taken as the eyepoint. If eye movements are ignored meaning that the computer-graphics eyepoints are fixed, then it was proposed that it is best to select the center of rotation of the eye as the eyepoint (Fry, 1969; Holloway, 1995). An in-depth analysis of this issue reveals that while the center of rotation yields higher accuracy in position, the center of the entrance pupil yields in fact higher angular accuracy (Vaissie & Rolland, 2000). Therefore, depending on the task involved, and whether angular accuracy or position accuracy is most important, the centers of rotation or the centers of the entrance pupil may be selected as best eyepoints location in HMDs.

Residual Optical Distortions   Optical distortion, an optical aberration that does not affect image sharpness, introduces warping of an image. It only occurs for optics including lenses or curved mirrors. If the optics

---

[3]Nodal points are conjugate points in an optical system that satisfy an angular magnification of 1. Two points are conjugate of each other if they are image of each other.

only includes plane mirrors, as in Peuchot's augmented reality system, there are no distortions. The outcome of such a mapping is errors in rendered depths. Distortion is an outcome of the location of the pupil of the user away from the nodal points of the optics. Moreover, it varies as a function of where the user looks through the optics. However, if the optics are well calibrated to account for the user's IPD, distortion will be fairly constant for typical eye movements behind the optics. Prewarping of the computer generated image can thus be conducted to compensate for the optical residual distortions (Robinett & Rolland, 1992; Rolland & Hopkins, 1993; Watson & Hodges, 1995).

Perceived Location of Objects in Depth   Once depths are accurately rendered according to a given computational model and the stereo images presented according to the computational model, the perceived locations of objects in depth and the perceived sizes of objects become an important issue for assessment of the technology and the model. Accuracy and precision can only be defined statistically. Given an ensemble of measured perceived locations of objects in depths, the depth percept will be accurate if objects appear in average at the location predicted by the computational model. Perceived location of objects in depth will be precise if objects appear within a small spatial zone around that average location. A strong component of rendering depth accurately is occlusion of overlapping objects. We shall thus distinguish between perceived locations of objects in depth of nonoverlapping and overlapping objects.

In the case of nonoverlapping objects, one may resort to depth cues other than occlusion. These include familiar sizes, stereopsis, perspective, texture, and motion parallax. A psychophysical investigation of perceived location of objects in depth in an optical see-through HMD using stereopsis and perspective as the visual cues to depth is given in Rolland, Ariely, & Gibson, (1995), Rolland & Arthur, (1997), and Rolland, Quinn, et al. (2000). The HMD is mounted on a bench to facilitate the calibration and the setting of system parameter (see Fig. 4.14).

In a first investigation, a systematic shift in the order of 50 mm in perceived location of objects in depth from predicted values was found (Rolland et al., 1995). Moreover, the precision of the measures varied significantly across subjects. As we learn more about the interface between the optics and the computational model used in the generation of the stereo image pairs, and as we improve the technology, we have since demonstrated errors in the order two millimeters (Rolland, Quinn, et al., 2000).

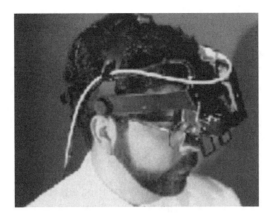

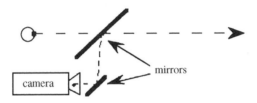

FIG. 4.14. A 10 degree FOV video see-through HMD: Dglasses developed at UNC-CH. Lipstick cameras and a double fold mirror arrangement were used to match the veiwpoints of the camera and user.

The technology is now ready to deploy for extensive testing in specific applications (e.g., the VRDA tool). Some measures of perceived size were conducted by Roscoe and colleagues in see-through HMDs and a main result was that objects seemed to be perceived smaller than they actually were (Roscoe, 1984, 1991). As the technology and the associated methods for image generation improve, follow-up experiments are required to assess the perception of objects in virtual environments.

Studies of perceived location of objects in depth, for overlapping objects in an optical see-through HMD, have been conducted by Ellis & Buchler (1994). They showed that the perceived location of objects in depth of a virtual object could be affected by the presence of a nearby opaque physical object. When a physical object was positioned in front of or at the initial perceived location of a 3D virtual object, the virtual object appeared to move closer to the observer. In the case where the opaque physical object was positioned substantially in front of the virtual object, human subjects often perceived the opaque object as transparent.

In the ODA Laboratory at UCF headed by Jannick Rolland, assessment of the technology through controlled psychophysical and human-factors studies has been an important component of the research program. The major difficulty we have encountered in conducting the assessment work is that of HMD calibration and maintenance of the calibration. The current calibration procedure is still tedious and future research should address quick calibration methods as well as maintenance of calibration over time.

3.2.3 Adaptation

When a system does not offer what the user ultimately wants, two paths may be taken: 1) Improve on the current technology or 2) study the ability of the human system to adapt to an imperfect technological unit and develop adaptation training when appropriate. This is possible because of the astonishing ability of the human visual and proprioceptive systems to adapt to new environments, as has been shown in multiple studies on adaptation (Rock, 1966).

Biocca and Rolland (1998) conducted a study of adaptation to visual displacement using a large FOV video see-through HMD. Users see the real world through two cameras that are located 62 mm higher and 165 mm forward from their natural eyepoints. Subjects showed evidence of perceptual adaptation to sensory disarrangement during the course of the study. This revealed itself as improvement in performance over time while wearing the see-through HMD and as negative aftereffects once they removed it. More precisely, the negative aftereffect manifested itself clearly as a large overshoot in a depth pointing task, as well as an upward translation in a lateral pointing task after wearing the HMD. Moreover, some participant experienced some early signs of cybersickness (Kennedy and Stanney, 1997).

The presence of negative aftereffects has some potentially disturbing practical implications for the diffusion of large FOV video see-through HMDs. Some of the intended earlier users of these HMDs are surgeons and other individuals in the medical profession. Hand' eye sensory recalibration for highly skilled users such as surgeons could have potentially disturbing consequences if the surgeon were to enter surgery within some period after use of a HMD. It is an empirical question how long the negative aftereffects might persist, and whether a program of gradual adaptation (Welch, 1994) or dual adaptation (Welch, 1993) might minimize the effect altogether. In any case, any shift in the camera eyepoints needs to be minimized as much as possible to facilitate the adaptation process that is taking

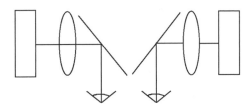

FIG. 4.15.   (a) Bench prototype head-mounted display with head-motion parallax developed in the VGILab at UCF (1997). (b) Schematic of the optical imaging from a top view of the setup.

place. This path has been taken in recent years at UNC-CH as a consequence of this investigation. As we learn more about these issues, we build devices with less error and the distance between using these systems and a pair of eyeglasses decreases so that adaptation takes less time and aftereffects decreases as well. Remaining issues are conflicts of accommodation and convergence in such displays. The issue can be solved at some cost (Rolland, Krueger, & Goon, 2000). For lower-end systems, a question of investigation concerns how users adapt to various settings of the technology. For high-end systems, much research is still needed in understanding the importance of perceptual conflicts and how to best minimize them.

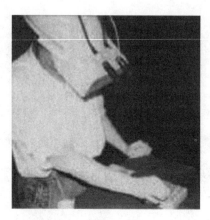

FIG. 4.16. Study of adaptation to visual displacement. A user wearing a video see-through HMD performs a localization task.

3.2.4 Peripheral FOV

Given that peripheral vision can be provided for both optical and video see-through systems, the next question is whether it is used as effectively for both systems. In optical see-through, there is almost no transition or discrepancy between the real scene captured by the see-through device and the peripheral vision seen on the side of the device.

For video see-through, the peripheral FOV has been provided by letting the user see around the device, as with optical see-through (Colucci & Chi, 1995). Especially in the latter case, it remains to be seen however whether the difference in presentation of the superimposed real scene and the peripheral real scene will cause discomfort or provide conflicting cues to the user. The issue is that the virtual displays call for a different accommodation for the user than the real scene in various cases.

3.2.5 Depth of Field

One important property of optical systems, including the visual system, is depth of field. Depth of field refers to the range of distances from the detector (e.g., the eye) in which an object appears to be in focus without the need for a change in the optics focus (e.g., eye accommodation). For the human visual system example, if an object is accurately focused monocularly, other objects somewhat nearer and further away are also seen clearly without any change in accommodation. Still nearer or further away objects are blurred. Depth of field reduces the necessity for precise accommodation and is markedly influenced by the diameter of the pupil. The larger the pupil, the smaller the depth of field. For a 2 and 4 mm pupil, the depths of field are $+/- 0.06$ and $+/- 0.03$ diopters, respectively. For a 4 mm pupil, for example, such a depth of field translates as a clear focus from 0.94 to 1.06 m for an object 1 m away, and from 11 to 33 m for an object 17 m away (Campbell, 1957; Moses, 1970). An important point is that accommodation plays an important role only at close working distance where depth of field is narrow.

With video see-through systems, the miniature cameras used for acquiring the real-scene images must provide a depth of field equivalent to the required working distance for a task. For a large range of working distances, the camera may need to be focused at the middle working distance. For closer distances, the small depth of field may require an autofocus instead of a fixed-focus camera.

With optical see-through systems, the available depth of field for the real scene is essentially that of the human visual system but for a larger pupil than would be accessible with unaided eyes. This can be explained by the brightness attenuation of the real scene by the half transparent mirror. As a result, the pupils are dilated (we assume here that the real and virtual scenes are matched in brightness). Therefore, the effective depth of field will be slightly less than with unaided eyes. This is only a problem if the user is working with nearby objects and the virtual images are focused outside of the depth of field required for nearby objects. For the virtual images and no autofocus capability for the 2D virtual images, the depth of field is imposed by the human visual system around the location of the displayed virtual images (Rolland, Krueger, & Goon, 2000).

When the retinal images are not sharp following some discrepancy in accommodation, the visual system is constantly processing somewhat blurred images and tends to tolerate blur up to the point at which essential detail is obscured. This tolerance for blur extends the apparent depth of field considerably, so that the eye may be as much as $+/- 0.25$ diopter out of focus without stimulating accommodative change (Moses, 1970).

3.2.6 Qualitative Aspects

The representation of virtual objects, and in some cases of real objects, is altered by see-through devices. Aspects of perceptual representation include the shape of objects, their color, brightness, contrast, shading, texture, and level of detail. In the case of optical see-through HMDs, folding the optical path by using a half-transparent mirror is necessary because it is the only configuration that leaves the real scene almost unaltered. A thin folding mirror will introduce a small apparent shift in depth of real objects precisely equal to $e(n-1)/n$ where e is the thickness of the plate and n is its index of refraction. This is in addition to a small amount of distortion of the scene at the edges of the FOV (e.g., <1% for a 60 degree FOV). Consequently, real objects are seen basically unaltered. Virtual objects, on the other hand, are formed from fusion of stereo images formed through magnifying optics. Each optical virtual image formed of the display associated with each eye is typically aberrated. For large FOV optics, astigmatism can be a limiting factor. Shenker (1994) proposes an objective performance estimate for evaluating visual performance in HMDs.

It must be noted that real and virtual objects in such systems may be seen sharply by accommodating in different planes under most visualization settings. This yields conflicts in accommodation for real and virtual imagery. For applications where the virtual objects are presented in a small working volume around some mean display distance (e.g., arm length visualization), the 2D optical images of the miniature displays can be located at that same distance to minimize conflicts in accommodation and convergence between real and virtual objects (Rolland, Ariely, & Gibson, et al., 1995). Another approach to minimizing conflicts in accommodation and convergence is multifocal-plane technology (Rolland, Krueger, and Goon, 2000).

Beside brightness attenuation and distortion, other aspects of objects representation are altered in video see-through HMDs. The authors' experience with at least one system is that the color and brightness of real objects are altered along with the loss in texture and levels of detail due to the limited resolution of the miniature video cameras and the wide-angle optical viewer (Biocca & Rolland, 1998). This alteration includes spatial, luminance, and color resolution. This is perhaps resolvable with improved technology but it currently limits the ability of the HMD user to perceive real objects as they would appear with unaided eyes. In wide FOV video see-through HMDs, both real and virtual objects call for the same accommodation; however, conflicts of accommodation and convergence are also present. As with optical see-through HMDs, these conflicts can be minimized if objects are perceived at a relatively constant depth near the plane of the optical images. In narrow FOV systems where the real scene is seen in large part outside the overlay imagery, conflicts in accommodation can also result between the real and computer generated scene.

For both technologies, an effective solution to these various conflicts in accommodation may be to allow autofocus of the 2D virtual images as a function of the location of the user gaze point in the virtual environment, or to implement multifocal planes (Rolland, Krueger, & Goon, 2000). Given eye-tracking capability, while certainly not the optimal appoach, autofocus could be provided because small displacements of the miniature display near the focal plane of the optics would yield large axial displacements of the 2D virtual images in the projected virtual space. The 2D virtual images would move in depth according to the user gaze point. Multifocal-plane approaches also allow autofocusing but with no need for eye tracking.

4. CONCLUSION

We have presented issues involving optical and video see-through head-mounted displays. In the authors'opinion, the most important issues are system latency, occlusion, the fidelity of the real-world view, and user acceptance. Optical see-through systems offer an essentially unhindered view of the real environment; they also provide an instantaneous real-world view that assures the synchronization of visual and proprioception information. Video systems give up the unhindered view in return for improved ability to see real and synthetic imagery simultaneously.

Some of us working with optical see-through devices strongly feel that providing the real scene through optical means is important for applications such as medical visualization where human lives are implicated. Others, working with video see-through devices feel that a video see-through device with a flip-up view is adequate for safety of the patient. Also, while how to perform occlusion is far from solved and is actively researched, the ability to selectively render occlusion of the real scene at given spatial locations may be important in various applications. Video see-through systems can also guarantee registration of the real and virtual scenes at the expense of a mismatch between vision and proprioception, which may or may not be perceived as a penalty if the human observer is able to adapt to such a mismatch.

Clearly, there is no "right" system for all applications: Each of the trade-offs discussed in this chapter must be examined with respect to specific applications and available technology to determine which type of system is most appropriate. A shared concern among scientists developing further technology is the lack of standards not only in the design, but also most importantly in the calibration and maintenance of HMD systems.

Acknowledgments This book chapter was significantly expanded from an earlier publication by Rolland, Holloway, and Fuchs (1995), and the authors would like to thank Rich Holloway for his earlier contribution to this work. We thank Myron Krueger from Artificial Reality Corporation for stimulating discussions on various aspects of the technology, and Martin Shenker from M.S.O.D. and Brian Welch from CAE Electronics for discussions on current optical technology. Finally we thank Bernard Peuchot, David Mizell, Steven Feiner, Derek Hill, and Andrei State for providing information about their research that has significantly contributed to the improvement of this document. We deeply thank our various sponsors not only for their financial support that has greatly facilitated our research in see-through devices but also for the stimulating discussions they have provided over the years. Contracts and grants include ARPA #DABT 63-93-C-0048, NSF Cooperative Agreement #ASC-8920219; "Science and Technology Center for Computer Graphics and Scientific Visualization," ONR #N00014-86-K-0680, ONR#N00014-94-1-0503, ONR#N000149710654, NIH #5-R24-RR-02170, NIH#1-R29-LM06322-OlAl, and DAAH04-96-C-0086. Industrial partners sponsorships include RSK-Assessment, Inc. and Artificial Reality Corporation.

REFERENCES

Allen, D. (1993). "A 1″ high resolution field sequential display for head-mounted applications." *Proceedings of IEEE Virtual Reality Annual International Symposium* (VRAIS'93), 364–370.

Argotti, Y., V. Otters, and J. P. Rolland, (in press, 2000). "Dynamics of superimposition of virtual objects on real objects," *Technical Report* TR2000-03, University of Central Florida.

Azuma, R., and G. Bishop (1994). "Improving static and dynamic registration in an optical see-through HMD." *Computer Graphics: Proceedings of SIGGRAPH '94*. Orlando, July 24–29, 197–204.

Baillot, Y., and J. P. Rolland (1998). "Modeling of a knee joint for the VRDA tool," *Proceedings of Medicine Meets Virtual Reality*, 366–367 (IOS Press).

Baillot, Y. (1999). "First Implementation of the Virtual Reality Dynamic Anatomy Tool," *Masters Dissertation*, University of Central Florida.

Baillot, Y., J. P. Rolland, and D. L. Wright (1999). "Automatic modeling of knee-joint motion for the virtual reality dynamic anatomy (VRDA) tool," *Proceedings of Medicine Meets Virtual Reality '99*, 30–37 (IOS Press).

Bajura, M., H. Fuchs, and R. Ohbuchi (1992). "Merging virtual objects with the real world," *Computer Graphics*, 26, 203–210.

Bajura, M., and U. Neumann (1995). "Dynamic registration correction in video-based augmented reality systems," *IEEE Computer Graphics and Applications*, 15(5), 52–60.

Baillot, Y., J. P. Rolland, K. Lin, and D. L. Wright (2000). "Automatic Modeling of Knee-Joint Motion for the Virtual Reality Dynamic Anatomy (VRDA) Tool," *Presence: Teleoperators and Virtual Environments* (MIT Press) 9(3), 223–235.

Barrette, R. E. (1992). "Wide field of view, full-color, high-resolution, helmet-mounted display," *SID '92 Symposium*, 69–72.

Biocca, F. A., and J. P. Rolland (1998). "Virtual eyes can rearrange your body: adaptation to virtual-eye location in see-thru head-mounted displays," *Presence: Teleoperators and Virtual Environments* (MIT Press), 7(3), 262–277.

Bishop, G., H. Fuchs, L. McMillan, and E. J. Scher-Zagier (1994). "Frameless rendering: double buffering considered harmful," *Proceedings of SIGGRAPH'94*, 175–176.

Bolt, R. A. (1981). "Gaze-orchestrated dynamic windows," *Computer Graphics*, 15(3), 109–119.

Brooks, F. P. (1992). "Walkthrough project: Final technical report to National Science Foundation Computer and Information Science and Engineering," *Technical Report*, TR92-026, University of North Carolina at Chapel Hill.

Bryson, S. (1991). "Interaction of objects in a virtual environment: a two-point paradigm," *Proceedings of SPIE* 1457, 180–187.

Buchroeder, R. A., G. W. Seeley, and D. Vukobratovich (1981). "Design of a Catadioptric VCASS helmet-mounted display," *Optical Sciences Center, University of Arizona, under contract to U.S. Air Force Armstrong Aerospace Medical Research Laboratory*, Wright-Patterson Air Force Base, Dayton, Ohio, AFAMRL-TR-81-133.

Campbell, F. W. (1957). "The depth of field of the human eye," *Optica Acta*, 4, 157–164.

Caudell, T. P., and D. W. Mizell (1992). "Augmented Reality: An application of heads-up display technology to manual manufacturing processes," *Proceedings of the 1992 IEEE Hawaii International Conference on Systems Sciences*, 659–669.

Colgate, T. P. (1968). "Reaction and response time of individuals reacting to auditory, visual, and tactile stimuli," *The Research Quarterly*, 39(3), 783–784.

Colucci, D., and V. Chi (1995). "Computer Glasses: A compact light weight and cost effective display for monocular and tiled wide field view systems," *Proceedings of SPIE Conference on Novel Optical Systems Design and Optimization*, 2537, 61–70.

Cutting, J. E., and P. M. Vishton (1995). "Perceiving the layout and knowing distances: the integration, relative potency, and contextual use of different information about depth," *Perception of Space and Motion*, ed. by W. Epstein and S. Rogers, Academic Press, 69–117.

Deering, M. (1992). "High resolution virtual reality," *Computer Graphics*, 26(2), 195–201.

Desplat, S. (November 1997). "Charactérization des éléments actuals et future de loculométre MétrovisionSextant," *Technical Report, Ecole Nationale Superieure de Physique de Marseilles.*

Dhome, M., M. Richetin, J. P. Lapreste, and G. Rives (1989). "Determination of the attitude of 3D objects from a single perspective view," *IEEE Trans. Pattern Analysis and Machine Intelligence*, 11(12), 1265–1278.

Dodge (1903). "Five types of eye movement in the horizontal meridian plane of the field of regard," *The American Journal of Physiology*, 8, 307–329.

Droessler, J. G., and D. J. Rotier (1990). "Tilted cat helmet-mounted display," *Optical Engineering*, 29(8), 849–854.

Edwards, E. K., J. P. Rolland, and K. P. Keller (1993). "Video see-through design for merging of real and virtual environments," *Proceedings of IEEE Virtual Reality Annual International Symposium (VRAIS'93)*, 223–233.

Edwards, P. J., D. J. Hawkes, D. L. G. Hill, D. Jewell, R. Spink, A. Strong, and M. Gleeson (1995). "Augmentation of reality using an operating microscope for otolaryngology and neurosurgical guidance," *J. Image Guided Surgery*, 1(3), 172–178.

Ellis, S. R., and U. J. Bucher (1994). "Distance perception of stereoscopically presented virtual objects optically superimposed on physical objects in a head-mounted see-through display," *Proceedings of the Human Factors and Ergonomics Society*, Nashville.

Feiner, S., B. Macintyre, and D. Seligmann (1993). "Knowledge-based augmented reality," *Communications of the ACM*, 36 (7), 53–62.

Feiner, S. B., B. Macintyre, H. Tobias, and A. Webster (1997). "A touring machine: prototyping 3D mobile augmented reality systems for exploring the urban environment," *Proceedings of ISWC'97*, 74–81.

Fernie, A. (1995). "Helmet-mounted display with dual resolution," *Proceedings of SID95, Applications Digest*, 37–40.

Ferrin, F. J. (1991). "Survey of helmet tracking technologies," *Proceedings of SPIE*, 1456, 86–94.

Fry, G. A. (1969). *Geometrical Optics*, Chilton Book Company.

Furness, T. A. (1986). "The super cockpit and its human factors challenges," *Proceedings of the Human Factors Society*, 30, 48–52.

Girolamo, H. (1991). "Notional helmet concepts: A survey of near-term and future technologies," US Army NATICK *Technical Report*, NATICK/TR-91/017.

Goldberg, M. E., H. M. Eggers, and P. Gouras (1991). "The Ocular Motor System," in *Principles of Neural Science*, 3rd Ed, ed. by E. R. Kandel, J. H. Schwartz, and T. M. Jessell, Appleton & Lange, Norwalk, CT.

Held, R., and N. Durlach (1987). "Telepresence, time delay and adaptation," NASA *Conference Publication*, 10032.

Hua, H., A. Girardot, C. Gao, and J. P. Rolland (2000). "Engineering of Head-Mounted Projective Displays," *Applied Optics*, 39(22), 3814–3824.

Holloway, R. (1995). "An Analysis of Registration Errors in a See-Through Head-Mounted Display System for Craniofacial Surgery Planning," *Ph.D. Dissertation*, University of North Carolina at Chapel Hill.

Jacobs, M. C., M. A. Livingston, and A. State (1997). "Managing latency in complex augmented reality systems," *Proceedings of 1997 Symposium on Interactive 3D Graphics, ACM SIGGRAPH*, 235–240.

Jacoby, R. H., and S. R. Ellis (1992). "Using virtual menus in a virtual environment," *Proceedings of SPIE*, 1668, 38–48.

Kaiser Electro-Optics (1994). Personal communication from Frank Hepburn of KEO, Carlsbad, CA. General description of Kaiser's VIM system ("Full immersion head-mounted display system") is available via ARPA's ESTO World-Wide Web home page: *http://esto.sysplan.com/ESTO/*.

Kancherla, A., J. P. Rolland, D. L. Wright, and G. Burdea (1995). "A novel virtual reality tool for teaching dynamic 3D anatomy," *Proceedings of CVRMed'95*, 163–169.

Kandebo, S. W. (1988). "Navy to evaluate Agile Eye helmet-mounted display system," *Aviation Week & Space Technology*, August 15, 94–99.

Kennedy, R. S., and K. M. Stanney (1997). "Aftereffects in virtual environment exposure: psychometric issues," in *Design of Computing*, ed. by M. J. Smith, G. Salvendy, and R. J. Koubek, Elsevier, Amsterdam.

Kojima, R., and T. Ojika (1997). "Transition between virtual environment and workstation environment with projective head-mounted display," *Proceedings of VRAIS'97*, 130–137.

Laveau, S., and O. Faugeras (1994). "3-D scene representations as a collection of images and fundamental matrices," *Institut National de Recherche en Informatique et en Automatique (INREA) Report* 2205, February.

Levine, M. D. (1985). *Vision in Man and Machine*, McGraw-Hill.

MacIntyre, B., and S. Feiner (1998). "A distributed 3D graphics library," *Proceedings of ACM SIGGRAPH 98*, 361–370.

Manhart, P. K., R. J. Malcom, and J. G. Frazee (1993). "Augeye: A compact, solid Schmidt optical relay for helmet mounted displays," *Proceedings of IEEE VRAIS'93*, 234–245.

Mine, M. (1993). "Characterization of end-to-end delays in head-mounted display systems," *Technical Report* TR93-001, University of North Carolina at Chapel Hill.

Moses, R. A. (1970). *Adler's Physiology of the Eye*, St. Louis, MO, Mosby.

Oster, P. J., and J. A. Stern (1980). "Measurement of Eye Movement," in *Techniques of Psychophysiology*, ed. by I. Martin, and P. H. Venables, Wiley & Sons.

Outters, V., Y. Argotti, and J. P. Rolland (1999). "Knee motion capture and representation in augmented reality," *Technical Report* TR99-006, University of Central Florida.

Parsons, J., and J. P. Rolland (1998). "A non-intrusive display technique for providing real-time data within a surgeons critical area of interest," *Proceedings of Medicine Meets Virtual Reality*, 246–251 (IOS Press).

Peuchot, B. (1993). "Camera virtual equivalent model: 0.01 pixel detectors," *Special issue on 3D Advanced Image Processing in Medicine in Computerized Medical Imaging and Graphics*, 17 (4/5), 289–294.

Peuchot, B. A. (1994). "Utilization de detecteurs subpixels dans la modelisation d'une camera-verification de l'hypothese stenope," *9ᵉ Congres AFCET, reconnaissance des formes et intelligence artificielle*, Tme1, 691–695 Paris, 11–14 January.

Peuchot, B., A. Tanguy, and M. Eude (1994). "Dispositif optique pour la visualization d'une image virtuelle tridimensionelle en superimposition avec an object notamment pour des applications chirurgicales," *Depot CNRS* #94106623, May 31.

Peuchot, B., A. Tanguy, and M. Eude (1995). "Virtual reality as an operative tool during scoliosis surgery," *Proceedings of CVRMed'95*, 549–554.

Rash, C. E. (1999). "Helmet-Mounted Displays": *Design Issues for Rotary-Wing Aircraft*, U.S. Government Printing Office, 735–164.

Robinett, W., and J. P. Rolland (1992). "A computational model for the stereoscopic optics of a head-mounted display," *Presence: Teleoperators and Virtual Environments* (MIT Press), 1(1), 45–62.

Rock, I. (1966). *The Nature of Perceptual Adaptation*, Basic Books.

Rolland, J. P., and T. Hopkins (1993). "A method for computational correction of optical distortion in head-mounted displays," *Technical Report*, TR93-045, University of North Carolina at Chapel Hill.

Rolland, J. P. (1994). "Head-mounted displays for virtual environments: the optical interface," presented at the International Optical Design Conference 94, *Proceedings of OSA*, 22, 329–333.

Rolland, J. P., D. Ariely, and W. Gibson (1995). "Towards quantifying depth and size perception in virtual environments," *Presence: Teleoperators and Virtual Environments*, 4(1), 24–49.

Rolland, J. P., R. L. Holloway, and H. Fuchs (1995). "Comparison of optical and video see-through, head-mounted displays," *Proceedings of SPIE*, 2351, 293–307.

Rolland, J. P., D. L. Wright, and A. R. Kancherla (1996). "Towards a Novel Augmented-Reality Tool to Visualize Dynamic 3D Anatomy," *Proceedings of Medicine Meets Virtual Reality*, 5, San Diego, CA (1997). *Technical Report*, TR96-02, University of Central Florida.

Rolland, J. P., and K. Arthur (1997). "Study of depth judgments in a see-through mounted display," *Proceedings of SPIE*, 3058, 66–75, AEROSENSE.

Rolland, J. P., A. Yoshida, L. Davis, and J. H. Reif (1998). "High-resolution inset head-mounted display," *Applied Optics*, 37(19), 4183–4193.

Rolland, J. P., J. Parsons, D. Poizat, and D. Hancock (1998). "Conformal optics for 3D visualization," *Proceedings of the International Lens Design Conference*, 760–764.

Rolland, J. P., A. Rapaport, and M. W. Krueger (1998). "Design of an anamorphic fisheye lens," *Proceedings of the International Lens Design Conference*, 274–277.

Rolland, J. P. (2000). "Wide angle, off-axis, see-through head mounted display," *Optical Engineering-Special issue on Pushing the Envelop in Optical Design Software*, 39(7) 1760–1767.

Rolland, J. P., M. Krueger, and A. Goon (2000). "Multi-focal planes in head-mounted displays," *Applied Optics*, 39(19) 3209–3215.

Rolland, J. P., A. Quinn, and K. Arthur, and E. Rinalducci (in press, 2000). "Accuracy of rendered depth in head-mounted displays: Comparison of two assessment technologies," *Technical Report*, TR2000-02, University of Central Florida.

Roscoe S. N. (1984). "Judgments of size and distance with imaging displays," *Human Factors*, 26(6), 617–629.

Roscoe S. N. (1991). "The eyes prefer real images," in *Pictorial Communication in Virtual and Real Environments*, ed. by Stephen R. Ellis, Taylor and Francis, New York, NY.

Scher-Zagier, E. J. (1997). "A human's eye view: motion, blur, and frameless rendering," *ACM Crosswords 97*.

Shenker, M. (1994). "Image quality considerations for head-mounted displays," *Proceedings of the OSA: International Lens Design Conference*, 22, 334–338.

Slater, M., and S. Wilbur (1997). "A framework for immersive virtual environments (FIVE): Speculations on the role of presence in virtual environments," *Presence: Teleoperators and Virtual Environments*, 6(6), 603–616.

State, A, D. Chen, C. Tector, A. Brandt, H. Chen, R. Ohbuchi, M. Bajura, and H. Fuchs (1994). Case study: Observing a volume rendered fetus within a pregnant patient," *Proceedings of Visualization '94*, Washington, DC, 364–373.

State, A., G. Hirota, D. T. Chen, W. E. Garrett, and M. Livingston (1996). "Superior augmented-reality registration by integrating landmark tracking and magnetic tracking," *Proceedings of SIG-GRAPH'96*, ACM SIGGRAPH, 429–438.

Sutherland, I. (1965). "The ultimate display," *Information Processing 1965: Proceedings of IFIP Congress*, 65, 506–508.

Sutherland, I. E. (1968). "A head-mounted three-dimensional display," *Fall Joint Computer Conference, AFIPS Conference Proceedings*, 33, 757–764.

Thomas, M. L., W. P. Siegmund, S. E. Antos, and R. M. Robinson (1989). "Fiber optic development for use on the fiber optic helmet-mounted display," in *Helmet-Mounted Displays*, J. T. Colloro, Ed., *Proceedings of the SPIE*, 1116, 90–101.

Tomasi, C., and T. Kanade (1991). "Shape and motion from image streams: a factorization method-Part 3: Detection and tracking of point features," *Carnegie Mellon Technical Report*, CMU-CS, 91–132.

Vaissie, L., and J. P. Rolland (in press, 2000). "Albertian errors in head-mounted displays: choice of eyepoints location," *Technical Report*, TR2000-01, University of Central Florida.

Volkman, F., L. A. Riggs, K. D. White, and R. K. Moore (1978). "Contrast sensitivity during saccadic eye movements," *Vision Research*, 18, 1193–1199.

Volkman, F. (1986). "Human visual suppression," *Vision Research*, 26, 1401–1416.

Watson, B., and L. F. Hodges (1995). "Using texture maps to correct for optical distortion in head-mounted displays," *Proceedings of VRAIS'95*, 172–178.

Welch, B., and M. Shenker (1984). "The fiber-optic Helmet-Mounted Display," *Image III*, 345–361.

Welch, R. B. (1993). "Alternating prism exposure causes dual adaptation and generalization to a novel displacement," *Perception and Psychophysics*, 54(2), 195–204.

Welch, R. B. (1994). "Adapting to virtual environments and teleoperators," *Unpublished Manuscript*, NASA-Ames Research Center, Moffett Field, CA.

Wright, D. L., J. P. Rolland, and A. R. Kancherla (1995). "Using virtual reality to teach radiographic positioning," *Radiologic Technology*, 66(4), 167–172.

5

Augmenting Reality Using Affine Object Representations

James Vallino
Rochester Institute of Technology

Kiriakos N. Kutulakos
University of Rochester

1. INTRODUCTION

An augmented reality system is a system that creates a view of a real scene that visually incorporates into the scene computer-generated images of three-dimensional (3D) virtual objects. As the user of such a system moves about the real scene the virtual objects appear as if they actually exist in the scene. One motivation for augmenting reality in this way is to enhance the performance of real-world tasks. The performance requirements for an augmented reality system are: (1) merge images of 3D virtual objects with images of the real environment, (2) generate a consistent view of those objects from all views of the real scene, and (3) perform these operations in real time to be interactive with the user. Augmented reality can be compared to the more commonly known virtual reality. Virtual reality systems immerse a user in an environment that is completely computer generated. Augmented reality systems, on the other hand, strive to maintain the user's immersion in the real environment. The rationale behind this is twofold. First, real environments contain a wealth of information, much of which is impossible to model and simulate by computer. Secondly, if the end goal is to enhance the performance of a real-world task the user will most naturally perform that task while looking at an augmented view of the real scene. Practical applications for augmented reality are described in other chapters of this book and include applications from the domains of manufacturing (Chapter 23), medicine (Chapter 21), and the military (Chapter 20).

Both virtual reality and augmented reality systems provide an interface that allows the user to operate in a natural 3D physical space while receiving a consistent set of sensory inputs for both the real and virtual worlds. The primary performance goal for a virtual reality system is to present visual stimuli that are consistent with the changes in body position sensed by the user. Any inconsistency perceived by the user results from a misregistration between the coordinate system the user is maintaining internally to describe body position and the coordinate system that describes the graphics system's viewpoint in the virtual scene. This can be contrasted to the primary performance goal for an augmented reality system, which is to render views of virtual objects that are consistent with the user's view of the real environment containing the objects. Any inconsistency, which manifests itself as a difference between two visual stimuli (i.e. the virtual and real images), derives from a misregistration between the coordinate system describing the user's viewpoint in the real scene and the graphics system's viewpoint in the virtual scene. The nature of this registration problem in augmented reality systems can be seen in Figure 5.1a. To create an image of the three-dimensional virtual objects that is consistent with the user's current view of the world and the object's placement in the real world requires the definition of the geometric relationships between the virtual and physical objects shown in Figure 5.1a. Any errors in the determination of these relationships appear to the user as inconsistencies in the appearance of the virtual objects in the real scene (Figures 5.1b,c). These errors in registering the two images are classified as either static or dynamic (Chapter 6). Static errors are perceived by the user as differences in the placement or appearance of the virtual objects when viewed from different viewpoints. The dynamic errors are caused by the system lagging behind due to not meeting its real-time requirements. The visual effect of these dynamic errors is a shift in the position of the virtual objects when there is motion in the system.

This chapter describes a method for solving the registration problem in augmented reality systems using affine object representations. The method defines a global non-Euclidean affine coordinate system and determines the relationships between that global coordinate system and all the coordinate systems in Figure 5.1a. Unlike other solutions to the augmented reality registration problem that require position sensing and calibrated

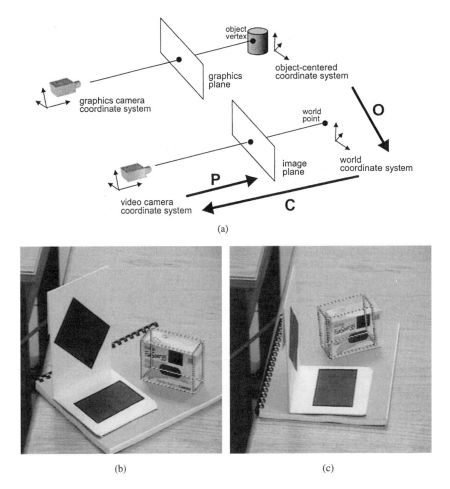

(a)

(b)   (c)

FIG. 5.1. Components in an augmented reality system. (a) The multiple coordinate systems that must be registered are shown. Several types of augmented reality systems exist (Chapter 2). This diagram depicts a system using a monitor-based display or video see-through head-mounted display. (b) View of an affine wireframe model correctly overlaid on a small box. (c) Example of a misregistration of a virtual object with the real scene. The virtual wireframe is not correctly registered on the small box.

cameras, this method relies solely on tracking four or more features in video images of the real scene using uncalibrated cameras. As shown in Figure 5.1a, our approach requires that a video camera view the real scene. This requirement favors operation with a monitor-based display or video see-through head-mounted display (Azuma 1997). This chapter describes working augmented reality systems that employ both those display types.

Operation with an optical see-through display, in addition to requiring the video camera, will require alignment of the camera with the see-through display (Janin, Mizell et al. 1993; Hoff, Nguyen et al. 1996).

2. THE REGISTRATION PROBLEM

The key requirement for creating an augmented reality image in which virtual objects appear to exist in the three-dimensional real scene is knowledge of the relationships among the object, world, and camera coordinate systems (Figure 5.1a). These relationships are determined by the object-to-world, $\mathbf{O}$, world-to-camera, $\mathbf{C}$, and camera-to-image plane, $\mathbf{P}$, transforms. The object-to-world transform specifies the position and orientation of a virtual object with respect to the world coordinate system that defines the real scene. The pose of the video camera that views the real scene is defined by the world-to-camera transform. The projection performed by the camera to create a 2D image of the 3D real scene is specified in the camera-to-image plane transform. Visually correct merging of virtual objects with the live video image requires computation of these relationships. Accurately performing this computation while maintaining real-time response and a low latency is the major challenge for an augmented reality system.

2.1 Augmenting Reality Using Pose Sensors

In many augmented reality systems, the problem of computing the transforms shown in Figure 5.1a is approached in a straightforward manner using sensing, calibration, and measurement to explicitly determine each transform (Feiner, MacIntyre et al. 1993; Ahlers, Breen et al. 1994; State, Chen et al. 1994). Sensors, based on mechanical, magnetic, or optical techniques are used to measure the position and angular orientation of the camera with respect to the world coordinate system. These two measurements together are termed the *pose* of the camera and determine the world-to-camera transform, $\mathbf{C}$. Quantifying the camera-to-image transform, $\mathbf{P}$, requires knowledge of the intrinsic parameters, such as focal length and aspect ratio, of the camera. These can be determined by performing a calibration procedure on the camera (Tsai 1987). The third transform, $\mathbf{O}$, is computed by simple measurement. The world coordinate system is a standard three-dimensional Euclidean space. The desired position and orientation for a virtual object can be measured in the real scene. Using the methods just described, all of the necessary transforms are known so that, at least in principle, virtual objects can be rendered and merged correctly with the live video.

These approaches do suffer from limitations. So far none of the pose sensors have been completely satisfactory for an augmented reality application (Azuma 1997). Mechanical sensors place limits on the range of the work space and require attachment to a restrictive linkage. Magnetic sensors are susceptible to disturbances in their generated magnetic field created by metal objects in the work space. Calibration for these distortions can reduce the errors (Byrson 1992). The magnetic sensors also have latencies that can only be improved with predictive estimation of pose (Azuma and Bishop 1994). Techniques for calibrating a camera to determine its intrinsic parameters are available (Tsai 1987) but calibration is a tedious process to perform. The intrinsic parameters of a camera may change over time requiring recalibration. In particular, zoom lenses, like those found on common consumer-grade video cameras, may change focal length with use either intentionally or with wear. Accurate sensing of zoom position is not commonly available, which would require recalibration with each change in focal length. Any errors introduced by incorrect pose sensing or camera calibration propagate through the system and will appear as misregistration in the final augmented reality image. The optical see-through head-mounted displays used in some augmented reality systems also must be calibrated even if the system does not use a video camera (Janin, Mizell et al. 1993). Any inaccuracy in the calibration of the display will result in misregistration in the augmented image.

2.2 Computer Vision for Augmented Reality

The initial approaches to augmented reality discussed in the previous section all overlook one source of significant information. Computer vision research has developed techniques for extracting information about the structure of a scene, the intrinsic parameters of the camera, and its pose from images of the scene. Recent augmented reality systems are applying computer vision methods to improve performance. Tuceryan et al. (1995) provide a careful analysis of procedures that rely on computer vision for calibrating a monitor-based augmented reality system. Image analysis of the video signal at runtime can also be beneficial. Several systems have been described in the literature that track fiducials in the scene at runtime.

Bajura and Neumann (1995) track LED fiducials to correct registration errors. Other systems (Hoff, Nguyen et al. 1996; Neumann and Cho 1996) use knowledge of the intrinsic camera parameters and tracking of fiducials placed in known locations in the scene to invert the camera projection operation and obtain an estimate of the viewer pose.

A hybrid method that uses fiducial tracking in combination with standard magnetic position tracking (State, Hirota et al. 1996) requires an initialization procedure to determine the intrinsic parameters of the cameras viewing the scene. Fiducials, whose location in the scene are known, are tracked in two video images. The position of the viewer is computed by inverting the projection operation. Position data obtained from a magnetic tracker aid in localization of the landmarks. This aid is particularly useful when large motions are encountered between two video frames. Algorithms that rely solely on vision-based tracking often cannot determine the interframe correspondences between fiducials when large motions occur between frames. The magnetic tracker position estimates are also used when occlusions prevent the vision system from seeing the required minimum number of fiducials.

Mellor (1995a,b) and Uenohara and Kanade (1995) describe two augmented reality systems that eliminate the need for explicit determination of the viewer's pose. For each viewpoint, Mellor (1995a,b) uses a linear method to solve for the camera's projection transform from the positions of tracked fiducials. The pose, specifying the 3D position and orientation of the camera, is not directly determined. Instead the computed projection transform is used to render the virtual objects. The camera must be calibrated at initialization time and a laser range finder provides the 3D positions of the fiducials in the scene. The second method for obtaining correct registration with neither position tracking nor camera calibration is presented by Uenohara and Kanade (1995). They track features in live video and represent the virtual points associated with planar overlays as the linear combination of feature points. The placement and rendering of three-dimensional virtual objects were not considered.

3. AUGMENTING REALITY USING AFFINE REPRESENTATIONS

The approach to augmenting reality described in the following sections is motivated by recent computer vision research that has determined structure for objects in a scene and the pose of the camera viewing it without knowledge of the object-to-world, world-to-camera, and camera-to-image plane transforms. The following observation was provided by Koenderink and van Doorn (1991) and Ullman and Basri (1991):

> Given a set of four or more non-coplanar 3D points, the projection of all points in the set can be computed as a linear combination of the projection of just four of the points.

This observation will be used to create a global coordinate system in which the coordinate systems diagrammed in Figure 5.1a can be expressed. Additionally, this global coordinate system will be defined solely from the locations of visible features in the real scene with no knowledge of the intrinsic parameters and pose of the camera.

3.1 Affine Camera Approximation

Accurate determination of where a point on a virtual object will project in the video image is essential for correct registration of the virtual and live-video images (Foley, van Dam et al. 1990; Tuceryan, Greer et al. 1995). In homogeneous coordinates, the projection, $[u \ v \ h]^T$ in the video image of a 3D point $[x \ y \ z \ w]^T$ can be expressed using the equation:

$$\begin{bmatrix} u \\ v \\ h \end{bmatrix} = \mathbf{P}_{3\times4}\mathbf{C}_{4\times4}\mathbf{O}_{4\times4} \begin{bmatrix} x \\ y \\ z \\ w \end{bmatrix}. \tag{1}$$

The transforms $\mathbf{P}_{3\times4}$, $\mathbf{C}_{4\times4}$, and $\mathbf{O}_{4\times4}$ are shown in Figure 5.1a and are the camera-to-image plane, world-to-camera, and object-to-world transforms respectively. Equation 1 assumes that object, world, and camera coordinate systems are independently defined. Previous approaches to augmented reality have been based on an explicit determination of each of the transforms that relate the coordinate systems. Our approach will represent the three coordinate systems in a single *non-Euclidean* coordinate system and express the projection operation with the equation:

$$\begin{bmatrix} u \\ v \\ h \end{bmatrix} = \Pi_{3\times4} \begin{bmatrix} x' \\ y' \\ z' \\ w' \end{bmatrix}, \tag{2}$$

where $[x' \ y' \ z' \ w']^T$ are the new coordinates for the point transformed from $[x \ y \ z \ w]^T$.

To fully exploit this simplification and the observation of Koenderink and van Doorn (1991) and Ullman and Basri (1991) we will use *weak perspective projection* to model the camera-to-image plane transform **P** (Shapiro, Zisserman et al. 1995). Under a weak perspective approximation, the projections in the image plane of 3D points are determined by first projecting the points along parallel rays orthogonal to the image plane. The entire image is then scaled by f/z_{avg}, where f is the camera's focal length and z_{avg} is the average distance of the points from the image plane. This approximation is commonly seen in the computer vision literature and holds when the front to back depth of objects along the viewing direction is small compared to the viewing distance (Thompson and Mundy 1987). While this does impose restrictions on the system it has also been shown that the weak perspective approximation can yield more accurate structure-from-motion computations (Boufama, Weinshall et al. 1994; Wiles and Brady 1996).

3.2   Global Affine Coordinate System

All points in this system are represented with an *affine representation* whose coordinate system is defined using the location of feature points in the image. This representation is invariant when an affine transform (i.e. translation, rotation, nonuniform scaling) is applied to all points. Transforms caused by the motion of a weak perspective camera viewing a scene will maintain this affine invariant representation. *Affine reprojection* or *transfer* (Barrett, Brill et al. 1992; Shashua 1993) is used to compute the projection of virtual objects placed into the real scene.

The affine representation for a collection of points, $p_0, \ldots, p_n$, is composed of: 1. the *affine basis points*, which are four non-coplanar points, one of which is specially designated as the *origin*, and 2. the *affine coordinates* of each point that define the point with respect to the affine basis points. The properties of affine point representation are illustrated in Figure 5.2.

Our affine augmented reality systems are based on two properties of affine representations (Koenderink and van Doorn 1991; Mundy and Zisserman 1992; Weinshall and Tomasi 1993):

Property 1 (Affine Reprojection Property) *The   projection,* $[u_p \ v_p]^T$, *of any point, p, represented with affine coordinates* $[x \ y \ z]^T$, *is expressed by the equation:*

$$\begin{bmatrix} u_p \\ v_p \end{bmatrix} = \underbrace{\begin{bmatrix} u_{p1} - u_{p0} & u_{p2} - u_{p0} & u_{p3} - u_{p0} \\ v_{p1} - v_{p0} & v_{p2} - v_{p0} & v_{p3} - v_{p0} \end{bmatrix}}_{\Pi_{2\times3}} \begin{bmatrix} x \\ y \\ z \end{bmatrix} + \begin{bmatrix} u_{p0} \\ v_{p0} \end{bmatrix},$$

(3)

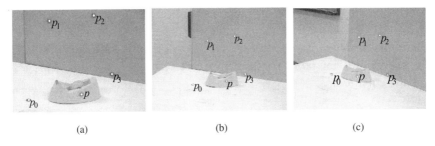

(a)   (b)   (c)

FIG. 5.2. Properties of affine point representations. The tracked
features p_0, p_1, p_2, p_3 define an affine coordinate frame within
which all world points can be represented: Point p_0 is the origin,
and points p_1, p_2, p_3 are the basis points. The affine coordinates
of a fifth point, p, are computed from its projection in (a) and (b)
using Property 2. Point p's projection in (c) can then be computed
from the projections of the four basis points using Property 1.

*where $[u_{pi} \ v_{pi}]^T, i = 0, \ldots, 3$ are the projections of the origin p_0 and
the three other basis points, p_1, p_2, and p_3, that define the affine coordinate system. This can be equivalently expressed in homogeneous
coordinates with the equation:*

$$\begin{bmatrix} u_p \\ v_p \\ 1 \end{bmatrix} = \underbrace{\begin{bmatrix} u_{p1} - u_{p0} & u_{p2} - u_{p0} & u_{p3} - u_{p0} & u_{p0} \\ v_{p1} - v_{p0} & v_{p2} - v_{p0} & v_{p3} - v_{p0} & v_{p0} \\ 0 & 0 & 0 & 1 \end{bmatrix}}_{\Pi_{3\times4}} \begin{bmatrix} x \\ y \\ z \\ 1 \end{bmatrix}.$$

(4)

Equation 4 provides an explicit definition for the projection matrix $\Pi_{3\times4}$
seen in Equation 2 and defines the projection of a 3D point in any new image
as a linear combination of the projections of the affine basis points in that
image. Equation 4 provides a method by which an augmented reality system
can calculate the projection of a point on a virtual object with knowledge
of only the location of the projections of the affine basis points and the
homogeneous affine coordinates for the virtual point. The affine basis points
will be defined by visually tracking features in the scene and determining
the projections in each new video image. The following property provides
the technique for determining the affine coordinates of any 3D point.

Property 2 (Affine Reconstruction Property) *The affine coordinates of any point can be computed from Equation 4 if its projection
in at least two views is known and the projections of the affine basis
points are also known in those views.*

This results in an overdetermined system of equations based on Equation 4. Given two views, I_1, I_2, of a scene in which the projections of the affine basis points, $p_0, \ldots, p_3$, are known then the affine coordinates $[x \ y \ z \ 1]^T$ for any point p can be recovered from the solution of the following equation:

$$
\begin{bmatrix} u_p^1 \\ v_p^1 \\ \ldots \\ u_p^2 \\ v_p^2 \end{bmatrix} = \begin{bmatrix} u_{p1}^1 - u_{p0}^1 & u_{p2}^1 - u_{p0}^1 & u_{p3}^1 - u_{p0}^1 & u_{p0}^1 \\ v_{p1}^1 - v_{p0}^1 & v_{p2}^1 - v_{p0}^1 & v_{p3}^1 - v_{p0}^1 & v_{p0}^1 \\ \ldots\ldots\ldots\ldots\ldots\ldots\ldots\ldots\ldots\ldots\ldots\ldots\ldots\ldots\ldots\ldots\ldots \\ u_{p1}^2 - u_{p0}^2 & u_{p2}^2 - u_{p0}^2 & u_{p3}^2 - u_{p0}^2 & u_{p0}^2 \\ v_{p1}^2 - v_{p0}^2 & v_{p2}^2 - v_{p0}^2 & v_{p3}^2 - v_{p0}^2 & v_{p0^2} \end{bmatrix} \begin{bmatrix} x \\ y \\ z \\ 1 \end{bmatrix}, \quad (5)
$$

where $[u_p^i \ v_p^i]^T$ and $[u_{p_j}^i \ v_{p_j}^i]^T$ are the projections of point p and affine basis point p_j, respectively, in image I_i.

3.3   Affine Augmented Reality

One performance goal for an augmented reality system is the ability to operate in real time, which will often limit the photorealism possible in rendering the virtual objects. As a minimum the very basic operation of hidden surface removal (Foley, van Dam et al. 1990) must be performed to have a correct visualization of any three-dimensional virtual object. The hidden surface removal algorithm uses a front-to-back ordering of surfaces to determine visibility. That ordering is obtained by assigning a depth to each rendered point that represents its distance from the viewpoint. The following two properties will extend the familiar notions of "image plane" and "viewing direction" to the affine representation. This extension will allow for the required front-to-back ordering of virtual surfaces and use of hardware-supported rendering via z-buffering. The image plane and viewing direction define a 3D coordinate system that describes the orientation of the camera. The graphics system can operate entirely within this global affine coordinate system and completely ignore the original object representation.

Property 3 (Affine Image Plane)   *Let χ and ψ be the homogeneous vectors corresponding to the first and second row of $\Pi_{2\times3}$, respectively. (1) The vectors χ and ψ are the directions of the rows and columns of the camera, respectively, expressed in the coordinate frame* *of the affine basis points. (2) The affine image plane of the camera is the plane spanned by the vectors χ and ψ.*

The unique direction in space along which all points project to a single pixel in the image defines the viewing direction of a camera under our model of weak perspective projection. In the affine case, this direction is expressed mathematically as the null-space of the matrix $\Pi_{2\times3}$:

Property 4 (Affine Viewing Direction) *When expressed in the co-ordinate frame of the affine basis points, the viewing direction, ζ, of the camera is given by the cross product*

$$\zeta = \chi \times \psi. \tag{6}$$

Property 4 guarantees that the set of points $\{p + t\zeta, t \in \Re\}$ that defines the line of sight of a point p will project to the same pixel under Equation 3. The z-buffer value needed for hidden surface removal that is assigned to every point is the dot product $[\zeta^T \ 0] \cdot p^T$. The actual magnitude of this value is irrelevant: The important characteristic is that the front-to-back order of virtual points rendered to the same pixel is correctly maintained along the camera viewing direction.

The affine image plane and viewing direction vectors define a 3D coordinate system that in general will not be an orthonormal reference frame in Euclidean 3D space. Despite this, correct visible surface rendering of any point $[x \ y \ z \ 1]^T$ defined in the global affine coordinate system can be performed by applying the transform:

$$\begin{bmatrix} u \\ v \\ w \\ 1 \end{bmatrix} = \underbrace{\begin{bmatrix} u_{p1} - u_{p0} & u_{p2} - u_{p0} & u_{p3} - u_{p0} & u_{p0} \\ v_{p1} - v_{p0} & v_{p2} - v_{p0} & v_{p3} - v_{p0} & v_{p0} \\ & \zeta^T & & 0 \\ & \mathbf{0} & & 1 \end{bmatrix}}_{\Pi_{4\times4}} \begin{bmatrix} x \\ y \\ z \\ 1 \end{bmatrix}, \tag{7}$$

where u and v are the graphic image coordinates of the point and w is its assigned z-buffer value.

$\Pi_{4\times4}$ has the same form as the viewing matrix that is commonly used in computer graphics systems to perform transforms of graphic objects. The structural similarity allows standard graphics hardware to be used for real-time rendering of objects defined in the affine coordinate system developed here. In our system, a Silicon Graphics Infinite Reality Engine is directly used to render virtual objects with hardware-supported hidden

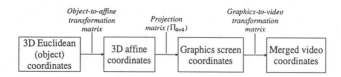

FIG. 5.3. Procedure for rendering virtual objects showing the coordinate systems and transforms involved.

surface removal. The graphics system not only renders the virtual objects, but it may also render affine models of real objects, as described in the next section when we discuss occlusions.

3.4   Rendering Virtual Objects

The affine projection matrix, $\Pi_{4\times4}$, will correctly render virtual objects for merging with the video image provided that those objects are represented in the global affine coordinate system. A method to define and place objects in this global coordinate system is needed. These operations will have to be performed at runtime since the structure of the coordinate system is not know a priori. The interactive methods are based on the *affine reconstruction property*. The resulting transform operations for rendering a virtual object are shown in Figure 5.3.

In order to use $\Pi_{4\times4}$ to render the virtual objects, those objects must be represented in the global affine coordinate system defined by the tracked basis points. The 3D object-centered Euclidean coordinate system that will commonly describe a virtual object must be transformed to the global affine coordinate system developed in the previous sections. The calculation of this object-to-affine transform will be done at runtime. In the simplest approach, the user interactively specifies this transform by placing a real object in the scene that defines a bounding box for the virtual object. In two separate views of the scene the user will specify the locations of four points defining the bounding box. The *affine reconstruction property* is then applied to determine the affine coordinates of the bounding box. The bounding box of the virtual object is used to compute the *object-to-world transform*, $O_{4\times4}$, from Figure 5.1a. In this case the world coordinate system is the common affine coordinate system. The computed transform handles both the change to the global coordinate and placement in the 3D scene.

The user can be further supported in the process of interactively placing virtual objects. By using results from stereo vision, constraints can be imposed on where the user is allowed to specify points representing physical locations in 3D space. Once a point has been specified in one image, the *epipolar constraint* (Shapiro, Zisserman et al. 1995) determines the line in the second image on which the projection of this point must lie. The user specification of the point in the second image can be "snapped" to the nearest point on this epipolar line. Additionally, techniques for constraining points to be collinear or coplanar with physical edges and surfaces are available (Kutulakos and Vallino 1996).

In an augmented view of the scene visual interactions between real and virtual objects must be considered. The affine projection matrix, $\Pi_{4\times4}$, will correctly handle hidden surface elimination within a virtual object (Figure 5.4a). It will also cause rendering algorithms to correctly occlude virtual objects that are behind other virtual objects (Figure 5.4b). Hidden surface removal does not occur when a real object occludes a virtual one (Figure 5.4c) because there is no information about the geometric relationship between these objects (Wloka and Anderson 1995). If an affine model of a real object is included as another virtual object and rendered in a key color the occlusions are resolved by chroma or luminance keying (Figure 5.4d). A method for directly creating an affine model for a real object is described in Section 5.

4.   THE UNIVERSITY OF ROCHESTER AUGMENTED REALITY SYSTEMS

Augmented reality systems based on affine representations have been developed in the Department of Computer Science at the University of Rochester. Conceptually, our augmented reality systems consist of tracking and graphics subsystems that work together. The major differences between the two systems that have been built to date are in feature tracking and the technology used for viewing the augmented reality image (Azuma 1997).

4.1   A Monitor-Based Augmented Reality System

A block diagram of our monitor-based augmented reality system is shown in Figure 5.5. The augmented reality technique described in this chapter requires the ability to track features in frames throughout a video sequence. It is not dependent on the particular type of feature being tracked. Our affine method does restrict the motion of the features to a rigid motion with respect to each other. Features on multiple objects that are moving relative to each other cannot be used. The tracking subsystem provides updates of the affine projection matrix to the graphics system and, as such, can be considered to be an "affine camera position tracking" system. We have

(a)   (b)

(c)   (d)

FIG. 5.4. Visible-surface rendering of affine virtual objects. The virtual towers were represented in OpenInventor™. Affine basis points were defined by the centers of the circular markers. The virtual towers were defined with respect to those points. (a) Initial augmented view. (b) Augmented view after a clockwise rotation of the object containing the affine basis points. (c) Hidden-surface elimination occurs only between virtual objects; correct occlusion resolution between physical and virtual objects requires information about the geometric relations between them (Wloka and Anderson 1995). (d) Real-time visible surface rendering with occlusion resolution between virtual and real objects. Visibility interactions between the virtual towers and the L-shaped object were resolved by first constructing an affine graphical model for the object. By painting the entire model a fixed background color and treating it as an additional virtual object, occlusions between that object and all other virtual objects are resolved via chroma or luminance keying. Affine models of real objects can be constructed using the interactive modeling technique of Section 5.

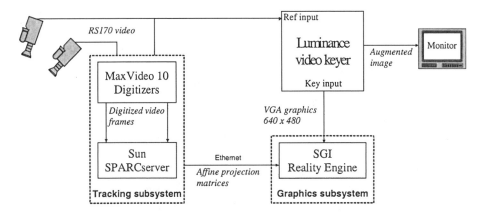

FIG. 5.5. Configuration of a monitor-based augmented reality system.

implemented trackers that use regions and lines as features (Kutulakos and Vallino 1996). The trackers use Datacube MaxVideo 10 boards as frame grabbers with two standard consumer camcorders (Sony TR CCD-3000) used to view the scene. The trackers are interactively initialized by the user selecting seed points in the regions to be tracked. A search for the boundary of each uniform intensity region proceeds from the seed point using a radially expanding coarse-to-fine search algorithm. The detected boundary points are grouped into linear segments using the polyline curve approximation algorithm (Ballard and Brown 1982) and lines are fitted to the segments using least squares. The region's vertices are located by intersecting the lines fitted to adjacent segments. The actual basis points used to define the affine coordinate system do not directly correspond to visible features. Instead they are the center of mass and the three principal components of the 3D set of vertices of the regions being tracked. The basis points are computed from more than the minimum four points needed to define the affine frame. The inclusion of more than the minimum number of points increases the robustness of the localization of the basis points (Reid and Murray 1996). A Kalman filter (Bar-Shalom and Fortmann 1988) is used for tracking the feature points. The output of the filters estimates the image position and velocity of the projections of the basis points.

The graphics subsystem is the second component of the system. It is based on a Silicon Graphics workstation with Infinite Reality graphics. The graphics rendering is performed using the OpenGL and Open Inventor graphics libraries. Communication between the tracker and graphics subsystems is via an Ethernet network. The rendered virtual objects are displayed in a graphics window on the workstation console. This image is used as the foreground element of a luminance keying operation (Jack 1993) in a Celect Translator keyer. The background element is the live video signal. The keyer will show live video in all areas of the image where the luminance value of the foreground element is below the key value. This value is set to key on the black background in the graphics image.

System operation is broken into initialization and runtime phases. There are three steps in the initialization phase: (1) graphics-to-tracker alignment, (2) affine basis initialization, and (3) placement of virtual objects. The graphics-to-tracker alignment is necessary due to differences in image coordinates used by the tracker and graphics subsystems. To establish the relationship between these two 2D coordinate systems the correspondence between three points in the two buffers must be known. The graphics system outputs an alignment pattern composed of three crosses on a black background. This pattern is then merged with the live video signal and digitized by the tracker. (During this alignment sequence the tracker digitizes the merged video signal, which is different than normal operation when it works with the original live video.) The user interactively specifies the location of the crosses in the digitized image. From these locations and the known position of the crosses in the graphics image the 2×3 *graphics-to-video transform* (Figure 5.3) is computed. The initialization of the affine coordinate system is performed interactively by the user when the region trackers are initialized as described above. Automatic tracking of these regions by the tracking subsystem commences after initialization. The next step is placement of virtual objects into the scene. This is accomplished interactively using the techniques specified in Section 3.4. With initialization complete the system enters its runtime phase. The tracking subsystem computes the affine projection matrix, $\Pi_{4 \times 4}$, and transmits updates to the graphics subsystem at rates between 30 and 60 Hz. Using the updated projection matrix the virtual scene is rendered and output to the luminance keyer to create the merged augmented reality image.

The static and dynamic performance of the system was measured. Static performance was measured to determine the misregistration errors caused by the affine approximation to perspective projection and any distortions introduced by the camera lens. The ground truth values were gotten from the image projections of vertices on a physical object. The projections were manually specified in the sequence of images from approximately 50 camera positions. The camera was moved in a roughly circular path around the object at multiple distances ranging to 5 m. Camera zoom was used to maintain a constant image size as the distance from the object increased. Four points were selected to define the affine basis through the entire set of images. From these points the affine projection matrix in each image was computed. The affine coordinates of the remaining vertices were calculated using the affine reconstruction property from two images in the sequence. The projections of the nonbasis points in the other images of the sequence were then computed and compared to the manually specified points. Misregistration errors range up to 15 pixels, with the larger errors seen for shorter distances to the object. This is as expected with the weak perspective approximation to perspective projection.

Dynamic performance is measured while the system is running and quantifies the misregistration error of overlays caused not only by the factors discussed in the previous paragraph but also by latencies in the system's real-time operation. The test was initialized by calculating the affine coordinates of a feature on a real object from its projection in two images. This feature was not one used for defining the affine coordinate system. The augmented image was a white dot that with perfect registration would align with the feature in the merged view. Two correlation based trackers provided independent measurements of the position of the feature in the live video and the white dot in the graphics image. The former was considered the ground truth and the Euclidean distance in image coordinates between the two positions was measured as the dynamic registration error. The test system was manually translated and rotated in an arbitrary fashion for approximately 90 seconds. The mean absolute registration error in the vertical and horizontal directions was 1.74 and 3.47 pixels, respectively.

4.2   Using a Video See-Through Display

The user's perception of being present in the augmented reality scene can be increased by using either a video or optical see-through display (Azuma 1997). The technique of augmenting reality with affine representations immediately lends itself to operation with a video see-through display. Since neither position information nor camera calibration parameters are needed, creating a video see-through display is simply a matter of attaching cameras to a standard virtual reality head-mounted display (HMD). We created our display by mounting two Panasonic miniature color CCD cameras each with a 7.5 mm lens on a Virtual Research VR4 HMD. To obtain a properly fused stereo view of the real scene the user manually adjusts the fixtures holding the two cameras to correctly orient them. This is the only additional system initialization step needed. The system block diagram is shown in Figure 5.6.

The features being tracked by cameras on the HMD can undergo large-scale shifts between video frames due to motion of the user's head (State,

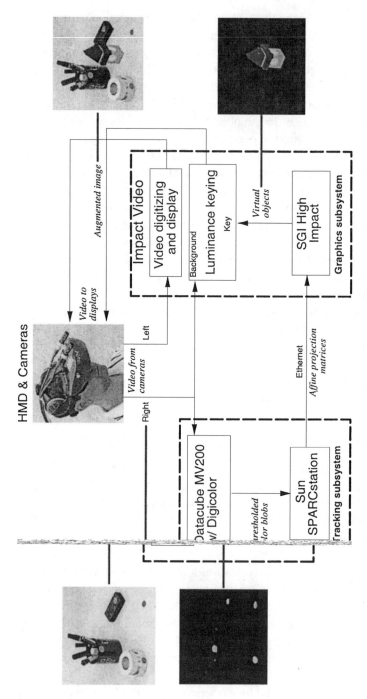

FIG. 5.6. Configuration using a video see-through display. The affine basis in the example images was defined by the four circular markers, which were tracked in real time using color segmentation. The markers were manually attached to objects in the environment and their 3D configuration was unknown.

174

Hirota et al. 1996). Because the region-based tracker had difficulty tracking these motions we implemented new trackers based on color blob detection. Color-based tracking also allows for increased flexibility in selecting the real scene because features are simply colored marker dots placed in the scene rather than the high-contrast regions used previously. The tracking subsystem is based on a Datacube MV200 image processing system with a Digicolor board for color frame grabbing. The video frames are digitized in the hue-saturation-value (HSV) colorspace where segmentation of color features is performed via table look-up. After performing blob coloring (Ballard and Brown 1982) and centroid calculations the tracker computes the affine projection matrices. These operations are performed at a 30 Hz rate. The registration errors and latencies in this implementation are comparable to those found on our original system using region-based trackers.

The HMD-based system incorporates a new graphics subsystem running on a Silicon Graphics Indigo2 workstation with High Impact graphics and Impact video. The video hardware provides a luminance keyer for merging the graphics on a single channel of live video. An interesting observation made during informal experiments in the laboratory is that the user was unaware of this limitation when the augmented view was presented to the user's dominant eye. Only when asked to close that eye did the user notice the missing augmentation in the nondominant eye.

5. AN AUGMENTED REALITY–BASED INTERFACE FOR INTERACTIVE MODELING

The ability to overlay virtual images on live video leads to an interactive method for building models that can be used as virtual objects. The approach uses the real object as a physical three-dimensional model. Instead of trying to solve the difficult problem of understanding the shape of an object from video images of it, the user uses a hand-held pointer to trace the surfaces of the object for which a model is desired.

First, the global affine coordinate system is computed from the projections of the tracked feature points. No placement of the virtual object is necessary. The user identifies a surface of the object by moving a hand-held pointer in contact with the surface as if painting it (Figure 5.7a). This motion is tracked from two viewpoints using a normalized correlation technique (Ballard and Brown 1982). Given the projection of the pointer in

(a)   (b)

(c)

FIG. 5.7. Interactive 3D affine modeling. Live video is provided
by two camcorders whose position and intrinsic parameters
were neither known in advance nor estimated. (a) An easily-
distinguishable hand-held pointer is moved over the surface of an
industrial part. The dark polygonal regions are tracked to establish the affine basis frame. The regions were only used to simplify
tracking and their Euclidean world coordinates were unknown.
(b) Visualizing the progress of 3D stenciling. The augmented display shows the user drawing a virtual curve on the object's surface in real time. (c) When the object is manually rotated in front of
the two cameras, the reconstructed points appear "locked" on the
object's surface, as though the curve traced by the pointer was
actually drawn on the object.

two images, the *affine reconstruction principle* is applied to compute affine
coordinates for that point on the object's surface. This turns the pointer into
the equivalent of a 3D digitizer (Foley, van Dam et al. 1990; Tebo, Leopold
et al. 1996). Feedback is given to the user by rendering small spheres at
the 3D points defined to be on the object's surface and merging this image
with the video image (Figure 5.7b). By looking at this augmented view
the user sees areas of the object that have not yet been modeled and those
that may require refinement. The augmentation is correctly rendered even

if the object is moved (Figure 5.7c), giving the appearance that the user has applied virtual paint on the surface of the object (Agrawala, Beers et al. 1995). We are currently developing techniques for incremental real-time triangulation, surface representations that can efficiently grow in an incremental fashion, and texture mapping of the object's video image onto the model.

6. DISCUSSION

The technique for augmenting reality using affine representations provides a method for merging virtual objects with live video without magnetic position tracking and camera calibration. While the elimination of these two inputs is an advantage, the technique comes with a set of new requirements and limitations, namely, the requirement for real-time tracking of features, the weak perspective approximation, and the use of a non-Euclidean reference frame.

Augmenting reality with affine representations is limited by the accuracy, response, and abilities of the tracking system. To date we have simplified our tracking problem by limiting it to tracking high-contrast objects and easily segmented color blobs. For each video image the tracking subsystem must compute a consistent global coordinate system. This becomes a problem when feature points become occluded or the tracker detects new features in the image. A promising approach to overcome this problem is the Variable State Dimension Filter (VSDF) (McLauchlan and Murray 1995). We are currently testing whether the application of the VSDF allows the tracker to maintain a stable affine coordinate system even when there are a variable number of feature points from one video frame to the next.

This entire system operates by defining all objects in a global affine coordinate system. The technique approximates the camera's perspective projection with a weak perspective model. The validity of this approximation is limited to regions close to the optical axis of the camera, and to objects whose front-to-back distance is small compared to the object's average distance from the camera. The weak perspective assumption can be eliminated if a common projective representation is used instead (Faugeras 1992). A projective representation requires a minimum of five feature points to be tracked instead of the four required by an affine representation. The relative merits of several methods to compute projective structure has been described in the literature (Zisserman and Maybank 1994; Hartley 1995; Rothwell, Csurka et al. 1995; Li, Brady et al. 1996). The weak perspective approximation also does not account for any distortion in the camera lens. Radial distortion is present to some degree in most lenses. Calculating an appropriate image warp (Mohr, Boufama et al. 1993) or estimating the distortion coefficients in conjunction with the tracking subsystem using a VSDF (McLauchlan and Murray 1996) can compensate for this.

Representing virtual objects in the global affine coordinate system imposes some constraints on system operation. Since the coordinate system in which the virtual objects will be represented is not defined until run time, the placement of virtual objects cannot be done beforehand. No metric information about the real 3D world is used, which eliminates the possibility of placing virtual objects based on measurements in the real scene. Instead virtual objects must be placed interactively after the global coordinate system has been determined via methods described in Section 3.4. The graphics system operates in the global affine coordinate system, which, in general, is not an orthonormal frame in space. Our system shows that projection computations, z-buffer determination of visible surface, and texture mapping can be performed within this affine representation. Other rendering algorithms, such as lighting computations, that require metric information in the form of measurement of angles, cannot be performed directly in affine space. Image-based methods can, in principle, provide correct rendering with lighting by linearly combining multiple shaded images of the objects that have been precomputed (Shashua 1992; Belhumeur and Kriegman 1996; Dorsey, Arvo et al. 1996).

There are also some limitations in our specific implementations that were described in Section 4. The latency in our current color trackers is one video frame for acquisition and one frame for blob coloring and centroid calculation. We have also measured a maximum delay of 90 ms (approximately three video frames) from transmission of a new projection update to rendering of the virtual objects. This results in a total latency on the order of 5 frames. We are investigating methods to perform tracking at 60 Hz to reduce the latency introduced by the tracking subsystem. We are also experimenting with predictive estimation of feature positions (Azuma and Bishop 1994) for mitigating the other latency in our system. Finally, our graphics subsystem only has the hardware capability to perform a single merging operation. For a monitor-based augmented reality system this is not a major limitation but in the HMD system it eliminates the possibility of the user viewing stereo virtual objects. The purchase of additional hardware to provide luminance keying on two video channels will overcome this limitation.

7. CONCLUSIONS AND FUTURE WORK

The primary challenge for an augmented reality system is to determine the proper rendering and registration of the virtual objects that will be merged with the view of the real scene. This requires computation of the relationships between multiple coordinate systems. Most augmented reality systems use methods that transform these coordinate systems into a common 3D Euclidean frame relying on position sensing, camera calibration, and knowing the 3D locations of fiducial markers in the scene. This chapter has presented an alternative method for rendering virtual objects that are registered with a live video image to create augmented reality images. The problem has been reduced to:

- real-time tracking of a set of four or more points that define a global affine coordinate system,
- representing the virtual objects in this coordinate system,
- computing the projections of virtual points in each video image as linear combinations of the projections of the affine basis points.

We are continuing to work in this area to overcome limitations present in the prototype. Affine representations are particularly well suited when a priori knowledge of the environment is not available. To work in this general setting, a tracking subsystem capable of tracking features in a natural setting is needed. The tracking subsystem should also handle temporary occlusions or permanent disappearance of features and the appearance of new features in images. We are investigating recursive estimation techniques that will compute a consistent global coordinate system while being robust to these perturbations in the feature set. The weak perspective approximation of the camera's perspective projection generates errors in rendering and registration when the system operates outside the range in which the approximation is valid. Representing objects in a common projective frame will remove the limitations of that approximation.

Acknowledgments   The authors would like to thank Chris Brown for many helpful discussions and for his constant encouragement and support throughout the course of this work. The financial support of the National Science Foundation under Grant No. CDA-9503996, of the University of Maryland under Subcontract No. Z840902, and of Honeywell under Research Contract No. 304931455 is also gratefully acknowledged.

REFERENCES

Agrawala, M., A. C. Beers and M. Levoy (1995), "3d painting on scanned surfaces," in *Proc. Symposium on Interactive 3D Graphics*, pages 145–150.

Ahlers, K., D. Breen, C. Crampton, E. Rose, M. Tuceryan, R. Whitaker and D. Greer (1994), "An augmented vision system for industrial applications," Technical Report ECRC-94-39, European Computer Industry Research Center (ECRC) 1994.

Azuma, R. T. (1997), "A survey of augmented reality," *Presence*, 6(4):355–385.

Azuma, R. and G. Bishop (1994), "Improving static and dynamic registration in an optical see-through hmd," in *Proceedings SIGGRAPH '94*, pages 197–204.

Bajura, M. and U. Neumann (1995), "Dynamic registration correction in video-based augmented reality systems," *IEEE Computer Graphics and Applications*, 15(5):52–60.

Ballard, D. H. and C. M. Brown (1982), *Computer Vision*, Englewood Cliffs: Prentice–Hall, Inc.

Barrett, E. B., M. H. Brill, N. N. Haag and P. M. Payton (1992), "Invariant linear methods in photogrammetry and model-matching," in J. L. Mundy and A. Zisserman, editors, *Geometric Invariance in Computer Vision*, 277–292, Cambridge, MA: The MIT Press.

Bar-Shalom, Y. and T. E. Fortmann (1988), *Tracking and Data Association*, Academic Press.

Belhumeur, P. N. and D. J. Kriegman (1996), "What is the set of images of an object under all possible lighting conditions," in *Proceedings IEEE Conference on Computer Vision and Pattern Recognition*, pages 270–277.

Boufama, B., D. Weinshall and M. Werman (1994), "Shape from motion algorithms: a comparative analysis of scaled orthography and perspective," in *Proceedings of the European Conference on Computer Vision*, pages 199–204.

Byrson, S. (1992), "Measurement and calibration of static distortion of position data from 3D trackers," in *Proceedings SPIE Vol. 1669: Stereoscopic Displays and Applications III*, pages 244–255.

Dorsey, J., J. Arvo and D. Greenberg (1996), "Interactive design of complex time dependent lighting," *IEEE Computer Graphics and Applications*, 15(2):26–36.

Faugeras, O. D. (1992), "What can be seen in three dimensions with an uncalibrated stereo rig?," in *Proceedings of Second Euopean Conference on Computer Vision*, pages 563–578.

Feiner, S., B. MacIntyre and D. Seligmann (1993), "Knowledge-based augmented reality," *Communications of the ACM*, 36(7):53–62.

Foley, J. D., A. van Dam, S. K. Feiner and J. F. Hughes (1990), *Computer Graphics Principles and Practice*, Reading, MA: Addison-Wesley Publishing Co.

Hartley, R. I. (1995), "In defence of the 8-point algorithm," in *Proceedings 1995 IEEE International Conference on Computer Vision*, pages 1064–1070.

Hoff, W. A., K. Nguyen and T. Lyon (1996), "Computer vision-based registration techniques for augmented reality," in *Proceedings SPIE Vol. 2904: Intelligent Robots and Computer Vision XV: Algorithms, Techniques, Active Vision, and Materials Handling*, pages 538–548.

Jack, K. (1993), *Video Demystified: A Handbook for the Digital Engineer*, Solana Beach, CA: HighText Publications Inc.

Janin, A. L., D. W. Mizell and T. P. Caudell (1993), "Calibration of head-mounted displays for augmented reality applications," in *Proceedings IEEE Virtual Reality Annual International Symposium '93*, pages 246–255.

Koenderink, J. J. and A. J. van Doorn (1991), "Affine structure from motion," *Journal of the Optical Society of America A*, 8(2):377–385.

Kutulakos, K. N. and J. R. Vallino (1996), "Affine object representations for calibration-free augmented reality," in *Proceedings of 1996 IEEE Virtual Reality Annual International Symposium*, pages 25–36.

Li, F., M. Brady and C. Wiles (1996), "Fast computation of the fundamental matrix for an active stereo vision system," in *Proceedings of the Fourth European Conference on Computer Vision*, pages 157–166.

McLauchlan, P. F. and D. W. Murray (1995), "A unifying framework for structure and motion recovery from image sequences," in *Proceedings of the 5th IEEE International Conference on Computer Vision*, pages 314–320.

McLauchlan, P. F. and D. W. Murray (1996), "Active camera calibration for a head-eye platform using the variable state-dimension filter," *IEEE Transactions on Pattern Analysis and Machine Intelligence*, 18(1):15–21.

Mellor, J. P. (1995a), *Enhanced Reality Visualization in a Surgical Environment*, Masters thesis, AI Lab, Massachusetts Institute of Technology.

Mellor, J. P. (1995b), "Realtime camera calibration for enhanced reality," in *Proceedings of Computer Vision, Virtual Reality, and Robotics in Medicine '95 (CVRMed '95)*, pages 471–475.

Mohr, R., B. Boufama and P. Brand (1993), "Accurate projective reconstruction," in J. L. Mundy, A. Zisserman and D. Forsyth, editors, *Applications of Invariance in Computer Vision*, 257–276, Berlin: Springer-Verlag.

Mundy, J. L. and A. Zisserman (1992), *Geometric Invariance in Computer Vision*, Cambridge, MA: The MIT Press.

Neumann, U. and Y. Cho (1996), "A self-tracking augmented reality system," in *Proceedings of ACM Symposium on Virtual Reality Software and Technology*, pages 109–115.

Reid, I. D. and D. W. Murray (1996), "Active tracking of foveated feature clusters using affine structure," *International Journal of Computer Vision*, 18(1):41–60.

Rothwell, C., G. Csurka and O. Faugeras (1995), "A comparison of projective reconstruction methods for pairs of views," in *Proceedings 1995 IEEE International Conference on Computer Vision*, pages 932–937.

Shapiro, L., A. Zisserman and M. Brady (1995), "3D motion recovery via affine epipolar geometry," *International Journal of Computer Vision*, 16(2):147–182.

Shashua, A. (1992), *Geometry and Photometry in 3D Visual Recognition*, PhD thesis, MIT.

Shashua, A. (1993), "Projective depth: a geometric invariant for 3D reconstruction from two perspective/orthographic views and for visual recognition," in *Proceedings 1993 IEEE International Conference on Computer Vision*, pages 583–590.

State, A., D. T. Chen, C. Tector, A. Brandt, H. Chen, R. Ohbuchi, M. Bajura and H. Fuchs (1994), "Case study: observing a volume rendered fetus within a pregnant patient," in *Proceedings of the 1994 IEEE Visualization Conference*, pages 364–368.

State, A., G. Hirota, D. T. Chen, W. F. Garrett and M. A. Livingston (1996), "Superior augmented reality registration by integrating landmark tracking and magnetic tracking," in *Proceedings of the ACM SIGGRAPH Conference on Computer Graphics*, pages 429–438.

Tebo, S. A., D. A. Leopold, D. M. Long, S. J. Zinreich and D. W. Kenedy (1996), "An optical 3D digitzer for frameless stereotactic surgery," *IEEE Computer Graphics and Applications*, 16(1):55–64.

Thompson, W. B. and J. L. Mundy (1987), "Three-dimensional model matching from an unconstrained viewpoint," in *Proceedings of the IEEE 1987 Robotics and Automation Conference*, pages 208–220.

Tsai, R. Y. (1987), "A versatile camera calibration technique for high-accuracy 3D machine vision metrology using off-the-shelf TV cameras and lenses," *IEEE Transactions of Robotics and Automation*, RA-3(4):323–344.

Tuceryan, M., D. S. Greer, R. T. Whitaker, D. E. Breen, C. Crampton, E. Rose and K. H. Ahlers (1995), "Calibration requirements and procedures for a monitor-based augmented reality system," *IEEE Transactions on Visualization and Computer Graphics*, 1(3):255–273.

Uenohara, M. and T. Kanade (1995), "Vision-based object registration for real-time image overlay," in N. Ayache, editor, *Computer Vision, Virtual Reality and Robotics in Medicine: CVRMed '95*, 14–22, Berlin: Springer-Verlag.

Ullman, S. and R. Basri (1991), "Recognition by linear combinations of models," *IEEE Transactions on Pattern Analysis and Machine Intelligence*, 13(10):992–1006.

Weinshall, D. and C. Tomasi (1993), "Linear and incremental acquisition of invariant shape models from image sequences," in *Proceedings 4th IEEE International Conference on Computer Vision*, pages 675–682.

Wiles, C. and M. Brady (1996), "On the appropriateness of camera models," in *Proceedings of the Fourth European Conference on Computer Vision*, pages 228–237.

Wloka, M. M. and B. G. Anderson (1995), "Resolving occlusion in augmented reality," in *Proceedings 1995 Symposium on Interactive 3D Graphics*, pages 5–12.

Zisserman, A. and S. J. Maybank (1994), "A case against epipolar geometry," in J. L. Mundy, A. Zisserman and D. Forsyth, editors, *Applications of Invariance in Computer Vision*, 69–88, Berlin: Springer-Verlag.

6

Registration Error Analysis for Augmented Reality Systems

Richard L. Holloway
University of North Carolina

1. INTRODUCTION

An important goal of most augmented reality systems is to display computer-generated objects so that they appear to be aligned with the real objects in the scene. For example, in a jet-engine maintenance application, we may want to display a computer-generated arrow that points to the faulty wire on the wiring harness; clearly, the arrow must point to the correct wire for the system to be useful at all. The correct positioning of the virtual objects with respect to the real objects is known as *registration*, and failure of the system to correctly align the virtual objects is called *registration error*. Figure 6.1 below shows a system with significant registration error and Figure 6.2 shows the same system with the error corrected.

Since no AR system is perfect, each has some amount of registration error. How much registration error is tolerable depends on the application: One would expect a missile-targeting system to be more tolerant of a one-centimeter registration error than a surgical system would be. What we would like, then, is a way of determining how much registration error to

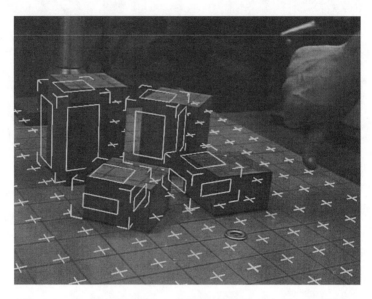

FIG. 6.1. Registration error. (Image courtesy of Gentaro Hirota and Andrei State, University of North Carolina at Chapel Hill © 1996.)

FIG. 6.2. Correctly registered real and virtual objects. (Image courtesy of Gentaro Hirota and Andrei State, University of North Carolina at Chapel Hill © 1996.)

expect from a given system so that we can determine whether the system will be usable, and, if not, how to change it so that it will be. Because AR systems are fairly complex, the analysis is nontrivial: It turns out that there are on the order of a dozen error sources in a typical AR system, and it is not always obvious which one is causing the observed misregistration.

This chapter presents an error model for AR systems that allows the system architect to determine

1. what the registration error sources are and which ones are the most significant contributors to the total error,
2. the sensitivity of the net registration error to input errors in each part of the system,
3. the nature of the distortions caused by each type of input error, and
4. the level of registration accuracy one can expect as a function of the input errors.

The chapter also provides insights on how to best calibrate the system.

In other words, the model tells the system architect where to spend his time and money in order to improve the system's registration, and also gives some idea of what level of registration he can expect for a given set of hardware and software.

The model presented applies to systems that use optical see-through head-mounted displays (STHMDs), although many of the results apply to video STHMDs and non–head-mounted systems as well.

The main results of the analysis conducted using the model are:

1. Even for moderate head velocities, system delay causes more registration error than all other sources combined. A rule of thumb for medical applications is that each millisecond of delay introduces a millimeter of registration error in the worst case, and $\frac{1}{3}$ mm/s in the average case. The only hope for good dynamic registration with optical see-through systems will be to use predictive head tracking.
2. Eye tracking is probably not necessary, since error due to eye rotation can be minimized by using the eye's center of rotation as the center of projection.
3. Tracker error is a significant problem both in head tracking *and* in system calibration, even when the tracker is calibrated for field distortion and other static errors.
4. The World (or reference) coordinate system adds error and should be omitted when possible.

TABLE 6.1
Error Sources and Associated Registration-Error Magnitudes

Rank	Error Source	Registration Error (mm)	Assumptions
1	Delay	20–60+	Max head velocities of 500 mm/s, 50°/s
2	Optical distortion	0–20 (in image plane)	11% distortion at corner of image, 4% at top of image, magnification = 6.0, image distance = 500 mm
3	World–Tracker calibration error	4–10+	Assumes head is ~500 mm from transmitter and viewed point is at arm's length (500 mm)
4	Tracker measurement error (static, dynamic, jitter)	1–7+	Assumes magnetic tracker w/source-sensor distance ≈500 mm
5	Acquisition/alignment error	1–3	Typical medical dataset w/voxel sizes of $1 \times 1 \times 3$ mm
6	Viewing error	0–2+	Virtual image at 500 mm, 5 mm of eye movement or calibration error, viewed point is ±200 mm from virtual image plane.
7	Display nonlinearity	1–2	$1''$ CRT with nonlinearity ≈1%, magnification = 6.0
8	Image misalignment, lateral color, aliasing	<1	Good calibration procedures Perceived image point is average of RGB images NTSC resolution

5. Computational correction of optical distortion may introduce more delay-induced registration error than the distortion error it corrects.

6. There are many small error sources that will make submillimeter registration almost impossible in an optical STHMD system.

Table 6.1 gives an approximate ranking of error sources with estimated error ranges and assumptions.

Holloway (1995) describes the model and the test system in full detail; the rest of this chapter gives a brief overview of the model and summarizes the most important results.

2. RELATED WORK

While there are many descriptions of STHMD systems in the literature (Sutherland (1968), Furness (1986), Bajura et al. (1992), and Feiner et al. (1993)), only a few papers have dealt with errors or models for HMD systems. Robinett and Rolland (1991) present a computational model for HMDs that identifies key parameters for characterizing a system, and Grinberg et al. (1994), Robinett and Holloway (1995), and Southard (1994) present models for correctly computing the complete viewing transformation once the system parameters are known. Janin et al. (1993) and Oishi and Tachi (1996) describe methods for calibration of STHMDs and discuss a number of registration-error sources and their effects. Deering (1992) describes a number of error sources encountered in creating a CRT-based AR tool. Azuma and Bishop (1994) give a brief listing of error sources and present methods for correcting some of the worst error sources via calibration procedures and predictive head tracking. Hodges and Davis (1993) discuss the geometry of stereoscopic viewing and list a number of error sources and their effects, but they stop short of a complete system analysis. Min and Jense (1994) also list several error sources and describe a user study to determine the optimal system parameters for each subject. State et al. (1994) describe problems encountered in attempting to register ultrasound data displayed with a video STHMD with a real patient. Bajura and Neumann (1995) present a model for video-based AR systems that dynamically corrects the registration error by forcing alignment of the real and virtual images.

3. ERROR MODEL

3.1 Registration Error Metrics

In the course of the analysis of registration-error sources, it became clear that there are several metrics for registration error and that each is useful for describing some aspect of the problem. This section describes the metrics used and when they are useful.

If two points that are supposed to be coincident are separated by some distance, one can describe the degree of separation or misregistration with a 3D error vector from one point to the other. *Linear registration error* is defined here to be the length of this error vector. While this is often a useful metric for registration error, generating the 3D error vector for a stereoscopic display requires knowledge of the projectors from both eyes in order to specify the location of the perceived point, and there are many cases in which we would like to examine the registration error for a single eye. Moreover, in cases where the projectors are nearly parallel, even tiny errors can cause the projectors to become parallel, inducing theoretically infinite linear registration error. To characterize such

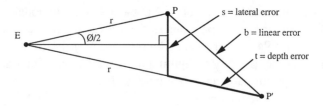

FIG. 6.3.   Registration error metrics.

a situation as having infinite registration error seems overly pessimistic and not very useful, since the projectors do pass near the point and may appear to converge at the point when coupled with other depth cues, such as head-motion parallax. All of this leads to the conclusion that registration errors in depth are somehow different from registration errors that cause a clear visual separation between the real and virtual points, particularly when a stereoscopic display is involved. For this reason, I will also describe the registration error using the three related metrics pictured in Figure 6.3.

In the figure, E is the eyepoint, P′ is the displayed point, and P is the real point (where we want P′ to appear). The *angular error* Ø is the visual angle subtended at the eyepoint by the line segment PP′. The *lateral error* *s* answers the question, "If P′ were at the same distance as P from the eye, how far apart would they be?" Mathematically, this is the length of a line from P to EP′ perpendicular to the bisector of the angle PEP′, and it is given by

$$s = 2 r \sin \frac{\emptyset}{2}. \tag{1}$$

Finally, the *depth error t* tells us how much closer or further P′ is relative to P and is given by

$$t = \|\mathbf{v}_{E\_P'}\| - \|\mathbf{v}_{E\_P}\| \tag{2}$$

where $\|\mathbf{v}_{E\_P'}\|$ and $\|\mathbf{v}_{E\_P}\|$ are the magnitudes of the vectors from E to P′ and P, respectively.

Clearly, all of these measures depend on the geometry defining the segment PP′, so a complete specification will depend on the situation being discussed. I will use these metrics for discussing viewing and display errors and the linear registration error for the analysis of head-tracking error and digitization/alignment error.

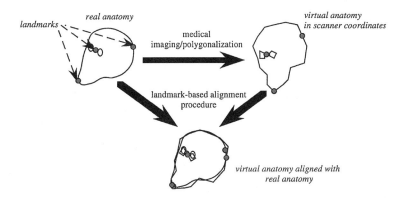

FIG. 6.4. Acquisition/alignment process.

3.2 System Operation

In a typical AR system, the virtual objects that are to be registered with their real counterparts are first acquired by some form of imaging equipment (e.g., a CT scanner) or modeled with some design tool (e.g., a computer-aided design package). This typically produces a virtual object defined in its own coordinate system (CS), which must then be aligned with the real object(s) in the laboratory or World CS. Figure 6.4 shows this process.

In the top part of the figure, the virtual object is created via a scanning or modeling process (indicated by the right-facing arrow). Because of errors in this process (such as scanning artifacts and approximation errors), the virtual object is only an approximation of the object it is intended to represent. In the next part of the figure, the real and virtual objects are aligned in World space via some alignment procedure. A typical method is to digitize landmarks on both objects and use an algorithm such as that described in Besl and McKay (1992) to rotate, translate, and scale the virtual object to be in a least-squares alignment with the real object. This is shown in the bottom part of the figure. Note that there is already some registration error at various points on the real and virtual objects and that this error is independent of the process of viewing the objects (which is discussed next).

In order to view the real and virtual objects, the system employs a see-through head-mounted display (STHMD), which superimposes the view of the virtual object onto the real object, as shown in Figure 6.5. The user sees the virtual image of the screen (created by the lens) reflected off of the beam splitter and can see the real environment as well. Figure 6.6

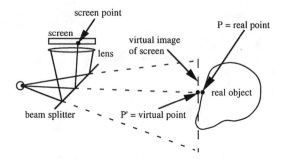

FIG. 6.5. STHMD operation for one eye.

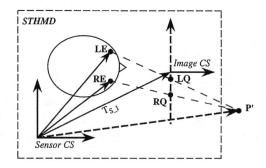

FIG. 6.6. Top view of binocular case showing perceived point.

gives a more abstract view from the top showing the situation for both eyes.[1]

As in the previous figure, P′ is the point displayed by the STHMD, and it is defined as the intersection of the projectors from the eyepoints (LE and RE) through the projected points (LQ and RQ). The reference coordinate system shown is that of the Sensor, which is the part of the tracker attached to the STHMD. The arrow labeled T_{S_I} in the figure represents the transformation between the Sensor and Image coordinate systems. The sensor's position and orientation is reported relative to the Tracker CS, as shown in Figure 6.7.

In this figure, the Tracker CS is the coordinate system defined by the tracker's base or transmitter, which is mounted somewhere in the World CS. The Sensor coordinate system is the reference CS for the STHMD, which displays the point P′. The World–Tracker transformation (represented by

---

[1]For clarity, errors are not shown in the figures in this section. See Holloway (1995) for a more thorough explanation.

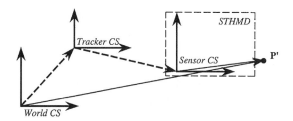

FIG. 6.7. System overview showing STHMD as a black box and all coordinate systems.

the heavy dashed line between these two CSs) is derived via a calibration procedure, and Tracker–Sensor transform is measured and reported by the tracker each frame.

3.3 Error Model Overview

Following the system overview given above, we can simplify the error analysis by dividing the registration error sources into four categories:

1. *Acquisition/alignment error:* Error in acquiring the data for the virtual anatomy and aligning it with the real patient in the laboratory. For this application, the error sources are CT scanning artifacts, approximations made in polygonalizing the resulting CT volume, and errors in the landmark-based alignment procedure.

2. *Head-tracking error:* Error in the World–Tracker and Tracker–Sensor transformations, which define where the STHMD is in World space. Error sources are tracker delay, static and dynamic tracker measurement error, and calibration error.

3. *Display error:* Error made in displaying the computed image. This includes optical distortion, miscalibration of the virtual images with respect to the tracker's sensor, aliasing, nonlinearity in the display devices (e.g., CRTs), and lateral color aberration.

4. *Viewing error:* Error in the modeled location of the user's eyepoints in the computer graphics model. Error sources are calibration error, rotation of the user's eyes, and slippage of the STHMD on the user's head.

The error model was derived by examining each of the four types of error and (where possible) deriving separate[2] analytical expressions for the

---

[2]To first order, these error sources can be treated as independent; the issue of interaction between the error sources is treated more thoroughly in Holloway (1995).

registration error as a function of the system parameters (e.g., viewing distance, sensor-to-transmitter separation, etc.) and the size of the input error (e.g., the magnitude of the translational error in the tracker measurement). Although the classic approach to error analysis is to use partial derivatives to determine a function's sensitivity to errors in its inputs, this approach yielded expressions so large they were useless. Instead, I derived error expressions by modeling the input errors explicitly in the geometry for each situation, which generally yielded smaller, more intuitive expressions. Moreover, while the partial-derivative approach is valid only for small errors, the geometric error model is valid for both small and large errors, which is important for examining the behavior of large error sources. Finally, the use of the error metrics discussed in Section 3.1 allowed the model to give finite error bounds whenever possible.

3.4 Description of System for Testing the Error Model

To test the error model, I conducted a set of experiments to verify that the model was both complete and accurate. That is, I wanted to verify that each error source contributed to the net registration error in the expected fashion *and* that I had not left out any significant error sources. I only checked the equations describing the major sources of error (discussed in the next section), that is, head-tracking error (due to delay, tracker error, and World-Tracker calibration error), optical distortion, and viewing error.[3] The behavior of the smaller error sources (image calibration error, aliasing, display nonlinearity, lateral color, and acquisition/alignment errors) was not tested (except to note the absence of any major effects due to these sources). The experimental results are reported in Holloway (1995) and will not be repeated here, since they add nothing to the discussion of the error expressions themselves. However, the test system itself may be of some interest, since it turned out to be a rather accurate AR system.

The system used for the error model test experiments was the UNC 30° STHMD connected to Pixel-Planes 5 (Fuchs et al. (1989)) and a Faro Industrial Metrecom mechanical tracker. The Faro arm is a very accurate tracker/digitizer[4] and was used so that the system could be calibrated

---

[3]This was tested not because it was a large error source, but rather because its behavior seemed complex enough to warrant at least a simple check.

[4]According to Faro (1993), the 2σ value for single-point repeatability is .3 mm, and the 2σ value for linear displacement is 0.5 mm. My experience is that it meets these specifications in real use.

FIG. 6.8. System overview.

accurately; however, due to its limited range and unwieldiness, it is not an ideal solution for a real surgical planning system. Figure 6.8 shows the experimental setup.

The general approach was to calibrate the system as well as possible and then deliberately introduce errors of each type and record their effect on the overall registration error. The setup used for most of these experiments was as follows: A small video camera (a Panasonic model GP-KS102 color CCD camera) was inserted into a Styrofoam head with the entrance-pupil center positioned roughly at the head's eyepoint. The system was calibrated and then errors from the list above were introduced and their corresponding registration errors measured. The test point in World space was a point on a sheet of graph paper surrounded by a ruled grid, which was used to measure the size of the error in World space. Because the system is calibrated and the errors are artificially introduced, the system has full knowledge of the errors in the transformations and can calculate and display both the correct location for the displayed point *and* the erroneous location, as shown in Figure 6.9.

In the figure, the large crosshair on the right is the point drawn using transformations containing error (in this case, error in the World–Tracker transform, which moves it 20 mm from its modeled location). The small crosshair aligned with the circle is the point drawn using the correct transformations. The box and crosshair to the left of the other two crosshairs

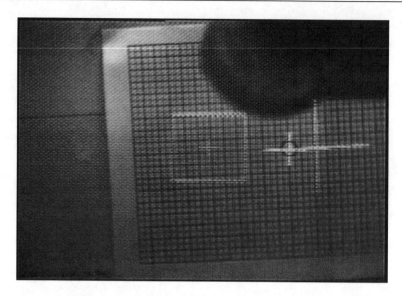

FIG. 6.9.   Camera image of virtual and real scene.

is just a reference marker fixed in the center of the virtual image. At any instant, the distance between the small and large crosshair gives the registration error due to the deliberately introduced error source, and this can be measured on the paper grid in World space or on a distortion-corrected grid in Image space (not shown). Since the location of the camera's "eyepoint" is also known via measurement, we can compute the angular error, the lateral error, and the depth error.

A side benefit to the model verification experiments is that the test system turned out to be a very accurate (albeit unwieldy) augmented reality system. After calibration, the system achieves static registration of 1–2 mm, which is the best of which I am aware. The main reason for the good registration with this system is that it uses an accurate mechanical tracker both for head-tracking and for digitizing calibration points, which (as we shall see) avoids some of the largest error sources. The disadvantages to this configuration are its unwieldiness and poor dynamic performance (the system latency is on the order of 300 ms).

The error model was quite useful during the calibration process, since it can be used to determine which parts of the system need to be calibrated carefully and which ones can be approximated. In particular, I found that the parameters whose values are difficult to measure are often those for which precise calibration is not necessary. For example, the z coordinate (i.e., the depth) for the Image coordinate system is difficult to measure with precision, but the net registration error is fairly insensitive to error in this parameter. In contrast, error in the x or y coordinate induces registration error directly and is therefore easy to detect and correct.

The next section discusses the most interesting results of the analysis and related experiments.

4. MAIN RESULTS

The results in this section follow the organization just presented, except that small error sources (including acquisition/alignment error) are treated together at the end. The section begins with delay and other head-tracking errors, then treats optical distortion, then viewing error, and finishes with a brief discussion of smaller error sources.

4.1 Delay Swamps Other Error Sources

It should come as no surprise to anyone who has used an AR system that the largest source of registration error is due to system delay. Even relatively slow head motions can induce large registration errors, which quickly kills the illusion that the virtual objects are fixed in the real environment. System delay is the sum of all the delays from the time the measurement of head position/orientation is made until the time that the image generated using that information is finally visible to the user and is discussed in Adelstein et al. (1992), Mine (1993), Olano et al. (1995) and Wloka (1995). Although many of the delays that contribute to this tracker-to-display latency are not specifically associated with the tracker, they each contribute to the discrepancy between the real and reported head position and orientation at display time. Mine (1993) gives a complete listing and analysis of delay sources for the UNC system; a similar list follows.

Delay Sources

- *Tracker delay:* This is the time required for gathering data, making calculations in order to derive position and orientation from the sensed data, and transmitting the result to the host. The Polhemus Fastrak is quoted at 4 ms of delay; Mine measured 11 ms at UNC, but this included transmission time and some host processing.
- *Host-computer delay:* This delay includes tasks such as fetching and massaging the tracker data, running host-based application code, and any operating-system tasks.

- *Image-generation delay:* This is the time to render the image corresponding to the current tracker report into the frame buffer. For UNC's Pixel-Planes 5 graphics engine, typical delay values for a small hardware configuration (13 graphics processors, 5 renderers) range from 75 ms for 4,000 primitives to 135 ms for 60,000 primitives. An experimental low-latency rendering system developed by Cohen and Olano (1994) reduced the delay to 17 ms, but only for very small data sets (100–200 triangles).
- *Video sync delay:* This is the delay while waiting for the next video frame to begin. The worst case for a 60 Hz refresh rate is 16.7 ms, and the best case is a synchronized system for which the delay is zero.
- *Frame delay:* Most raster devices paint the image sequentially from top to bottom. For a 60 Hz noninterlaced display, the delay is roughly 17 ms between the display of the upper left pixel and the lower right pixel.
- *Internal display delay:* Some display devices add additional delays due to processing within the display device itself. For example, Mine (1993) reports that the LCDs in an HMD in use at UNC added an additional field time (about 17 ms) of delay. This could be due to the display having a different resolution from the input signal and having to resample the input before it can display the current frame.

For immobile objects, the amount of registration error due to delay is determined by the amount the head moves from the time the tracker makes its measurement until the image is scanned out. If we use a simple first-order model for head motion, the general expression for bounding the delay-induced error is

$$b_{\text{delay}} = \|\dot{\mathbf{v}}_{\text{head}}\Delta t\| + 2 \cdot \sin\frac{|\dot{\varnothing}_{\text{head}}\Delta t|}{2} \cdot \|\mathbf{v}_{\text{S\_P'}}\| \qquad (3)$$

where $\dot{\varnothing}_{\text{head}}$ and $\dot{\mathbf{v}}_{\text{head}}$ are the angular and linear velocity of the user's head, and Δt is the net delay (which in the worst case is the sum of all the delays listed above). The values for $\dot{\varnothing}_{\text{head}}$ and $\dot{\mathbf{v}}_{\text{head}}$ will clearly be application dependent: One would expect fighter pilots to have higher velocity values than surgeons, for example.

To get an idea of representative head velocities for surgery planning, I measured the angular and linear velocities of a physician's head (with a Fastrak magnetic tracker) while he conducted a simulated planning session. Most of the head movements were slower than about 50 deg/s and 500 mm/s, and the average velocities were 164 mm/s and 20 deg/s. This is consistent with data collected by Azuma (1995) for naive users in a demo application: The linear velocities peaked at around 500 mm/s, and most of the angular velocities were below 50 deg/s (although the peak velocities did get as high as 120 deg/s in some cases).

If we take 500 mm/s and 50 deg/s as fairly conservative upper bounds for head movement and plug them into the expression for b_{delay} for the minimum delay number for the normal Pixel-Planes rendering system (65 ms), we get

$$b_{T\_S} = 500 \text{ mm/s} \cdot 0.065 \text{ s} + 2 \cdot \sin \frac{50 \text{ deg/s} \cdot 0.065 \text{ s}}{2} \cdot 500 \text{ mm}$$
$$= 28.4 + 32.5 = 60.9 \text{ mm},$$

which is clearly a very large error. If we plug in the mean velocities, we get 22 mm. This is still quite a large error and gives an indication of just how serious a problem delay-induced registration error is. Note that, at least for this application, the linear and angular terms contribute equally to the net registration error. Note also that these delay values have not included Δt_{frame}, the time to draw a full NTSC field, which adds up to 17 ms for the last pixel scanned out.

Using these numbers, a simple rule of thumb for this application is that *we can expect about 1 mm of registration error for every millisecond of delay in the worst case and $\frac{1}{3}$ mm/ms in the average case.* Note the significance of this result: If our goal is registration to within 1 mm, unless we do predictive head tracking, the system will only have (in the worst case) 1 millisecond to read the tracker, do its calculations, and update the displays! Even the most aggressive strategies for reducing system delay cannot hope to achieve this level of performance. *The only hope for good dynamic registration will be to use predictive head tracking.*

In summary, delay is clearly the largest error source in our current system and is likely to be a problem for the foreseeable future. For the maximum head velocities and typical system delays, delay-induced registration error is greater than all other registration errors combined. Angular velocity seems to dominate for applications in which the user is surrounded by the data (e.g., a building walkthrough), while translational velocity is more of a factor in applications where a single object is being studied (as in surgery planning). Azuma and Bishop (1994) were able to reduce dynamic registration errors by a factor of 5 to 10 by using predictive filtering with rate gyroscopes and linear accelerometers, but the problem is far from solved. One of their results is that prediction errors increase at greater than linear rates with respect to increasing prediction intervals, which means that prediction may not be effective for long delays (which for their data was

> 80 ms). Thus, systems must be optimized for low latency (see Olano et al. (1995)), which is in direct conflict with the need for high throughput. In addition, techniques to synchronize the rendering process with the display scanout (such as beginning-of-frame synchronization, frameless rendering (Bishop et al. 1994), and just-in-time pixels (Mine and Bishop 1993)) will also be essential for reducing delay-induced error.

4.2 Beware the World Coordinate System

It is common practice to have a user- or system-defined World CS as a reference. An example of when we need a World CS is when we have a special digitizer (such as a camera measurement system) for precisely locating points in W, but which is not suitable for head tracking. We then have W as the reference CS and T is expressed relative to it. To understand the error this causes, let us examine the process of aligning virtual points within the real environment. If we want the virtual point P' to coincide with the real point P, we must express the location of P in some coordinate system known to the system. In one method, we measure P relative to some World CS, which we have defined for convenience, and then transform the vector from W to P into Sensor space for viewing in the STHMD via the transformation:

$$\mathbf{v}_{S\_P} = T_{S\_T} \cdot T_{T\_W} \cdot \mathbf{v}_{W\_P}, \tag{4}$$

where T_{S_T} is the inverse of the Tracker–Sensor transform (reported by the head tracker), T_{T_W} is the inverse of the World–Tracker transform (which expresses the Tracker CS in W), and $\mathbf{v}_{W\_P}$ and $\mathbf{v}_{S\_P}$ are the vectors to P from W and S, respectively.

A second method is to measure P with respect to the Tracker CS directly by digitizing the point via a number of measurements in Tracker space. This reduces the previous equation to

$$\mathbf{v}_{S\_P} = T_{S\_T} \cdot \mathbf{v}_{T\_P} \tag{5}$$

This method has better error properties (as we shall see) but is not always an option, since accurate digitizing trackers are often ill-suited for head tracking (as mentioned in Section 3.3).

The problem with the first approach lies in its error propagation behavior. In the absence of other errors, the linear registration error due to error in T_{W_T} is given by

$$b_{W\_T} = \|\delta\mathbf{v}_{W\_T}\| + 2\cdot\sin\frac{|\delta\varnothing_{W\_T}|}{2}\cdot(\|\mathbf{v}_{T\_S}\| + \|\mathbf{v}_{S\_P}\|). \tag{6}$$

Here, δv_{W_T} is the error in positioning the tracker's origin in W, $\delta \varnothing_{W\_T}$ is the orientation error of T in W, v_{T_S} is the vector from the origin of T to the origin of S, and v_{S_P} is the vector from the Sensor CS to P. Note that while the translational error in T_{W_T} just adds to the net error, the rotational error term is scaled by the magnitudes of two vectors which may be rather large.

As reported in Janin et al. (1993), the origin and orientation of magnetic trackers (such as the Polhemus Fastrak) are difficult to measure directly with any accuracy, since the origin is inside of a transmitter. Therefore, a common approach for orienting and positioning the tracker source in World space involves taking tracker readings and deducing the T_{W_T} transform from them. If we assume that the determination of $T_{W\_T'}$ is limited by the static accuracy of the tracker, we can use the specifications for the tracker to get another estimate of the net error. The quoted specifications for the Polhemus Fastrak for distances up to 760 mm are $0.15°$ RMS for static angular accuracy and 1.3 mm static translational accuracy (Polhemus (1993)). If we assume that the user's head is 500 mm from the tracker origin and that P is 500 mm from the sensor, we get a linear registration error of 4 mm *just from error in locating the tracker in World space.* If the tracker–head distance and sensor-to-point distances go up to 1000 mm, the error reaches 6.6 mm. Again, this is independent of error in measuring the head position and orientation, which will add even more error.

The heart of the problem is that angular errors in orienting the tracker precisely in W are magnified by the "moment arm" of the tracker-to-point distance, which can be quite large. If we can eliminate this transform from the system by measuring point locations relative to the tracker, the net error should go down. For systems that require a separate digitizer for aligning the virtual objects in the real environment (and therefore a World CS), it may be possible to use the scaling behavior of this error source to calibrate it out of the system; that is, by using large head-to-tracker and head-to-point distances, we might be able to use this moment arm to our advantage to reduce δv_{W_T} to a negligible level.

4.3 Tracker Measurement Error

Apart from the delay error and World–Tracker error already discussed, there is the problem of error in the measurement of head position as reported by the tracker. We can break the tracker error into three categories:

1. *Static field distortion:* This is any systematic, repeatable distortions of the measurement volume, such as the warping seen in the presence

of metal for magnetic trackers. This distortion can be corrected via calibration to the extent that it is repeatable and systematic.

2. *Nonrepeatable tracker error (or jitter):* This is error that cannot be calibrated out of the system and includes both short-term variations due to noise and long-term variations that cause readings to change from one day to the next.

3. *Dynamic tracker error:* This is any error that is a function of the sensor's motion. For example, systems that assume the sensor's motion is negligible with respect to their measurement interval will have some amount of error for moving objects.

The problem with quantifying tracker error is that it is very dependent on the tracker technology and the environment in which the tracker is used. For example, magnetic trackers are sensitive to metal and electromagnetic fields in their operating environment, yet most AR setups are in labs chock full of electronic equipment and have significant amounts of metal in walls, floors, etc. In such hostile environments, the error in the tracker measurements may exceed the manufacturer's specifications by an order of magnitude or more.

As indicated above, there are two questions for static tracker accuracy: 1) How much noise (or jitter) is there in the tracker readings over time? and 2) Can we calibrate the tracker so that the average accuracy is roughly equal to the average jitter?

The work of Bryson (1992) addresses both questions for a Polhemus Isotrak. They measured the accuracy of a Polhemus Isotrak and reported that for source–sensor distances of up to 760 mm, readings taken on two different days varied as much as 1–2" (25–51 mm), even though the standard deviation of the readings in any one-second period was less than 3 mm (all errors increased with the source–sensor distance). They tried several calibration methods and were able to calibrate the tracker to within 1–2" for separations of around 30". In short, they were able to calibrate the tracker to within the long-term jitter, but the net error was still about ten times the standard deviation of the short-term jitter (25–50 mm).

I measured the jitter of the Polhemus Fastrak and the Ascension Flock of Birds in the UNC laboratory; summary plots for both are given in Figures 6.10 and 6.11.

For the Fastrak for transmitter–sensor separations of less than 500 mm,[5] the translation sigma values are 0.25 mm or lower, and the orientation

---

[5]In the head-motion study cited earlier, I also measured the range of head motion for the surgeon and found that most of the time his head was within 500 mm of the patient; therefore a centrally located transmitter could keep the transmitter–sensor separation to 500 mm or less most of the time.

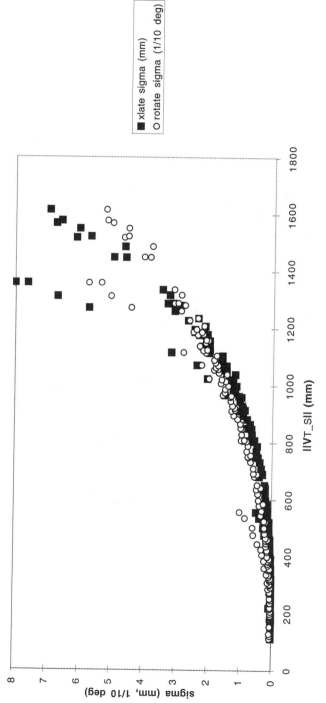

FIG. 6.10   Translation (in mm) and orientation (in tenths of degrees) jitter sigma values for Fastrak vs. transmitter-to-sensor distance (in mm).

201

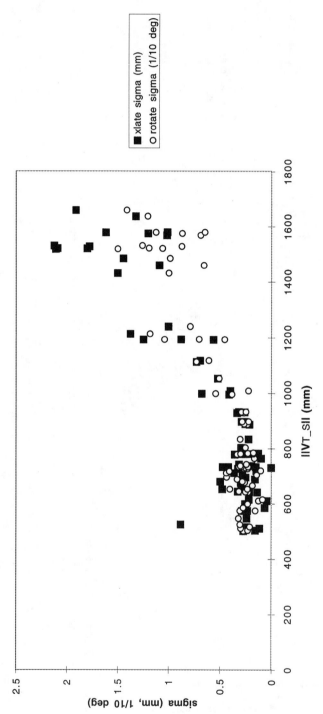

FIG. 6.11 Translation (in mm) and orientation (in tenths of degrees) jitter sigma values for Flock of Birds vs. transmitter-to-sensor distance (in mm).

sigmas are all below 0.05°. At 1000 mm, the translation sigma rises to 1.3 mm and the orientation is 0.15°. The readings were taken over intervals of a few seconds (although intervals of a few minutes showed no significant difference); longer intervals were not tested.

For the Flock of Birds with the extended range transmitter, transmitter-to-sensor separations of less than 500 mm led to saturation of the sensor inputs and therefore readings were not taken in this region. The jitter was generally less than 0.5 mm and 0.05° for separations of 500–1000 mm, and then rose to about 2 mm and 0.15° at 1500 mm. As with the Fastrak, the jitter appears to be a function of the square of the source–sensor separation (because of the falloff of the magnetic field with distance).

Efforts at calibrating the Flock at UNC have not come close to the measured short-term jitter values. Livingston and State (1995) report that they were able to calibrate the Flock to an average error of 5 mm (in a volume roughly equal to one-half a cubic meter) for translation error (down from an average error of 42 mm before calibration). This is similar to what Bryson reported: His calibration reduced the error to about 10 times the short-term jitter standard deviation. Thus, while calibration can reduce tracker error significantly, calibrating trackers to the one-sigma level looks like a nontrivial task.

Turning now to the error model, we find that the sensitivity of overall registration error to tracker measurement error is given by

$$b_{\text{tracker}} = \|\delta \mathbf{v}_{T\_S}\| + 2 \cdot \sin \frac{|\delta \varnothing_{T\_S}|}{2} \cdot \|\mathbf{v}_{S\_P}\|. \tag{7}$$

This shows that translation error just adds to the registration error, but rotational error is magnified by the distance to the point. For $\|\mathbf{v}_{S\_P}\| = 500$ mm, each tenth of a degree of angular error yields about a millimeter of registration error.

If we use the specified static accuracy for the Polhemus Fastrak (1.3 mm, 0.15°), we get 2.6 mm of error for $\|\mathbf{v}_{S\_P}\| = 500$ mm. If the error is as bad as 10 times the measured jitter value at 500 mm (i.e., 2.5 mm and 0.5°), we get about 7 mm of registration error.

As for dynamic error, recent work by Adelstein et al. (1995) indicates that for the Fastrak and the Flock, there does not seem to be any additional error for a moving sensor for normal volitional head motion.

The upshot is that tracker calibration is a difficult but important task and that systems using magnetic trackers should do as much as possible to reduce the source–sensor distance in order to reduce the nonrepeatable tracker errors. Equation 7 gives an idea of the quality of registration one can expect for a given amount of nonrepeatable tracker error.

4.4 Optical Distortion

Optical distortion (hereafter referred to as *distortion*) is the most significant optical aberration for most HMD systems and has been analyzed in Robinett and Rolland (1991) and in Rolland and Hopkins (1993). Distortion is a lateral shift in the position of the imaged points, which can be approximated to third order by the following equation:

$$r_i = m \cdot r_s + k(m r_s)^3, \tag{8}$$

where r_i is the radial distance from the optical axis to the point in image space, m is the linear magnification, k is the third-order coefficient of optical distortion, and r_s is the radial distance to the point in screen space. If k is positive, the magnification increases for off-axis points, and the aberration is called *pincushion distortion*; if k is negative, the magnification decreases, and it is called *barrel distortion*. Pincushion distortion is more common in HMD systems and is pictured in Figure 6.12.

Since this error does not vary with time, it can be corrected by prewarping the image prior to display so that it appears undistorted when viewed through the optics (several approaches for this are given in the above references). While optical and electronic methods for predistortion exist, they are not always feasible for various reasons and many systems can only predistort in the rendering process. Currently, though, predistortion is so computationally intensive that it may induce more system-latency error than the warping error it corrects. For example, on Pixel-Planes 5, Lastra

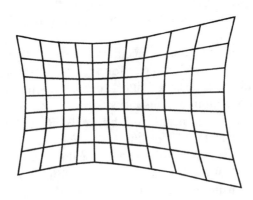

FIG. 6.12. Pincushion distortion.

(1994) reports that he was able to achieve 20 frames per second with predistortion, but only by adding it as a stage in the rendering pipeline. This added a frame of delay, or about 50 ms, which corresponds to about 50 mm of error for the head velocities observed in this application. This is a much larger registration error than that introduced by the distortion itself, and leads to the conclusion that predistortion in the rendering process is not a quick and simple fix for all systems. It is therefore useful to examine the effect of uncorrected distortion in order to compare it with other error sources.

The general term for lateral display error is

$$s_{display} \approx (d - z)\frac{q}{d},\tag{9}$$

where q is the magnitude of the display error in the projection plane. Note that lateral display error is zero at the eyepoint (where $z = d$) and increases linearly as z approaches negative infinity. Assuming that we have correctly modeled the linear magnification, the q value for distortion (in the absence of other errors) is

$$q_{dist} = k(m\, r_s)^3.\tag{10}$$

The plot in Figure 6.13 shows the distortion error for the current UNC STHMD, for which $m = 6.0$, $k = 2.66 \times 10^{-6}$ mm^{-2}, and the screens are 54.7×41 mm. x_n is the normalized x coordinate in screen space (similarly

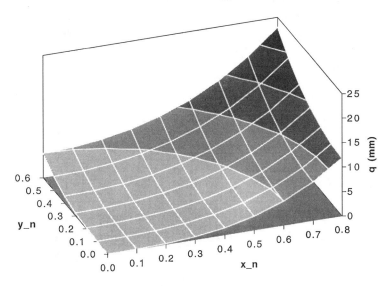

FIG. 6.13. Distortion error.

for y_n) and we have used the previously calculated values for k and m. Because of the 4:3 aspect ratio, the normalized screen-space coordinates have maxima in x and y at 0.8 and 0.6, respectively, and the corner is at unit screen-space radius. It is clear from the plot that the error in image space becomes quite significant in the corners of the image, where the distortion error is 23 mm (corresponding to 11.2% distortion). At the center of right edge, the distortion error gets up to 11.8 mm (7.2%), and at the center of the top edge it reaches 5 mm (4%). For points 200 mm beyond the screen image, the lateral error is 32 mm at the corner, 17 mm at the left/right edge, and 7 mm at the top/bottom.

In general, the distortion error scales linearly with k and as the cube of $m \cdot r_G$. Thus, for a given value of k, using a larger display device to increase the system FOV will also increase the distortion error significantly at the edges of the larger virtual image. Because distortion is systematic, we can look at the binocular case to see what sort of warpings in depth it is likely to cause.

Figure 6.14 shows a top view of the warping of a square caused by distortion. The points LI and RI denote the centers of the two coplanar images and therefore the centers for distortion in each image. Thus, the projectors for the points A and B for RE pass through RI and are not distorted (since $r_s = 0$), whereas all the other displayed points have nonzero r_s values and are moved accordingly. The distorted projectors are shown for A* and D*; note how much more the projectors from LE are moved than those from RE, since the projected points for LE are much further from LI. Distortion tends to cause peripheral objects to "wrap around" the user; that is, it tends to move points further into the periphery but shifted inward toward the user.

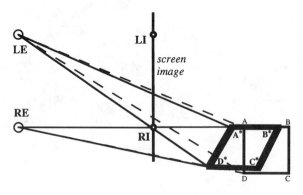

FIG. 6.14. Warping due to optical distortion.

In general, distortion is a small error source in the center of the images but increases rapidly in the periphery and can become quite large. For objects that fill the field of view, the misregistration may well be unacceptable. Moreover, because distortion is an image-space error, the amount of warping will be a function of where the object is drawn within the field of view, which means that the object will seem to change shape as the user's head moves. Finally, because the eyes converge to look at an object, one or both eyes will typically be looking at image points that are not at the exact center of the image, which means that some amount of distortion error will be present even in the best of cases.

4.5 Is Eye Tracking Unnecessary?

Another source of registration error is *viewing error*, which is the error in the modeled eyepoint locations. This is not usually a large source of registration error, but one of the byproducts of the analysis was the realization that there is a method for calibrating systems such that eye tracking may not be necessary in order to eliminate the small error that eye rotation causes. This section begins with a discussion of viewing error in general and then moves on to the issues of eye tracking and system calibration.

For this discussion, I use the center of the entrance pupil[6] E as the eyepoint, following Rolland et al. (1995) rather than the first nodal point[7] N as in Deering (1992). E is approximately 11 millimeters forward of the center of rotation for the eye[8] (vs. 6 mm for the first nodal point), as shown Figure 6.15.

The eyepoints deviate from their modeled locations for two reasons: calibration error and eye movement. That is, a calibration procedure is used to derive the eyepoint locations, but the actual eyepoint E will deviate from the modeled eyepoint E' because of eye movement[9] and error in the procedure. The approximate bound for the resulting lateral error is

$$s_{\text{view}} \approx e \cdot \frac{|z|}{d}, \tag{11}$$

---

[6]The *entrance pupil* of the eye is the image of the pupil seen through the cornea.

[7]A ray passing through the first nodal point will emerge at the same angle from the second nodal point.

[8]This value was calculated using data from Longhurst (1957). The entrance pupil is about 0.5 mm forward of the pupil itself.

[9]If an eye tracker is used, there will still be residual error attributed to eye movement.

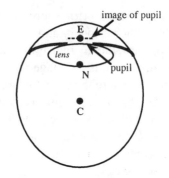

FIG. 6.15. Simple schematic of an eye.

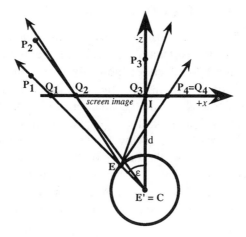

FIG. 6.16. Viewing error for modeled eyepoint at center of rotation.

where e is the magnitude of the viewing error, d is the distance from E′ to the screen image, and z is the distance *from the screen image (or projection plane)* to P′. The first observation is that the registration error due to viewing error goes to zero for points in the projection plane[10] (like P_4 in Figure 6.16) since $P = Q$ and $z = 0$, which suggests that we should position the screen's virtual image in the center of the working volume (preferably dynamically) in order to minimize the effects of viewing error. Also, this property should also be of use in system calibration, since it helps distinguish viewing error from other error sources.

---

[10]Oishi and Tachi (1996) also noted this property and use it to make their calibration method more accurate.

While it might seem that letting d approach infinity (i.e., using collimated images) would reduce viewing error to zero, the situation is somewhat more complex. z is measured relative to the virtual image (not the eye) so that when d becomes large, z also becomes large (for relatively close points). Thus, moving the projection plane to infinity will make $z/d \rightarrow 1$ for near points, inducing lateral errors approaching e. For close work with shallow depths of field, this is probably not desirable. However, for applications requiring a large range of depths, putting the virtual image at infinity has the advantage of capping the lateral error at a value equal to e, while a smaller value for d can induce large lateral errors for $z/d \gg 1$.

Let us now examine the case where all the error is due to eye rotation. That is, we assume a perfect calibration procedure that identifies E when the eye is looking straight ahead, and then we examine the error as the eye rotates. If we designate the measured, straight-ahead eyepoint by E' and rotate E by an angle ε about C, the viewing error magnitude is given by

$$e = \|\mathbf{v}_{C\_E} - \mathbf{v}_{C\_E'}\| = 2 \cdot \sin \frac{|\varepsilon|}{2} \cdot \|\mathbf{v}_{C\_E}\|. \qquad (12)$$

For a 60° monocular FOV, we would expect ε to range from $-30°$ to $+30°$, corresponding to an eyepoint movement of ± 5.7 mm in the worst case. If we use $d = 500$ mm and $e = 5$ mm, we find that the lateral errors for points in the range $0 < |z| < 500$ mm vary linearly from 0 to 5 mm (i.e., the lateral error increases by 1 mm every 100 mm). Depending on the precision required by the application, it would seem that eye tracking would be the only way to reduce this error. Fortunately, there is reason to hope that eye tracking will not be necessary if the point C can be located with precision. That is, it turns out that using C as the modeled eyepoint may reduce the viewing error in this case to a negligible level, even without eye tracking. This is because C is always aligned with the true eyepoint for a point in the center of the eye's field of view, as shown in Figure 6.16.

The figure shows four points, P_1–P_4, projected using E' = C as the center of projection. As the eye rotates about C to fixate on P1, E comes into alignment with E' (= C), Q_1, and P_1, and thus there is no registration error for P_1. While the eye is fixated on P_1 (and Q_1), there is a slight registration error for P_2, a larger error for P_3, but none for P_4 since it is in the projection plane. Similarly, if the eye rotates to fixate on P_3 for example, E, E', Q_3, and P_3 will all fall on the same line, and the registration error for P_3 will then be zero. Thus, when the eye is not looking at a point, it will have some amount of registration error, but when it turns to look at the point, its error goes to zero.

The next question is: How much viewing error is induced in the nonfixated points? The answer depends on the viewing parameters but appears to be fairly small in general. For $d = 500$ mm, the angular errors for a 60° FOV HMD are less than 1.5° for points from 200 mm from the eye on out to infinity (points closer than 200 mm can have much larger angular errors, but these points are closer than the *near point*, or the closest point of comfortable focus for an adult (Longhurst 1957)). Points lying along or near the gaze direction and points within or near the screen image will have zero or very small angular registration error. Although the angular errors calculated here correspond to large lateral errors for distant points, because the angles corresponding to the errors are small, it is unlikely that the human eye would detect such errors due to the falloff in acuity for nonfoveal vision. Thus, although it remains to be confirmed by user studies, there is reason to hope that eye tracking will not be necessary for systems that can accurately locate the center of rotation of the user's eyes. Another benefit of this result is that in some cases it is easier to find C than E, since calibration procedures often require the eye to swivel in order to align itself with two or more World-space vectors (as in Azuma and Bishop (1994) and one method used in Janin et al (1993)); such procedures identify the center of rotation rather than the eyepoint.[11]

As for eye calibration error, if the center of rotation of the eye is used as the center of projection, the lateral error is bounded by

$$s_{\text{view}} \approx c \cdot \frac{|z|}{d}, \tag{13}$$

where c is the magnitude of the calibration error. For $d = 500$ mm, 5 mm of calibration error will induce lateral errors of about 5 mm for points 500 mm from the screen image (and 0 mm at the screen image). Clearly, the error is linear in c, so halving the calibration error will halve the lateral error, etc.

Finally, if we examine the binocular case, we can characterize and quantify some of the distortions induced by viewing error. Two cases of particular interest are rigid-pair motion, where the eyes translate together relative to the screen image, and inter-pupillary distance (IPD) error, in which the modeled IPD is different from the actual IPD. As noted in Hodges and

---

[11]Note that if the user is wearing eyeglasses, the point to use is the image of C as seen through the glasses, which is exactly the point located by these calibration procedures.

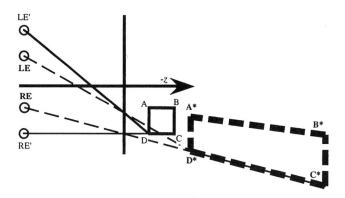

FIG. 6.17. IPD-error distortion.

Davis (1993) and Rolland et al. (1995), rigid-pair motion induces a shear distortion for horizontal/vertical motion, and a compression/elongation distortion for in/out movements. According to the error model, the deviation of each point is equal to $\frac{|z|}{d}$ e for a motion of size e.

IPD error can induce gross registration errors in depth,[12] as shown by the exaggerated case depicted in Figure 6.17. In this top view, LE′ and RE′ are the modeled eyepoints and LE and RE are the actual eyepoints. The heavy dashed line indicates the theoretically perceived version of the square ABCD. In general, the error vector between the theoretically perceived point P′ and the real point $P = (x, y, z)$ is

$$\mathbf{v}_{P'\_P} = \frac{1}{1 + \frac{z}{d}\left(\frac{i'}{i} - 1\right)} \cdot \begin{bmatrix} x \\ y \\ (z - d) \end{bmatrix}, \tag{14}$$

where i' is the modeled IPD and i is the actual IPD. Note that for suitable values of the parameters, the denominator can go to zero, leading to an infinite-length error vector, corresponding to the case where the projectors for a point become parallel for the perceived point.

In summary, viewing error is not likely to be a large registration error source, but its effects can be minimized by using the center of the eye's rotation as the center of projection, calibrating the system carefully for each user, and setting the screen image in the center of the working volume (since

---

[12]That is, the mathematical model shows gross errors in depth based on the intersection of the projectors from each eye; since the process of visual perception is based on multiple depth cues, it is somewhat unlikely that the depth error predicted by the model will be a reliable predictor of *perceived* depth error.

the registration error due to viewing error is zero there). Systems displaying objects at different depths may need to have an automatic adjustment for screen-image distance, which would also make the accommodation distance for the virtual and real objects the same.

4.6 Other Error Sources

This section will briefly discuss each of the remaining, small error sources. While these are small in comparison to the other error sources, it is worth noting that they would have to be dealt with in a system requiring submillimeter accuracy, and they might prove difficult to correct as well.

- *Virtual object alignment/scanning:* These are errors accrued in modeling or scanning the virtual objects and aligning them in the real environment. The amount of error is clearly a function of the accuracy of the scanning/modeling method and the accuracy of the alignment process. Since both of these are very application dependent, it is very hard to generalize about this error source. For 3D registration of medical data sets, the literature seems to indicate that mean-squared error values of less than 1 mm can be achieved, although larger errors (2–8 mm) are not uncommon depending on the alignment method and scan accuracy (Udupa and Herman (1991)). A general registration algorithm by Besl and McKay (1992) reports good results at matching 3D shapes, often to within 0.1% of the shape size, which for a head-sized data set would correspond to less than a millimeter. Holloway (1995) examines the behavior of this alignment algorithm in the presence of errors in picking the landmarks.
- *Virtual-image calibration errors:* In order to properly render the stereo projections of the virtual objects, the transformation between the Sensor CS and the Image CSs must be determined precisely. The approximate image-space error q for image alignment error is

$$q_{\text{im}} \approx \|\delta \mathbf{v}_{\text{S\_I}}\| + 2 \sin \frac{|\delta \varnothing_{\text{S\_I}}|}{2} \|\mathbf{v}_{\text{I\_P'}}\|, \qquad (15)$$

which is just rigid-body transformation error of the projected points. Various calibration methods and their results have been reported (Janin et al (1993) and Azuma and Bishop (1994)).

The critical observation regarding virtual-image misalignment is that the resulting error is fixed in image space and is therefore independent of viewing direction, etc. Moreover, certain misalignments are readily detectable (and therefore correctable). For example, if the screen image is shifted in x or y relative to its modeled location by 2 mm, *all* of the points in the scene will be shifted by that amount. Rotation about the z axis will displace points as a function of their distance from the rotation axis and can easily be detected with the use of a crosshair or grid. Errors in the remaining three degrees of freedom (z translation, x and y axis rotation) are less easily detected, *but that is precisely because they do not induce much registration error unless the error is severe.* That is, these errors for the most part move points along their projectors, which, because of the projection operation, has very little effect on q_{im}. If the errors are severe, they will be systematic and can therefore be distinguished from other error sources and corrected. In summary, errors in this transformation that induce noticeable registration error can be corrected, and those that do not induce noticeable registration error can be ignored. Based on these observations and the experience with the prototype system, I estimate that this error can be reduced to 1 mm or less.

- *Aliasing:* If antialiasing is not done (e.g., for performance reasons), the worst-case error is for all of the edge pixels for a primitive to be shifted by half a pixel in x and y. In this case, the primitive's center of mass will shift by $\sqrt{2}/2$ times the pixel spacing. Assuming the display does not resample the signal from the frame buffer, the net error is just this shift magnified by the optics, or

$$q_{al} = m \cdot \frac{\sqrt{2}}{2} p, \qquad (16)$$

where p is just the screen width divided by the horizontal resolution (for square pixels), and m is the linear magnification. For a 52 mm-wide LCD screen with 640 pixels/line and a magnification of 6.0, this amounts to 0.4 mm of error. Aliasing should not be a major error source for most systems.

- *Display device nonlinearity:* Certain displays (CRTs in particular) have nonlinearities that cause the final screen display to deviate from a regular rectangular grid. For CRTs, nonlinearities in the beam deflection process can distort the final image. This nonlinearity is quoted as a percentage of the screen size; values range from 0.1% to 3%. While values of 3% can induce enough error to be troublesome (on the order of 5 mm), values of 1% (corresponding to 1–2 mm of error) are more common. Moreover, any prewarping implemented for distortion correction could be modified to compensate for this problem as well. Finally, one can always use more elaborate (and expensive) drive electronics to further reduce this error source.

5. CONCLUSIONS AND FUTURE WORK

Most of the major error sources are associated with the tracker in some way. The tracker is a critical component for making augmented reality work, and any error or delay in its data causes serious flaws in the illusion. The errors associated with the tracker are due to delay in displaying the tracker data (due to delay in the entire system), error in the tracker measurement, error in locating the Tracker CS in the World CS, and errors in tracker readings used for system calibration.

Another result of this analysis is that eye tracking is probably unnecessary if the eye's center of rotation is used as the center of projection, since it gives the correct projection for fixated points and small errors for nonfixated points. The last of the major error sources is optical distortion, which can cause large errors in the image periphery; unfortunately, inverting the distortion in the rendering process may cause more delay error than the warping error it corrects. Finally, there are several other small error sources, each of which may add a small amount of registration error.

The analysis shows that submillimeter registration is not likely any time soon since there are many error sources on the order of a millimeter; but it does seem probable that we will achieve 5–10 millimeter dynamic accuracy (and 1–2 mm for static accuracy) with the use of predictive tracking, synchronized display methods, and careful calibration. The progress toward submillimeter registration error will probably be asymptotic, with increasing effort and expense required to gain each small increase in precision.

The error model discussed here was tailored to a particular application (surgery planning) and as such was not thoroughly explored for different systems and different applications. In particular, this model could easily be expanded to analyze video STHMD systems, CAVE systems,[13] and opaque HMDs. For video STHMDs, the model for viewing error would have to be changed, but much of the rest of the model would work as is. In CAVE systems, many of the problems of head tracking disappear (since the images

---

[13]CAVE systems use several rear-projection screens and head-tracked stereo glasses to immerse the user in a 10′ × 10′ virtual environment (Ghazisaedy et al. (1995)).

are fixed in the environment), but the analysis of viewing error would be rather useful, especially since multiple users often view a scene that is only correct for the user wearing the head-tracker. Finally, opaque HMDs do not have the strict requirements for registering real and virtual objects (since the real objects are not visible), but nevertheless they suffer from the apparent swimming of virtual objects due to delay and tracker error, as well as the visual distortions from viewing error and optical distortion.

For the most part, however, the future work suggested by this research is not in the area of extending the work presented here, but rather in addressing the problems that it describes. In particular, more work needs to be done in the following areas: tracker and system calibration methods, low-latency rendering, synchronized rendering with just-in-time incorporation of tracker measurements, predictive and hybrid tracking methods, and feedback methods such as those used in video STHMDs.

Acknowledgments   This work was my dissertation research while at the Department of Computer Science at the University of North Carolina at Chapel Hill and was supported by the following grants: ARPA DABT63-94-C-0048, the NSF/ARPA Science and Technology Center for Computer Graphics and Visualization (NSF prime contract 8920219), and ONR N00014-86-K-0680.

I thank Fred Brooks, Jannick Rolland, Vern Chi, Stephen Pizer, Henry Fuchs, and Jefferson Davis for their help over the course of this project. Thanks also to Bernard Adelstein, Dave Allen, Ron Azuma, Gary Bishop, Vincent Carrasco, Jerry Cloutier, Jack Goldfeather, David Harrison, Linda Houseman, John Hughes, Kurtis Keller, Anselmo Lastra, Mark Livingston, Dinesh Manocha, Mark Mine, Warren Robinett, Andrei State, Russ Taylor, Kathy Tesh, Greg Turk, Fay Ward, Mary Whitton, Steve Work, Terry Yoo, and to the various reviewers for their comments.

REFERENCES

Adelstein, B., E. Johnston, S. Ellis. 1992. A testbed for characterizing dynamic response of virtual environment spatial sensors. *Proceedings, UIST '92, Fifth Annual ACM Symposium on User Interface Software and Technology*. Monterey, CA, pp. 15–22.

Adelstein, B., E. Johnston, S. Ellis. 1995. Dynamic response of electromagnetic spatial displacement trackers. Presence 5:3 pp. 302–318, 1996 Summer.

Azuma, R., and G. Bishop. 1994. Improving static and dynamic registration in an optical see-through HMD. *Proceedings of SIGGRAPH '94*. Orlando, July 24–29, pp. 197–204.

Bajura, M., and U. Neumann. 1995. Dynamic registration correction in AR systems. *Proc IEEE Virtual Reality Annual International Symposium (VRAIS)*.

Bajura, M., H. Fuchs, and R. Ohbuchi. 1992. Merging virtual objects with the real world. *Computer Graphics*. 26: 203–10.

Besl, P., and N. McKay. 1992. A method for registration of 3-D shapes. *IEEE Trans. on Pat. Anal and Mach. Int.* 14:2.

Bryson, S. 1992. Measurement and calibration of static distortion of position data from 3D trackers. *Proc. SPIE Vol. 1669: Stereoscopic Displays and Applications III*.

Cohen, J., and M. Olano. 1994. Low latency rendering on Pixel-Planes 5. UNC technical report TR 94-028.

Deering, M. 1992. High resolution virtual reality. *Computer graphics*. 26(2):195.

Faro Technologies. 1993. Industrial Metrecom Manual. pp. 73–76. Lake Mary, FL.

Feiner, S., B. Macintyre, and D. Seligmann. 1993. Knowledge-based augmented reality. *Communications of the ACM*. 36(7): 53–62.

Fuchs, H., J. Poulton, J. Eyles, T. Greer, J. Goldfeather, D. Ellsworth, S. Molnar, G. Turk, B. Tebbs, and L. Israel. 1989. Pixel-Planes 5: A heterogeneous multiprocessor graphics system using processor-enhanced memories. *Computer Graphics: Proceedings of SIGGRAPH '89*. 23:3.

Furness, T. 1986. The super cockpit and its human factors challenges. *Proceedings of the Human Factors Society*. 30: 48–52.

Ghazisaedy, M., D. Adamczyk, D. Sandin, R. Kenyon, and T. DeFanti. 1995. Ultrasonic calibration of a magnetcic tracker in a virtual reality space. *Proceedings of Virtual Reality Annual International Symposium*. pp. 179–88.

Grinberg, V., G. Podnar, and M. Siegel. 1994. Geometry of binocular imaging. *Proc. SPIE Stereoscopic Displays and VR Systems*. February. Vol. 2177.

Hodges, L., and E. Davis. 1993. Geometric considerations for stereoscopic virtual environments. *Presence*. 2:1.

Holloway, R. 1995. Registration errors in augmented reality systems. PhD dissertation. University of North Carolina at Chapel Hill. Also UNC technical report TR95-016.

Janin, A., D. Mizell, and T. Caudell. 1993. Calibration of head-mounted displays for augmented reality applications. *Proc. IEEE Virtual Reality Annual International Symposium*.

Lastra, A. 1994. University of North Carolina. Personal communication.

Livingston, M., and A. State. 1995. Improved registration for augmented reality systems via magnetic tracker calibration. UNC technical report TR95-037.

Longhurst, R. 1957. *Geometrical and Physical Optics*. Longmans. New York.

Min, P., and H. Jense. 1994. Interactive stereoscopy optimization for head-mounted displays. *Proc. SPIE Stereoscopic Displays and VR Systems*. February.

Mine, M. 1993. Characterization of end-to-end delays in head-mounted display systems. UNC technical report TR93-001.

Mine, M., and G. Bishop. 1993. Just-in-time pixels. UNC technical report 93-005.

Oishi, T., and S. Tachi. 1996. Methods to calibrate projection transformation parameters for see-through head-mounted displays. *Presence*. 5:1.

Olano, M., J. Cohen, M. Mine, and G. Bishop. 1995. Combatting rendering latency. *Proceedings of 1995 Symposium on Interactive 3D Graphics*. Monterey, CA, April 9–12.

Polhemus. 1993. Fastrak specifications sheet and personal communication. Colchester, VT.

Robinett, W., and R. Holloway. 1995. The visual display transformation for virtual reality. *Presence*. 4:1.

Robinett, W., and J. Rolland. 1991. A computational model for the stereoscopic optics of a head-mounted display. *Presence*. 1:1.

Rolland, J. P., D. Ariely, and W. Gibson. 1995. Towards quantifying depth and size perception in 3D virtual environments. *Presence*. 4:1.

Rolland, J. P., and T. Hopkins. 1993. A method of computational correction for optical distortion in head-mounted displays. UNC Technical Report #TR93-045.

Southard, D. 1994. Viewing model for stereoscopic head-mounted displays. *Proc. SPIE Stereoscopic Displays and VR Systems*. February.

State, A., D. T. Chen, C. Tector, A. Brandt, H. Chen, R. Ohbuchi, M. Bajura, and H. Fuchs. 1994. Case study: Observing a volume-rendered fetus within a pregnant patient. *Proceedings of IEEE Visualization '94*, edited by R. D. Bergeron and A. E. Kaufman, IEEE Computer Society Press, Los Alamitos, CA. pp. 364–368.

Sutherland, I. 1968. A head-mounted three dimensional display. *Proc. Fall Joint Computer Conference*.

Udupa, J., and G. Herman, eds. 1991. *3D Imaging in Medicine*. CRC Press. Boca Raton, FL.

Wloka, M. Lag in multiprocessor VR. Presence 4:1. Winter 1995. pp. 50–63.

7

Mathematical Theory for Mediated Reality and WearCam-Based Augmented Reality

Steve Mann
University of Toronto

This chapter provides the mathematical framework for WearCam-based augmented reality and mediated reality (which, by definition, is WearCam-based).

Traditionally video has been either part of the environment, as with video surveillance cameras mounted on or inside a building or video conferencing systems based on fixed cameras within a special room, or it has been the domain of large organizations such as broadcast television stations. Recently, however, a new field of research called "Personal Imaging," has emerged. Personal Imaging systems are typically characterized by video from a truly first-person perspective. This first-person perspective arises from a head-mounted camera and display together with an image processing computer worn on the body of the user. The possibilities afforded by Personal Imaging include a personal safety device for crime reduction, a new kind of videoconferencing system for computer supported collaboration, as well as a new tool for photojournalism. This chapter briefly describes the mathematical framework for Personal Imaging, as used, for example, in computer supported collaborative wireless video.

1. WHY WEARCAM?

Implicit within mediated reality is the need for a camera [Mann, 1994]. However, there are actually two classes of apparatus that will provide video from a first-person perspective and are of interest within the context of this chapter:

1. Mediated Reality (MR)
2. Camera-based augmented reality

In the case of MR, the camera is an essential component of the reality mediator and therefore may be used, for example, for head-tracking, in addition to being or as part of the means by which light is absorbed and quantified for purposes of altering the visual perception of reality. In the case of augmented reality, a camera may be added to the system if it does not already have one, so that head-tracking can be done using the camera added thereupon.

1.1 Why Camera-Based Head-Tracking?

A goal of Personal Imaging is to facilitate the use of the WearCam [Mann, 1997] apparatus in ordinary everyday situations, not just on a factory assembly line "workcell," or other restricted space. Thus it is desired that the apparatus have a head-tracker that need not rely on any special apparatus in the environment.

Accordingly, most embodiments of the WearCam invention incorporate, at the very least, camera-based head-tracking (sometimes in addition to tracking by inertial system, compass, GPS, etc.).

Moreover, the camera goes beyond functioning just as a head-tracker, and the setup suggests a symbiosis in which the wearer becomes both a producer and consumer of information.

2. INFINITE SCREEN RESOLUTION WITH CAMERA-BASED HEAD-TRACKING

As previously stated, the reality mediator, by definition [Mann, 1994], must have a camera in order for the wearer to see. Accordingly, other uses of this camera are proposed in this section. These other uses are (1) as a head-tracker and (2) to make the wearer into a producer as well as a consumer of virtual information.

The proposed methodology (MR) is quite different from other related work where the assumption is often that there is a controlled environment such as the assembly line of a factory, where VR headsets might be used to make employees more productive, and where head-tracking and the like may therefore be done quite easily with a special device fixed in the environment. By tethered cables, workers are, in effect, imprisoned in their workcells, unable to roam freely without taking off their VR headsets.

Quite the opposite is true with Personal Imaging, where there is no assumption regarding a fixed location. Indeed, the goal of Personal Imaging is that the apparatus function in nearly any environment, with no special preparation of the environment being required.

Accordingly, a new method of head-tracking based on the use of a video camera has been proposed [Mann, 1997] and is based on the VideoOrbits algorithm [Mann and Picard, 1995]. The VideoOrbits algorithm performs head-tracking, visually, based on a natural environment, and it works without the need for object recognition. Instead it is based on algebraic projective geometry and a featureless means of estimating the change in spatial coordinates arising from movement of the wearer's head, as illustrated in Figure 7.1.

3. HISTORICAL CONTEXT FOR PERSONAL IMAGING: FROM "PAINTING WITH LIGHT" TO "PAINTING WITH LOOKS"

In the early days of Personal Imaging, a specific location was selected from which a measurement space or the like was constructed. From this single vantage point, a collection of differently illuminated/exposed images was constructed using the wearable computer and associated illumination apparatus. However, this approach was often facilitated by transmitting images from this single specific location (base station) back to the wearable computer, and vice versa. Thus, when I developed the eyeglass-based computer display/camera system, it was natural to exchange viewpoints with another person (namely the operator of the base station). This mode of operation ("seeing eye-to-eye") made the notion of perspective very apparent, and thus projective geometry is at the heart of personal imaging.

Personal Imaging situates the camera such that it provides a unique first-person perspective, that is, in the case of the eyeglass-mounted camera, the machine captures the world from the same perspective as its host (human).

In this chapter, I emphasize the importance of projective geometry, and I present some new results that are germane to the principles of Personal

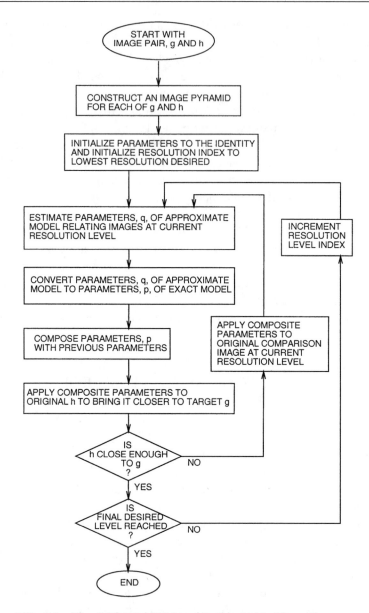

FIG. 7.1. The "VideoOrbits" head-tracking algorithm: The new head-tracking algorithm requires no special devices installed in the environment. The camera in the Personal Imaging system simply tracks itself based on its view of objects in the environment. The algorithm is based on algebraic projective geometry and provides an estimate of the true projective coordinate transformation, which, for successive image pairs is composed using the

Imaging and are used in such applications as "painting with looks" (building environment maps by looking around), wearable tetherless computer-mediated reality, and the new genre of personal documentary that arises from this new perspective.

4. WHY PROJECTIVE GEOMETRY?

I present direct featureless methods for estimating the 8 parameters of an "exact" projective (homographic) coordinate transformation to register pairs of images, together with the application of seamlessly combining a plurality of images of the same scene, resulting in a single image (or new image sequence) of greater resolution or spatial extent. The approach is "exact" for two cases of static scenes: (1) images taken from the same location of an arbitrary 3-D scene, with a camera that is free to pan, tilt, rotate about its optical axis, and zoom, or (2) images of a flat scene taken from arbitrary locations. The featureless projective approach generalizes interframe camera motion estimation methods, which have previously used an *affine* model (which lacks the degrees of freedom to "exactly" characterize such phenomena as camera pan and tilt) and/or which have relied upon finding points of correspondence between the image frames. The featureless projective approach, which operates directly on the image pixels, is shown to be superior in accuracy and ability to enhance resolution. The proposed methods work well on image data collected from both good-quality and poor-quality video under a wide variety of conditions (sunny, cloudy, day, night). These new fully automatic methods are also shown to be robust to deviations from the assumptions of static scene and no parallax.

Many problems require finding the coordinate transformation between two images of the same scene or object. Whether to recover camera motion between video frames, to stabilize video images, to relate or recognize photographs taken from two different cameras, to compute depth within a 3-D

---

projective group [Mann and Picard, 1995]. Successive pairs of images may be estimated in the neighbourhood of the identity coordinate transformation of the group, while absolute head-tracking is done using the exact group by relating the approximate parameters q to the exact parameters p in the innermost loop of the process. The algorithm typically runs at 5–10 frames per second on a general-purpose computer but the simple structure of the algorithm makes it easy to implement in hardware for the higher frame rates needed for full-motion video.

scene, or for image registration and resolution enhancement, it is important to have both a precise description of the coordinate transformation between a pair of images or video frames, and some indication as to its accuracy.

Traditional *block matching* (e.g., as used in *motion estimation*) is really a special case of a more general *coordinate transformation*. In this chapter I demonstrate a new solution to the motion estimation problem using a more general estimation of a coordinate transformation and propose techniques for automatically finding the 8-parameter projective coordinate transformation that relates two frames taken of the same static scene. I show, both by theory and example, how the new approach is more accurate and robust than previous approaches, which relied on affine coordinate transformations, approximations to projective coordinate transformations, and/or the finding of point correspondences between the images. The new techniques take as input two frames and automatically output the 8 parameters of the "exact" model, to properly register the frames. They do not require the tracking or correspondence of explicit features, yet are computationally easy to implement.

Although the theory I present makes the typical assumptions of static scene and no parallax, I show that the new estimation techniques are robust to deviations from these assumptions. In particular, I apply the direct featureless projective parameter estimation approach to image resolution enhancement and compositing, illustrating its success on a variety of practical and difficult cases, including some that violate the nonparallax and static scene assumptions.

5. BACKGROUND ON MOTION ESTIMATION

Hundreds of papers have been published on the problems of motion estimation and frame alignment. (For review and comparison, see Barron and Fleet [1994].) In this section I review the basic differences between coordinate transformations and emphasize the importance of using the "exact" 8-parameter projective coordinate transformation.

5.1 Coordinate Transformations

A coordinate transformation maps the image coordinates, $\mathbf{x} = [x, y]^T$, to a new set of coordinates, $\mathbf{x}' = [x', y']^T$. The approach to "finding the coordinate transformation" relies on assuming it will take one of the forms in

TABLE 7.1

Image Coordinate Transformations Discussed in This Chapter

Model	Coordinate Transformation from $\mathbf{x}$ to $\mathbf{x}'$	Parameters
Translation	$\mathbf{x}' = \mathbf{x} + \mathbf{b}$	$\mathbf{b} \in \mathbb{R}^2$
Affine	$\mathbf{x}' = \mathbf{A}\mathbf{x} + \mathbf{b}$	$\mathbf{A} \in \mathbb{R}^{2\times2}, \mathbf{b} \in \mathbb{R}^2$
Bilinear	$x' = q_{x'xy}xy + q_{x'x}x + q_{x'y}y + q_{x'}$ $y' = q_{y'xy}xy + q_{y'x}x + q_{y'y}y + q_{y'}$	$bf q_* \in \mathbb{R}$
Projective	$\mathbf{x}' = \dfrac{\mathbf{A}\mathbf{x}+\mathbf{b}}{\mathbf{c}^T\mathbf{x}+1}$	$\mathbf{A} \in \mathbb{R}^{2\times2}, \mathbf{b}, \mathbf{c} \in \mathbb{R}^2$
Relative-projective	$\mathbf{x}' = \dfrac{\mathbf{A}\mathbf{x}+\mathbf{b}}{\mathbf{c}^T\mathbf{x}+1} + \mathbf{x}$	$\mathbf{A} \in \mathbb{R}^{2\times2}, \mathbf{b}, \mathbf{c} \in \mathbb{R}^2$
Pseudo-perspective	$x' = q_{x'x}x + q_{x'y}y + q_{x'} + q_\alpha x^2 + q_\beta xy$ $y' = q_{y'x}x + q_{y'y}y + q_{y'} + q_\alpha xy + q_\beta y^2$	$\mathbf{q}_* \in \mathbb{R}$
Biquadratic	$x' = q_{x'x^2}x^2 + q_{x'xy}xy + q_{x'y^2}y^2 + q_{x'x}x + q_{x'y}y + q_{x'}$ $y' = q_{y'x^2}x^2 + q_{y'xy}xy + q_{y'y^2}y^2 + q_{y'x}x + q_{y'y}y + q_{y'}$	$bf q_* \in \mathbb{R}$

Table 7.1 and then estimating the parameters (2 to 12 parameters depending on the model) in the chosen form. An illustration showing the effects possible with each of these forms is shown in Fig. 7.2.

The most common assumption (especially in motion estimation for coding and optical flow for computer vision) is that the coordinate transformation between frames is translation. Tekalp, Ozkan, and Sezan [Tekalp et al., 1992] have applied this assumption to high-resolution image reconstruction. Although translation is the least constraining and simplest to implement of the seven coordinate transformations in Table 7.1, it is poor at handling large changes due to camera zoom, rotation, pan, and tilt.

Zheng and Chellappa [Zheng and Chellappa, 1993] considered the image registration problem using a subset of the affine model—translation, rotation and scale. Other researchers [Irani and Peleg, 1991; Teodosio and Bender, 1993] have assumed affine motion (6 parameters) between frames. For the assumptions of static scene and no parallax, the affine model exactly describes rotation about the optical axis of the camera, zoom of the camera, and pure shear, which the camera does not do, except in the limit as the lens focal length approaches infinity. The affine model cannot capture camera pan and tilt, and therefore it cannot properly express the "keystoning" and "chirping" we see in the real world. (By "chirping" I mean the effect of increasing or decreasing spatial frequency with respect to spatial location,

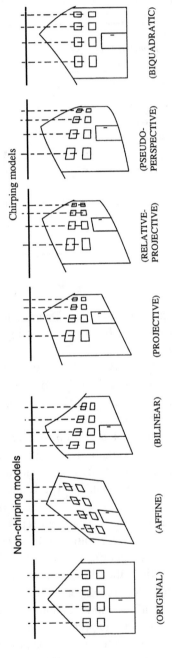

FIG. 7.2. Pictorial effects of the six coordinate transformations of Table 7.1, arranged left to right by number of parameters. Note that translation leaves the ORIGINAL house figure unchanged, except in its location. Most importantly, only the four rightmost coordinate transformations affect the periodicity of the window spacing (inducing the desired "chirping," which corresponds to what we see in the real world). Of these four, only the PROJECTIVE coordinate transformation preserves straight lines. The 8-parameter PROJECTIVE coordinate transformation "exactly" describes the possible image motions ("exact" meaning under the idealized zero-parallax conditions).

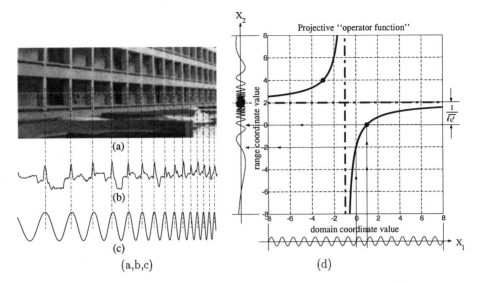

FIG. 7.3. The "projective chirping" phenomenon. (a) A real-world object that exhibits periodicity generates a projection (image) with "chirping"—"periodicity-in-perspective." (b) Center raster of image. (c) Best-fit projective chirp of form $\sin(2\pi((ax + b)/(cx + 1)))$. (d) Graphical depiction of exemplar 1-D projective coordinate transformation of $\sin(2\pi x_1)$ into a "projective chirp" function, $\sin(2\pi x_2) = \sin(2\pi((2x_1 - 2)/(x_1 + 1)))$. The range coordinate as a function of the domain coordinate forms a rectangular hyperbola with asymptotes shifted to center at the *vanishing point* $x_1 = -1/c = -1$ and "exploding point," $x_2 = a/c = 2$, and with "chirpiness" $c' = c^2/(bc-a) = -1/4$.

as illustrated in Fig. 7.3.) Consequently, the affine model attempts to fit the wrong parameters to these effects. Even though it has fewer parameters, I find that the affine model is more susceptible to noise because it lacks the correct degrees of freedom needed to properly track the actual image motion.

The 8-parameter *projective* model gives the desired 8 parameters that exactly account for all possible zero-parallax camera motions; hence, there is an important need for a featureless estimator of these parameters. To the best of my knowledge, the only algorithms proposed to date for such an estimator are those by Mann [1993], and shortly after, Szeliski and Coughlan [1994]. In both of these, a computationally expensive nonlinear optimization method was presented. In the earlier work [Mann, 1993], a direct method was also proposed. This direct method uses simple linear algebra and is noniterative insofar as methods such as Levenberg-Marquardt and the like are in no way required. The proposed method instead uses repetition with the correct law of composition on the projective group, going from one pyramid level to the next by application of the group's law of composition. Because the parameters of the projective coordinate transformation had traditionally been thought to be mathematically and computationally too difficult to solve, most researchers have used the simpler affine model or other approximations to the projective model. Before I propose and demonstrate the featureless estimation of the parameters of the "exact" projective model, it is helpful to discuss some approximate models.

Going from first order (affine), to second order, gives the 12-parameter *biquadratic* model. This model properly captures both the chirping (change in spatial frequency with position) and converging lines (keystoning) effects associated with projective coordinate transformations, but it does not constrain chirping and converging to work together (the example in Fig. 7.2, being chosen with zero convergence yet substantial chirping, illustrates this point). Despite its larger number of parameters, there is still considerable discrepancy between a projective coordinate transformation and the best-fit biquadratic coordinate transformation. Why stop at 2nd order? Why not use a 20-parameter *bicubic* model? While an increase in the number of model parameters will result in a better fit, there is a trade-off, where the model begins to fit noise. The physical camera model fits exactly in the 8-parameter projective group; therefore, we know that "eight is enough." Hence, it seems reasonable to have a preference for approximate models with exactly eight parameters.

The 8-parameter bilinear model is perhaps the most widely used [Wolberg, 1990] in the fields of image processing, medical imaging, remote sensing, and computer graphics. This model is easily obtained from the biquadratic model by removing the four x^2 and y^2 terms. Although the resulting bilinear model captures the effect of converging lines, it completely fails to capture the effect of chirping.

The 8-parameter *pseudo-perspective* model [Adiv, 1985] and an 8-parameter *relative-projective* model both do, in fact, capture both the converging lines and the chirping of a projective coordinate transformation. The pseudo-perspective model, for example, may be thought of as first, removal of two of the quadratic terms ($bf q_{x'y^2} = q_{y'x^2} = 0$), which results in a 10-parameter model (the "q-chirp" of Navab and Mann [1994]), and then constraining the four remaining quadratic parameters to have two degrees of freedom. These constraints force the "chirping effect" (captured by $\mathbf{q}_{x'x^2}$ and $\mathbf{q}_{y'y^2}$) and the "converging effect" (captured by $\mathbf{q}_{x'xy}$ and $\mathbf{q}_{y'xy}$) to work together in the "right" way to match, as closely as possible, the effect of a projective coordinate transformation. By setting $\mathbf{q}_\alpha = \mathbf{q}_{x'x^2} = \mathbf{q}_{y'xy}$, the chirping in the x direction is forced to correspond with the converging of parallel lines in the x direction (and likewise for the y direction).

Of course, the desired "exact" 8 parameters come from the projective model, but they have been perceived as being notoriously difficult to estimate. The parameters for this model have been solved by Tsai and Huang [Tsai and Huang, 1981], but their solution assumed that features had been identified in the two frames, along with their correspondences. The main contribution of this chapter is a simple featureless means of automatically solving for these 8 parameters.

Other researchers have looked at projective estimation in the context of obtaining 3-D models. Faugeras and Lustman [Faugeras and Lustman, 1988], Shashua and Navab [Shashua and Navab, 1994], and Sawhney [Sawhney, 1994] have considered the problem of estimating the projective parameters while computing the motion of a rigid planar patch, as part of a larger problem of finding 3-D motion and structure using parallax relative to an arbitrary plane in the scene. Kumar et al. [Kumar et al., 1994] have also suggested registering frames of video by computing the flow along the *epipolar* lines, for which there is also an initial step of calculating the gross camera movement assuming no parallax. However, these methods have relied on feature correspondences and were aimed at 3-D scene modeling. My focus is not on recovering the 3-D scene model but on aligning 2-D images of 3-D scenes. Feature correspondences greatly simplify the problem; however, they also have many problems. The focus of this chapter is simple featureless approaches to estimating the projective coordinate transformation between image pairs.

5.2 Camera Motion: Common Assumptions and Terminology

Two assumptions are typical in this area of research. The first assumption is that the scene is constant—changes of scene content and lighting are small between frames. The second assumption is that of an ideal pinhole camera—implying unlimited depth of field with everything in focus (infinite resolution) and implying that straight lines map to straight lines.[1] Consequently, the camera has three degrees of freedom in 2-D space and eight

---

[1] When using low cost wide-angle lenses, there is usually some barrel distortion, which we correct using the method of Campbell and Bobick [1995].

TABLE 7.2

The two "no parallax" cases for a static scene. Note that the first situation has 7
degrees of freedom (yaw, pitch, roll, translation in each of the 3 spatial axes, and
zoom), while the second has 4 degrees of freedom (pan, tilt, rotate, and zoom).
Both, however, are represented within the 8 scalar paramters of the projective
group of coordinate transformations

	Scene Assumptions	Camera Assumptions
Case 1:	arbitrary 3-D	free to zoom, rot., pan, and tilt, fixed COP
Case 2:	planar	free to zoom, rot., pan, and tilt, free to trans.

degrees of freedom in 3-D space: translation (X, Y, Z), zoom (scale in each
of the image coordinates x and y), and rotation (rotation about the optical
axis, pan, and tilt). These two assumptions are also made in this chapter.

In this chapter, an "uncalibrated camera" refers to one in which the
principal point[2] is not necessarily at the center (origin) of the image and
the scale is not necessarily isotropic.[3] I assume that the zoom is continually
adjustable by the camera user and that we do not know the zoom setting,
or whether it changed between recording frames of the image sequence. I
also assume that each element in the camera sensor array returns a quantity
that is linearly proportional to the quantity of light received.[4] With these
assumptions, the exact camera motion that can be recovered is summarized
in Table 7.2.

5.3   Video Orbits

Tsai and Huang [Tsai and Huang, 1981] pointed out that the elements of
the projective *group* give the true camera motions with respect to a planar surface. They explored the group structure associated with images of
a 3-D rigid planar patch, as well as the associated *Lie algebra*, although
they assume that the correspondence problem has been solved. The solution presented in this chapter (which does not require prior solution of

---

[2]The principal point is where the optical axis intersects the film.

[3]Isotropic means that magnification in the x and y directions is the same. Our assumption facilitates
aligning frames taken from different cameras.

[4]This condition can be enforced over a wide range of light intensity levels, by using the Wyckoff
principle [Wyckoff, 1962; Mann and Picard, 1994a].

correspondence) also relies on projective group theory. I briefly review the basics of this theory, before presenting the new solution in the next section.

5.3.1 Projective Group in 1-D Coordinates

A group is a set upon which there is defined an associative law of composition (*closure, associativity*), which contains at least one element (*identity*) who's composition with another element leaves it unchanged, and for which every element of the set has an *inverse*.

A *group* of operators together with a *set* of operands form a *group operation*.[5]

In this chapter, coordinate transformations are the operators (group), and images are the operands (set). When the coordinate transformations form a group, then two such coordinate transformations, p_1 and p_2, acting in succession, on an image (e.g., p_1 acting on the image by doing a coordinate transformation, followed by a further coordinate transformation corresponding to p_2, acting on that result) can be replaced by a single coordinate transformation. That single coordinate transformation is given by the *law of composition* in the group.

The *orbit* of a particular element of the set, under the group operation [Artin, 1991] is the new set formed by applying to it all possible operators from the group.

In this chapter, the orbit is a collection of pictures formed from one picture through applying all possible projective coordinate transformations to that picture. I refer to this set as the "video orbit" of the picture in question. Image sequences generated by zero-parallax camera motion on a static scene contain images that all lie in the same video orbit.

For simplicity, I review the theory first for the projective coordinate transformation in one dimension.[6]

Suppose we take two pictures, using the same exposure, of the same scene from fixed common location (e.g., where the camera is free to pan, tilt, and zoom between taking the two pictures). Both of the two pictures capture the same pencil of light,[7] but each one projects this information differently onto the film or image sensor. Neglecting that which falls beyond

---

[5] Also known as a *group action* or *G-set* [Artin, 1991].

[6] In this 2-D world, the "camera" consists of a center of projection (pinhole "lens") and a line (1-D sensor array or 1-D "film").

[7] We neglect the boundaries (edges or ends of the sensor) and assume that both pictures have sufficient field of view to capture all of the objects of interest.

the borders of the pictures, each picture captures the same information about the scene but records it in a different way. The same object might, for example, appear larger in one image than in the other, or it might appear more squashed at the left and stretched at the right than in the other. Thus we would expect to be able to construct one image from the other, so that only one picture should need to be taken (assuming its field of view covers all the objects of interest) in order to synthesize all the others. We first explore this idea in a make-believe "Flatland" where objects exist on the 2-D page, rather than the 3-D world in which we live, and where pictures are real-valued functions of one real variable, rather than the more familiar real-valued functions of two real variables.

For the two pictures of the same pencil of light in Flatland, I define the common COP at the origin of our coordinate system in the plane. In Fig. 7.4 I have depicted a single camera that takes two pictures in succession as

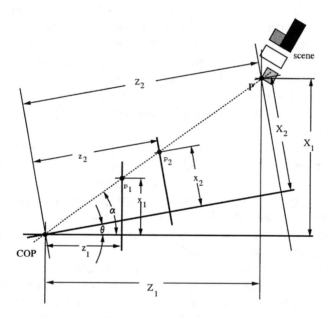

FIG. 7.4.   Camera at a fixed location. An arbitrary scene is photographed twice, each time with a different camera orientation, and a different principal distance (zoom setting). In both cases the camera is located at the same place (COP) and thus captures the same pencil of light. The dotted line denotes a ray of light traveling from an arbitrary point, P, in the scene, to the COP. Heavy lines denote both camera optical axes in each of the two orientations as well as the image sensor in each of its two pan and zoom positions. The two image sensors (or films) are in front of the camera to simplify mathematical derivations.

two cameras shown together in the same figure, one having been rotated through on angle of θ with respect to the other. Let Z_k, $k \in \{1, 2\}$ represent the distances, along each optical axis, to an arbitrary point in the scene, P, and let X_k represent the distances from P to each of the optical axes. The principal distances are denoted z_k. In the example of Fig. 7.4, we are *zooming in* (increased magnification) as we go from frame 1 to frame 2.

Let $\alpha = \arctan(x_1/z_1)$ then the geometry of Fig. 7.4 defines a mapping from x_1 to x_2, given by [Mann, 1992, 1994]:

$$x_2 = z_2 \tan(\arctan(x_1/z_1) - \theta), \quad \forall x_1 \neq o_1$$
$$= (ax_1 + b)/(cx_1 + 1), \quad \forall x_1 \neq o_1, \tag{1}$$

where $a = z_2/z_1$, $b = -z_2 \tan(\theta)$, and $c = \tan(\theta)/z_1$, and where $o_1 = z_1 \tan(\pi/2 + \theta) = -1/c$ is the location of the singularity in the domain ("appearing point" [Mann, 1994]). I should emphasize here that if we set $c = 0$ we arrive at the affine group. (Recall, also, that c, the degree of perspective, has been given the interpretation of a chirp rate [Mann, 1992].)

Let $\mathbf{p} \in \mathbf{P}$ denote a particular mapping from x_1 to x_2, governed by the three parameters $\mathbf{p}' = [z_1, z_2, \theta]$, or equivalently by a, b, and c from (1).

Proposition 1   *The set of all possible operators, $\mathbf{P}_1$, given by the coordinate transformations (1), $\forall a \neq bc$, acting on a set of 1-D images, neglecting the situation when ($\theta = \pi/2$ cameras are at right angles) forms a group operation.*

Proof: A pair of images related by a particular camera rotation and change in principal distance (depicted in Fig. 7.4) correspond to an operator that takes any function g on image line 1, to a function h on image line 2:

$$h(x_2) = g(x_1) = g((-x_2 + b)/(cx_2 - a)), \quad \forall x_2 \neq o_2$$
$$= g \circ x_1 = g \circ \mathbf{p}^{-1} \circ x_2, \tag{2}$$

where $\mathbf{p} \circ x = (ax + b)/(cx + d)$ and $o_2 = a/c$, $d = 1$ neglecting the case $d = 0$. As long as $a \neq bc$, each operator, $\mathbf{p}$, has an inverse, namely that given by composing the inverse coordinate transformation:

$$x_1 = (b - x_2)/(cx_2 - a), \quad \forall x_2 \neq o_2 \tag{3}$$

with the function $h(\ )$ to obtain $g = h \circ \mathbf{p}$. The identity operation is given by $g = g \circ e$, where e is given by $a = 1$, $b = 0$, and $c = 0$.

In complex analysis (see, for example, Ahlfors [Ahlfors, 1979]) the form $(az+b)/(cz+d)$ is known as a linear fractional transformation. Although our mapping is from $\mathbb{R}$ to $\mathbb{R}$ (as opposed to theirs from $\mathbb{C}$ to $\mathbb{C}$), I can still borrow the concepts of complex analysis. In particular, a simple group representation is provided using the 2×2 matrices, $\mathbf{p} = [a, b; c, 1] \in \mathbb{R}^2 \times \mathbb{R}^2$. Closure[8] and associativity are obtained by using the usual laws of matrix multiplication followed with dividing the resulting vector's first element by its second element.   □

Proposition 1 says that an element of the $(ax + b)/(cx + 1)$ group (neglecting $d = 0$) can be used to align any two frames of the (1-D) image sequence provided that the COP remains fixed.

Proposition 2   *The set of operators that take nonsingular projections of a straight object to one another form a group,* $\mathbf{P}_2$.

A "straight" object is one which lies on a straight line in Flatland.[9]

Proof: Consider a geometric argument. The mapping from the first (1-D) frame of an image sequence, $g(x_1)$, to the next frame, $h(x_2)$, is parameterized by the following: camera translation perpendicular to the object, t_z; camera translation parallel to the object, t_x; pan of frame 1, θ_1; pan of frame 2, θ_2; zoom of frame 1, z_1; and zoom of frame 2, z_2. (See Fig. 7.5.) We want to obtain the mapping from x_1 to x_2. Let's begin with the mapping from X_2 to x_2:

$$x_2 = z_2 \tan(\arctan(X_2/Z_2) - \theta_2) = \frac{a_2 X_2 + b_2}{c_2 X_2 + 1},$$

$$\text{neglecting } \theta = \frac{\pi}{2} \tag{4}$$

which can be represented by the matrix $\mathbf{p}_2 = [a_2, b_2; c_2, 1]$, so that $x_2 = \mathbf{p}_2 \circ X_2$. Now $X_2 = X_1 - t_x$ and it is clear that this coordinate transformation is inside the group, for there exists the choice of $a = 1$, $b = -t_x$, and $c = 0$ that describe it: $X_2 = \mathbf{p}_t \circ X_1$, where $\mathbf{p}_t =$

---

[8] Also know as *law of composition* [Artin, 1991].

[9] An important difference to keep in mind, with respect to pictures of a flat object, is that in Flatland, if you take a picture of a picture, that is equivalent to a single picture for an equivalent camera orientation and position. However, with 2-D pictures in a 3-D world, a picture of a picture is, in general, not necessarily a simple perspective projection (however, if you continue taking pictures you do not get anything new beyond the second picture). However, the 2-D version of the group representation contains both cases.

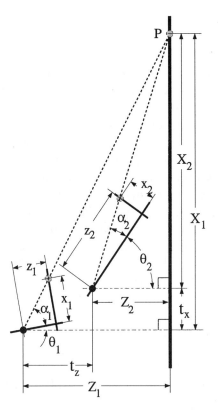

FIG. 7.5. Two pictures of a flat (straight) object. The point P is imaged twice, each time with a different camera orientation, a different principal distance (zoom setting), and different camera location (resolved into components parallel and perpendicular to the object).

$[1, -t_x; 0, 1]$. Finally, $x_1 = z_1 \tan(\arctan(X_1/Z_1) - \theta) = \mathbf{p}_1 \circ X_1$. Let $\mathbf{p}_1 = [a_1, b_1; c_1, 1]$. Then $\mathbf{p} = \mathbf{p}_2 \circ \mathbf{p}_t \circ \mathbf{p}_1^{-1}$ is in the group by the law of composition. Hence, the operators that take one frame into another, $x_2 = \mathbf{p} \circ x_1$, form a group. $\qquad\square$

Proposition 2 says that an element of the $(ax + b)/(cx + 1)$ group can be used to align any two images of linear objects in Flatland, regardless of camera movement.

Proposition 3 *The two groups $\mathbf{P}_1$ and $\mathbf{P}_2$ are isomorphic; a group representation for both is given by the 2×2 square matrix $[a, b; c, 1]$, neglecting $\theta = \pi/2$.*

Isomorphism follows because $\mathbf{P}_1$ and $\mathbf{P}_2$ have the same group representation.[10] The $(ax + b)/(cx + 1)$ operators in the above propositions form
the *projective group* $\mathbf{P}$ in Flatland.

The affine operator that takes a function space G to a function space
H may itself be viewed as a function. Let us now construct a similar plot
for a member of the group of operators, $\mathbf{p} \in \mathbf{P}$, in particular, the operator
$\mathbf{p} = [2, -2; 1, 1]$, which corresponds to $\mathbf{p}' = \{1, 2, 45°\} \in \mathbf{P}_1$. We have
also depicted the result of mapping $g(x_1) = \sin(2\pi x_1)$ to $h(x_2)$. When G
is the space of Fourier analysis functions (harmonic oscillations), then H
is a family of functions known as P-chirps [Mann, 1992], adapted to a
particular *vanishing point*, o_2 and normalized chirp rate, $c' = c^2/(bc - a)$
[Mann, 1994]. Figure 7.6(b) is a *rectangular hyperbola* (e.g., $x_2 = \frac{1}{c'x_1}$)
with an origin that has been shifted from $(0, 0)$ to (o_1, o_2).

A member of this group of coordinate transformations, $x' = (ax + b)/$
$(cx + d)$, $\forall ad \neq bc$ (where the images are functions of one variable,
x) is denoted by $p_{a,b,c,d}$ and has inverse $p_{-d,b,c,-a}$. The law of composition is given by $p_{e,f,g,h} \circ p_{a,b,c,d} = p_{ae+cf,be+df,ag+cd,bg+d^2}$. In almost all
practical engineering applications, $d \neq 0$, so I will divide through by
d and denote the coordinate transformation $x' = (ax + b)/(cx + 1)$ by
$x' = p_{a,b,c} \circ x$. This is what I mean by neglecting $\theta = \pi/2$. When $a \neq 0$
and $c = 0$, the projective group becomes the affine group of coordinate
transformations, and when $a = 1$ and $c = 0$, it becomes the group of
translations.

Of the coordinate transformations presented in the previous section, only
the projective, affine, and translation operations form groups.

The equivalent two cases of Table 7.2 for this hypothetical Flatland world
of 2-D objects with 1-D pictures correspond to the following. In the first
case a camera is at a fixed location and is free to zoom and pan. In the second
case, a camera is free to translate, zoom, and pan, but the imaged object
must be flat (i.e., lie on a straight line in the plane). The resulting two (1-D)
frames taken by the camera are related by the coordinate transformation
from x_1 to x_2, given by (1)

$$x_2 = z_2 \tan(\arctan(x_1/z_1) - \theta), \quad \forall x_1 \neq o_1$$

$$= (ax_1 + b)/(cx_1 + 1), \quad \forall x_1 \neq o_1, \tag{5}$$

---

[10]For 2-D images in a 3-D world, the isomorphism no longer holds. However, the group still *contains*
and therefore represents both cases.

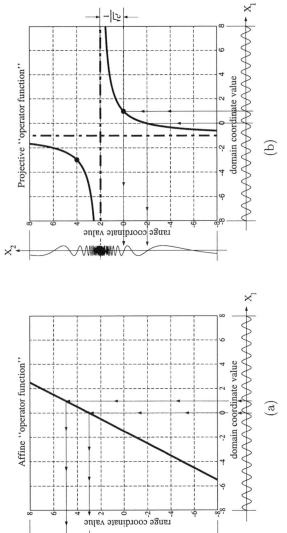

FIG. 7.6. Comparison of 1-D affine and projective coordinate transformations, in terms of their operator functions, acting on a sinusoidal image. Note that whether the function is enlarged, or "chirped," the same information remains present but is simply recorded in a different way by the new function. (a) Orthographic projection is equivalent to affine coordinate transformation, $y = ax + b$. In this example, $a = 2$ and $b = 3$. (b) Perspective projection for a particular fixed value of $p' = [1, 2, 45°]$. Note that the plot is a rectangular hyperbola like $x_2 = 1/(c'x_1)$ but with asymptotes at the shifted origin $(-1, 2)$. Here $g(x_1) = \sin(2\pi x_1)$. The arrows indicate how a chosen cycle of this sine wave is mapped to the corresponding cycle of the P-chirp, $h(x_2)$.

5.3.2 Projective Group in 2-D Coordinates

The theory for the projective, affine, and translation groups also holds for the familiar 2-D images taken of the 3-D world. The "video orbit" of a given 2-D frame is defined to be the set of all images that can be produced by applying operators from the 2-D projective group to the given image. Hence, I restate the coordinate transformation problem: Given a set of images that lie in the same orbit of the group, I wish to find, for each image pair, that operator in the group which takes one image to the other image.

If two frames, say, f_1 and f_2, are in the same orbit, then there is an group operation $\mathbf{p}$ such that the mean-squared error (MSE) between f_1 and $f_2' = \mathbf{p} \circ f_2$ is zero. In practice, however, I find which element of the group takes one image "nearest" the other, for there will be a certain amount of parallax, noise, interpolation error, edge effects, changes in lighting, depth of focus, etc. Figure 7.7 illustrates the operator $\mathbf{p}$ acting on frame f_2, to move it nearest to frame f_1. (This figure does not, however, reveal the precise shape of the orbit, which occupies an 8-D space.)

Summarizing, the 8-parameter projective group captures the exact coordinate transformation between pictures taken under the two cases of Table 7.2. The primary assumptions in these cases are that of no parallax and of a static scene. Because the 8-parameter projective model is

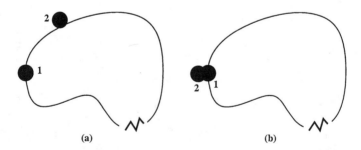

(a)   (b)

FIG. 7.7. Video orbits. (a) The orbit of frame 1 is the set of all images that can be produced by acting on frame 1 with any element of the operator group. Assuming that frames 1 and 2 are from the same scene, frame 2 will be close to one of the possible projective coordinate transformations of frame 1. In other words, frame 2 "lies near the orbit of" frame 1. (b) By bringing frame 2 along its orbit, we can determine how closely the two orbits come together at frame 1.

"exact," it is theoretically the right model to use for estimating the coordinate transformation. Examples presented in this chapter demonstrate that it also performs better in practice than the other proposed models.

6. FRAMEWORK: MOTION PARAMETER ESTIMATION AND OPTICAL FLOW

To lay the framework for my new results, I will review existing methods of parameter estimation for coordinate transformations. This framework will apply to both existing methods as well as the new methods. The purpose of this review is to bring together a variety of methods that appear quite different, but which actually can be described in a more unified framework, which I present here.

The framework I give breaks existing methods into two categories: feature-based and featureless. Of the featureless methods, I consider two subcategories: 1) methods based on minimizing MSE (generalized correlation, direct nonlinear optimization) and 2) methods based on spatiotemporal derivatives and optical flow. Note that variations such as *multiscale* have been omitted from these categories; multiscale analysis can be applied to any of them. The new algorithms I propose in this chapter (with final form given in Sec. 7) are featureless and based on (multiscale if desired) spatiotemporal derivatives.

Some of the descriptions of methods below will be presented for hypothetical 1-D images taken of 2-D "scenes" or "objects." This simplification yields a clearer comparison of the estimation methods.

The new theory and applications will be presented subsequently for 2-D images taken of 3-D scenes or objects.

6.1 Feature-Based Methods

Feature-based methods [Tsai and Huang, 1984; Huang and Netravali, 1984] assume that point correspondences in both images are available. In the projective case, given at least three correspondences between point pairs in the two 1-D images, I will find the element $\mathbf{p} = \{a, b, c\} \in \mathbf{P}$ that maps the second image into the first. Let $x_k, k = 1, 2, 3, \ldots$ be the points in one image, and let x'_k be the corresponding points in the other image. Then $x'_k = (ax_k + b)/(cx_k + 1)$. Rearranging yields $ax_k + b - x_k x'_k c = x'_k$, so that a,

b, and c can be found by solving $k \geq 3$ linear equations in three unknowns:

$$[x_k \quad 1 \quad -x_k' x_k][a \quad b \quad c]^T = [x_k'] \qquad (6)$$

using least squares if there are more than three correspondence points. The extension from 1-D "images" to 2-D images is conceptually identical; for the affine and projective models, the minimum number of correspondence points needed in two dimensions is three and four respectively.

A major difficulty with feature-based methods is finding the features. Good features are often hand-selected, or computed, possibly with some degree of human intervention [Navab and Shashua, 1994]. A second problem with features is their sensitivity to noise and occlusion. Even if reliable features exist between frames (e.g., line markings on a sports playing field) these features may be subject to signal noise and occlusion (e.g., running players blocking a feature). The emphasis in the rest of this chapter will be on robust featureless methods.

6.2 Featureless Methods Based on Generalized Cross-Correlation

The purpose of this subsection is for completeness. We'll consider first what is perhaps the most most obvious approach (generalized cross-correlation in 8-D parameter space) in order to motivate a different approach provided in Sec 6.3, the motivation arising from ease of implementation and simplicity of computation.

Cross-correlation of two frames is a featureless method of recovering translation model parameters. Affine and projective parameters can also be recovered using generalized forms of cross-correlation.

Generalized cross-correlation is based on an inner-product formulation, which establishes a similarity metric between two functions, say, g and h, where $h \approx \mathbf{p} \circ g$ is an approximately coordinate-transformed version of g, but the parameters of the coordinate transformation, $\mathbf{p}$, are unknown.[11] We can find, by exhaustive search (applying all possible operators, $\mathbf{p}$, to h), the "best" $\mathbf{p}$ as the one that maximizes the inner product:

$$\int_{-\infty}^{\infty} g(x) \frac{\mathbf{p}^{-1} \circ h(x)}{\int_{-\infty}^{\infty} \mathbf{p}^{-1} \circ h(x)\, dx}\, dx, \qquad (7)$$

---

[11] In the presence of additive white Gaussian noise, this method, also known as "matched filtering," leads to a maximum likelihood estimate of the parameters [Van Trees, 1968].

where I have normalized the energy of each coordinate-transformed h before making the comparison. Equivalently, instead of maximizing a similarity metric, we can minimize some distance metric, such as MSE, given by $\int_{-\infty}^{\infty}(g(x) - \mathbf{p}^{-1} \circ h(x))^2 - Dx$. Solving (7) has an advantage over finding MSE when one image is not only a coordinate-transformed version of the other but is also an amplitude-scaled version, as generally happens when there is an automatic gain control or an automatic iris in the camera.

In one dimension, the orbit of an image under the affine group operation is a family of *wavelets* (assuming the image is that of the desired "mother wavelet," in the sense that a wavelet family is generated by 1-D affine coordinate transformations of a single function) while the orbit of an image under the projective group of coordinate transformations is a family of "projective chirplets" [Mann and Haykin, 1995],[12] the objective function (7) being the cross-chirplet transform. A computationally efficient algorithm for the cross-wavelet transform has been presented [Young, 1993]. (See Weiss [1993] for a good review on wavelet-based estimation of affine coordinate transformations.)

Adaptive variants of the chirplet transforms have been previously reported in the literature [Mann and Haykin, 1992]. However, there are still many problems with the adaptive chirplet approach; thus, for the remainder of this chapter, we consider featureless methods based on spatiotemporal derivatives.

6.3 Featureless Methods Based on Spatiotemporal Derivatives

6.3.1 Optical Flow (Translation Flow)

When the change from one image to another is small, optical flow [Horn and Schunk, 1981] may be used. In one dimension, the traditional optical flow formulation assumes each point x in frame t is a translated version of the corresponding point in frame $t + \Delta t$ and that Δx and Δt are chosen in the ratio $\Delta x / \Delta t = u_f$, the translational flow velocity of the point in question. The image brightness $E(x, t)$ is described by:

$$E(x, t) = E(x + \Delta x, t + \Delta t), \quad \forall (x, t), \tag{8}$$

---

[12]Symplectomorphisms of the time-frequency plane [Berthon, 1989; Grossmann and Paul, 1984] have been applied to signal analysis [Mann and Haykin, 1995], giving rise to the so-called q-chirplet [Mann and Haykin, 1995], which differs from the projective chirplet discussed here.

where u_f is the translational flow velocity of the point. In the case of pure translation, u_f is constant across the entire image. More generally, though, a pair of 1-D images are related by a quantity $u_f(x)$ at each point in one of the images.

Expanding the right-hand side of (8) in a Taylor series, and canceling 0th order terms, gives the well-known optical flow equation: $u_f E_x + E_t + h.o.t. = 0$, where E_x and E_t are the spatial and temporal derivatives respectively and $h.o.t.$ denotes higher order terms. Typically, the higher order terms are neglected, giving the expression for the optical flow at each point in one of the two images:

$$u_f E_x + E_t \approx 0. \tag{9}$$

6.3.2 Weighing the Difference between "Affine Fit" and "Affine Flow"

A comparison between two similar approaches is presented, in the familiar and obvious realm of linear regression versus direct affine estimation, highlighting the obvious differences between the two approaches. This difference, in weighting, motivates new weighting changes, which will later simplify implementations pertaining to the new methods.

Given the optical flow between two images, g and h, I wish to find the coordinate transformation to apply to h to register it with g. We now describe two approaches based on the affine model:[13] (1) finding the optical flow at every point, and then fitting this flow with an affine model ("affine fit"), and (2) rewriting the optical flow equation in terms of an affine (not translation) motion model ("affine flow").

"Affine Fit"   Wang and Adelson have proposed fitting an affine model to the optical flow field [Wang and Adelson, 1994a] between two 2-D images. I briefly examine their approach with 1-D images; the reduction in dimensions simplifies analysis and comparison to affine flow. Denote coordinates in the original image, g, by x, and in the new image, h, by x'. Suppose that h is a dilated and translated version of g, so $x' = ax + b$ for every corresponding pair (x', x). Equivalently, the affine model of velocity (normalizing $\Delta t = 1$), $u_m = x' - x$, is given by $u_m = (a - 1)x + b$. We can

---

[13]The 1-D affine model is a simple yet sufficiently interesting (non-Abelian) example selected to illustrate differences in weighting.

expect a discrepancy between the flow velocity, u_f, and the model velocity, u_m, due to either errors in the flow calculation or to errors in the affine model assumption, so I apply linear regression to get the best least-squares fit by minimizing:

$$\varepsilon_{\text{fit}} = \sum_x (u_m - u_f)^2 = \sum (u_m + E_t/E_x)^2. \tag{10}$$

The constants a and b that minimize ε_{fit} over the entire patch are found by differentiating (10), and setting the derivatives to zero. This results in what I call the "affine fit" equations:

$$\begin{bmatrix} \sum_x x^2, \sum_x x \\ \sum_x x, \sum_x 1 \end{bmatrix} \begin{bmatrix} a-1 \\ b \end{bmatrix} = - \begin{bmatrix} \sum_x x E_t/E_x \\ \sum_x E_t/E_x \end{bmatrix}. \tag{11}$$

"Affine Flow" Alternatively, the affine coordinate transformation may be directly incorporated into the brightness change constraint equation (8). Bergen et al. [Bergen, Burt, Hingorini, and Peleg, 1990] have proposed this method, which I will call "affine flow," to distinguish it from the "affine fit" model of Wang and Adelson (11). Let us show how affine flow and affine fit are related. Substituting $u_m = (ax + b) - x$ directly into (9) in place of u_f and summing the squared error:

$$\varepsilon_{\text{flow}} = \sum_x (u_m E_x + E_t)^2 \tag{12}$$

over the whole image, differentiating, and equating the result to zero gives a linear solution for both a and b:

$$\begin{bmatrix} \sum_x x^2 E_x^2, \sum_x x E_x^2 \\ \sum_x x E_x^2, \sum_x E_x^2 \end{bmatrix} \begin{bmatrix} a-1 \\ b \end{bmatrix} = - \begin{bmatrix} \sum_x x E_x E_t \\ \sum_x E_x E_t \end{bmatrix}. \tag{13}$$

To see how this result compares to the affine fit I rewrite (10) as

$$\varepsilon_{\text{fit}} = \sum_x \left(\frac{u_m E_x + E_t}{E_x} \right)^2 \tag{14}$$

and observe, comparing (12) and (14), that affine flow is equivalent to a weighted least-squares fit, where the weighting is given by E_x^2. Thus the affine flow method tends to put more emphasis on areas of the image that are spatially varying than does the affine fit method. Of course, one is free to separately choose the weighting for each method in such a way that affine fit and affine flow methods both give the same result. Both my intuition and our practical experience tend to favor the affine flow weighting, but, more generally, perhaps we should ask "What is the best weighting?" Lucas and Kanade [Lucas and Kanade, 1981], among others, have considered weighting issues, though the rather obvious difference in weighting between fit and flow doesn't appear to have been pointed out previously in the literature. The fact that the two approaches provide similar results, yet have drastically different weightings, suggests that we can exploit the choice of weighting. In particular, we will observe in Sec 6.3.3 that we can select a weighting that makes the implementation easier.

Another approach to the affine fit involves computation of the optical flow field using the multiscale iterative method of Lucas and Kanade and *then* fitting to the affine model. An analogous variant of the affine flow method involves multiscale iteration as well, but in this case the iteration and multiscale hierarchy are incorporated directly into the affine estimator [Bergen, Burt, Hingorini, and Peleg, 1990]. With the addition of multiscale analysis, the "fit" and "flow" methods differ in additional respects beyond just the weighting. My intuition and experience indicate that the direct multiscale affine flow performs better than the affine fit to the multiscale flow. Multiscale optical flow makes the assumption that blocks of the image are moving with pure translational motion, and then, paradoxically, the affine fit refutes this pure-translation assumption. However, "fit" provides some utility over "flow" when it is desired to segment the image into regions undergoing different motions [Wang and Adelson, 1994b], or to gain robustness by rejecting portions of the image not obeying the assumed model.

6.3.3 "Projective Fit" and "Projective Flow": New Techniques

Analogous to the affine fit and affine flow of the previous section, I now propose two new methods: "projective fit" and "projective flow." For the 1-D affine coordinate transformation, the graph of the range coordinate as a function of the domain coordinate is a straight line; for the projective coordinate transformation, the graph of the range coordinate as a function of the domain coordinate is a rectangular hyperbola (Fig. 7.2(d)). The affine fit case used linear regression; however, in the projective case I use hyperbolic regression. Consider the flow velocity given by (9) and the model velocity:

$$u_m = x' - x = \frac{ax + b}{cx + 1} - x \qquad (15)$$

and minimize the sum of the squared difference as was done in (10):

$$\varepsilon = \sum_x \left(\frac{ax + b}{cx + 1} - x + \frac{E_t}{E_x} \right)^2. \tag{16}$$

As discussed earlier, the calculation can be simplified by judicious alteration of the weighting; in particular, multiplying each term of the summation (16) by $(cx + 1)$, and solving, gives:

$$\left(\sum_x \phi(x) \phi^T(x) \right) [a, b, c]^T = \sum_x (x - E_t/E_x) \phi(x), \tag{17}$$

where the *regressor* is $\phi = [x, 1, xE_t/E_x - x^2]^T$.

For projective-flow (p-flow), I substitute $u_m = \frac{ax+b}{cx+1} - x$ into (12). Again, weighting by $(cx + 1)$ gives:

$$\varepsilon_w = \sum (axE_x + bE_x + c(xE_t - x^2E_x) + E_t - xE_x)^2 \tag{18}$$

(the subscript w denotes weighting has taken place) resulting in a linear system of equations for the parameters:

$$\left(\sum \phi_w \phi_w^T \right) [a, b, c]^T = \sum (xE_x - E_t) \phi_w, \tag{19}$$

where $\phi_w = [xE_x, E_x, xE_t - x^2E_x]^T$. Again, to show the difference in the weighting between projective flow and projective fit, we can rewrite (19):

$$\left(\sum_x E_x^2 \phi \phi^T \right) [a, b, c]^T = \sum E_x^2 (xE_x - E_t) \phi, \tag{20}$$

where ϕ is that defined in (17).

6.3.4 The Unweighted Projectivity Estimator

If we do not wish to apply the ad hoc weighting scheme, we may still estimate the parameters of projectivity in a simple manner, still based on solving a linear system of equations. To do this, we write the Taylor series of u_m:

$$u_m + x = b + (a - bc)x + (bc - a)cx^2 + (a - bc)c^2x^3 + \cdots \tag{21}$$

and use the first three terms, obtaining enough degrees of freedom to account for the three parameters being estimated. Letting $\varepsilon = \sum(-h.o.t.)^2 = \sum((b + (a - bc - 1)x + (bc - a)cx^2)E_x + E_t)^2$, $\mathbf{q}_2 = (bc - a)c$, $\mathbf{q}_1 = a - bc - 1$, and $\mathbf{q}_0 = b$, and differentiating with respect to each of the three parameters of $\mathbf{q}$, setting the derivatives equal to zero, and verifying with the second derivatives gives the linear system of equations for "unweighted projective flow":

$$
\begin{bmatrix} \sum x^4 E_x^2 & \sum x^3 E_x^2 & \sum x^2 E_x^2 \\ \sum x^3 E_x^2 & \sum x^2 E_x^2 & \sum x E_x^2 \\ \sum x^2 E_x^2 & \sum x E_x^2 & \sum E_x^2 \end{bmatrix} \begin{bmatrix} q_2 \\ q_1 \\ q_0 \end{bmatrix} = - \begin{bmatrix} \sum x^2 E_x E_t \\ \sum x E_x E_t \\ \sum E_x E_t \end{bmatrix} . \tag{22}
$$

In Sec. 7 I will extend this derivation to 2-D images.

7. MULTISCALE IMPLEMENTATIONS IN TWO DIMENSIONS

In the previous section, two new techniques, projective-fit and projective-flow were proposed. Now I describe these algorithms for 2-D images. The brightness constancy constraint equation for 2-D images [Horn and Schunk, 1981], which gives the flow velocity components in the x and y directions, analogous to (9) is:

$$
\mathbf{u_f}^T \mathbf{E_x} + E_t \approx 0. \tag{23}
$$

As is well-known [Horn and Schunk, 1981] the optical flow field in two dimensions is underconstrained.[14] The model of *pure translation* at every point has two parameters, but there is only one equation (23) to solve; thus it is common practice to compute the optical flow over some neighborhood, which must be at least two pixels but is generally taken over a small block, 3×3, 5×5, or sometimes larger (e.g., the entire image, as in this chapter).

Our task is not to deal with the 2-D translational flow, but with the 2-D projective flow, estimating the eight parameters in the coordinate transformation:

$$
\mathbf{x'} = \begin{bmatrix} x' \\ y' \end{bmatrix} = \frac{\mathbf{A}[x, y]^T + \mathbf{b}}{\mathbf{c}^T [x, y]^T + 1} = \frac{\mathbf{A}\mathbf{x} + \mathbf{b}}{\mathbf{c}^T \mathbf{x} + 1}. \tag{24}
$$

---

[14]Optical flow in one dimension did not suffer from this problem.

The desired eight scalar parameters are denoted by $\mathbf{p} = [\mathbf{A}, \mathbf{b}; \mathbf{c}, 1]$, $\mathbf{A} \in \mathbb{R}^{2 \times 2}$, $\mathbf{b} \in \mathbb{R}^{2 \times 1}$, and $\mathbf{c} \in \mathbb{R}^{2 \times 1}$.

Analogous to (14), we have, in the 2-D case:

$$\varepsilon_{\text{flow}} = \sum \left(\mathbf{u_m}^T \mathbf{E_x} + E_t \right)^2 = \sum \left(\left(\frac{\mathbf{Ax} + \mathbf{b}}{\mathbf{c}^T \mathbf{x} + 1} - \mathbf{x} \right)^T \mathbf{E_x} + E_t \right)^2,$$

(25)

where the sum can be weighted as it was in the 1-D case:

$$\varepsilon_w = \sum \left((\mathbf{Ax} + \mathbf{b} - (\mathbf{c}^T \mathbf{x} + 1)\mathbf{x})^T \mathbf{E_x} + (\mathbf{c}^T \mathbf{x} + 1)E_t \right)^2. \quad (26)$$

Differentiating with respect to the free parameters $\mathbf{A}$, $\mathbf{b}$, and $\mathbf{c}$ and setting the result to zero gives a linear solution:

$$\left(\sum \phi \phi^T \right) [a_{11}, a_{12}, b_1, a_{21}, a_{22}, b_2, c_1, c_2]^T = \sum (\mathbf{x}^T \mathbf{E_x} - E_t)\phi,$$

(27)

where

$$\phi^T = [E_x(x, y, 1), E_y(x, y, 1), x E_t - x^2 E_x$$
$$- xy E_y, y E_t - xy E_x - y^2 E_y].$$

7.1 Unweighted Projective Flow

As with the 1-D images, we make similar assumptions in expanding (24) in its own Taylor series, analogous to (21). If we take the Taylor series up to 2nd order terms, we obtain the biquadratic model mentioned in Sec. 5.1. As mentioned in Sec. 5.1, by appropriately constraining the 12 parameters of the biquadratic model we obtain a variety of 8-parameter approximate models. In my algorithms for estimating the "exact unweighted" projective group parameters, I use one of these approximate models in an intermediate step.[15]

The Taylor series for the bilinear case gives:

$$u_m + x = q_{x'xy} xy + (q_{x'x} + 1)x + q_{x'y} y + q_{x'},$$
$$v_m + y = q_{y'xy} xy + q_{y'x} x + (q_{y'y} + 1)y + q_{y'}. \quad (28)$$

---

[15]Use of an approximate model that doesn't capture chirping or preserve straight lines can still lead to the true projective parameters as long as the model captures at least eight degrees of freedom.

Incorporating these into the flow criteria yields a simple set of eight linear equations in eight unknowns:

$$\left(\sum_{x,y}(\phi(x, y)\phi^T(x, y))\right) \mathbf{q} = -\sum_{x,y} E_t\phi(x, y), \qquad (29)$$

where $\phi^T = [E_x(xy, x, y, 1), E_y(xy, x, y, 1)]$.

For the relative-projective model, ϕ is given by

$$\phi^T = [E_x(x, y, 1), E_y(x, y, 1), E_t(x, y)] \qquad (30)$$

and for the pseudo-perspective model, ϕ is given by

$$\phi^T = [E_x(x, y, 1), E_y(x, y, 1), (x^2 E_x + xy E_y, xy E_x + y^2 E_y)]. \quad (31)$$

In order to see how well the model describes the coordinate transformation between two images, say, g and h, one might *warp*[16] h to g, using the estimated motion model, and then compute some quantity that indicates how different the resampled version of h is from g. The MSE between the reference image and the warped image might serve as a good measure of similarity. However, since we are really interested in how the *exact model* describes the coordinate transformation, we assess the goodness of fit by first relating the parameters of the approximate model to the exact model, and then we find the MSE between the reference image and the comparison image after applying the coordinate transformation of the exact model. A method of finding the parameters of the exact model, given the approximate model, is presented in Sec 7.1.1.

7.1.1 Four-Point Method for Relating Approximate Model to Exact Model

Any of the approximations above, after being related to the exact projective model, tend to behave well in the neighborhood of the identity, $\mathbf{A} = \mathbf{I}, \mathbf{b} = \mathbf{0}, \mathbf{c} = \mathbf{0}$. In one dimension, I explicitly expanded the model Taylor series about the identity; here, although I do not explicitly do this, I shall assume that the terms of the Taylor series of the model correspond to those taken about the identity. In the 1-D case we solve the three linear equations in three unknowns to estimate the parameters of the approximate motion

---

[16]The term *warp* is appropriate here, since the approximate model does not preserve straight lines.

model and then relate the terms in this Taylor series to the exact parameters, a, b, and c (which involves solving another set of three equations in three unknowns, the second set being nonlinear, although very easy to solve).

In the extension to 2-D, the estimate step is straightforward, but the relate step is more difficult, because we now have eight nonlinear equations in eight unknowns, relating the terms in the Taylor series of the approximate model to the desired exact model parameters. Instead of solving these equations directly, I now propose a simple procedure for relating the parameters of the approximate model to those of the exact model, which I call the four-point method:

1. Select four ordered pairs (e.g., the four corners of the bounding box containing the region under analysis, or the four corners of the image if the whole image is under analysis). Here suppose, for simplicity, that these points are the corners of the unit square: $\mathbf{s} = [s_1, s_2, s_3, s_4] = [(0,0)^T, (0,1)^T, (1,0)^T, (1,1)^T]$.

2. Apply the coordinate transformation using the Taylor series for the approximate model (e.g., (28)) to these points: $\mathbf{r} = \mathbf{u}_m(\mathbf{s})$.

3. Finally, the correspondences between $\mathbf{r}$ and $\mathbf{s}$ are treated just like features. This results in four easy to solve linear equations:

$$\begin{bmatrix} x'_k \\ y'_k \end{bmatrix} = \begin{bmatrix} x_k, y_k, 1, 0, 0, 0, -x_k x'_k, -y_k x'_k \\ 0, 0, 0, x_k, y_k, 1, -x_k y'_k, -y_k y'_k \end{bmatrix}$$
$$\times \, [a_{x'x}, a_{x'y}, b_{x'}, a_{y'x}, a_{y'y}, b_{y'}, c_x, c_y]^T, \tag{32}$$

where $1 \le k \le 4$. This results in the exact eight parameters, $\mathbf{p}$.

We remind the reader that the four corners are *not* feature correspondences as used in the feature-based methods of Sec. 6.1, but, rather, they are used so that the two featureless models (approximate and exact) can be related to one another.

It is important to realize the full benefit of finding the exact parameters. While the "approximate model" is sufficient for small deviations from the identity, it is not adequate to describe large changes in perspective. However, if we use it to track small changes incrementally, and each time relate these small changes to the exact model (24), then we can accumulate these small changes using the *law of composition* afforded by the group structure. This is an especially favorable contribution of the group framework. For example, with a video sequence, we can accommodate very large accumulated

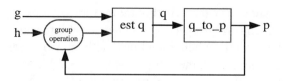

FIG. 7.8. Method of computation of eight parameters p between
two images from the same pyramid level, *g* and *h*. The approximate model parameters q are related to the exact model parameters p in a feedback system.

changes in perspective in this manner. The problems with cumulative error
can be eliminated, for the most part, by constantly propagating forward the
true values, computing the residual using the approximate model, and each
time relating this to the exact model to obtain a goodness-of-fit estimate.

7.1.2 Algorithm for Unweighted Projective Flow: Overview

Below is an outline of the algorithm; details of each step are in subsequent
sections.

Frames from an image sequence are compared pairwise to test whether
or not they lie in the same orbit:

1. A Gaussian pyramid of three or four levels is constructed for each
    frame in the sequence.
2. The parameters **p** are estimated at the top of the pyramid, between
    the two lowest-resolution images of a frame pair, *g* and *h*, using the
    iterative method depicted in Fig. 7.8.
3. The estimated **p** is applied to the next higher-resolution (finer) image
    in the pyramid, **p** ∘ *g*, to make the two images at that level of the
    pyramid nearly congruent before estimating the **p** between them.
4. The process continues down the pyramid until the highest-resolution
    image in the pyramid is reached.

7.2 Multiscale Iterative Implementation

The Taylor-series formulations I have used implicitly assume smoothness;
the performance is improved if the images are blurred before estimation.
To accomplish this, I do not downsample critically after low-pass filtering
in the pyramid. However, after estimation, I use the original (unblurred)
images when applying the final coordinate transformation.

The strategy I present differs from the multiscale iterative (affine) strategy of Bergen et al. in one important respect beyond simply an increase from six to eight parameters. The difference is the fact that we have two motion models, the "exact motion model" (24) and the "approximate motion model," namely the Taylor series approximation to the motion model itself. The approximate motion model is used to iteratively converge to the exact motion model, using the algebraic *law of composition* afforded by the exact projective group model. In this strategy, the exact parameters are determined at each level of the pyramid and passed to the next level. The steps involved are summarized schematically in Fig. 7.8 and described below:

1. Initialize: Set $h_0 = h$ and set $\mathbf{p}_{0,0}$ to the identity operator.
2. Iterate ($k = 1 \ldots K$):
    (a) *ESTIMATE:* Estimate the eight or more terms of the approximate model between two image frames, g and h_{k-1}. This results in approximate model parameters $\mathbf{q}_k$.
    (b) *RELATE:* Relate the approximate parameters $\mathbf{q}_k$ to the exact parameters using the four-point method. The resulting exact parameters are $\mathbf{p}_k$.
    (c) *RESAMPLE:* Apply the *law of composition* to accumulate the effect of the $\mathbf{p}_k$s. Denote these composite parameters by $\mathbf{p}_{0,k} = \mathbf{p}_k \circ \mathbf{p}_{0,k-1}$. Then set $h_k = \mathbf{p}_{0,k} \circ h$. (This should have nearly the same effect as applying $\mathbf{p}_k$ to h_{k-1}, except that it will avoid additional interpolation and antialiasing errors you would get by resampling an already resampled image [Wolberg, 1990]).

Repeat until either the error between h_k and g falls below a threshold, or until some maximum number of iterations is achieved. After the first iteration, the parameters $\mathbf{q}_2$ tend to be near the identity since they account for the residual between the "perspective-corrected" image h_1 and the "true" image g. We find that only two or three iterations are usually needed for frames from nearly the same orbit.

A rectangular image assumes the shape of an arbitrary quadrilateral when it undergoes a projective coordinate transformation. In coding the algorithm, I pad the undefined portions with the quantity NaN, a standard IEEE arithmetic value, so that any calculations involving these values automatically inherit NaN without slowing down the computations. The algorithm (in Matlab on an HP 735) takes about six seconds per iteration for a pair of 320×240 images.

7.3 Exploiting Commutativity for Parameter Estimation

There is a fundamental uncertainty [Wilson and Granlund, 1984] involved in the simultaneous estimation of parameters of a noncommutative group, akin to the Heisenberg uncertainty relation of quantum mechanics. In contrast, for a commutative[17] group (in the absence of noise), we can obtain the exact coordinate transformation.

Segman [Segman et al., 1992] considered the problem of estimating the parameters of a commutative group of coordinate transformations, in particular, the parameters of the affine group [Segman, 1992]. His work also deals with noncommutative groups, in particular, in the incorporation of scale in the Heisenberg group[18] [Segman and Schempp, 1993].

Estimating the parameters of a commutative group is computationally efficient (e.g., through the use of Fourier cross-spectra [Girod and Kuo, 1989]). I exploit this commutativity for estimating the parameters of the noncommutative 2-D projective group by first estimating the parameters that commute. For example, we improve performance if we first estimate the two parameters of translation, correct for the translation, and then proceed to estimate the eight projective parameters. We can also simultaneously estimate both the isotropic-zoom and the rotation about the optical axis by applying a log-polar coordinate transformation followed by a translation estimator. This process may also be achieved by a direct application of the Fourier–Mellin transform [Sheng et al., 1988]. Similarly, if the only difference between g and h is a camera pan, then the pan may be estimated through a coordinate transformation to cylindrical coordinates, followed by a translation estimator.

In practice, I run through the following "commutative initialization" before estimating the parameters of the projective group of coordinate transformations:

1. Assume that h is merely a translated version of g.
    (a) Estimate this translation using the method of Girod [Girod and Kuo, 1989].
    (b) Shift h by the amount indicated by this estimate.

---

[17] A commutative (or *Abelian*) group is one in which elements of the group commute, for example, translation along the x axis commutes with translation along the y axis, so the 2-D translation group is commutative.

[18] While the Heisenberg group deals with translation and frequency-translation (modulation), some of the concepts could be carried over to other more relevant group structures.

    (c)   Compute the MSE between the shifted h and g, and compare to the original MSE before shifting.

    (d)   If an improvement has resulted, use the shifted h from now on.

2. Assume that h is merely a rotated and isotropically zoomed version of g.

    (a)   Estimate the two parameters of this coordinate transformation.

    (b)   Apply these parameters to h.

    (c)   If an improvement has resulted, use the coordinate-transformed (rotated and scaled) h from now on.

3. Assume that h is merely an "x-chirped" (panned) version of g, and, similarly, "x-dechirp" h. If an improvement results, use the x-dechirped h from now on. Repeat for y (tilt).

Compensating for one step may cause a change in choice of an earlier step. Thus it might seem desirable to run through the commutative estimates iteratively. However, my experience on lots of real video indicates that a single pass usually suffices and, in particular, will catch frequent situations where there is a pure zoom, a pure pan, a pure tilt, etc., both saving the rest of the algorithm computational effort, as well as accounting for simple coordinate transformations such as when one image is an upside-down version of the other. (Any of these pure cases corresponds to a single parameter group, which is commutative.) Without the "commutative initialization" step, these parameter estimation algorithms are prone to get caught in local optima, and thus never converge to the global optimum.

8. PERFORMANCE AND APPLICATIONS

Figure 7.9 shows some frames from a typical image sequence captured by Wearable Wireless Webcam [Mann, 1997]. Figure 7.10. shows the same frames brought into the coordinate system of frame (c), that is, the middle frame was chosen as the *reference frame*.

    Given that we have established a means of estimating the projective coordinate transformation between any pair of images, there are two basic methods we use for finding the coordinate transformations between all pairs of a longer image sequence. Because of the group structure of the projective coordinate transformations, it suffices to arbitrarily select one frame and find the coordinate transformation between every other frame and this frame. The two basic methods are:

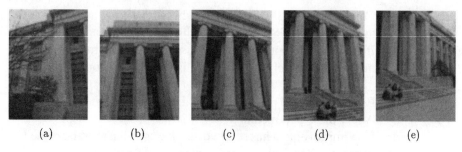

(a)   (b)   (c)   (d)   (e)

FIG. 7.9. Frames from original image orbit, sent from my personal imaging apparatus. Note that the camera is mounted sideways so that it can "paint" out the image canvas with a wider "brush," when sweeping across for a panorama. Thus the visual field of view that I experienced was rotated through 90 degrees. Much like George Stratton did with his upside-down glasses, I adapted, over an extended period of time, to experiencing the world rotated 90 degrees. (Adaptation experiments will be covered in Chapter 9.)

(a)   (b)   (c)   (d)   (e)

FIG. 7.10. Frames from original image video orbit after a coordinate transformation to move them along the orbit to the reference frame (c). The coordinate-transformed images are alike except for the region over which they are defined. Note that the regions are not parallelograms; thus, methods based on the affine model fail.

1. **Differential parameter estimation:** The coordinate transformations between successive pairs of images, $\mathbf{p}_{0,1}$, $\mathbf{p}_{1,2}$, $\mathbf{p}_{2,3}$, ... , estimated.

2. **Cumulative parameter estimation:** The coordinate transformation between each image and the reference image is estimated directly. Without loss of generality, select frame zero (E_0) as the reference frame and denote these coordinate transformations as $\mathbf{p}_{0,1}$, $\mathbf{p}_{0,2}$, $\mathbf{p}_{0,3}$, ...

Theoretically, the two methods are equivalent:

$$E_0 = p_{0,1} \circ p_{1,2} \circ \ldots \circ p_{n-1,n} E_n \quad \text{(differential method)},$$
$$E_0 = p_{0,n} E_n \quad \text{(cumulative method)}. \tag{33}$$

However, in practice, the two methods differ for two reasons:

1.   *Cumulative error:* In practice, the estimated coordinate transformations between pairs of images register them only approximately, due to violations of the assumptions (e.g., objects moving in the scene, center of projection not fixed, camera swings around to bright window and automatic iris closes, etc.). When a large number of estimated parameters are composed, cumulative error sets in.

2.   *Finite spatial extent of image plane:* Theoretically, the images extend infinitely in all directions, but, in practice, images are cropped to a rectangular bounding box. Therefore, a given pair of images (especially if they are far from adjacent in the orbit) may not overlap at all; hence it is not possible to estimate the parameters of the coordinate transformation using those two frames.

The frames of Fig. 7.9 were brought into register using the differential parameter estimation, and "cemented" together seamlessly on a common canvas. "Cementing" involves piecing the frames together, for example, by median, mean, or trimmed mean, or combining on a subpixel grid [Mann and Picard, 1994b]. (Trimmed mean was used here, but the particular method made little visible difference.) Figure 7.11 shows this result ("projective/projective"), with a comparison to two nonprojective cases. The first comparison is to "affine/affine" where affine parameters were estimated (also multiscale) and used for the coordinate transformation. The second comparison, "affine/projective," uses the six affine parameters found by estimating the eight projective parameters and ignoring the two "chirp" parameters $\mathbf{c}$ (which capture the essence of tilt and pan). These six parameters $\mathbf{A}, \mathbf{b}$ are more accurate than those obtained using the affine estimation, as the affine estimation tries to fit its shear parameters to the camera pan and tilt. In other words, the affine estimation does worse than the six affine parameters within the projective estimation. The affine coordinate transform is finally applied, giving the image shown. Note that the coordinate-transformed frames in the affine case are parallelograms.

8.1   Extreme Violations of the Underlying Assumptions

To show that the algorithms are robust to large amounts of "noise" (violations of the underlying assumptions), we consider a nonstatic scene containing some people (Figure 7.12). Images were acquired by the author, using an embodiment of the WearCam [Mann, 1997] invention. Because

projective/projective

affine/projective

affine/affine

FIG. 7.11. Frames of Fig. 7.10 "cemented" together on single image "canvas," with comparison of affine and projective models. Note the good registration and nice appearance of the projective/projective image despite the noise in the amateur television receiver, wind-blown trees, and the fact that the rotation of the camera was not actually about its center of projection. Note also that the affine model fails to properly estimate the motion parameters (affine/affine), and even if the "exact" projective model is used to estimate the affine parameters, there is no affine coordinate transformation that will properly register all of the image frames.

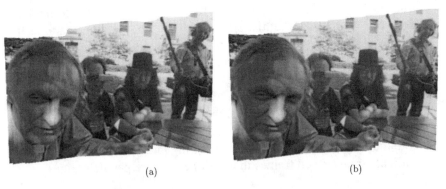

(a)   (b)

FIG. 7.12. PIC made from scene in which the assumptions are violated. The nonstatic scene (with television crew) also involved some parallax from nearby parts of the scene (especially leftmost person), changes in lighting resulting from sun shining through leaves which were blowing in the wind, etc. (a) These violations in the assumptions resulted in striping or banding artifacts, mainly due to the rapid fluctuations in illumination. However the estimator still performed quite well. (b) With a feather radius of 21 pixels, frames were blended into each other, resulting in a seamless PIC.

the camera is worn some distance from the center of rotation (neck), and there are elements in the scene quite close (in particular, the leftmost person was about arms-length away), there was considerable parallax. The robustness of the algorithms is quite evident.

9. SUMMARY

I have proposed and demonstrated featureless estimation of the projective coordinate transformation between successive pairs of images captured from an eyeglass-based WearCam system. Not just one method, but various methods were proposed, among these, "projective fit" and "projective flow," which estimate the projective (homographic) coordinate transformation between pairs of images, taken with a camera that is free to pan, tilt, rotate about its optical axis, and zoom. The new approach was also formulated and demonstrated within a multiscale iterative framework. Applications to seamlessly combining images in or near the same orbit of the projective group of coordinate transformations were also presented. The proposed approach solves for the 8 parameters of the "exact" model (the projective group of coordinate transformations).

The proposed method was found to work well on image data collected from both good-quality and poor-quality video under a wide variety of conditions (sunny, cloudy, day, night). It has been tested with a head-mounted wireless video camera and performs successfully even in the presence of noise, interference, scene motion (such as people walking through the scene), lighting fluctuations, and parallax (due to movements of the wearer's head). It remains to be shown which variant of the proposed approach is optimal, and under what conditions.

Acknowledgments   The author would like to thank many individuals for suggestions and encouragement, including Rosalind Picard, Simon Haykin, Shawn Becker, Charles Wyckoff, John Wang, Kris Popat, Nassir Navab, Ujjaval Desai, and Chris Graczyk.

The author would also like to thank Lee Campbell for help with the correction of barrel distortion (enabling really cheap plastic lenses to be used in the eyeglass-based personal imaging systems).

Thanks also to Alex Drukarev, Jeanne Wiseman, and Paul and Claire Hubel of HP Labs, Palo Alto.

Some of the software to implement the p-chirp models was developed in collaboration with Shawn Becker.

Additional thanks to Kodak, Digital Equipment Corporation, Compaq, Kopin, VirtualVision, and HP labs.

REFERENCES

[Adiv, 1985] Adiv, G. (July 1985). Determining 3D motion and structure from optical flow generated by several moving objects. *IEEE Trans. Pattern Anal. Machine Intell.*, pages 304–401.

[Ahlfors, 1979] Ahlfors, L. (1979). *Complex Analysis*. International Series in Pure and Applied Mathematics. McGraw Hill, Inc., 3rd edition.

[Artin, 1991] Artin, M. (1991). *Algebra*. Prentice-Hall.

[Barron, 1994] Barron, J. L., Fleet, D. J., and Beauchemin, S. S. (1994). Systems and experiment performance of optical flow techniques. *International Journal of Computer Vision*, Vol 12, No. 1, pages 43–77.

[Bergen, Burt, Hingorini, and Peleg, 1990] Bergen, J., Burt, P., Hingorini, R., and Peleg, S. (1990). Computing two motions from three frames. In *Proc. Third Int'l Conf. Comput. Vision*, pages 27–32, Osaka, Japan.

[Berthon, 1989] Berthon, A. (1989). Operator groups and ambiguity functions in signal processing. In Combes, J., editor, *Wavelets: Time-Frequency Methods and Phase Space*. Springer-Verlag.

[Campbell and Bobick, 1995] Campbell, L. and Bobick, A. (1995). Correcting for radial lens distortion: A simple implementation. TR 322, M.I.T. Media Lab Perceptual Computing Section, Cambridge, Massachusetts.

[Faugeras and Lustman, 1988] Faugeras, O. D. and Lustman, F. (1988). Motion and structure from motion in a piecewise planar environment. *International Journal of Pattern Recognition and Artificial Intelligence*, 2(3):485–508.

[Girod and Kuo, 1989] Girod, B. and Kuo, D. (1989). Direct estimation of displacement histograms. *OSA Meeting on Image Understanding and Machine Vision*.

[Grossmann and Paul, 1984] Grossmann, A. and Paul, T. (1984). Wave functions on subgroups of the group of affine cannonical tranformations. Lecture notes in physics, No. 211: *Resonances—Models and Phenomena*, pages 128–138. Springer-Verlag.

[Horn and Schunk, 1981] Horn, B. and Schunk, B. (1981). Determining optical flow. *Artificial Intelligence*, Vol. 17, pages 185–203.

[Huang and Netravali, 1984] Huang, T. S. and Netravali, A. (1984). Motion and structure from feature correspondences: a review. *Proc. IEEE*, Feb. 1984, Vol. 82, No. 2, pages 252–268.

[Irani and Peleg, 1991] Irani, M. and Peleg, S. (1991). Improving resolution by image registration. *CVGIP*, 53:231–239.

[Kumar et al., 1994] Kumar, R., Anandan, P., and Hanna, K. (1994). Shape recovery from multiple views: a parallax based approach. *ARPA Image Understanding Workshop*.

[Lucas and Kanade, 1981] Lucas, B. D. and Kanade, T. (1981). An iterative image-registration technique with an application to stereo vision. In *Image Understanding Workshop*, pages 121–130.

[Mann, 1992] Mann, S. (1992). Wavelets and chirplets: Time-frequency perspectives, with applications. In Archibald, P., editor, *Advances in Machine Vision, Strategies and Applications*. World Scientific, Vol. 32 in World Scientific Series in Computer Science.

[Mann, 1993] Mann, S. (1993). Compositing multiple pictures of the same scene. In *Proceedings of the 46th Annual IS&T Conference*, Cambridge, Massachusetts. The Society of Imaging Science and Technology.

[Mann, 1994] Mann, S. (1994). Mediated Reality. TR 260, M.I.T. Media Lab Perceptual Computing Section, Cambridge, Massachusetts, http://wearcam.org/mr.htm.

[Mann, 1997] Mann, S. (1997). Wearable computing: A first step toward personal imaging. *IEEE Computer; http://wearcam.org/ieeecomputer.htm*, 30(2).

[Mann and Haykin, 1992] Mann, S. and Haykin, S. (1992). Adaptive "chirplet" transform: an adaptive generalization of the wavelet transform. *Optical Engineering*, 31(6):1243–1256.

[Mann and Haykin, 1995] Mann, S. and Haykin, S. (1995). The chirplet transform: Physical considerations. *IEEE Trans. Signal Processing*, 43(11).

[Mann and Picard, 1994a] Mann, S. and Picard, R. (1994a). Being 'undigital' with digital cameras: Extending dynamic range by combining differently exposed pictures. Technical Report 323, M.I.T. Media Lab Perceptual Computing Section, Boston, Massachusetts. Also appears in *IS&T's 48th Annual Conference*, pages 422–428, May 1995.

[Mann and Picard, 1994b] Mann, S. and Picard, R. W. (1994b). Virtual bellows: Constructing high-quality images from video. In *Proceedings of the IEEE First International Conference on Image Processing*, Austin, Texas.

[Mann and Picard, 1995] Mann, S. and Picard, R. W. (1995). Video orbits of the projective group; a simple approach to featureless estimation of parameters. TR 338, Massachusetts Institute of Technology, Cambridge, Massachusetts. Also appears in *IEEE Trans. Image Proc.*, Sept 1997, Vol. 6 No. 9.

[Navab and Mann, 1994] Navab, N. and Mann, S. (1994). Recovery of relative affine structure using the motion flow field of a rigid planar patch. *Mustererkennung 1994, Tagungsband.*

[Navab and Shashua, 1994] Navab, N. and Shashua, A. (1994). Algebraic description of relative affine structure: Connections to Euclidean, affine and projective structure. *MIT Media Lab Memo No. 270.*

[Sawhney, 1994] Sawhney, H. (1994). Simplifying motion and structure analysis using planar parallax and image warping. *ICPR*, 1. 12th IAPR.

[Segman, 1992] Segman, J. (1992). Fourier cross correlation and invariance transformations for an optimal recognition of functions deformed by affine groups. *Journal of the Optical Society of America, A*, 9(6):895–902.

[Segman et al., 1992] Segman, J., Rubinstein, J., and Zeevi, Y. Y. (1992). The canonical coordinates method for pattern deformation: Theoretical and computational considerations. *IEEE Trans. on Patt. Anal. and Mach. Intell.*, 14(12):1171–1183.

[Segman and Schempp, 1993] Segman, J. and Schempp, W. (1993). *Two methods of incorporating scale in the Heisenberg group.* JMIV special issue on wavelets.

[Shashua and Navab, 1994] Shashua, A. and Navab, N. (1994). Relative affine: Theory and application to 3D reconstruction from perspective views. *Proc. IEEE Conference on Computer Vision and Pattern Recognition.*

[Sheng et al., 1988] Sheng, Y., Lejeune, C., and Arsenault, H. H. (1988). Frequency-domain Fourier–Mellin descriptors for invariant pattern recognition. *Optical Engineering*, Vol. 27, No. 5, pages 354–7, May 1998.

[Szeliski and Coughlan, 1994] Szeliski, R. and Coughlan, J. (1994). Hierarchical spline-based image registration. *CVPR*, pages 194–201.

[Tekalp et al., 1992] Tekalp, A., Ozkan, M., and Sezan, M. (1992). High-resolution image reconstruction from lower-resolution image sequences and space-varying image restoration. In *Proc. of the Int. Conf. on Acoust., Speech and Sig. Proc.*, pages III–169, San Francisco, CA. IEEE.

[Teodosio and Bender, 1993] Teodosio, L. and Bender, W. (1993). Salient video stills: Content and context preserved. *Proc. ACM Multimedia Conf.*

[Tsai and Huang, 1981] Tsai, R. Y. and Huang, T. S. (1981). Estimating three-dimensional motion parameters of a rigid planar patch. *Trans. Accoust., Speech, and Sig. Proc.*, ASSP(29), pages 1147–1152.

[Tsai and Huang, 1984] Tsai, R. Y. and Huang, T. S. (1984). Multiframe image restoration and registration. *ACM.*

[Van Trees, 1968] Van Trees, H. L. (1968). *Detection, Estimation, and Modulation Theory (Part I).* John Wiley and Sons.

[Wang and Adelson, 1994a] Wang, J. Y. and Adelson, E. H. (1994a). Spatio-temporal segmentation of video data . In *SPIE Image and Video Processing II*, pages 120–128, San Jose, California.

[Wang and Adelson, 1994b] Wang, J. Y. A. and Adelson, E. H. (1994b). Representing moving images with layers. *Image Processing Spec. Issue: Image Seq. Compression*, 12(1).

[Weiss, 1993] Weiss, L. G. (1993). Wavelets and wideband correlation processing. *IEEE Signal Processing Magazine*, pages 13–32, Jan. 1994.

[Wilson and Granlund, 1984] Wilson, R. and Granlund, G. H. (1984). The Uncertainty Principle in image processing. *IEEE Transactions on Pattern Analysis and Machine Intelligence*.

[Wolberg, 1990] Wolberg, G. (1990). *Digital Image Warping*. IEEE Computer Society Press, 10662 Los Vaqueros Circle, Los Alamitos, CA. IEEE Computer Society Press Monograph.

[Wyckoff, 1962] Wyckoff, C. W. (1962). An experimental extended response film. *S.P.I.E. Newsletter*, pages 16–20.

[Young, 1993] Young, R. K. (1993). Wavelet theory and its applications. Kluwer Academic Publishers, Boston.

[Zheng and Chellappa, 1993] Zheng, Q. and Chellappa, R. (1993). A computational vision approach to image registration. *IEEE Transactions Image Processing*, pages 311–325.

III

Augmented Reality

8

Studies of the Localization of Virtual Objects in the Near Visual Field

Stephen R. Ellis
NASA Ames Research Center

Brian M. Menges
San José State University Foundation

ABSTRACT

Errors in the localization of nearby virtual objects presented via see-through, helmet-mounted displays are examined as a function of viewing conditions and scene content in four experiments using a total of 38 subjects. Monocular, biocular or stereoscopic presentation of the virtual objects, accommodation (required focus), subjects' age, and the position of physical surfaces are examined. Nearby physical surfaces are found to introduce localization errors that differ depending upon the other experimental factors. These errors apparently arise from the occlusion of the physical background by the optically superimposed virtual objects. But they are modified by subjects' accommodative competence and specific viewing conditions. The apparent physical size and transparency of the virtual objects and physical surfaces respectively are influenced by their relative position when superimposed. The implications of the findings are discussed in terms of display design and potential applications.

INTRODUCTION

The study of observers interacting with virtual images dates at least from the first human recognition that the mirrored image produced by a smooth surface on a pond has a different call on reality than the objects which it mirrors. Though the structure in such an image can match the visual appearance of objects with high fidelity, attempts to touch the virtual image reveal it to be merely a chimera, albeit well detailed. Furthermore, it is slavishly linked to the real objects that are the source of the image and devoid of most normal visual interactivity.

Nevertheless, the virtual image seen in a mirror has a very distinct apparent depth, which may be manipulated by changing the binocular convergence needed to see it without diplopia, the accommodation needed to focus it clearly, or the binocular disparity needed to see it in depth. Indeed, when a mirror is used to reflect a real scene, each distinct element of the scene provides its own distinct binocular stimulus to convergence, accommodation, and disparity and, when viewed in isolation, has correspondingly easily calculable apparent distances, especially for objects less than a meter distant.

Technical advances in miniature visual display technology and electronic position measurement have made possible the creation of a new class of virtual images, viewed through stereoscopes, which can interact with operators' changing viewpoints as if they were real objects: They are visual virtual objects. Increasingly, rich sensory display devices are promising to add to them auditory and tactile elements. However, because the visual components of these virtual objects are generally presented via stereoscopes, the binocular stimuli that they present are compromises. Usually only one fixed plane of accommodation is presented, though disparity and convergence may vary widely. Consequently, the apparent distance of these virtual objects may be ambiguous, especially if they are viewed in a see-through format that optically superimposes them on a viewer's visual world.

Until the past several years, most studies concerning the design of see-through, virtual image displays have only considered systems that present virtual images focused at large distances from the users' eyes (>2 m). Research on the design of cockpit Heads Up Displays (HUDs) is an example of such work (Weintraub and Ensing, 1992). The recent development of inexpensive, head-mounted, see-through displays, however, has extended their range of possible application to include many cases in which users manually interact with nearby virtual images. Possible applications include previewing mechanical assembly, surgical planning and training, and visualization of virtual objects produced by computer-aided design systems. Computer-generated images, which behave in these applications like real objects, are often described as "virtual objects" and raise questions about optical design and information formatting not previously confronted. The studies in this chapter examine several such questions concerning some cues to distance that influence operator interaction with objects within arms reach.

The perceptual cues to space have been classically separated in terms of the presence or absence of movement or binocular information. More recent analyses of depth perception focusing on the behavioral affordances of vision have more usefully reclassified the classical depth cues into three categories: those primarily important with respect to personal space (e.g., stereopsis; (2 m ~ 1–2 eye heights), those mainly relevant for action space (e.g., self-motion parallax; 3–30 m ~ 2–20 eye heights), and those especially relevant for vista space (e.g., aerial perspective; >30 m ~ >20 eye heights) (Cutting and Vishton, 1995).

This reclassification of sources of information concerning the spatial layout surrounding a viewer is particularly useful since it focuses attention on what vision is to be used for in each of these distinct regions of space. Not surprisingly, observers' sensitivity to detecting changes in depth varies across the different cues and the different cues have varying importance for different regions (Nagata, 1993). In particular, binocular convergence, accommodation (required focus) and associated reflexes play roles mainly relevant for the personal space associated with coordinated, manipulative activity. In this region accommodation itself was classically not thought to be a direct or potent influence on perceived depth (Graham, 1965). However, through the accommodation–vergence reflex (Krishnan, Phillips & Stark, 1973; Ciuffreda, 1991; 1992; Semlow & Hung, 1983), accommodation can indirectly, but significantly influence perceived depth by causing a binocular-vergence-associated rescaling of perceptual space. And other results suggest that some observers may be able to use accommodation itself as a depth cue (Fisher & Ciuffreda, 1988). Consequently, investigations of the localization of nearby virtual objects need to consider the role of accommodation and its associated reflexes.

Understanding the interaction of these physiological reflexes on depth perception will have growing importance as head-mounted displays of virtual objects are introduced into the industrial and medical work place. Recent applications of such displays are designed to present to their users nearby, spatially conformal, computer-generated virtual objects for medical and manufacturing applications (Rolland, Ariely, & Gibson, 1995; Janin, Mizell & Caudell, 1993; Azuma & Bishop, 1994). This work has focused attention initially on precise calibration and alignment of the displays but now needs to be expanded to include the study of perceptual

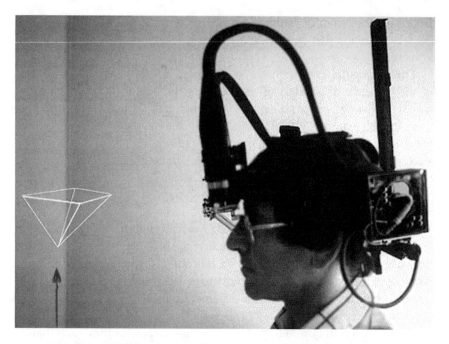

FIG. 8.1.   Side view 1996 of the head-mounted haploscope used
in the experiments summarized in this chapter. It was designed to
allow independent left and right eye bore-sight alignment through
the use of rotatable prisms and electronic adjustments. Alignment
accuracy was 3–5 arcmin. Focus was adjusted by either rotating
the lens barrel or introducing an optometric trial lens between the
collimating optics and the semisilvered mirror. Focus was verified
with a callibrated camera. Schematic overlays illustrate a virtual
inverted pyramid and object and real pointer, which was used to
localize.

phenomena that might degrade operator performance even in well-
calibrated systems.

The following four experiments explore such phenomena by examining
subjects' ability to adjust the distance of a physical pointer to match that
of a nearby virtual object. The object is generated by a high performance,
computer graphics system and presented by a head-mounted, see-through
display (Figure 8.1). This localization task was selected since it is close to
the visual-manual manipulation expected of users of virtual objects and its
precision and accuracy can be easily and reliably measured. Preliminary
testing, for example, showed that our subjects could set our mechanically
displaced, physical pointer to match the distance of physical targets with
several millimeter accuracy and that this accuracy corresponded to their

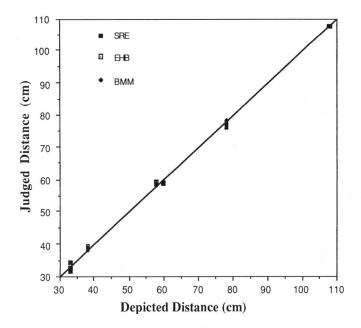

FIG. 8.2. The absolute stereoscopic calibration of the haplo-scope is verified by three subjects' adjustments of the physical cursor to match stereoscopically presented virtual object targets, the inverted tetrahedra described in the text, at several different distances. Each subject made 5 or 10 independent settings of the cursor using the method of adjustment to localize virtual targets at each depicted distance. Many data points are invisible because of overplotting. The level of performance indicated in this figure requires correct adjustment of the display optics and computer graphics eyepoints to match the subjects' intrapupilary distances with an accuracy of one millimeter. The correct display magnifi-cation was measured with an accuracy of 2%. Naive or inexperi-enced subjects generally do not show such precise localization but their average accuracy is comparable.

ability to match target distances with their fingers (Figure 8.2). Addition-ally, the use of our pointer allowed examination of targets just beyond arms reach.

The initial experiment below examines the effects of three different view-ing conditions on the subjects' accuracy of placement of the physical pointer under a virtual object. The conditions are: 1) monocular, 2) biocular or 3) stereoscopic viewing. These represent a range of cost and image fidelity (i.e., the completeness with which an objects' physical characteristics are presented). The monocular condition presents a virtual image of an object much as monocular helmet-mounted sights do and represents a minimal

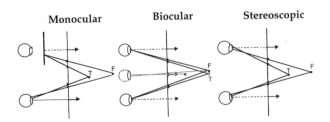

FIG. 8.3. Viewing geometry for each viewing condition for a target at T and fixation at F.

hardware/software rendering cost for such virtual image displays. But the monocular virtual images it presents are subject to visual suppression due to binocular rivalry. The biocular condition, which presents two identical virtual images to the subjects' left and right eyes, can avoid the rivalry problem. The images it presents are projected from a cyclopean position between the viewers eyes but are shifted to allow unstressed fusion at a selected distance (see Figure 8.3). This condition halves the rendering cost with respect to stereo displays but doubles the head-mounted display hardware requirements with respect to monocular displays. It presents a pattern of disparity approximating that of a transparent flat display surface and represents an intermediate cost system with potentially more stable image brightness. The third condition, conventional stereo display with parallel viewing vectors, presents the highest spatial fidelity but doubles both the rendering and hardware display cost compared to the monocular condition.

Because the monocular and biocular viewing conditions degrade the fidelity with which distance is presented, we expect that the stereoscopic display should support the most accurate localization. The biocular display may, however, provide a competitive alternative for virtual objects with relatively little internal depth by avoiding the potential problem of binocular rivalry. The first experiment provides a descriptive study of relative localization accuracy of virtual objects presented via the alternative viewing conditions. Since observer age and accommodative demand (required focus) could be expected to interact and influence localization accuracy, these display characteristics are also examined in the first study to develop designer guidelines for the adjustment of focus and selection of personnel to use virtual object displays. Subsequent studies consider the effect of the introduction of a nearby physical surfaces on the localization of the virtual object, identify a phenomenon that introduces errors into such localization, and explore two alternative explanations for the phenomenon.

EXPERIMENT 1

Methods

Apparatus

All experiments reported in this chapter used a custom, head-mounted, see-through display, called an electronic haploscope by the authors, capable of presenting a 20–30° diameter circular monocular field to each eye with variable monocular overlap. In the following experiments the system was used at 100% overlap and 20° field of view. The display system used two vertically mounted Citizen 1.5′ 1000 line miniature Cathode Ray Tube's (CRT) in National Television Standards Committee (NTSC) mode, which were driven by a Silicon Graphics (SGI) computer (4D/210GTXB) through custom video conditioning circuits. For the simple 3D imagery used in the following experiments, the computer could maintain a 15 Hz graphics update rate. The CRT images were infinity collimated by standard glass telescope eyepieces (e.g. Erfl 32 mm and Ploessl 42 mm) mounted directly under the CRTs. After the signal transformation from the RGB to NTSC, individual pixels, which corresponded in the current configuration to 3 arcmin horizontal resolution measured from subjects' eyes, were easily discriminated. The collimated light could be modified by lenses and rotating prisms from a standard optometric trial lens set that allowed precise positioning with at least 5 arcmin resolution of the separate left and right images and allowed variation of the accommodative demand for each eye. The images were relayed to the subject's eyes by custom, partially silvered (15%) polycarbonate mirrors mounted at 45° directly in front of each eye. The left and right viewing channels could be mechanically adjusted between 55 mm and 71 mm separations for different subject's interpupilary distances. The video signal conditioning also allowed lateral adjustment of the video frame. Consequently, the display system can precisely position the center of each graphics viewport directly in front of the eyes of all subjects for bore-sight alignment.

The entire display system (Figure 8.1), built around a snug fitting, rigid headband, is intended to be worn by a freely moving subject and weights between 0.77 and 1.1 kg depending upon configuration. In the lightest configuration the moments of inertia have been measured when mounted on an erect head to be: 0.0782 kg-m^2 vertical axis, 0.0644 kg-m^2 longitudinal axis, and 0.0391 kg-m^2 lateral axis. In all of the experiments described below, the band was fitted to each subjects head and then supported by a special pivoted mount at the end of a 1.8 m table. This mount restricted horizontal movement but allowed some pitch movement. Subjects sat at this end during the course of each experiment. The mount and chair were adjusted so that the virtual objects could be presented at eye level. Lateral head movement was restricted during all the following experiments but a residual pitch of ±10° was allowed for subject comfort. In practice the subjects were monitored by the experimenters to keep their heads approximately level at an individually selected orientation during the course of the experiments.

Stimuli

A monocular, biocular, or stereoscopic virtual image of an upside-down, axially rotating (∼3 rpm) pyramid was presented at a distance of 58 cm away from the subjects' eyes by a head-mounted see-through display. It was seen against a grey, cloth covered wall 2.2 m from the subjects. Preliminary experiments examining varying the rotation rate of the pyramid for each trial showed that such variation had no effect on the localization of the virtual image. The reference distance of 58 cm was chosen for the experiment because it corresponded to a possible working distance for several industrial applications of interest to the authors. All displays were operated under moderate indoor artificial illumination (approximately 50 lux). The virtual image was presented with either 2 diopters accommodative relief or at optical infinity. The stereo display was, however, calibrated (see below) over a range of 30 to 110 cm. The monocular display was simply the stereo channel that corresponded to the subjects' dominant eye. The biocular display was produced by positioning the graphics eyepoint midway between the subject's eyes. The left and right images were identical copies of this view but were shifted laterally so that when the subject's eyes converged to the centers of each viewport they would have 0 disparity relative to the reference convergence point of 58 cm. The plane of 0 disparity was thus set so that the subjects could easily fuse the images when converged at 58 cm. This technique was used in general for all biocular stimuli at different depths, which were experimentally interjected as described below. Though no keystone correction was applied for the several arcmin of distortion caused by the image shift, the disparity pattern produced in this biocular image closely approximates that of a flat image of the target as if it were drawn on a transparent projection surface at the simulated convergence distance.

The wire-frame pyramid had a nominal 10 cm base and 5 cm height. The width of the wire-frame lines were about 9 arcmin. The depicted size of the presented virtual object was randomly scaled from 70 to 130% of its nominal size for each trial to interfere with subjects' possible use of angular size as a depth cue. The lines of the wire frame an all other computer-generated lines had a luminance of about 65 cd/m^2 and were seen against the gray cloth background of 2.9 cd/m^2 with visible vertical seams. While the presented luminance did not approach that used in aircraft heads up displays, which must be visible against a 30,000 cd/m^2 background, it was just adequate for indoor work against most colored surfaces and is at least 3 times brighter than the luminance available in off-the-shelf, see-through head-mounted displays such as the IGlasses[TM] formerly made by VIO.

Four out of every thirty judgments were based on unanalyzed random variations in the depicted depth of the pyramid. This variation was introduced to help ensure that the subjects did not notice that the same depicted depth was repeated. However, the major factor masking the repetition of the depicted depth were the perceptual effects causing changes in the apparent depth with different viewing conditions. Since the viewing conditions could be unobtrusively intermixed and no feedback was given to the subjects, there was no way for them to tell that the depicted distance was not in fact changing. In fact, no naive subject in any of the following experiments reported noticing the repetition of the depicted target depth.

Task

The subjects' task was to use a method of adjustment to position the binocularly visible, physical cursor, a yellow-green light emitting diode (LED) (about 20 cd/m^2) pointer, shaped like a pyramid (base 0.5 cm, height 1 cm) into vertical alignment with the apex of the inverted pyramidal virtual object. The physical cursor was moved on a rail by a chain and gear system and positioned about 2 degrees of visual angle below the virtual object. The distance to the pointer was automatically recorded through use of a shaft-encoder interfaced to the display computer. The adjustment was self-paced, but subjects were encouraged to take between 15 and 30 seconds for each adjustment and were allowed to take breaks at one-half to one hours intervals as needed. As part of the standard procedure for use of human subjects, all subjects were informed they could terminate the experiment at anytime if they experienced any undo discomfort and were asked at the end if they experienced any viewing difficulties seeing the virtual objects.

Stereo Calibration

The haploscope display system was adjusted by monocular superimposition of reference virtual images on an 18 cm diameter circumscribed circle presented at a distance of 2.2 m. In addition to position adjustment this allowed adjustment of the field of view angle of the graphics system to match the total magnification of the system. This was done separately for each eye and for each subject before the experiment was started. Thus, we could account for any changes due to variation of accommodative demand and corrective lenses that might be worn by the subjects. The subjects' interpupilary distances (IPDs) were measured with a binocular-type viewing device with digital readout (Varilux Model: Digital CRP). All displays and algorithms were adjusted to reflect the measured IPD values.

Preliminary test results for virtual targets placed between 33 and 108 cm showed that within the full range of adjustment used for the experiment subjects using a stereo display could align the cursor within ±0.3 cm of the depicted virtual object target depth (Ellis & Menges, 1997a). The distance responses were completely linear, unbiased, and unskewed and were conducted in the same full room illumination as the experiment. Similar tests of the biocular viewing condition showed equally linear responses. Tests in the monocular condition showed, expectedly, inconsistent behavior. (Figure 8.4) In further examination of the localization technique pilot subjects were asked to use the pointer to match the depth of physical targets. These tests showed linear, unbiased, unskewed estimates with statistical ranges of ±0.15 cm about depicted physical distances for targets used in the experiment.

Subjects

Ten subjects, five young (15–29 yrs.) and five older (38–47 yrs.), participated in the experiment. Subjects in the older group could be presumed by population data to be at least early presbyopes (Moses, 1987). All but one young and one older subject (i.e., the authors) were naive with respect to the purpose of the experiment. The others were either paid subjects recruited through the Ames Bionetics contractor or were laboratory personnel. All subjects were screened on the Bauch & Lomb Orthorater stereo tests for

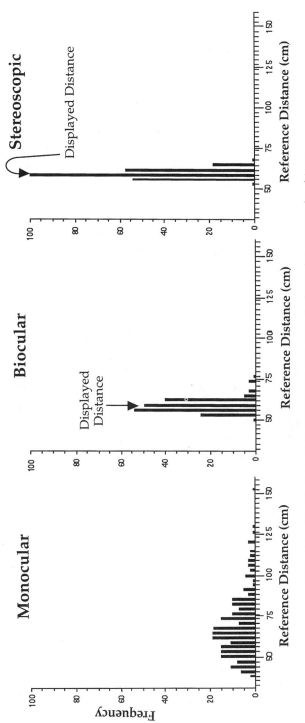

FIG. 8.4. Distribution of localizations of a virtual object for monocular, biocular, and stereo viewing conditions. Results from 25 subjects collected from current and previous experiments illustrate the relative variability of localization for monocular, biocular, and stereoscopic viewing. Five of the subjects were laboratory personnel, the remaining were naive subjects selected by a laboratory contractor (Ellis & Menges, 1995). Accommodative demand was set at 2 diopters (50 cm) for all judgments. Since the size of the displayed object was varied sufficiently to preclude size as a distance cue and no lateral head movement was allowed, the clustering of judgments for the monocular viewing condition is due to the accommodative demand and response biases inherent in the measurement conditions. In fact, the monocular objects do not have a displayed distance.

273

stereoacuity better than 1 arcmin. Subjects who normally wore prescription spectacles were allowed to wear them during the screening test and during the experiment. During pilot testing for the experiment, inadvertent errors of 0.1–0.2 cm in modeling of subjects' interpupilary distances in the graphics simulation produced easily noticeable artifacts. Precise stereo or biocular presentation of virtual objects evidently requires measurement and modeling of interpupilary distance with an accuracy on the order of ±0.1 cm.

Design

Viewing Conditions (monocular, biocular, and stereoscopic) were crossed with Accommodation (0 or 2 diopters) and nested within Age groups. The experiment used a blocked design in which blocks of 5 replications of a given condition were presented for each of the 3 viewing conditions producing uninterrupted 15 judgment sequences. The sequence of viewing conditions was randomly assigned to each subject and thereafter systematically permuted after each set of 3 viewing conditions were presented. In general, it was possible to switch the viewing conditions solely through software. Thus, the subjects were generally unaware which viewing condition was presented and the perceptual variation in apparent depth caused by variation in view condition was readily interpreted by them as variation in depicted depth. The viewing conditions were blocked for a given accommodative demand, which was switched by interrupting the experiment after every 15 trials to change the viewing lenses. The order of presentation of accommodative demand was permuted within subjects and balanced across subjects.

Results

Analysis of variance (ANOVA) showed that the viewing conditions had a major effect on the bias of the subjects distance judgments ($F = 15.580$; $df = 2, 16$; $p < .001$). The mean stereoscopic and biocular localizations were almost completely correct, but a judgment bias appeared as an overestimate when the stereo depth cues associated with the virtual object were removed by monocular viewing. This effect interacted with accommodative demand and age, as indicated in Figure 8.5 ($F = 7.76$; $df = 2, 16$; $p < .004$). All other effects are related to this three-way interaction and will not be discussed individually. No subjects reported any difficulties seeing the virtual objects during the experiment.

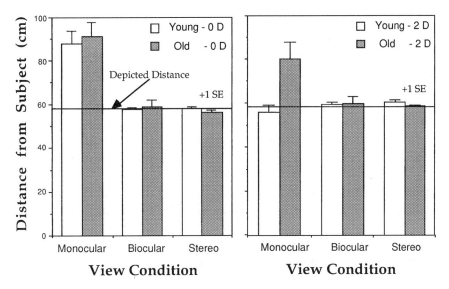

FIG. 8.5.   Effect of Age and Accommodative demand on localization.

Discussion

The results of the ANOVA plotted in Figure 8.2 show that when depth cues to the virtual object are degraded to monocular conditions, judgments of its distance drop back toward the distance of the background wall at 2.2 m from the subjects. A phenomenon that could explain the increased judged target distance in the monocular condition is the specific distance effect, which causes visual targets of unknown physical size presented with weak or ambiguous depth cues to tend to appear in visually impoverished environments about 2–3 meters away (Gogel & Tietz, 1973; Foley, 1993). This distance effect is also associated with tonic accommodation and vergence, which relax to approximately 1–2 meters in the absence of distance cues (Owens & Liebowitz, 1976). Changes of convergence to a more distant resting position could cause the localization of the virtual object to recede from the viewer. But since these effects are generally seen when targets appear against featureless backgrounds they are not likely important for the present results.

Another phenomenon probably more relevant is the direct effect of change in ocular convergence on perceived depth (Owens & Liebowitz, 1983; Zuber, 1965). It is the kind of result to be expected if the subject's actual convergence was driven or attracted by the wall, which provided a visually sharp, textured cloth with vertical seams at 2.2 m. In the monocular condition, the only source of information that the virtual object is located any particular distance in front of the background is provided by accommodative demand. It is therefore not surprising that when 2 diopters of accommodative demand was provided, only the subjects young enough to respond to this cue were able to localize the virtual object approximately correctly. In these subjects the accommodative-convergence reflex allowed them to maintain convergence on the virtual object. Older subjects unable to respond to accommodative cues only have the disparity information provided by the background cloth to control their vergence. Thus, they would still diverge to fuse the background and the monocularly presented virtual objects would still appear toward or on the background wall.

That the monocular virtual objects are not judged to be exactly on the wall reflects the response bias within the experiment originating from constraints on movement of the physical cursor, unavoidable guesses the observers may have made about the approximate size of the object, and the possibility that the observer converged to other closer objects such as the pointer and its supports, which, though darkened to be less conspicuous, were still visible against the background. In fact, if the observers were to hold their eyes completely still, the monocular viewing situation is very similar to that of viewing a monocular afterimage in a demonstration of Emmert's Law, the classic observation that an afterimage often appears at the distance of a physical surface against which it is projected (Brown, 1965). The readily available correct disparity information in the biocular and stereo conditions, however, provide the missing cue that allowed all observers to correctly judge the distance to the virtual object.

Finally, since no difficulties seeing the virtual objects were reported during the experiment, we find no evidence that binocular rivalry interfered with subjects' ability to see the virtual objects in the monocular condition. The finding is consistent with all observations we have made of the monocular wire-frame virtual objects during preparation for the experiment. Apparently, the high contrast and motion of the objects we have examined easily overcome any binocular rivalry that might be present.

EXPERIMENT 2

Experiment 1 examined the effect of different viewing conditions on the localization of virtual objects superimposed on a physical surface 2.2 m distant. But since new uses of virtual objects are likely to bring them closer to physical surfaces, Experiment 2 examines the effect of introduction of a much closer physical surface. In view of the interacting roles of accommodative and convergence in the discussion of Experiment 1, one could reasonably expect the introduction of a nearby real surface to cause the observers to localize monocularly viewed virtual objects at the same distance of introduced surface. If the accommodative demand for the virtual object is already matched to its displayed distance, one would expect that this improvement in the accuracy of localization would be larger for older observers than younger ones who already would have accommodative cues to the virtual object distance (Ellis & Menges, 1997b). Accordingly, subjects of different age groups were used while correct accommodative demand to the virtual object was provided. In order to study the effects of introduction of a physical surface, subjects must first judge the distance of the virtual image by itself. This first judgment is identical to those made with the viewing conditions in Experiment 1 with 2D accommodative demand and provides a chance to replicate that part of the experiment.

Methods

Stimuli

The virtual image stimuli used in Experiment 2 were identical to those in Experiment 1 for the 2 diopter accommodation condition but a new physical stimulus was introduced. This physical surface was a slowly, irregularly rotating checkerboard (~ 2 rpm) made of xeroxed paper glued on foam-core and was mechanically introduced along the line of sight to the pyramid as illustrated in Figure 8.6. Motion of the checkerboard was introduced because preliminary testing showed that changes in localization that it produced were enhanced by motion. The checkerboard was a disk 29 cm in diameter with 5 cm black and white checks having either 1.3 cd/m^2 or 17.8 cd/m^2 luminance. It was positioned so that the virtual image of the pyramid could be seen against the lower rim of the disk in order to allow the subjects to adjust the physical cursor to the apparent distance of the virtual image in the presence of the disk. Care was taken to be sure the physical cursor was below the bottom of the virtual object and the edge of the disk as in Experiment 1. As before, subjects viewed the virtual objects with monocular, biocular, or stereoscopic view conditions.

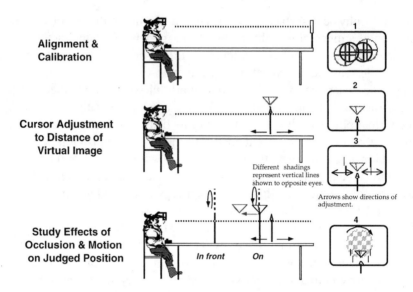

FIG. 8.6. Experimental procedure illustration for Experiments 2, 3, and 4. Top: alignment, magnification, and interpupilary adjustment. Middle: Initial localization of virtual object depth. Bottom: Testing conditions representing the "on" or "in front" placement of the rotating checkerboard and the second localization of the virtual object depth. The rightmost panels 3 and 4 represent use of nonius lines to detect relative convergence in Experiment 3.

Subjects

Thirteen subjects, seven young (15–29 yrs.) and six older (38–47 yrs.) participated in the experiment. All but one young and one older subject (the authors) were naive with respect to the purpose of the experiment. The others were either recruited through the Ames Bionetics contractor or were laboratory personnel. All subjects were screened for stereo vision as in Experiment 1.

Task

The first part of the subjects' task was to mechanically place the yellow-green LED pointer under the nadir of the slowly rotating, wire-frame virtual pyramid, which varied randomly in size for each trial as in Experiment 1. The second part of the task involved an adjustment of the pointer to match the pyramid's distance after the slowly, irregularly rotating, opaque checkerboard was introduced along the line of sight to the pyramid. The checkerboard was introduced at the previously judged distance of the apex of the virtual pyramid. This fact was unknown to all the naive subjects and remained unnoticed throughout the experiment. Though the virtual pyramid was also presented a second time at the same distance as the first localization, the experimental variations generally concealed this fact from the naive subjects who were led to believe each trial, with or without the checkerboard, involved a potentially different depicted depth. As in Experiment 1, the occasional introduction of unanalyzed sham targets at different depths enforced the naive subjects' belief that the virtual image possibly could be displaced variously in depth for every localization.

Results

Analysis of the subjects' first localization of the virtual object under the three viewing conditions with 2 diopters of accommodative demand closely replicates the findings of Experiment 1. The basic result is a significant two-way interaction of view condition and age ($F(2, 42) = 19.160, p < 0.0009$) in which age variation affects the judged distance only for the monocular viewing condition, with the younger subjects (Figure 8.4).

Analysis of the offset of the mean judged distance to the virtual object associated with the introduction of the physical surface also showed a main effect of viewing condition ($F(2, 26) = 91.340, p < .0001$) and a significant interaction between viewing condition and age ($F(2, 26) = 21.921, p < .0001$) (Figure 8.4). These effects modulated the overall significant offset ($F(1, 13) = 90.623, p < .0001$) of the judged distance to the virtual object toward the viewer that was caused by introduction of the physical surface. This effect appears for all viewing conditions as a closer localization of the target after interposition of the physical surface.

Discussion

The first virtual object localization shown in Figure 8.7 does replicate the 2D viewing conditions in Experiment 1 showing that the older subjects are unable to use the accommodative information to estimate the virtual object distance and consequently localize the virtual object erroneously toward the background wall. Interestingly, interposition of the checkerboard at the judged distance of the monocularly viewed virtual object causes a substantial forward movement of the judged virtual object position (Figure 8.7). This change is what would be expected if insertion of the disparity and other cues to nearness of the checkerboard were to cause relative convergence with respect to the position of eyes before its introduction.

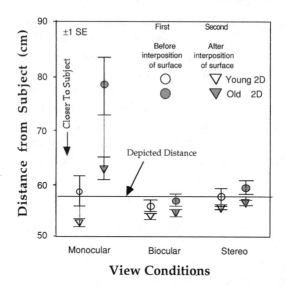

FIG. 8.7.   Interaction of Age and Viewing Condition.

It is important to note that the forward movement of the localization is from the *initial judged* position for each appearance of the isolated virtual object. Since the older subjects tended to initially judge the virtual object to be too far away, the tendency for the checkerboard to bring the judge distance closer was corrective (see Figure 8.7). On the other hand, the younger subjects viewing the virtual object monocularly did not significantly misjudge the virtual object distance, as evident from the error bars. For them introduction of the checkerboard was detrimental, causing the virtual object to be judged too near.

In fact, the checkerboard insertion caused the virtual object to generally appear too near for the other two viewing conditions, which otherwise supported correct localization of the isolated virtual object. This was generally true for both age groups and suggests that introduction of the physical surface, the checkerboard, could be causing a small relative convergence under these conditions as well. Though disparity information for correct convergence to the virtual object is available under these two conditions, the virtual object providing this information is not of high visual quality. Its presentation corresponds roughly to 20/100 visual acuity. Under such conditions, convergence based on stereoscopic cues might not be precise and could exhibit a fixation disparity. If this error were an exofixation disparity (i.e., it would tend to the distant side), introduction of the high visual fidelity checkerboard could correct it, causing a relative convergence and associated decrease of the judged distance to the virtual object. If this small corrective convergence were incompletely compensated due to the breakdown of distance constancy, errors of localization could be expected. Thus, the change in localization after introduction of the checkerboard could be due to a change in static convergence.

EXPERIMENT 3

Experiment 3 explicitly tests for such a change associated with the change in judged distance. Attention is focused on the stereoscopic viewing condition to see if the closer judged distance of the virtual object associated with introduction of the physical surface could be associated with an increase in static convergence. Since the amount of expected change is small, for example a change of 3 cm from 58 cm to 55 cm, a sensitive measure of convergence is needed. Angular changes of monocular position of only about 3 arcmin would be expected if the change in localization were explainable by vergence change alone. Such a measurement is difficult to make without encumbering eye-tracking technology that preserves a clear visual field, but it is conveniently just at the display resolution of the display configuration used. Therefore, a technique using nonius lines on the display itself was adopted (Ellis, Bucher & Menges, 1995). A nonius line is a line that is broken into two line segments, each of which is visible by only one eye. Such lines have typically been used to measure equivalent oculocentric directions to determine the position of the stereoscopic horopter. When the two segments are moved laterally during a period of constant convergence so that they appear to be collinear, their positions may be used to record a specific convergence position.

Methods
Stimuli

Since pilot studies had suggested that longer distances enhance the effect of introduction of the checkerboard, the virtual pyramid was presented at 108 cm rather than 58 cm away from the subjects' eyes. Such an enhancement was deemed helpful to increase the detectability of any relative convergence. This display, otherwise similar to that used in Experiment 2, was operated under normal room illumination with 1 diopter accommodative relief for the virtual image. Flanking nonius lines (lower right panels of Figure 8.6) of the same luminance and line width as the pyramid described in the Task section below were also occasionally presented to detect changes in static convergence.

Subjects

Five men and one woman with measured stereo resolution of better than 1 arcmin participated in the experiment. Some subjects had vision corrected by contact lenses or glasses and were able to wear their corrections during the experiment. Subject ages ranged from 17 to 47 and included laboratory personnel as well as paid subjects recruited by a contractor at Ames. Because of the computer control of the experiment it was possible to conduct this experiment double blind.

Task

The subject's task had three basic parts:

1. Localization of an isolated virtual pyramid and measurement of associated static ocular convergence
2. Relocalization of the virtual pyramid in the presence of a real surface either at or in front of the pyramid's apparent distance
3. Measurement of changes in static convergence associated with the relocalization

The first part of the subject's task was to mechanically place the LED pointer under the nadir of the slowly rotating, wire-frame pyramid. After aligning the pointer, the subjects were presented with two sets of vertical nonius lines just flanking the pyramid (Figure 8.6, right panel #3). These lines were then adjusted to appear vertically collinear (i.e., to have equal visual directions on each side of the pyramid). This adjustment was made by moving the lower left and right segments with a joystick control (see Figure 8.6) and effectively recorded the subjects' static convergence during this part of the experiment. Subsequent brief presentations of the nonius line will accordingly show how static convergence may have changed by revealing vertical misalignments. The second part of the task involved another adjustment of the pointer to the pyramid's depth after the slowly, irregularly rotating checkerboard was introduced along the line of sight to the pyramid. The pyramid was then presented a second time at the same depicted depth in this new configuration. As before, the experimental variations generally concealed this fact from the subjects so that they believed each trial, with or without the checkerboard, involved a potentially different depicted depth. And as before, unanalyzed trials with random variation in distance were introduced to maintain the subjects' uncertainty.

After the second judgment of the pyramid's depth, the nonius lines were flashed briefly (ca 250 ms) next to the pyramid while the subjects fixated it (Figure 8.6, right panel #4). Then the subjects made a forced choice indicating whether the upper or lower pair of the flashed nonius lines appeared closer laterally. The eye assignments of each segment of the nonius lines were randomly selected so that the meaning of the alternative possibilities in terms of convergence or divergence varied randomly across the trials. The assignment of the lower part of the left nonius line and the upper part of the right line to one eye, and the other upper-lower pair to the other eye, produced a differential effect doubling the relative misalignment for any given vergence change and increasing the sensitivity of the technique for detecting changes in convergence. The subject reported, by a button press, which of the paired nonius lines were closer.

In fact, three different experimental conditions were used in the second part of the experiment because of the need for a control case. In the "on" condition the checkerboard was mechanically introduced at the judged depth of the virtual pyramid object so that the pyramid appeared "on" the checkerboard. For the "in front" condition the checkerboard was introduced 30 cm in front of the judged depth. In the control condition the second judgment was a replication of the first judgment in that the subject made a second judgment of the depth of the virtual object. But this time the subject made the forced-choice judgment of the nonius lines alignment without the addition of the checkerboard. Thus, the control was identical to the experimental conditions except the checkerboard was not introduced into the line of sight. Therefore, this control provides an individual baseline for subjects' judgment biases and changes of their convergence during the course of a trial. Each condition was repeated 15 times for each subject in a randomized block design in which blocks of 5 replications of each condition were repeated. The 6 possible orders of the 3 conditions were distributed randomly across the 6 subjects in the experiment.

The change in judged distance of the virtual object was analyzed in a single factor repeated measures ANOVA. Chi-square analyses were conducted on each individual subject's distribution of judgments of convergence/divergence for each of the 3 experimental conditions. Taking the control condition as a baseline, the relative strength of convergence could be measured by a ratio of the probability of convergence in each experimental condition to the probability of convergence in each subject's individual control. This ratio allows control for the possibility that subjects might have an individual bias to converge or diverge simply because of a repeated presentation of the virtual object.

Results

Single factor repeated measures analysis of the effect of superposition of the checkerboard and virtual images replicated the previous observations that the virtual object was moved closer to the viewer ($F(2, 10) = 7.549$, $p <$ 0.01). Individual data are shown in Figure 8.8. This effect was somewhat stronger for the "on" condition than for the "in front" case and varied in strength across the six subjects. One subject interestingly showed essentially no effect.

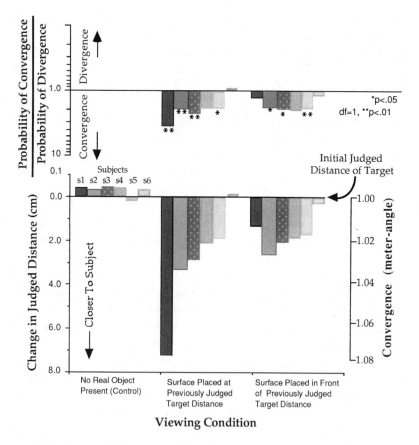

FIG. 8.8. Relative Convergence versus Change in Judged Distance.

TABLE 8.1

Frequency of Convergence to Divergence during Depth Judgments
of Virtual Objects

	Convergence	Divergence	Total Judgments	Ratio of Convergence to Divergence
On	84	21	105	1.58
In Front	70	35	105	1.32
Control	53	52	105	1.00

The cause of this individual subject's result is illuminated by considering all subjects' tendency to relatively converge during judgment of the depth of the virtual object in the presence of the checkerboard. This tendency is summarized for the experiment in Table 8.1, which displays the frequency of convergence or divergence indicated by the nonius judgments for all subjects in the three experimental conditions (chi-square $= 20.37, df = 2$, $p < .001$). The control case shows the expected 50 : 50 break, collapsing across all subjects, while the other two conditions show clear convergence, the "on" condition being somewhat stronger.

For further analysis each subject's individual tendency to converge was computed separately as the ratio of their probability of convergence in an experimental condition to their probability of convergence in the control. These ratios are plotted in Figure 8.8 for each subject. Since the control was used as the reference, all ratios for the control condition are 1. A 2×2 chi square contingency was also computed to compare the distribution of convergence and divergence for each experimental condition to that of the control condition. This was done separately for each subject for whom statistically significant differences in distributions are indicated by asterisks (Figure 8.8).

Discussion

The individual subject's localization errors in Figure 8.5 are sorted by the size of the change in the judged position of the virtual object for the "on" condition. These results can then be compared with the ratio of the convergence probabilities. As is clear from the figure, the two measurements are almost perfectly correlated across the subjects. The only subject not to show a displacement of the virtual object caused by the checkerboard also is the only one to show essentially no relative convergence. The subject showing the largest displacement due to introduction of the checkerboard is also the one with the strongest tendency to converge. The results for the "in front" condition show a weaker apparent displacement of the virtual image but also show a correlation of convergence tendencies and changes in localization. The correlation of relative convergence with magnitude of displacement for the "on" and "in front" conditions across subjects and conditions is, in fact, $r = 0.894$ ($df = 10; t = 6.31, p < .002$). These results generally support the supposition that the change in judged depth could be related to a change in convergence, but the mechanism underlying this change remains to be clarified. Correction of a fixation disparity, for example, by introduction of a high-resolution physical stimulus, for example, is not the only possible mechanism.

One other possibility is that change in convergence is due to so-called perspective (Enright, 1991) or proximal vergence (Cuiffreda, 1992). These phenomena are changes in convergence due to changes in the apparent nearness of objects. They provide evidence that spatial interpretations of the distance of a visual image themselves can simulate the vergence system. Accordingly, the results from the present experiment, while showing that there is a clear oculomotor response associated with the error in judged depth, does not resolve its cause. A change in the apparent nearness of the virtual object due to its appearing to occlude the nearby checkerboard could be the cause of the measured convergence rather than its consequence. This question can only be resolved experimentally.

EXPERIMENT 4

One approach to analyzing whether the oculomotor effect (i.e. the convergence) observed in Experiment 3 is caused by the superposition of the virtual object on the background is to devise a stimulus condition that on one hand strengthens the oculomotor cues to convergence but on the other hand weakens the visual evidence for occlusion, thus reducing the likelihood of convergence caused by proximal vergence.

We have attempted to create such a stimulus by cutting an annular slot 8 cm wide out of our rotating checkerboard so that the virtual pyramid would be just able to "fall through" the resulting hole (see Figure 8.9). The

FIG. 8.9.   Slotted physical surface.

outer rim of the checkerboard was supported by thin radial wire matching the color of the background wall and therefore being invisible to the observers. This stimulus triples the number of moving edges that provide the strong disparity discontinuity, which could be the stimulus to convergence that could be the cause of the change in static convergence observed in Experiment 3. If the better stimulus to convergence provided by the checkerboard were the cause of the change in static convergence, this stimulus should strengthen the effect. On the other hand, the slotted hole virtually eliminates the visual evidence for occlusion. If proximal vergence triggered by occlusion were the cause of the change in judged distance, one would expect not to find a change in the judged distance of the virtual object when it is presented in the slot.

Methods

Subjects

Nine subjects, aged 23–47, who were either laboratory personnel or paid subjects provided by Bionetics were used in Experiment 4.

This experiment was conducted using a methodology equivalent to that of Experiment 2 for stereoscopic virtual objects. Two different depicted distances of the virtual objects, 83 cm and 108 cm, were randomly ordered into blocks of 20 runs. In each block either a solid checkerboard or a slotted checkerboard was introduced along the line of sight of the virtual object.

After introduction of the checkerboard, the change in the judged distance to the virtual object was measured. Block types were alternated for all subjects. All but one subject, from whom half of his data were lost, experienced 4 blocks, making a total of 80 judgments per subject. The order of presentation of the two checkerboard types was counterbalanced across subjects.

Results

Analysis of variance showed that while the solid disk caused a previously observed offset of the judged virtual object distance toward the observer (mean: 2.80 cm; SE: ±0.75 cm), the slotted disk caused only a mean 0.72 cm (SE ±0.49 cm) change. This difference was statistically significant ($F(1, 8) = 19.605$, $p < 0.002$). The offset for the slotted disk condition, though quite close to 0, is statistically significantly less than 0 ($t = -2.59, df = 8, p < .04$)

The size of the offset was larger for the greater depicted distance. The offset for the 83 cm virtual object was 1.49 cm (SE ±0.43) and 2.03 cm (SE ±0.57) for the 108 cm object. This difference was just significant ($F(1, 8) = 5.349$, $p < 0.05$). There was no statistical interaction ($F(1, 8) = 3.068$, $p > 0.05$), so the effects of placing the virtual object in the slot was statistically indistinguishable for the two presentation distances. Accordingly, the data from the two presentation distances may be collapsed as in Figure 8.10. This figure shows the full distribution of all the subjects responses illustrating the effect of the presence of the slot on the judged distance to the virtual object.

Discussion

As is clear from Figure 8.10, introduction of the slot in the checkerboard that removed the occlusion between the virtual object and the checkerboard greatly reduced the offset in judged virtual object position produced by the checkerboard introduction. This reduction makes the oculomotor explanation of the change less likely since the binocular depth cues of the checkerboard would be expected, if anything, to strengthen any vergence response. Since the overlap of the virtual object contours and those of the checkerboard seems to be the key feature causing the shift in its judged position, the occlusion cue placing the virtual object in front of the checkerboard seems to be the best explanation for the nearer localization of the virtual objects. The change in static convergence, thus, appears to be a consequence of proximal vergence and resembles effects in recent reports

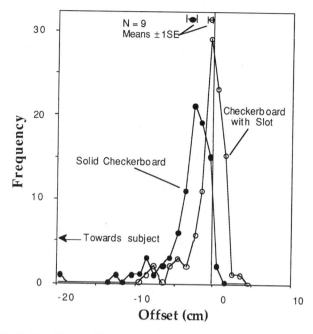

FIG. 8.10. Effect of hole on judged distance to virtual objects.

of convergence being driven by the kinetic depth effects that produce a perception of nearness (Ringach, Hawken & Shapley, 1997).

There is, however, another possible interpretation of the results, which could be based on the proposal that the alternative explanations of the offset of the judged distance of the virtual object are not mutually exclusive (W. Shebilske, personal communication, 1997). In this view the introduction of the slotted checkerboard stimulus could introduce a proximal vergence tendency to fixate farther away since any objects seen through the slot would necessarily be more distant than the slot. Were this tendency for a more distant fixation to occur, it would oppose any tendency of the visual information to reduce oculomotor bias. The net effect of the opposition could account for the almost negligible forward displacement of the virtual object observed while subjects were exposed to the slotted checkerboard.

One way to dissociate the two influences could be to study the individual oculomotor biases of each subject, measuring for example their phoria with a Maddox Rod test or equivalent. To the extent subjects exhibit exophoria, one could expect a exofixation disparity while viewing the virtual object. Its removal by presentation of the checkerboards could be associated with the change in judged distance. An explanation of the observed effects of the checkerboards solely in terms of proximal vergence would not predict a correlation between the phoria tests and the subjects' individual phoria variations. Accordingly, future investigations could examine individual subject's oculomotor biases as a technique to more precisely determine the cause of the offset of the change in judged distance associated with superimposition of virtual objects against physical surfaces.

DESIGN CONSIDERATIONS

1.   The present results were observed with a static eyepoint and can be expected to change when significant lateral head-movement producing motion parallax is introduced. Nevertheless, it is important to realize that since many of the new applications of head-mounted see-through displays, in fact, will involve relatively static viewing, the conditions used remain practically relevant.

2.   Since weight and cost considerations may argue for the use of monocular displays, they are likely to be initial candidates for many applications. Accordingly, such displays should have a variable focus control to appropriately direct the convergence of users younger than 40. Designers and supervisors should be aware that operators over 40 may not substantially benefit from the variable focus adjustment. For these viewers the effect of introducing real objects near the part of the visual field containing virtual objects will be much stronger if the virtual objects are presented monocularly.

3.   Biocular and stereo displays should be used with a bore-sighting procedure in which focus is adjusted to a reference target so as to correct for any errors in depth due to inappropriate vergence.

4.   Computer-generated or other targets presented binocularly should have individually reduced binocular parallax to correct their spatial localization so as to compensate for the tendency of virtual objects to appear to float in front of the surfaces that they are seen against.

General Appearance of Nearby Virtual Objects Close to Physical Surfaces

In addition to affecting the localization of a virtual object, a proximate physical surface can markedly affect its appearance. For example, when a fixed size, monocularly viewed virtual object is projected against physical surfaces at different distances, its apparent size appears to grow and shrink as the distance of the surface increases or decreases. The head restraint, discrete nature of specific testing conditions, and the size randomization used in the present experiments prevented our subjects from noticing this effect during testing. But it can be very prominent and it was readily noticed during pilot testing and when subjects were allowed to continuously view virtual objects while their heads were unrestrained.

Another prominent phenomenal effect occurred during the presentation of the opaque physical surface in front of the location of the stereoscopic virtual object in Experiment 3. In this case, the surface appears to change its physical properties and become transparent in the regions where the virtual object overlaps. The placement of the surface 30 cm in front of the virtual object was selected to ensure that this transparency would be induced for all subjects. The effect is a very striking one because its onset can be very sudden and can depend on the distance the surface is placed in front of the virtual object. It is most strikingly demonstrated with a moving surface that is initially behind a virtual object. As the surface is moved forward, the object first is "pushed" along in front. But after a critical distance varying from subject to subject, the virtual object will abruptly "fall back" behind the surface and in doing so impart a sense of transparency to the surface. This transformation is quite striking, especially when the surface is an observer's hand. When the demonstration is conducted close to the observer so that changes in their ocular convergence is easily visible, a divergence associated with the falling back of the virtual object is easily seen. As is shown in Figure 8.5 from Experiment 3, the localization of the virtual object is nevertheless still brought closer to the observer by the surface, even when the surface becomes transparent. This observation thus underscores the particular difficulty faced if a designer of an optically overlaid virtual object wishes to place the object precisely inside or behind a physical surface or object. The resultant distortions in the virtual object's localization may be correctable by reduction in its disparity, but this technique will have to be verified by future research.

Acknowledgments   Part of the head-mounted display used in this experiment was based on a mechanical design by Ramon Alarcon. Experiment 4 arose from discussions with Charles Neveu. Preliminary reports of the experiments described in this paper were made at the 1995 and 1996 annual meetings of the Human Factors and Ergonomics Society, the 1995 meeting of the Psychonomics Society, and the 1995 meeting of the International Federation of Automatic Control. Most of this chapter is based on a paper in *Human Factors* (Ellis & Menges, 1998).

REFERENCES

Azuma, R. & Bishop, G. (1994) Improving static and dynamic registration in an optical see-through HMD. *Proceedings of SIGGRAPH '94* (pp. 197–204), July 24–29, Orlando, Fl., New York: ACM.

Brown, J. L. (1965) Afterimages. In C. H. Graham (Ed.), *Vision and Visual Perception* (p. 485), New York: Wiley

Cuiffreda, K. J. (1991) Accommodation and its anomalies. In *Visual Optics and Instrumentation, Vol 1, Vision and Visual Dysfunction* (pp. 231–279), W. N. Charman (Ed.), New York: MacMillan Press.

Cuiffreda, K. J. (1992) Components of clinical near vergence testing. *Journal of Behavioral Optometry*, *3*(1), 3–13.

Cutting, J. E. & Vishton, P. M (1995) Perceiving layout and knowing distances: The integration, relative potency and contextual use of different information about depth. In W. Epstein & S. Rogers (Eds.), *Handbook of Perception and Cognition, Vol. 5.*, New York: Academic Press.

Ellis, S. R., Bucher, U. J. & Menges, B. M. (1995) The relationship of binocular convergence to error in the judged distance of virtual objects. Proceedings of the International Federation for Automatic Control (pp. 297–301), Boston, June 26–27.

Ellis, S. R. & Menges, B. M. (1995) Judged distance to virtual objects in the near visual field. *Proceedings, 39th Annual Meeting of the Human Factors and Ergonomics Society* (pp. 1400–1404), Santa Monica, CA. Human Factors and Ergonomics Society.

Ellis, S. R. & Menges, B. M. (1997a) Judged distance to virtual objects in the near visual field. *Presence*, *6*(4), 452–460.

Ellis, S. R. & Menges, B. M. (1997b) Effects of age on the judged distance to virtual objects in the near visual field. In W. A. Rogers (Ed.), *Designing for an Aging Population: Ten Years of Human Factors/Ergonomics Research* (pp. 15–19), Santa Monica, CA: Human Factors and Ergonomics Society.

Ellis, S. R. & Menges, B. M. (1998) Localization of virtual objects in the near visual field. *Human Factors*, *40*(3), 415–431.

Enright, J. T. (1991) Paradoxical monocular stereopsis and perspective vergence. In S. R. Ellis, et al. (Eds.), *Pictorial Communication in Virtual and Real Environments* (pp. 567–576), London: Taylor and Francis.

Fisher, S. K. & Ciuffreda, K. J. (1988) Accommodation and apparent distance. *Perception*, *17*, 609–621.

Foley, J. M. (1993) Stereoscopic distance perception. In S. R. Ellis, et al. (Eds.), *Pictorial Communication in Virtual and Real Environments* (pp. 558–566), London: Taylor and Francis.

Gogel, W. C. & Tietz, J. D. (1973) Absolute motion parallax and the specific distance tendency. *Perception and Psychophysics*, *13*, 284–292.

Graham, C. H. (1965) Space perception. In C. H. Graham (Ed.), *Vision and Visual Perception*, (pp. 519–20), New York: Wiley.

Janin, A. L., Mizell, D. W. & Caudell, T. P. (1993) Calibration of head-mounted displays for augmented reality applications. *Proceedings of IEEE VRAIS '93* (pp. 246–255), Seattle, WA, New York: IEEE.

Krishnan, V. V., Phillips, S. & Stark, L. (1973) Frequency analysis of accommodation, accommodative vergence, and disparity vergence. *Vision Research*, *13*, 1545–1554.

Moses, R. A. (1987) Accommodation. In R. A. Moses & W. M. Haret Jr. (Eds.), *Adler's Physiology of the Eye* (pp. 291–310), Washington D. C.: Mosby.

Nagata, S. (1993) How to reinforce perception of depth in single two-dimensional pictures. In S. R. Ellis, et al. (Eds.), *Pictorial Communication in Virtual and Real Environments* (pp. 527–545), London: Taylor and Francis.

Owens, D. A. & Liebowitz, H. W. (1976) The specific distance tendency. *Perception and Psychophysics*, *20*, 2–9.

Owens, D. A. & Liebowitz, H. W. (1983) Perceptual and motor consequences of tonic vergence. In K. J. Ciuffreda & C. M. Shor (Eds.), *Vergence Eye Movements: Basic and Clinic Aspects* (p. 50), Boston: Butterworths.

Ringach, D. L., Hawken, M. J. & Shapley, R. (1997) Binocular eye movements caused by the perception of three dimensional structure from motion. *Vision Research, 36*(10), 1479–1492.

Rolland, J. P., Ariely, D. & Gibson, W. (1995) Towards quantifying depth and size perception in 3D virtual environments. *Presence, 4*(1), 24–49.

Roscoe, S. (1991) The eyes prefer real images. In S. R. Ellis, et al. (Eds.), *Pictorial Communication in Virtual and Real Environments* (pp. 577–585), London: Taylor and Francis.

Semlow, J. L & Hung, G. K. (1983) The near response: Theories of control. In Schor, C. M. and Ciuffreda, K. J. (Eds.), *Vergence Eye Movements: Basic and Clinic Aspects* (pp. 175–196), Boston: Butterworths.

Weintraub, D. J. & Ensing, M. (1992) *Human Factors Issues in Head-Up Display Design: The Book of HUD*. Wright Patterson AFB, Ohio: CSERIAC.

Zuber. B. (1965) Physiological control of eye movements in humans (p. 103), Ph.D. thesis, Cambridge, Mass: MIT.

9

Fundamental Issues in Mediated Reality, WearComp, and Camera–Based Augmented Reality

Steve Mann
University of Toronto

1. INTRODUCTION

This chapter introduces the notion of Personal Imaging and wearable tetherless computer mediated reality.

Personal Imaging embodies a camera–based augmented reality experience, or a mediated reality experience (which, by definition is camera–based), that runs on a wearable tetherless computer system equipped with image processing and computer graphics capabilities.

Specific applications, motivating historical context, and the underlying mathematical framework of Personal Imaging and associated results in the areas of algebraic projective geometry and photoquantigraphic imaging are presented in followup chapters later in this book.

1.1 The Original Motivating Goal of Personal Imaging

The primary motivating goal behind *Personal Imaging* is best captured in the following quote:

The technologies for recording events lead to a curious result . . . [a] vicarious experience, even for those who were there. In this context "vicarious" means to experience an event through the eyes (or the recording device) of another. Yet here we have the real experiencer and the vicarious experiencer being the same person, except that the real experiencer didn't have the original experience because of all the activity involved in recording it for the latter, vicarious experience . . . we are so busy manipulating, pointing, adjusting, framing, balancing, and preparing that the event disappears . . . . But there is a positive side to the use of recording devices: situations where the device intensifies the experience. Most of the time this takes place only with less sophisticated artifacts: the sketch pad, the painter's canvas . . . Those who benefit from these intensifying artifacts are usually artists . . . with these artifacts, the act of recording forces us to look and experience with more intensity and enjoyment than might otherwise be the case. [Norman, 1992]

The original goal of Personal Imaging was to create a portable tetherless photographic environment that, although technically sophisticated, would function more in the spirit of the less sophisticated artifacts such as the artist's sketch pad or painter's canvas to which Norman (a leading HCI researcher and defender of humanstic concerns in the machine age) refers. This goal of Personal Imaging was to create tools that would intensify and augment the experience of seeing, through the embodiment of a photographically mediated visual experience used in conjunction with tools similar in many ways to the artist's paintbrush and canvas. These tools were based on (or inspired, depending on one's point of view) a framework consisting of "wearable computing" and "mediated reality."

2.   WHAT IS "WEARCOMP"?

The tools developed during the 1970s and early 1980s to intensify the experience of seeing, which led to personal imaging, were based on (or one might say, they defined) a concept called "wearable computing," which is based on realizing some combination of three basic modes of operation.

2.1 The Three Fundamental Operational Modes of WearComp

The three operational modes in this new interaction between human and computer, are:

1. Constancy (not just operational constancy but also interactional constancy—it does not need to be turned on or even opened up prior to use);
2. Augmentation (the notion that computing is *not* the primary task—the user will be doing something else at the same time as doing the computing);
3. Mediation (it can encapsulate or partially encapsulate the user and thus allows the user to control inbound informational flow for solitude and outbound informational flow for privacy).

These three basic modes of operation, characterized by their signal flow paths (signals between human and computer, as well as signals between the human + computer and the environment) are illustrated in Figure 9.1. Note also that constancy is an attribute of the other two (e.g., constancy is a special case of the other two). Moreover, the three basic modes of operation are a special case of the architecture presented in Figure 9.2.

2.2 The Six Attributes (Signal Flow Paths) of WearComp

Each of the six information channels (signal flow paths) of Figure 9.2 corresponds to one attribute of WearComp. These six attributes are listed in Table 9.1.

2.3 Other Properties that Result from the Six Attributes of WearComp

The six basis properties described in Table 9.1 may be used to derive additional properties, some of which are listed below:

- **CONSTANT:** If a system is constantly OBSERVABLE and CONTROLLABLE (e.g., is interactionally constant) it must also be operationally constant. Thus it must be "always ready." It may have "sleep modes" but it is never "dead." Unlike a laptop computer, which must be opened up, switched on, and booted up before use, it is always on and always running.
- **PERSONAL:** Human and computer are inextricably intertwined. PROSTHETIC: You can adapt to it so that it acts as a true extension of mind and body; after time you forget that you are wearing it.

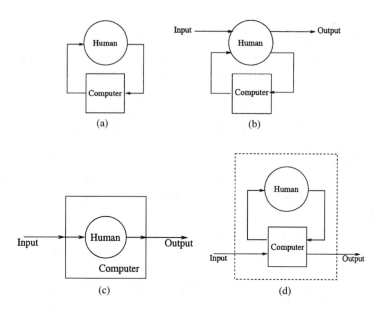

(a)

(b)

(c)

(d)

FIG. 9.1. The three basic operational modes of WearComp. (a) Signal flow paths for a computer system that runs continuously, constantly attentive to the user's input, and constantly providing information to the user. Over time, constancy leads to a symbiosis in which the user and computer become part of each other's feedback loops. (b) Signal flow path for augmented intelligence and augmented reality. Interaction with the computer is secondary to another primary activity, such as walking, attending a meeting, or perhaps doing something that requires full hand–to eye coordination, like running down stairs or playing volleyball. Because the other primary activity is often one that requires the human to be attentive to the environment as well as unencumbered, the computer must be able to operate in the background to augment the primary experience, for example, by providing a map of a building interior, or providing other information, through the use of computer graphics overlays superimposed on top of the real world. (c) The wearable computer can be used like clothing, to encapsulate the user and function as a protective shell, whether to protect us from cold, protect us from physical attack (as traditionally facilitated by armour), or to provide privacy (by concealing personal information and personal attributes from others). In terms of signal flow, this encapsulation facilitates the possible mediation of incoming information to permit solitude and the possible mediation of outgoing information to permit privacy. It is not so much the absolute blocking of these information channels that is important; it is the fact that the wearer can control to what extent, and when, these channels are blocked, modified, attenuated, or amplified, in various degrees, that makes WearComp much more empowering to the user than other similar forms of portable computing. (d) An equivalent depiction of encapsulation (mediation) redrawn to give it a similar form to that of (a) and (b), where the encapsulation is understood to comprise a separate protective shell.

298

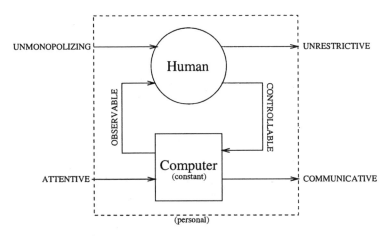

FIG. 9.2. The architecture of Humanistic Intelligence as provided by WearComp. The six signal flow paths between the human, the computer, and the environment each define one of the six attributes of WearComp, as described in Table 9.1.

ASSERTIVE: It can have a barrier to prohibition or to requests by others for removal during times when you wish such a barrier. This is in contrast to a laptop computer in a briefcase or bag that could be separated from you by a "please leave all bags and briefcases at the counter" policy of a department store or library.

PRIVATE: Others can't observe or control it unless you let them. Others can't determine system status unless you want them to (e.g., the clerk at the refund counter in a department store where photography is prohibited can't tell if you are taking a picture, in contrast to camcorder where it is obvious you are taking a picture when you hold it up to your eye). (Note that privacy rights of the observed person versus the observer (wearer of the WearComp apparatus) have been addressed elsewhere in the literature [Mann, 1996]). It has been argued that, especially in the many situations where we are placed under surveillance by an establishment, we should also have our own record of the interaction. It is the very establishments (department stores, gambling casinos, etc.) that tend to prohibit photography in which patrons are more often placed under the greater surveillance. Diffusionism [Mann, 1998a] argues that such forms of surveillance (e.g., that match the classic definition of "totalitarian" [Mann, 1998a]) are the most likely to give rise to abuse. A

TABLE 9.1

The Six Attributes (Signal Flow Paths) of WearComp

Attribute	Explanation
UNMONOPOLIZING of the user's attention:	It does not cut you off from the outside world like a reality game or the like. You can attend to other matters while using the apparatus. In fact, ideally, it will provide enhanced sensory capabilities. It may, however, mediate (augment, alter, or deliberately diminish) the sensory capabilities.
UNRESTRICTIVE to the user: ambulatory, mobile, roving,	You can do other things while using it, e.g. you can type while jogging, etc.
OBSERVABLE by the user:	It can get your attention continuously if you want it to. The output is almost–always–observable.
CONTROLLABLE by the user: Responsive.	You can grab control of it at any time you wish. Even in automated processes you can manually override to break open the control loop and become part of the loop at any time you want to. The input is infinitely–often–controllable.
ATTENTIVE to the environment:	It is environmentally aware, multimodal, and multisensory. Thus it ultimately gives the user increased situational awareness.
COMMUNICATIVE to others:	It can be used as a communications medium when you want it to. It is expressive: It allows the wearer to be expressive through the medium, whether as a direct communications medium to others or as means of assisting the production of expressive media (artistic or otherwise).

full discussion of the philosophical underpinnings of this work is beyond the scope of the article. However, a good understanding of Diffusionism may be attained by first understanding Reflectionism [Mann, 1998a].

- **CONNECTED:** Insofar as technically feasible, wireless communication for a network connection is desirable. If it is ATTENTIVE (e.g., can receive inputs from other systems and networks) and COMMUNICATIVE (can send messages to other systems and networks) then it has connectivity.

- **ANTICIPATORY** (PROACTIVE): That it is ATTENTIVE and CONSTANT in its processing of incoming information implies that it cannot only respond to events as they happen but can also anticipate events so that the apparatus can "learn" from the environment in a way that is influenced by prior conditions. Moreover, that it can be constantly OBSERVABLE means that it can take an initiative and "make things happen" (e.g., it can get your attention). That it is COMMUNICATIVE means that it can also take initiative with respect to interacting with others. By "others," what is meant is other people as well as other entities. For example, it may take the initiative to send a message to the heater in the building to switch off when it senses you're on the verge of a "sweatdown" (sweat-induced computer shutdown resulting from excessive sweat adversely affecting system components that have not been sealed from the effects of moisture).
- **RETROACTIVE:** That it is ATTENTIVE and CONTROLLABLE through processes that are CONSTANT means that it may continually collect data that might be of use later. Thus it may "learn" upon subsequent recall.
- **HUMANISTIC:** It embodies human intellect as a part of the feedback loop of its processing. This arises from the fact that it is CONTROLLABLE and OBSERVABLE. In this way the intended definition distinguishes itself from entities such as the obedience cuffs [Hoshen et al., 1995] worn by prisoners around their ankles, many of which even have some form of OBSERVABLE property (e.g., in the form of pain-giving electric shocks that some obedience cuffs produce). What the obedience cuff lacks is the CONTROLLABLE attribute (in the sense that the prisoner does not have control of the apparatus). That it is **HUMANISTIC** means also that automatic (or autonomous) aspects of the system may be overridden, and therefore it is possible for the human operator to "step inside" the feedback loop of a process when and if desired.

Note that these are affordances, rather than requirements. For example PRIVATE and COMMUNICATIVE are in some ways opposite, yet together they facilitate an expanded range of options to move further in one direction or the other on a continuum between these two opposites, or to selectively embody both. The important aspect here is choice for the user to move in an increased affordances space rather than a specific definition of how the user should act while using the apparatus.

Implicit in UNRESTRICTIVE and UNMONOPOLIZING is that use of the apparatus will be a secondary rather than a primary task (as in the "Augmenting" mode of operation).

2.4 Comparison of Formal Definition of WearComp to Other Definitions of Wearable Computing

2.4.1 Comparison to Previously Published Definition of Wearable Computing

These six attributes (as described in Table 9.1) are equivalent to the three attributes of wearable computing, *Eudaemonic*, *Existential*, and *Ephemeral*, previously published [Mann, 1997a]:

- *Ephemeral* interaction arises from a CONSTANT user interface (OBSERVABLE and CONTROLLABLE), where constancy is made possible through the fact that it is UNMONOPOLIZING and UNRESTRICTIVE.
- *Existential* properties of self-determination and mastery over one's own destiny are realized by a CONTROLLABLE device.
- *Eudaemonic* implies that it is PERSONAL.

2.4.2 Comparison to the Oranchak Definition of Wearable Computing

According to the more recent Oranchak definition, wearable computing is regarded as a third hemisphere of the brain and is defined therefore in terms of specialization (e.g., the left side of the brain is good at some things and the right side good at others, so too the third hemisphere will have its own areas of specialty). The definition proposed in the previous subsections captures the spirit of this "third hemisphere" definition. Moreover, attributes such as the PRIVATE characteristic of PERSONAL become all the more evident, in the sense that the brain be regarded as one's own personal space, protected by the U.S. Constitution and the like (e.g., that methods such as torture or involuntary dissection, of extracting, against one's will, information from the brain, are regarded as unacceptable). Thus the "third hemisphere" also becomes a personal space, like a personal diary that is protected from subpoena or other involuntary disclosure. Such protection is necessary if it is to become a true extension of the mind in the fullest sense.

2.4.3 The Solipsist Definition of Wearable Computing

Strictly speaking, ATTENTIVE and COMMUNICATIVE are not absolute requirements (e.g., a device to help you remember cards in a game of blackjack could function even if its only input was from the human and its only output to the human and would be a good example of wearable computing according to this previous definition). Thus the definition of wearable computing provided here is not specifically meant to exclude certain kinds of computing, but, rather, it is presented in the spirit of trying to define a core concept around which other equally or possibly more valid forms of computation have and may continue to emerge, perhaps at its periphery as much as at its core.

2.4.4 The "Eudaemonic" Definition

The ASSERTIVE aspect of the PERSONAL attribute (which is at the core of the *Eudaemonic* [Mann, 1997a] property) has been the least understood property of wearable computing over the past twenty years. Recently, a personal anecdote put forth by a leading researcher who commented on the above definition of wearable computing helps clarify this property and suggests a possible design methodology for wearable computers of the future:

> I think of a black trench coat I used to have—when I was 18 and living in Salzburg, Austria—I could go to the university library and go into the stacks if I was wearing my coat. While all the other students were stopped at the front and never permitted to go into the stacks [with their coats on], I was treated like a priest (which is what the librarians thought I looked like) and could go into the stacks [with my coat on]. They would never have ask me to take it off—as it was part of my 'uniform' from their point of view and it would have been inconceivable that I would *not* wear a long coat. (Gerald Q "Chip" Maguire Jr., 14 Feb 1998)

3. AUGMENTED/MEDIATED REALITY

The intent of Augmented Reality (AR) is to *add* virtual objects to the real world [Caudell and Mizell, 1992]. A typical AR apparatus consists of a video display with partially transparent visor, beam splitter, or the like upon which computer-generated information is *overlayed* on top of a see–through view of the real world.

A more general framework called Mediated Reality (MR), of which AR is a special case, is proposed. The intent of MR, like typical AR, includes *adding* virtual objects to visual reality but also includes the ability to *take away*, *alter*, deliberately *diminish*, and *significantly alter* the perception of visual reality. MR attempts to visually "mediate" real objects, using a body-worn apparatus where both the *real* and *virtual* objects are placed on an equal footing, in the sense that both are presented together via a synthetic medium.

Successful wearable, tetherless implementations of MR have been realized by *viewing* the real world using a head-mounted display (HMD) fitted with video camera(s), together with a wearable computer system and/or bidirectional wireless communications system. This untethered approach is quite distinct from other related research where various head–trackers are installed in a fixed environment, and the wearer of a virtual reality headset must remain tethered to a fixed location such as an operating room [Rolland et al., 1995]. The primary difference between early Personal Imaging systems as described in Mann [1994] and some of the similar more recent work in augmented reality [Rolland et al., 1995] is that the earlier Personal Imaging systems were characterized by little or no reliance on any special environment, so that they could be used nearly anywhere. This portability enabled various forms of the Personal Imaging apparatus to be tested extensively in everyday circumstances, such as while riding the bus, shopping, banking, etc., so that what emerged was a form of *Computer Supported Cooperative Living*, which included not only working (work being what is emphasized by the field of *Computer Supported Cooperative Work*) but all other facets of life as well.

The proposed approach shows promise in applications where it is desired to have the ability to reconfigure reality. For example, color may be deliberately diminished or completely removed from the real world at certain times when it is desired to highlight parts of a virtual world with graphic objects having unique colors. The fact that vision may be *completely* reconfigured also suggests utility to the visually handicapped.

3.1   A Brief Introduction to Mediated Reality (MR)

Ivan Sutherland, a pioneer in the field of computer graphics, described a head-mounted display with half-silvered mirrors so that the wearer could see a virtual world superimposed on reality [Earnshaw et al., 1993; Sutherland, 1968], giving rise to "Augmented Reality (AR)."

Others have adopted Sutherland's concept of a Head-Mounted Display (HMD) but generally without the see-through capability. An artificial environment in which the user cannot see through the display is generally referred as a Virtual Reality (VR) environment. One of the reasons that Sutherland's approach was not more ubiquitously adopted is that he did not merge the virtual object (a simple cube) with the real world in a meaningful way. Feiner's group was responsible for demonstrating the viability of AR as a field of research, using sonar (Logitech 3D trackers) to track the real world so that the real and virtual worlds could be registered [Feiner et al., 1993a,b]. Other research groups [Fuchs et al.; Caudell and Mizell, 1992] also contributed to this development. Some research in AR also arises from work in telepresence [Drascic and Milgram, 1996]. (See also Mizell's chapter of this book.)

An important goal of AR is to *add* computer graphics or the like to the real world. A typical AR apparatus does this with beam splitter(s) so that the user sees directly through the apparatus while simultaneously viewing a computer screen.

A goal of this chapter is to consider a wireless (untethered) apparatus worn over the eyes that, in real time, computationally *reconfigures* reality in addition to adding to it. This mediation of reality may be thought of as a *filtering* operation applied to reality and then a combining operation to insert *overlays* (Figure 9.3(a)). Equivalently, the addition of

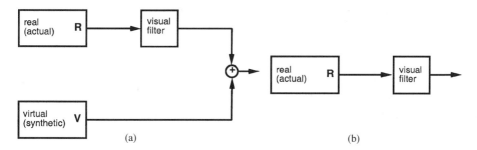

FIG. 9.3. Two equivalent interpretations of mediated reality (MR): (a) In addition to the ability to add computer-generated (synthetic) material to the wearer's visual world, there is potential to alter reality, if desired, through the application of a "visual filter." The coordinate transformation embodied in the visual filter may either be inserted into the virtual channel as well, or the graphics may be rendered in the coordinate system of the filtered reality channel, so that the real and virtual channels are in register. (b) The visual filter need not be a *linear system*. In particular, the visual filter may itself embody the ability to create computer-generated objects and therefore subsume the "virtual" channel.

computer-generated material may be regarded as arising from this filtering operation itself (Figure 9.3(b)).

A means of *mediating* (augmenting, deliberately diminishing, or otherwise altering) reality, in real time, through an apparatus worn over the eyes, will first be described using an idealized implementation, and later in terms of practical implementations that were actually built. The entire apparatus will be referred to as a Reality Mediator (RM).

3.2   Idealized Reality Mediator

Suppose that we were to make sunglasses from a special kind of glass, known as "lightspace" glass [Ryals, 1995] with a capability to absorb and produce light rays endowing it with the ability to determine the complete measurement space of how a scene or object responds to light. This hypothetical lightspace glass would be totally opaque but could absorb and quantify every ray of light incident upon it and send this information to a wearable computer system concealed inside or near the glasses, so that the glasses became a complete wearable multidimensional light measurement instrument. Suppose also that the glass could generate any desired ray of light under program control of the wearable computer. That is, the computer could specify where any given ray of light would emerge from, and in what direction it would go, of what wavelength it would be, and its intensity as a function of time. Suppose also that any number of such processes could be run simultaneously so that it could produce any desired superposition of such light rays. The combination of these special sunglasses and the wearable computer would form what is called an "idealized reality mediator."

Clearly, the apparatus could be used as a VR display, because its ability to produce any collection of rays of light means that it could function as a video display. (See Figure 9.4 (VR).)

Although the glass would be totally opaque, it would have the capability of functioning like an ordinary window in the sense that its ability to absorb and quantify all the rays of light incident upon it, and then to generate new rays of light, could be used to send exactly the same rays of light out the other side of the glass. The ideal nature of this glass permits it to perfectly sustain the illusion of transparency.

This illusion of transparency would have many uses of its own. Obviously the apparatus could be used to function as a pair of ordinary sunglasses by darkening rays of light coming out the other side. Because of the computer control, the darkening could even vary, in accordance with some gradient that would be darker up where the sun was, resulting in "smart"

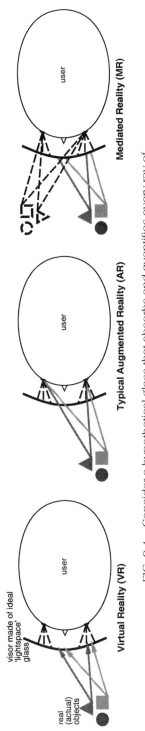

FIG. 9.4. Consider a hypothetical glass that absorbs and quantifies every ray of light that hits it and is also capable of generating any desired rays of light. Sunglasses made from such glass could produce a virtual reality (VR) experience by ignoring all rays of light from the real world and generating rays of light that simulate a virtual world. Rays of light from real (actual) objects are indicated by solid shaded lines; rays of light from the display device itself are indicated by dashed lines. The device could also produce a typical augmented reality (AR) experience by creating the illusion of transparency and also generating rays of light to make computer-generated "overlays." Furthermore, it could mediate the visual experience, allowing the perception of reality itself to be altered. In this figure, a non useful (except in the domain of psychophysical experiments) but illustrative example is shown: Objects are *left-right reversed* before being presented to the viewer.

307

sunglasses. The smart sunglasses would use machine vision to track the sun and adjust the position of the darkening mask. Such motion–stabilized overlays have been implemented [Mann and Picard, 1995] in video reality mediators and these will be described later.

Many conventional sunglasses have a fixed gradient, typically being darker at the top than at the bottom, which creates an annoying artificial percept of motion when the wearer tilts his or her head back or forward, even though nothing in the scene is moving. Smart sunglasses would eliminate this problem by fixing the darkening mask with respect to the scene rather than the wearer.

Now in addition to creating the illusion of allowing light to pass right through, the RM could also create new rays of light, having nothing to do with the rays of light coming into it. The combined illusion of transparency and the new light could provide the wearer with a typical AR experience (Figure 9.4 (AR)).

In an AR environment, graphics sometimes fail to stand out from the real objects. For example, when looking through the glasses at a brightly colored scene, there may not exist a unique color to use for the overlays. Suppose, however, that the sunglasses, instead of creating an illusion of transparency, create an illusion of being *achromat transparent*. Being achromat transparent means that each incoming ray of light is absorbed and quantified, while its wavelength is ignored. A ray from the same location is sent out in the same direction, at the same time, but with a flat (grey) spectrum. This would make the user colorblind to real objects, making the real world appear less "busy" when combined with some colorful computer-generated overlays where color could be used, more effectively, to accentuate the virtual objects. This would prevent computer-generated objects from being "lost" in the clutter of the real world.

With practical approximations to the idealized reality mediator described above, it has been found that color-reduced reality mediation is quite useful, for example, when comfortably seated on a commercial airline or commuter train and one wishes to read text on the virtual screen created by the sunglasses (e.g., to read email), one can "tone down" the surroundings so that they take on a lesser role. However, it is not desirable to be completely blind to the surroundings, as is someone who is reading a real paper newspaper (newspapers can easily end up covering most of a person's visual field). Thus the proposed framework has unique and useful features.

This form of reality mediation allows one to adjust the degree to which one's attention is devoted to the virtual world, which might, for example, be comprised of email, a computer source file, and other miscellaneous work, running in emacs19, with colorful text, where the text colors are chosen so that no black, white, or grey text (text colors that would get lost in the new reality) is used. One is then completely aware of the world behind a virtual newspaper or the like, but it does not distract from one's ability to read the newspaper.

Alternatively, the real world could be left in color, but the color mediated slightly so that unique and distinct colors could be reserved for virtual objects and graphics overlays. In addition to this chromatic mediation, other forms of mediated reality are often useful, such as those that might assist the visually challenged correct for a visual defect, or allow an artist or photographer to experience a diminished dynamic range for purposes of attaining a heightened sense of awareness of the range and color of light in various scenes.

3.3 Practical Issues

3.3.1 Registration between Real and Virtual Worlds

For many situations, alignment of the real and virtual worlds is very important, yet it has been very difficult to obtain. (See, for example, Azuma's chapter of this book.)

The problem with many implementations of AR is that even once registration is attained, if the glasses slip down the wearer's nose, ever so slightly, the real and virtual worlds will not generally remain in perfect alignment.

Using the illusory transparency approach, the illusion of transparency is perfectly coupled with the virtual world once the signals (e.g., video) corresponding to the real and virtual worlds are put into register and combined into one signal. Not all applications lend themselves to easy registration at the signal level, but there are a good many that do. One example of an application that works a lot better in mediated reality than in augmented reality is the finger–tracking mouse [Mann, 1997b]. In this and other applications in which mediated reality could provide perfect registration (to within subpixel accuracy), it was found that when the glasses slipped down the nose a little (or a lot for that matter), both the real and virtual worlds continued to move together, within subpixel accuracy, in a unified way, and remained in almost perfect register. Since they were both the same medium (e.g., video) once registration was attained between the real and virtual video signals themselves, the registration problem remained solved regardless of how the glasses moved around on the wearer, or how the wearer's eyes were focused or positioned with respect to the glasses [Mann, 1994].

Another important point is that even with perfect registration, when using a see-through visor (with beam splitter or the like), real objects may lie in a variety of different depth planes, while virtual objects are generally flat (in each eye that is), to the extent that their distance is at a particular focus (apart from the variations in binocular disparity of the virtual world). This is not so much of a problem when all of the objects are far away as is often the case in aircraft (e.g., in the fighter jets using HUDs), but in many other applications (such as in a typical building interior) the differing depth planes destroy the illusion of unity between real and virtual worlds. This loss of unity can be detrimental to certain tasks, so there are situations in which it would be desirable to maintain such unity. (The issue of binocular convergence and errors in judged distance to virtual objects has also been considered by Ellis [Ellis et al., 1995], in the context of see–through helmet mounted displays.)

Within a mediated reality environment (e.g., with the illusory transparency approach), the real and virtual worlds exist in the same medium and therefore are not only registered in location but also in depth, since the depth limitations of the display device affect both the virtual and real environments in exactly the same way.

3.3.2 Transformation of the Perceptual World

Even without graphics overlays, mediated realities are still interesting and useful. For example, the colorblinding glasses in themselves might be useful to an artist trying to study relationships between light and shade. While it is certain that the average person might not want any part of this experience, especially given the cumbersome nature of the current realization of the RM, without question, there are at least a small number of users who would be willing to wear an expensive and cumbersome apparatus in order to see the world in a different light. Consider, for example, the artist who travels halfway around the world to see the morning light in Italy. As the cost and size of the RM decreases, no doubt there would be a growing demand for glasses that alter (enhance or diminish) tonal range, allowing artists to manipulate contrast, color, and the like.

MR glasses could (in principle) be used to synthesize the effect of ordinary glasses, but with a computer-controlled prescription that would modify itself automatically, while conducting automatically scheduled eye tests on the user.

The RM might also, for example, reverse the direction of all outgoing light rays to allow the wearer to live in an "upside-down" world (Figure 9.4 (MR)), perhaps being useful for experiments in developmental psychology.

The vast majority of RM users of the future will no doubt have no desire to live in an upside-down, left-right-reversed, tonally reversed, or sideways rotated world, except, perhaps in certain situations, such as when an American is driving a car in Britain, or reading Da Vinci's notebooks (with backwards writing), or evaluating photographic negatives (reversed tonality). However, despite their lack of apparent practical utility, these altered visual worlds serve as illustrative examples of extreme forms of reality mediation.

In his 1896 paper [Stratton, 1896], George Stratton reported on experiments in which he wore eyeglasses that inverted his visual field of view. Stratton argued that since the image upon the retina was inverted, it seemed reasonable to examine the effect of presenting the retina with an "upright image."

His "upside-down" glasses consisted of two lenses of equal focal length, spaced two focal lengths apart, so that rays of light entering from the top would emerge from the bottom, and vice versa. Stratton, upon first wearing the glasses, reported seeing the world upside-down, but, after an adaptation period of several days, he was able to function completely normally with the glasses on.

Dolezal [Dolezal, 1982] (page 19) describes "various types of optical transformations," such as the *inversion* explored by Stratton, as well as *displacement*, *reversal*, *tilt*, *magnification*, and *scrambling*. Kohler [Kohler, 1964] also discusses "transformation of the perceptual world."

Each of these "optical transformations" could be realized by selecting a particular *linear time-invariant system* as the visual filter in Figure 9.3. (A good description of *linear time-invariant systems* may be found in a communications or electrical engineering textbook such as Heykin [1983].)

The optical transformation to greyscale, described earlier, could also be realized by a "visual filter" (Figure 9.3(a)) that is a linear time-invariant system, in particular, a *linear integral operator* [Arfken, 1985, page 669] that, for each ray of light, collapses all wavelengths into a single quantity giving rise to a ray of light, having a flat spectrum, emerging from the other side.

The visual filter of Figure 9.3(b) may not, in general, be realized through a *linear system*, but there exists an equivalent *nonlinear* filter arising from incorporating the generation of virtual objects into the filtering operation.

3.4   Video Transparency: Practical Embodiments of Mediated Reality

The "idealized reality mediator" does not exist in practice, since one would require an infinitely dense lattice of spectral photometers or the like with infinitely fast response in order to absorb and quantify every possible incoming ray of light. Even a discrete realization made from a dense array of miniature video cameras is costly and bulky. The corresponding display is even more costly and bulky.

However, since the visor may be relatively well fixed with respect to the wearer, there is not really a great need for full parallax holographic video. The fixed nature of the visor conveniently prevents the wearer from having *look-around* with respect to the visor itself (e.g., look-around is accomplished when the user and the visor move together to explore the space). Thus two views, one for each eye, suffice to create a reasonable illusion of transparency. A practical color stereo reality mediator (RM) may be made from video cameras and display. One example, made from a stereo display, each eye having approximately the same resolution as a standard NTSC television, is depicted in Figure 9.5.

It is desired to have the maximum possible visual bandwith, even if the RM is going to be used, each eyeto conduct experiments on *diminished reality*. For example, the apparatus of Figure 9.5 was used to experience colorblindness and reduced resolution by applying the appropriate visual filter to select the desired degree of degradation in a controlled manner that could also be automated by computer. For example, color was gradually reduced over the course of a day, to experiment with the extent that one can become colorblinded yet adapt to it to the extent that one would not even realize the loss of color.

The cameras of the apparatus in Figure 9.5 (Sony cameras later updated to Elmo CN401 cameras) were mounted the correct interocular distance apart, and cameras that had the same field of view as the display devices were used. In general, with the cameras connected directly to the displays, the illusion of transparency will be realized to some degree, at least to the extent that each ray of light entering the apparatus (e.g., absorbed and quantified by the cameras) will appear to emerge at roughly the same angle (by virtue of the display).

Others [Fuchs et al.; Drascic, 1993; Nagao, 1995] have also explored video-based illusory transparency, augmenting it with virtual overlays. Nagao, in the context of his hand-held TV set with single camera [Nagao, 1995] calls it "video see-through."

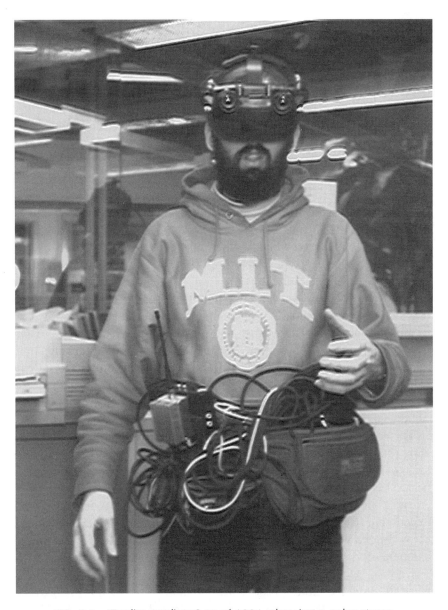

FIG. 9.5. "Reality mediator" as of 1994, showing a color stereo head-mounted display (VR4) with two cameras mounted to it. The inter-camera distance and field of view match approximately the author's interocular distance and field of view when the apparatus is removed. The components around the author's waist comprise radio communications equipment (video transmitter and receiver). The antennas are located at the back of the head-mount to balance the weight of the cameras, so that the unit is not front-heavy.

It is worth noting that whenever illusory transparency is used, as in the work of Fuchs et al.; Drascic [1993], and Nagao [1995], reality will be mediated, whether or not that mediation was intended. At the very least this mediation takes on the form of limited dynamic range and color gamut, as well as some kind of distortion, which may be modeled as a 2D coordinate transformation. Since this mediation is inevitable, it is worthwhile to attempt to exploit it, or at least plan for it, in the design of the apparatus. A visual filter may even be used to attempt to mitigate the distortion.

A first step in using a reality mediator is to wear the system for a while to become accustomed to its characteristics. Unlike in typical beam-splitter implementations of augmented reality, in mediated reality, transparency, if desired, is synthesized and therefore is only as good as the components used to make the apparatus.

The apparatus was worn by the author in identity map configuration (cameras connected directly to the displays) for several days. It was easy to walk around the building, up and down stairs, through doorways, to and from the lab, etc. There was, however, difficulty in scenes of high dynamic range or poor contrast, and also in reading fine print, and particular difficulty when both spatial and tonal resolution were needed, such as in reading a restaurant menu or a department store receipt printed in faint ink when the ribbon was near the end of its useful life.

The unusual appearance of the apparatus was itself a hindrance in my daily activities (for example when worn to a formal dinner), but after some time, over a long-term adaptation process beginning in 1994, people appeared to become accustomed to seeing the author this way.

The attempt to create an illusion of transparency was itself a useful experiment because it established some working knowledge of what can be performed when vision is *diminished* or *degraded* to RS170 resolution and field of view is somewhat limited by the apparatus.

Knowing what can be performed when reality is mediated (e.g., diminished) through the limitations of a particular HMD (e.g., the VR4) would be useful to researchers who are designing VR environments for that HMD, because it establishes a sort of *upper bound* on how good a VR environment could ever hope to be when presented through that particular HMD. A reality mediator may also be useful to those who are really only interested in designing a traditional beam-splitter-based AR system because RM could be used as a development tool and could also be used to explore new conceptual frameworks.

3.5 Wearable, Tetherless Computer–Mediated Reality

Once the display and mediation apparatus is worn long enough to be comfortable with the illusory transparency, mediation of the reality can begin.

The computer power required to perform general-purpose manipulation of color video streams is too unwieldy to be worn in a backpack (although body-worn video processing systems and other hardware to facilitate very limited forms of reality mediation have been constructed by the author). In particular, a large processing system with good video-processing capability may be accessed remotely by establishing a full-duplex video communications channel between the RM and the host video processor(s). In particular, the full–duplex communications link comprises a high-quality communications link called the inbound-channel, which is used to send the stereo video signal from the wearable cameras to the remote computer(s), while a lower quality communications link called the outbound channel is used to carry the processed signal from the computer back to the HMD. This apparatus is depicted in a simple diagram (Figure 9.6). Ideal channels would

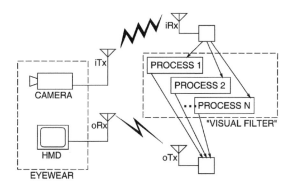

FIG. 9.6. Simple implementation of a reality mediator (RM). A stereo video signal is sent to a network of remote computers over a high-quality microwave communications link, called the inbound channel, comprising transmitter iTx and receiver iRx. The high bandwidth of the link ensures that there is not any appreciable delay, as might be the case with slow scan television, or other such narrowband systems. The remote computer systems send back the processed video stream over a UHF communications link called the outbound channel, comprising transmitter oTx and receiver oRx. The term visual filter refers to the processes that mediate the visual reality and possibly insert "virtual" objects into the reality stream.

be of high-quality, but the machine-vision algorithms were found to be much more susceptible to noise than was the author's own vision (ability to see through an apparatus that provided only a noisy illusion of transparency), and, due to a need to separate the frequency bands, an inferior frequency band being necessary for one of the channels, it was decided that the higher quality channel should be the inbound channel. Since the illusion of transparency is degraded by the total noise level of inbound plus outbound artifacts, it was found to be equally poor regardless of the order of these two degradation artifacts. Therefore it was decided to at least allow the remote video processing hardware to receive the best quality picture.

The use of broad bandwidth communications channels ensured that there was negligible delay, so that the main delay was in processing the video. The entire pipeline delay was kept to less than one frame (1/30 second). Of course delay was added, when desired, in order to experiment with the effects of delay, but it was found that even a small amount of delay was not tolerable.

To a very limited extent, looking through a camcorder provides a mediated reality experience, because we see the real world (usually in black and white, or if it is in color it has a very limited color fidelity) together with virtual text objects, such as shutter speed and other information about the camera superimposed on top of the image. Therefore, to the extent that this simple apparatus may be modeled by regarding it as a reality mediator, the implicit visual filter at work here is the colorblindness or loss of color fidelity, as well as the nonorthoscopic viewing condition, shift of center of projection from the eye into the camera, etc., which one experiences while looking through the viewfinder with the other eye closed. This visual filter is present whether intended by the manufacturer or not. Any optical path (even ordinary sunglasses) may be modeled by a reality mediator with the appropriate visual filter. Quite likely the manufacturer of the camcorder would rather have provided a full-color viewfinder, but a black and white viewfinder or reduced color viewfinder is used to reduce cost. This is a very trivial example of a mediated reality environment where the filtering operation is *unintentional* but nevertheless always present in any system so modeled.

3.6 The Reconfigured Eyes

Using the reality mediator, the author repeated the classic studies like those of Stratton [Stratton, 1896; 1897] and Anstis [Anstis, 1992] (e.g., living in an upside-down or negated world), as well as some new studies, such as

FIG. 9.7. Living in a "Rot 90" world: In order to fully adapt to a 90 degree rotated field of view, over a week long experiment, it was found to be necessary to rotate both the cameras together as a single unit (so that one was on top of the other). It was found not to be possible to properly adapt to having each camera rotated 90 degrees separately (e.g., rotated and mounted side-by-side). The use of stereo cameras and HMD rather than prisms allowed the author greater flexibility in psychophysical experiments such as this one. The parallax in the up-down direction which the author experienced after adapting to this apparatus was found to afford a similar sense depth perception that we normally experience with eyes spaced from left to right.

learning to live in a world rotated 90 degrees. However, in this sideways world, it was found that one could not adapt to having each of the images rotated by 90 degrees separately, but the cameras had to rotate together (Figure 9.7).

The video-based RM (e.g., Figure 9.5) permits one to experience any coordinate transformation that can be expressed as a mapping from a 2D domain to a 2D range. Such processing can be done in real time (30 frames/s = 60 fields/s) in full color, through the use of special–purpose video processing hardware. The RM allowed the author to experiment with various computationally generated coordinate transformations both indoors and outdoors, in a variety of different practical situations. Examples of some useful coordinate transformations appear in Figure 9.8.

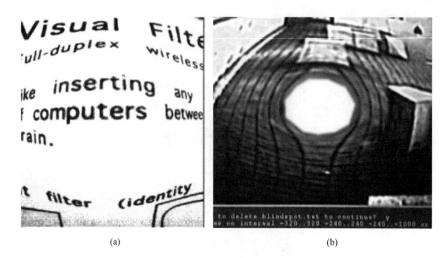

(a)   (b)

FIG. 9.8. Living in coordinate-transformed worlds: Color video images are transmitted, coordinate-transformed, and then received back at 30 frames per second—the full frame-rate of the VR4 display device. (a) This visual filter would allow a person with very poor vision to read (due to the central portion of the visual field being hyper-foveated for a very high degree of magnification in this area), yet still provide good peripheral vision (due to a wide visual field of view arising from demagnified periphery). (b) This visual filter would allow a person with a *scotoma* (a blind or dark spot in the visual field) to see more clearly, once having learned the mapping. Note the distortion of the square cobblestones on the ground and the cube-shaped outdoor stone sculptures. Visual filters also typically provide edge enhancement and contrast enhancement in addition to the coordinate transformation.

Researchers at Johns Hopkins University have been experimenting with the use of cameras and head-mounted displays for helping the visually handicapped. Their approach has been to use the optics of the cameras for magnification, together with the contrast adjustments of the video display to increase apparent scene contrast [Wilmer, 1995]. They also talk about using *image remapping* in the future:

One of the most exciting developments in the field of low vision is the Low Vision Enhancement System (LVES). This is an electronic vision enhancement system that provides contrast enhancement .... *Future enhancements* to the device include text manipulation, autofocus and *image remapping*.

(quote from their WWW page [Wilmer, 1995], emphasis added). This research effort suggests the utility of the real-time visual mappings (Figure 9.8) successfully implemented using the apparatus of Figure 9.5.

3.7 Extremely Poor Alignment Better Than Almost Orthoscopic

The purpose of this subsection is to provide background for explaining the rationale for the author's continued desire to live in a coordinate-transformed world over a period of many years. The usefulness of such a transformed world extends beyond psychophysical adaptation experiments and appears to have practical utility in the domains of photojournalism, photography, and cinematography.

As previously described, the idea of living in a coordinate-transformed world has been explored extensively by other authors [Kohler, 1964; Dolezal, 1982], using optical methods (such as prisms and the like). Much could also be written about the author's own experiences in various electronically coordinate-transformed worlds, but a detailed account of all of the various experiences is beyond the scope of this book chapter. Of note, however, the author observed that visual filters differing slightly from the identity, such as small shifts, or rotation by just a few degrees, had a more lasting impression upon removal of the apparatus than visual filters that were far from the identity, such as the "upside–down" mapping (rotation by 180 degrees). The lasting aftereffect that the author experienced, upon removing the glasses, even after a short period of wearing them, left the author far more incapacitated when the mapping had been close to the identity than when it had been quite far from the identity. Furthermore, visual filters close to the identity tended to leave an opposite aftereffect (e.g., the author would consistently reach too high after taking off the RM where the images had been translated down slightly, or reach too far clockwise after removing the RM that had been rotating images a few degrees counterclockwise). Visual filters far from the identity (such as reversal or upside-down mappings) did not leave an opposite aftereffect. For example, it was found that the world would *not* be seen as being upside down upon removing upside-down glasses to which adaptation had taken place. Moreover, when two (or more) adaptation spaces were sufficiently distinct, for example, in the case of the identity map and the rotation operation (rot 90), the author could sustain a dual adaptation space and switch back and forth between the "portrait" orientation of the identity operator and "landscape" orientation of the rot 90 operator without one causing lasting aftereffects in the other.

3.8 Why Living in a Coordinate Transformed World Makes Practical Sense

Regardless of how much care is taken in creating the illusion of transparency, there will be a variety of flaws, not the least of which is limited resolution, lack of dynamic range, limited color (mapping from the full spectrum of visible light to three responses of limited color gamut), and improper alignment and placement of the cameras. In Figure 9.5, for example, the cameras are mounted *above* the author's eyes. Even if they are mounted in front of the eyes, they will extend out from the eyes (except in the case of a blind person in the future when sufficient technological advances permit eyes to be removed and replaced by cameras in the exact same location as the eyes were). Thus some adaptation is needed. For purposes of this article, reality mediators are assumed to require an adaptation process; see "WearCam (The Wearable Camera)" [Mann, 1998b] for description of a class of reality mediators that do not need adaptation.

After wearing the apparatus for an extended period of time, the author eventually adapted, despite its flaws, whether these were unintended (e.g., limited dynamic range, limited color gamut, etc.) or intended (e.g., deliberately presenting an upside–down or sideways view of the world). In some sense the visual reconfiguration induced by the apparatus is subsumed into the brain, so that the apparatus and human host act as a single unit.

3.8.1 Slowglasses: Delayed Transparency

Delay is a serious problem in MR as much as it is in AR, except that the effects are different. In AR delay causes virtual objects to lag behind the real world, while in MR everything is delayed, both real and virtual. In order to better understand the effects of delay, a *delayed transparency* was created using the reality mediator (Figure 9.5) with a video delay.

As is found in any poor simulation of virtual reality, wearing "slowglasses" induced a similar dizziness and nausea. After experimenting with various delays one will develop an appreciation of the importance of moving the information through the RM in a timely fashion to avoid this unpleasant delay [Mann, 1994].

3.8.2   Edgertonian Eyes

Instead of a fixed delay of the video signal, the author experimented by applying a repeating freeze-frame effect to it (with the cameras' own shutters set to 1/10000 second). With this video *sample and hold*, it was found that nearly periodic patterns would appear to freeze at certain speeds. For example, while looking out the window of a car, periodic railings that were a complete blur without the RM would snap into sharp focus while looking through the RM. Slight differences in each strut of the railing would create interesting patterns that would dance about revealing slight irregularities in the structure. If one thinks of any nearly periodic structure as being a superposition of two components, a periodic component plus a difference signal, the difference signal is what gave rise to the interesting patterns. Looking out at other cars, traveling at 110 or 120 km/hour (approximately the same speed as the car the author was in), the writing on the tires could be easily read, and the number of bolts on the wheel rims could be easily counted.

While seated in an airplane that was flying in formation with other airplanes, the number of blades on the spinning propellers of the other planes could be easily counted. Depending on the sampling rate of the RM, the blades would appear to rotate slowly backwards or forwards, in much the same way as objects do under the stroboscopic lights of Harold Edgerton [Edgerton, 1979]. *What was learned from these simple observations was that by manually adjusting the processing parameters of the RM, many things that escape normal vision could be seen.*

A sufficiently capable RM may implement the Wyckoff principle [Mann, 1993], [Mann and Picard, 1994]. The Wyckoff principle, introduced in Mann and Picard [1994], is a means by which differently exposed pictures are used to determine the unknown nonlinear response function of a camera, and to compensate for the unknown nonlinearity, so that a photoquantigraphic measurement array can be constructed from each of the images. Implementation of this principle involves automatically capturing pictures at a variety of different exposures to allow the wearer to see in darkness, while at the same time see extremely bright objects. Thus the user may, for example, be able to look at a welder's flame in a dark room and see details in the structure of the flame without eye damage, and at the same time see details around the room that would be too dark to see with the naked eye. In this way the user would be able to see both greater dynamic range and subtle differences in intensity (e.g., through the use of a Wyckoff color scale, also known as a "pseudocolor" scale and invented by Charles Wyckoff in creating pictures that conveyed no color information but instead used color to display extended greyscale response [Wyckoff, 1962]). Such a sufficiently capable RM would, of course, be able to recognize patterns in motion that might escape normal vision due to their speed, regularity, or the like.

3.8.3   Virtual "Smart" Strobe

By applying machine vision (some rudimentary intelligence) to the incoming video, the RM could decide what sampling rate to apply. For example, it could recognize a nearly periodic or cyclostationary signal and adjust the sampling rate to lock onto the signal much like a phase–locked loop. A sufficiently advanced RM with eye tracking and other sensors might make inferences about what the wearer would like to see, and, for example, when looking at a group of airplanes in flight would freeze the propeller of the one that the wearer was concentrating on. The author's original goal in researching the psychophysical effects of freeze–frame, reduced frame rates, and Edgertonian effects was to take a first step toward inventing an apparatus that would enhance the wearer's photographic awareness. What is meant by photographic awareness is the general mindset that an experienced photographer has, combined with the enhanced sense of awareness of light and shade that exceptionally talented visual artists often have.

3.9   Human-Centered and Camera-Centered Computing

The main reason for which the MR paradigm was formulated was to make the camera take on a primary and central role in the context of wearable computing. While AR does not require a camera, a camera may be used in AR to recognize objects and insert overlays or other information into the visual reality stream. However with MR a camera *must* be present. Moreover, since the camera is the only means by which the wearer of the RM can see (since the apparatus covers the eyes completely), the camera actually becomes the wearer's eye, for all practical purposes. Thus if, for example, the camera is not adjusted properly, the wearer will be unable to see properly. In this way, the wearer is part of the feedback loop that keeps the camera adjusted properly. Therefore, whether through conscious effort (manual adjustment of the camera) or subconscious effort (choice of gaze angle, positioning of the head, and general conduct and behavior) the camera will always end up being situated for best picture. In this way the synergy between human and machine operate as a system that always seeks best picture. The wearer's entire body posture, mannerisms, and general conduct become part of a natural control system that constantly adjusts itself to obtain the clearest image upon the screen, with a side effect that when the video signal is observed by any other entity (be it other humans or other machine vision algorithms) the synergism of human and camera creates a single unit that also adjusts itself for the clearest image at the remote site. This characteristic gives rise to the MR genre of documentary cinematography to be described later, as well as a new form of computer supported collaboration.

3.10   Partially Mediated Reality

A camera and a display device completely covering only one eye (Figure 9.9(a)) can be used to create a partially mediated reality. In the apparatus of Figure 9.9(a) the author's right eye sees a green image (processed NTSC on a VGA display) while the left eye is unobstructed. (The brain fuses the unobstructed (full-color) view through the left eye with the green mediated video to create the appearance of a doubled vision of the same subject matter.)

Often the mediated and unmediated zones are in poor register and the brain cannot fuse them. Therefore, often the poor register is generated deliberately, as described earlier, so that the mediated and unmediated

(a)   (b)

FIG. 9.9.   Partially mediated reality: (a) Half MR: Author's right eye is *completely* immersed in a mediated reality environment arising from a camera on the right, while the author's left eye is free to see the unmediated real world. (b) Substantially less than half MR: Author's left eye is *partially* immersed in a mediated reality environment arising from a camera on the left.

zones remain visually distinct. The author accomplishes this most often by having one eye see a rotated (rot 90) view even though this means that one cannot see in stereo in a meaningful way. However, after a long-term adaptation of more than one year wearing the apparatus on a regular day–to–day basis it was found that this rotation made it easier to switch concentration back and forth between direct and mediated vision. In this way, the author was able to more easily selectively decide to concentrate on one or the other of these two worlds.

A Virtual Vision television set [Quint and Robinson, 1996] was also used as the basis for making a reality mediator with partial mediation. The structure of the Virtual Vision television set permitted a mediation of even lesser scope to take place. Not only does it play into just one eye, but the field of view of the display only covers part of that eye (Figure 9.9(b)). The visor is transparent so that both eyes can see the real world. In the original Virtual Vision system, the left eye was partially blocked. The author also experimented with the use of beam splitters so that the partial mediation was also partially transparent. With these glasses, one sees an object with both eyes, through the transparent visor, and may then look over to the "mediation zone" where the left eye sees, "through" the illusion of transparency in the display. Again, it was found that long–term adaptation to a 90 degree rotated field of view made it easier to switch mental attention back and forth between the mediated reality and ordinary vision. The result was again a double–vision effect (e.g., when looking at someone's face through the glasses of Figure 9.9(b), the author would see two replicas of the person's face, the one that was mediated, and the one that was not). This doubling effect, due to either the deliberately imperfect registration between the mediated and unmediated zones (as in the rot 90 mapping), or the imperfections that arise even if one attempts to align the two, may or may not be a problem depending on how the RM is used. For example, it was found that if the mediated world was presented in grey rather than in color, it remained distinct from the unmediated world, and the author was able to mentally switch back and forth between seeing directly and seeing in the mediated world, even when the two over-lapped almost exactly (as when they are closely aligned with mappings near the identity). Thus colorblinding served a similar role to rotation by 90 degrees.

Thus, in order to maintain either a dual adaptation space or another form of distinctness between mediated and unmediated zones, depending on the application or intent, there may be desire to either register or to deliberately misregister the possibly overlapping direct and mediated zones.

3.11 Computer Supported Cooperative
Work + Play + Living

Barfield defines "presence" as the feeling of "being there" or of a person or thing "being here," and he has established a comprehensive research program to investigate the relationships among presence, situational awareness, and performance within virtual environments [Barfield and Hendrix, 1995].

By having a remote site (such as a desktop computer) through which one or more remote participants can interact with the wearer of the MR, it is possible to share, remotely, a sense of presence, to allow a remote viewer to vicariously share in the experience of the wearer of the apparatus. Thus the wireless communications system already developed for the RM also facilitates a new form of computer supported collaborative telepresence. Because of the tetherless nature of the apparatus, it may be worn during all facets of life. The capabilities include those of the related field of Computer Supported Cooperative Work (CSCW), but they go beyond just facilitating work (e.g., specific tasks in an employer's structured "workcell" or assembly line or the like). It will be seen later, for example, that such an apparatus is also useful while shopping (for example, to collaborate visually with a spouse at a remote location) or during play and leisure activities.

With two reality mediators of the kind depicted in Figure 9.9(b), it was found that the author could wirelessly transmit the video output from his cameras to the screen of another person and she could transmit the output of her cameras to his screen. In this way, someone else would see through the author's eyes and the author would see through the other person's eyes. The Virtual Vision glasses allowed for concentration mainly on what was in one's own visual field of view (because of the transparent visor), but at the same time it provided a general awareness of the other persons's visual field. This "seeing eye-to-eye," as it is called, allowed for an interesting form of collaboration. Seeing eye-to-eye through the apparatus of Figure 9.5 requires a *picture in picture* process (unless one wishes to endure the nauseating experience of looking *only* through the other person's eyes), usually having the wearer's own view occupy most of the space, while using the apparatus of Figure 9.9(b) does not require any processing at all.

Usually when we communicate (e.g., by voice or video) we expect the message to be received and concentrated on. However seeing eye-to-eye is a new form of communication where there is not the expectation of constant interaction with the other person. It affords a general presence, where serendipity is important. Each of the two people would sometimes pay attention and sometimes not, depending on the relative level of interest between visual material present locally (one's own visual world) and visual material from the remote site (the other person's visual world).

Acknowledgments   The author wishes to thank Simon Haykin, Rosalind Picard, Steve Feiner, Don Norman, Charles Wyckoff, Chuck Carter, Kent Nickerson, Kris Popat, Jonathan Rose, and Chris Barnhart for much in the way of useful feedback, constructive criticism, etc. as this work has evolved. Gerald Q. "Chip" Maguire Jr., Vaughan Pratt, and Bruce Macdonald made valuable suggestions toward arriving at a clear definition of wearable computing. Moreover, this work would not have likely fallen into this state of final order were it not for the help of Woodrow Barfield who, through his editorial assistance, has made this work take its final finished form.

Additional thanks are offered to Kodak, Digital Equipment Corporation, Compaq, Kopin, VirtualVision, HP labs, Chris Barnhart, Miyota, Chuck Carter, Bob Kinney, Thought Technologies Limited, and the Institute for Interventional Informatics.

REFERENCES

[Anstis, 1992] Anstis, S. (1992). Visual adaptation to a negative, brightness-reversed world: some preliminary observations. In Carpenter, G. and Grossberg, S., editors, *Neural Networks for Vision and Image Processing*, pages 1–15. MIT Press.

[Arfken, 1985] Arfken, G. (1985). *Mathematical Methods for Physicists*. Academic Press, Orlando, Florida, third edition.

[Barfield and Hendrix, 1995] Barfield, W. and Hendrix, C. (1995). The effect of update rate on the sense of presence within virtual environments. *Virtual Reality: Research, Development, and Application*, 1(1):3–15.

[Caudell and Mizell, 1992] Caudell, T. and Mizell, D. (1992). Augmented reality: An application of heads-up display technology to manual manufacturing processes. *Proc. Hawaii International Conf. on Systems Science*, 2:659–669.

[Dolezal, 1982] Dolezal, H. (1982). *Living in a World Transformed*. Academic Press Series in Cognition and Perception. Academic Press, Chicago, Illinois.

[Drascic, 1993] Drascic, D. (1993). David Drascic's papers and presentations. http://vered.rose. utoronto.ca/people/david_dir/Bibliography.html.

[Drascic and Milgram, 1996] Drascic, D. and Milgram, P. (1996). Perceptual issues in augmented reality. *SPIE Volume 2653: Stereoscopic Displays and Virtual Reality Systems III*, pages 123–134.

[Earnshaw et al., 1993] Earnshaw, R. A., Gigante, M. A., and Jones, H. (1993). *Virtual Reality Systems*. Academic Press, Orlando, Florida.

[Edgerton, 1979] Edgerton, H. E. (1979). *Electronic Flash, Strobe.* MIT Press, Cambridge, Massachusetts.

[Ellis et al., 1995] Ellis, S. R., Bucher, U. J., and Menges, B. M. (1995). The relationship of binocular convergence and errors in judged distance to virtual objects. *Proceedings of the International Federation of Automatic Control.*

[Feiner et al., 1993a] Feiner, S., MacIntyre, B., and Seligmann, D. (1993a). Karma (knowledge-based augmented reality for maintenance assistance). http://www.cs.columbia.edu/graphics/projects/karma/karma.html.

[Feiner et al., 1993b] Feiner, S., MacIntyre, B., and Seligmann, D. (1993b). Knowledge-based augmented reality. *Communications of the ACM*, 36(7).

[Fuchs et al.,] Fuchs, H., Bajura, M., and Ohbuchi, R. (1993). Teaming ultrasound data with virtual reality in obstetrics. http://www.ncsa.uiuc.edu/Pubs/MetaCenter/SciHi93/1c.Highlights-BiologyC.html.

[Haykin, 1983] Haykin, S. (1983). *Communication Systems.* Wiley, second edition, New York.

[Hoshen et al., 1995] Hoshen, J., Sennott, J., and Winkler, M. (1995). Keeping tabs on criminals. *IEEE SPECTRUM*, pages 26–32.

[Kohler, 1964] Kohler, I. (1964). *The Formation and Transformation of the Perceptual World*, volume 3(4) of *Psychological Issues*. International University Press, 227 West 13 Street. monograph 12.

[Mann, 1993] Mann, S. (1993). "Compositing Multiple Pictures of the Same Scene." The Society of Imaging Science and Technology. *Proceedings of the 46th Annual IS&T Conference.* Cambridge, Massachusetts, pages 50–52, May 9–14.

[Mann, 1994] Mann, S. (1994). "Mediated reality." TR 260, M.I.T. Media Lab Perceptual Computing Section, Cambridge, Massachusetts, http://wearcam.org/mr.htm.

[Mann, 1996] Mann, S. (1996). "Smart clothing": Wearable multimedia and "personal imaging" to restore the balance between people and their intelligent environments. *Proceedings, ACM Multimedia 96*, pp. 163–174, Boston, MA; http://wearcam.org/acm-mm96.htm.

[Mann, 1997a] Mann, S. (1997a). Smart clothing: The wearable computer and wearcam. *Personal Technologies.* Volume 1, Issue 1.

[Mann, 1997b] Mann, S. (1997b). Wearable computing: A first step toward personal imaging. *IEEE Computer; http://wearcam.org/ieeecomputer.htm*, 30(2).

[Mann, 1998a] Mann, S. (1998a). Reflectionism and diffusionism. *Leonardo*, 30(2):93–102.

[Mann, 1998b] Mann, S. (1998b). "WearCam" (the wearable camera). In *IEEE ISWC-98*, Pittsburgh, Pennsylvania.

[Mann and Picard, 1994] Mann, S. and Picard, R. (1994). Being 'undigital' with digital cameras: Extending dynamic range by combining differently exposed pictures. Technical Report 323, M.I.T. Media Lab Perceptual Computing Section, Boston, Massachusetts. Also appears in *IS&T's 48th Annual Conference*, pages 422–428, May 1995.

[Mann and Picard, 1995] Mann, S. and Picard, R. W. (1995). Video orbits of the projective group; a simple approach to featureless estimation of parameters. TR 338, Massachusetts Institute of Technology, Cambridge, Massachusetts. Also appears in *IEEE Trans. Image Proc.*, Sept 1997, Vol. 6 No. 9.

[Nagao, 1995] Nagao, K. (1995). Ubiquitous talker: Spoken language interaction with real world objects. http://www.csl.sony.co.jp/person/nagao.html.

[Norman, 1992] Norman, D. (1992). *Turn Signals are the Facial Expressions of Automobiles.* Addison-Wesley.

[Quint and Robinson, 1996] Quint, J. L. and Robinson, J. W. (1996). Head mounted display system with light blocking structure. U.S. Patent 5,546,099.

[Rolland et al., 1995] Rolland, J. P., Biocca, F., Barlow, T., and Kancherla, A. R. (1995). Quantification of adaptation to virtual-eye location in see-thru head-mounted displays. *Proceedings of the Virtual Reality Annual International Symposium*, pages 55–66.

[Ryals, 1995] Ryals, C. (1995). Lightspace: A new language of imaging. *PHOTO Electronic Imaging*, 38(2):14–16. http://www.peimag.com/ltspace.htm.

[Stratton, 1896] Stratton, G. M. (1896). Some preliminary experiements on vision. *Psychological Review*.

[Stratton, 1897] Stratton, G. M. (1897). Vision without inversion of the retinal image. *Psychological Review*.

[Sutherland, 1968] Sutherland, I. (1968). A head-mounted three dimensional display. In *Proc. Fall Joint Computer Conference*, pages 757–764.

[Wilmer, 1995] Wilmer (1995). Lions vision research and rehabilitation center. http://www.wilmer.jhu.edu/low_vis/ low_vis.htm.

[Wyckoff, 1962] Wyckoff, C. W. (1962). An experimental extended response film. *S.P.I.E. Newsletter*, pages 16–20.

10

STAR: Tracking for Object-Centric Augmented Reality

Ulrich Neumann

University of Southern California

1. INTRODUCTION

This chapter distinguishes between two classes of augmented reality applications, those that require world-centric tracking and those that require object-centric tracking. Emphasis is placed on object-centric approaches that are suitable for applications in which annotated objects may move within an environment. These application classes impart differing requirements upon the tracking systems that enable the presentation of augmented reality media in spatial relationships to objects. The tracking approaches are contrasted, and an instance of an object-centric approach is detailed to illustrate the research and implementation issues. The example system overcomes a limitation of many object-centric tracking systems through its ability to sense and integrate new features into its tracking database, thereby extending the tracking region semiautomatically.

Tracking has been at the center of research and development in augmented reality (AR) since it's inception in the 1960s [1]. Many systems have been developed to track the six degree of freedom (DOF) pose of an object (or person) relative to a fixed coordinate frame in the environment

[2, 3, 4, 5, 6, 7, 8, 9]. These tracking systems employ a variety of sensing technologies, each with unique strengths and weaknesses, to determine a world-centric pose measurement that facilitates the rendering of graphics in a virtual or augmented reality. In a virtual reality, a fixed world coordinate frame is appropriate as the basis for tracking the user's viewing pose and positioning all the elements of the virtual world. Augmented reality, however, differs from virtual reality in that the virtual data or media are often linked to real objects in the environment. Tracking in a fixed frame of reference, therefore, can present a limitation for augmented realities, since it implies that objects in the environment are calibrated to the tracker's frame of reference and, after calibration, they do not move.

The assumption that objects are calibrated and fixed within the environment may be valid for applications such as augmented architectural visualization [10, 11] where the real walls, floors, and doors form a rigid structure whose coordinates can be measured, or are already known from the design. A rigid body transformation calibrates the structure's coordinates to the tracking system. An application may add virtual elements (e.g., furniture) placed within the building and the tracking system's range. Figure 10.1 illustrates an example with virtual chairs and a virtual lamp registered to a world coordinate frame. The lamp and chair positions and occlusions are correct relative to the real table because the virtual objects,

FIG. 10.1. Architectural augmented reality visualization showing a real desk, a virtual lamp, and two virtual chairs. (Courtesy of ECRC.)

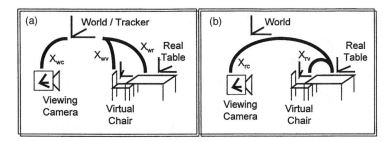

FIG. 10.2. World-centric (a) and object-centric (b) tracking approaches to the scene shown in Figure 10.1. The object-centric approach preserves the relative positions of the table, lamp, and chairs for any (unknown world) table position. ($X_{\alpha\beta}$ denotes the transformation between coordinate frames α and β.)

the real world, and the tracking system are all related by known static transformations.

Figure 10.2a illustrates the coordinate frames and transformations needed to construct the augmented reality room scene (Figure 10.1) using a world-centric tracking approach. The arcs show the needed transformations. (The world and tracker coordinate frames are shown as a single entity, although an additional transformation may exist between them.) The tracker dynamically measures the viewing pose for a camera or human observer. The virtual chair and the real world table are both calibrated to the tracker.

Figure 10.2a shows clearly that if the real table moves, it must be tracked to make the chair and lamp move with it. If we must track the table anyway, why not simply track *from* the table? This is the basic idea behind object-centric tracking and it is illustrated in Figure 10.2b where the table's pose within the world is unknown (and not needed). The viewpoint is tracked relative to the table, and all the virtual objects are placed relative to the table as well. One immediate advantage is that the table can be anywhere in a room, a building, or even outside without a need for a tracking system to cover the entire space of possibilities. Another advantage is that tracking the camera directly from the table (X_{rc}) reduces the error propagation [12, 13] caused by concatenating two measured and possibly dynamic transformations (X_{wr} and X_{wc}).

A number of AR applications provide annotation on objects whose positions in a room or the world may vary freely without impact on the desired AR media display. Figure 10.3 illustrates an example of some annotation that may appear on a door. Regardless of whether the door is open or closed, the annotation appears correctly aligned. Numerous AR applications in

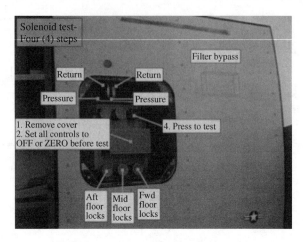

FIG. 10.3.   Example of AR annotation supporting an object-centric
maintenance task. (Courtesy of The Boeing Corp.)

manufacturing, maintenance, and training [10, 14, 15] require virtual annotations that provide task guidance and specific component indications on
subassemblies or portions of structure. These applications are well suited
to object-centric tracking, and useful approaches, based on viewing objects
[16, 17, 18, 19, 20, 21, 22, 23], are provided by the pose estimation methods
developed in the fields of computer vision and photogrammetry [24, 25].
A summary of these methods is presented in Section 2.

World-centric trackers can be used to calibrate movable objects [15, 21,
26, 27], but this generally entails placing and calibrating tracking elements
on each object of interest and operating within range of the shared tracking
infrastructure (e.g., magnetic fields or active beacons [4, 9]). These requirements may make it difficult or expensive to calibrate moving objects
with current world-centric tracking approaches. A summary of the qualitative contrasts between the world- and object-centric tracking approaches
is presented in Table 10.1.

When used for object-centric tracking, vision-based pose accuracy tends
to improve with proximity to an object. Image sensors measure the world in
pixel-units that have a variable relationship to absolute world dimensions,
depending on the lens system and the viewing pose. A distant view of
an object may produce significant pose error because image pixels cover
a relatively large world-dimension on the object. A close-up macro lens
may measure microscopic dimensions and provide commensurate accuracy
and resolution of pose. A camera and lens combination should match the
tracking accuracy required for a given application and setting.

TABLE 10.1
Contrasting World-Centric and Object-Centric Tracking System Characteristics

World-Centric Tracking	Object-Centric Tracking
Assumes that objects are calibrated to infrastructure and remain fixed	Permits tracked objects to move freely in environment
Tracking may cover large contiguous regions	Tracking regions are local to an object
Permits substantial tracking infrastructure that may be a permanent part of environment	Objects carry minimal tracking infrastructure that has little impact on environment
Resolution and accuracy are in fixed world units relative to tracking system components (e.g., centimeters or inches over the working area)	Resolution and accuracy are relative to camera view of objects (pixel-units of error vary over a range of world dimensions)
Tracking extensions and occlusions may be difficult to accommodate	Tracking extensions and occlusions may be semiautomatically accommodated

A drawback of vision-based methods is that tracking is only possible in a constrained set of views for which certain features of the scene are visible. In Section 2.3 we describe how to reduce this limitation in many cases, by allowing users to dynamically expand the range of tracked camera views as the AR system is operating.

2. VISION-BASED TRACKING

Vision-based tracking is the problem of calibrating a camera's pose relative to an object, given one or more images. Many tracking methods have been developed and they can be coarsely grouped into the following categories, based on their input requirements:

- Three or more known 3D points must be visible in a single image [25, 28, 29, 10, 22, 23, 30, 17, 8].
- A sequence of images with correspondences must be available from a moving camera, where 3D point positions may be known or unknown [31, 32, 33, 9, 20].
- A 3D model of the scene or image templates are available for matching to a single image [34, 16].

The first class of approaches utilizes a calibrated camera to provide constraints, so only a few known points within a single image are needed for pose determination. These methods are successfully employed in AR tracking systems, including the example architecture described in the remainder of this chapter.

The points needed for tracking may be natural features such as corners and holes, or they can be intentionally designed and applied targets or fiducials. We selected the latter option for our system since naturally occurring features can be difficult to recognize due to their variety and unpredictable characteristics. Also, objects do not always have features where they are needed for tracking; large regions of surfaces are often indistinguishable when viewed without context. Fiducials have the advantage that they can be designed to maximize the AR system's ability to detect and distinguish between them, they can be inexpensive, and they can be placed almost arbitrarily on objects. The design and detection of fiducials are important topics in themselves [7, 35], but for this chapter we assume that fiducials have some basic characteristics such as color and shape. Colored stickers are good examples of fiducials.

2.1   Camera Calibration

The camera's internal parameters for focal length and lens distortion must be determined [36, 37]. Our method for focal length determination (Figure 10.4) uses a planar target with a known grid pattern and spacing. Multiple images are taken at measured offsets D along the viewing direction. For several pairs of images the focal length f is computed using the equations and geometry in Figure 10.4. The final estimate is the average of all pair results. Grid points closest to the center of the image are used since they are least affected by lens distortions.

The lens distortion calibration is a variation of the DLTEA-II algorithm [38]. Rather than use an arbitrary camera orientation with seven independent points, no four of which can be coplanar, a planar grid target and perpendicular view pose are employed, as described above. The image

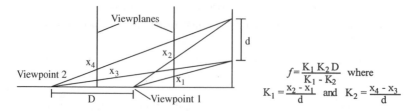

FIG. 10.4.   Two camera positions suffice for calculating focal length f.

coordinates (u,v) of a 3D point (x, y, z) can be represented by

$$
\begin{bmatrix} U \\ V \\ 1 \end{bmatrix} = \lambda \begin{bmatrix} f_x & 0 & U_0 \\ 0 & f_y & V_0 \\ 0 & 0 & 1 \end{bmatrix} [R][I, \ -T] \begin{bmatrix} x \\ y \\ z \\ 1 \end{bmatrix},
\tag{1}
$$

where λ is a scale factor relating to perspective projection, f_x and f_y are effective focal lengths along the u and v axes, respectively, and (U_0, V_0) is the center of lens distortion in the image. R and T are components of the view transformation, and due to our camera setup, R is approximated as an identity and all calibration points are on the $z = 0$ plane. Thus

$$
\begin{bmatrix} U \\ V \\ 1 \end{bmatrix} = \lambda \begin{bmatrix} f_x & 0 & U_0 \\ 0 & f_y & V_0 \\ 0 & 0 & 1 \end{bmatrix} \begin{bmatrix} I, & \begin{matrix} -X_0 \\ -Y_0 \\ -Z_0 \end{matrix} \end{bmatrix} \begin{bmatrix} x \\ y \\ 0 \\ 1 \end{bmatrix}.
\tag{2}
$$

Allowing for second-order lens distortion, we have

$$
\begin{aligned}
U + \Delta U &= U + f_x (k_1 U_r D^2 + t_1 (D^2 + 2U_r^2) + 2t_2 U_r V_r) \\
&= \frac{f_x X_0 + Z_0 U_0 - f_x x}{Z_0},
\end{aligned}
\tag{3}
$$

$$
\begin{aligned}
V + \Delta V &= V + f_y (k_1 V_r D^2 + t_2 (D^2 + 2V_r^2) + 2t_1 U_r V_r) \\
&= \frac{f_y Y_0 + Z_0 V_0 - f_y y}{Z_0},
\end{aligned}
\tag{4}
$$

where k_1 and t_1, t_2 are the coefficients of radial distortion and tangential distortion, respectively, as described in [38]. We thus have

$$
U_r = \frac{(U - U_0)}{f_x}, \quad V_r = \frac{(V - V_0)}{f_y}, \quad D = \sqrt{U_r^2 + V_r^2}.
\tag{5}
$$

The calculation of X_0, Y_0, k_1, t_1, and t_2 is an iterative minimization of the least-square distortion residual.

The remaining calibration issue is to specify how the fiducials and annotations will be calibrated to the actual assembly and each other. We calibrate the initial known points with a digitizer probe. In the manufacturing setting, jigs that support many assemblies have precision landmarks. We anticipate using those, perhaps with a digitizing probe, to fix an initial set of points on the surface of the assembly. Alternatively, CAD information can supply the positions of features that are designed into the assembly, or the autocalibration methods described in Section 2.3 may be employed. The best solution is application dependent, and we recognize a need for an array of flexible options and tools that can be applied as needed in each case.

2.2 Pose Estimation

Pose determination is computed based on three visible points [25]. In practice, we find the method stable over a wide range of viewing conditions. The method involves the solution to a quartic polynomial, so up to four solutions may exist. In many cases the number of solutions collapses to two that are distant from each other. Discrimination between multiple solutions is performed in several ways. Assemblies often have a front-side that is exposed so viewpoints behind the assembly are culled. If more than three points are visible, the screen location of the fourth point must agree with its projection under the correct pose. In successive frames, proximity to the previous frame's position is meaningful. The placement of fiducials is not critical, but unstable poses are known to exist for any three-point geometry. The singularities occur in practice, but users quickly learn to position themselves for stable views. The numerical characteristics of several three-point pose solution methods are described in Ref. [39].

2.3 Extendible Tracking

As noted earlier, a limitation of object-centric tracking is the need for fiducials to be in view at all times. Ideally, a wide range of camera motion should provide tracked viewing from a minimal number of fiducials that must be applied to an object a priori. A worst case occurs when all regions of a large object are equally likely to be viewed and therefore a dense distribution of fiducials is needed to support tracking over the entire object. Covering large objects (e.g., airplanes and vehicles) with calibrated fiducials is impractical. Our approach is an alternative analogous to "lazy-evaluation" in algorithm design; fiducials are only placed, and their positions computed, as the need for them arises. An initial set of fiducials is strategically placed and calibrated on an object or a fixture rigidly connected to an object. As regions of the object require additional fiducials to support tracking, users simply add new fiducials to those regions and allow the system to automatically calibrate them (as described in the next section). Once calibrated, the new fiducials are added to the database of known fiducials and they are used for tracking.

Adding fiducials is practical when a limited region of the object needs to be viewed and tracked for a given task, but that region or task is just one of many that may be of interest. Assembly, training, and maintenance tasks often require information that relates to a local region, but the specific task is only one of many that may be performed by different people, at different times, in different places. Figure 10.3 illustrates an example of information display that requires tracking in only a limited region around an access panel within a larger structure.

3. EXTENDIBLE OBJECT-CENTRIC TRACKING SYSTEM ARCHITECTURE

Figure 10.5 depicts a system architecture for extendible object-centric tracking. A single camera provides real-time video input, and a user observes the camera images and the overlaid AR media on a display that may be desktop, hand-held, or head-mounted. The shaded areas (Figure 10.5) indicate functions that provide the new extendible tracking capability.

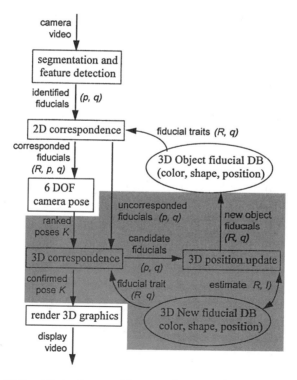

FIG. 10.5. AR system architecture for extendible vision-based tracking. Functions for new fiducial integration are shown on gray background.

3.1   Segmentation and Feature Detection

The first stage of the system (Figure 10.5) is responsible for robust fiducial detection in the video images. Fiducials are represented by a 2D-screen position p and a type q that encodes characteristics such as color and shape. Our fiducial design is a colored circle or triangle [17], but other designs such as concentric circles or coded squares are equally valid [7, 10]. The three primary and three secondary colors, along with the triangle and circle shapes, produce twelve unique fiducial types. Fiducials are detected by segmenting the image into regions of similar intensity and color and testing the regions for color and geometric properties that make them likely fiducials. Detection strategies are often dependent on the characteristics of the fiducials [18, 35, 40, 41].

3.2   2D Correspondence

The 2D fiducials (p, q) must be corresponded to elements in a database of known 3D fiducial positions R and types q. The result of a successful match is a correspondence (R, p, q) between detected fiducials (p, q) and known 3D fiducials (R, q). Fiducials that are detected in the image but do not correspond to the world database are passed on as uncorresponded fiducials and potential new points to estimate.

Computing correspondences is a hard problem in the general case but trivial if the fiducial types q are unique for each element in the database, as in Refs. [7, 10, 17, 18]. When multiple instances of a fiducial type q are possible, clusters of fiducials must be recognized to establish correspondences. Techniques for cluster recognition include the use of geometric and projective constraints [23, 41, 42, 43, 44], template matching [34], and aspect graphs [19, 45]. These and other approaches are active research topics, and a scalable solution to this problem is vital for vision-based tracking to perform well in large-scale applications.

3.3   Camera Pose

Three or more corresponded fiducials allow the camera pose K to be determined, as related in Section 2.2. Multiple pose solutions occur frequently [30], so rather than selecting only one pose at this point, the result of this function is an array of poses, ranked by their probability [46]. The new point functions (shown over gray in Figure 10.5) further refine the ranking of each pose by comparing the uncorresponded fiducial positions p against the projections of new fiducials of the same type. The pose that produces the best agreement between new fiducials and uncorresponded fiducials is taken as the confirmed pose of the camera.

3.4 New Point Estimation

This section details the unique aspects of this architecture that provide the interactive extension of the tracked viewing range. Given the camera pose and image coordinates of the uncorresponded features, these features' position estimates are updated in one or two possible ways.

3.4.1 Initial Estimates

For uncorresponded fiducials (p, q), the intersection[1] of two lines connecting the camera positions and the fiducial locations in the image create the initial estimate of the 3D position of the fiducial. The intersection threshold is scaled to the expected size of the fiducials. For example, if the radius of the fiducials is 0.5 inch, as in our case, the threshold is 1.0 inch. Our selection is based on the minimum distance that two distinct fiducials can be placed without overlap. Lines with a closest point of approach less than 1.0 inch are considered to be intersecting.

3.4.2 Extended Kalman Filter (EKF)

The Extended Kalman Filter (EKF) has been adapted to many applications, and details of the method can be found in, for example, Refs. [9, 32, 47]. Inputs to the EKF are the current camera pose and the image coordinates of the fiducial. The state of the EKF is the current estimate of the fiducial's 3D position. The 3D positions of the fiducials are constant over time, so no dynamics are involved in the EKF equations. Parameters of the EKF, including the current state and its covariance matrix, are explained below:

c_k: camera pose at kth time step,

p_c: intrinsic camera parameters including focal length,

z_k (measurement): image coordinate of the fiducial at kth time step,

---

[1]Since two lines in 3D space may not actually intersect, the point midway between the points of closest approach is used as the intersection.

$\hat{z}_k$ (measurement prediction): predicted measurement
at kth time step (see Eq. 11),

$\tilde{z}_k$ (residual): $\tilde{z}_k = z_k - \hat{z}_k$,   (6)

x_{k-1}: real value of 3D fiducial position;
$$x_k = x_{k-1} = x_{k-2} = \cdots = x_1.$$

Given $Z_{k-1} = (z_1 \, z_2 \, \cdots \, z_{k-1})$, $C_{k-1} = (c_1 \, c_2 \, \cdots \, c_{k-1})$, and p_c:

$\hat{x}_{k-1}$ (state): filter's estimate of the state value at $(k-1)$th
time step,

$\hat{x}_{k-1} = E(x_{k-1} \mid Z_{k-1}, C_{k-1}, p_c)$,   (7)

$\hat{x}_k^-$ (state prediction): predicted state estimate at kth time
step given measurements up to $(k-1)$th step,

$\hat{x}_k^- = E(x_k \mid Z_{k-1}, C_{k-1}, p_c)$,   (8)

Q (process noise): a 3×3 covariance matrix,

R (measurement noise): a 2×2 covariance matrix.

Let U_3 be 3×3 identity matrix; then $Q = 10^{-5} \cdot U_3$. Let U_2 be 2×2
identity matrix; then $R = 2 \cdot U_2$, assuming an measurement error variance
of 2, and no correlation between x and y coordinates in the image.
Other parameters are:

P_{k-1} (state uncertainty): 3×3 uncertainty covariance matrix
at $(k-1)$th time step,

$P_{k-1} = E[(x_{k-1} - \hat{x}_{k-1})(x_{k-1} - \hat{x}_{k-1})^T]$,   (9)

P_k^- (state uncertainty prediction): predicted state uncertainty
at kth time step given measurement up to $(k-1)$th
time step,

$P_k^- = E[(x_k - \hat{x}_k^-)(x_k - \hat{x}_k^-)^T]$,   (10)

$\vec{h}$ (measurement function): returns the projection
(measurement estimate) of the current position estimate
given the current camera pose and camera parameters,

$$\hat{z}_k = \vec{h}(\hat{x}_k^-, c_k, p_c), \tag{11}$$

H_k (Jacobian): Jacobian matrix of $\vec{h}$

K_k (Kalman Gain): 3×2 matrix (see Eq. 15).

The EKF process is composed of two groups of equations: *predictor* (time update) and *corrector* (measurement update). The *predictor* updates the previous $(k-1)$th state and its uncertainty to the predicted values at the current kth time step. Since the 3D fiducial position does not change with time, the predicted position at the current time step is the same as the position of the previous time step:

$$\hat{x}_k^- = \hat{x}_{k-1}, \tag{12}$$

$$P_k^- = P_{k-1} + Q, \tag{13}$$

$$\hat{z}_k = \vec{h}(\hat{x}_k^-, c_k, p_c). \tag{14}$$

Corrector equations correct the predicted state value $\hat{x}_k^-$ based on the residual of actual measurement z_k and measurement estimate $\hat{z}_k$. The Jacobian matrix linearizes the nonlinear measurement function:

$$K_k = P_k^- H_k^T (H_k P_k^- H_k^T + R)^{-1}, \tag{15}$$

$$\tilde{z}_k = z_k - \hat{z}_k, \tag{16}$$

$$\hat{x}_k = \hat{x}_k^- + K_k \cdot \tilde{z}_k, \tag{17}$$

$$P_k = (I - K_k H_k) P_k^-. \tag{18}$$

3.4.3 Recursive Average of Covariances (RAC) Filter

The RAC filter models each measurement as a 3D line through the current camera position and the fiducial location in the image. The 3D position estimate (X) of the fiducial is updated based on the measurement line (l). The uncertainty covariance matrix of the estimate is used in computing the update direction vector and the update magnitude of the estimate, and then it is recursively averaged with the uncertainty covariance matrix of the line.

To obtain the update direction of the position estimate, a vector $\vec{v}$ from the current estimate (X) to a point on the line (l) that is closest to X is computed first. The update direction vector $\vec{q}$ is a version of the vector

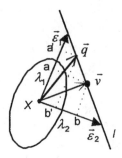

FIG. 10.6. 2D analogy of the RAC algorithm for computing the update direction.

$\vec{v}$ scaled by the uncertainty of X, to move the estimate in the direction of larger uncertainty. Figure 10.6 shows the 2D analogy of the update direction computation. The vector $\vec{v}$ is scaled in directions of $\vec{\varepsilon}_1$ and $\vec{\varepsilon}_2$ by λ_1 and λ_2 respectively to produce $\vec{q}$.

The update direction vector $\vec{q}$ is computed by first finding parameters (a, b, c) satisfying $a \cdot \vec{\varepsilon}_1 + b \cdot \vec{\varepsilon}_2 + c \cdot \vec{\varepsilon}_3 = \vec{v}$. These coefficients represent the vector from the current estimate to the closest point on the new line in the uncertainty eigenvector coordinate system. Then we compute (a', b', c') by scaling (a, b, c) by $(\lambda_1 \lambda_2 \lambda_3)$. Finally, the update direction $\vec{q}$ is obtained using the scaled components (a', b', c'):

$$[\vec{\varepsilon}_1 \quad \vec{\varepsilon}_2 \quad \vec{\varepsilon}_3] \begin{bmatrix} a \\ b \\ c \end{bmatrix} = \vec{v}, \tag{19}$$

$$\begin{bmatrix} a \\ b \\ c \end{bmatrix} = [\vec{\varepsilon}_1 \quad \vec{\varepsilon}_2 \quad \vec{\varepsilon}_3]^{-1} \cdot \vec{v}, \tag{20}$$

$$[a' \quad b' \quad c'] = [l_1 \cdot a \quad l_2 \cdot b \quad l_3 \cdot c], \tag{21}$$

$$\vec{q} = a' \cdot \vec{\varepsilon}_1 + b' \cdot \vec{\varepsilon}_2 + c' \cdot \vec{\varepsilon}_3. \tag{22}$$

The update magnitude m of the position estimate is

$$m = \min(d_u, d_l) \tag{23}$$

where d_u is the uncertainty of the current estimate in the direction of $\vec{q}$ and d_l is the scaled distance from X to l in the direction of $\vec{q}$ (i.e., $|\vec{q}|$).

The uncertainty of the position estimate is represented by a 3×3 covariance matrix. Each line has a constant uncertainty L_k, which is narrow and long along the direction of the line. Uncertainty is updated by recursively averaging the covariance matrices, similar to the process used in the Kalman Filter. In the Kalman Filter, the uncertainty covariance matrix is updated by performing weighted-averaging with the 2D-measurement error covariance as below:

$$K_k = P_k^- H_k^T (H_k P_k^- H_k^T + R_k)^{-1}, \qquad (24)$$

$$P_k = (I - K_k H_k) P_k^-. \qquad (25)$$

In the RAC filter, the measurement is modeled as a line that is already in the same 3D space the estimate is in. Therefore we can eliminate the Jacobian matrix and its linearization approximation. This is an advantage of the RAC filter over the EKF, simplifying and reducing the computational overhead. Let P_k^- be the uncertainty covariance matrix of the current estimate and L_k be that of the new line; then computation of the updated uncertainty covariance matrix P_k as simplified as below. The initial value of P_k is obtained in the same way by replacing P_k^- with L_k, resulting from the two initial line uncertainties.

$$K_k = P_k^- (P_k^- + L_k)^{-1},$$

$$P_k = (I - K_k) P_k^- \qquad (26)$$

$$= (I - P_k^- (P_k^- + L_k)^{-1}) P_k^- \qquad (27)$$

$$= ((P_k^- + L_k)(P_k^- + L_k)^{-1} - P_k^- (P_k^- + L_k)^{-1}) P_k^-$$

$$= (L_k (P_k^- + L_k)^{-1}) P_k^-. \qquad (28)$$

4.   EXPERIMENT AND RESULT

Both filters appear stable in practice. The EKF is known to have good characteristics under certain conditions [32]; however, the RAC gives comparable results, and it is simpler, operating completely in 3D world-space with 3D lines modeling measurements. The RAC approach eliminates the Jacobian matrices required as the EKF linear approximation.

Synthetic data sets were created, based on real camera poses captured by real camera movements. White noise of 0.5 and 2.0 peak pixel error were added to the measurements of the image coordinates of the fiducials. Figures 10.7 and 10.8 show some results of our experiments with synthetic data sets.

In a real-world test, the EKF and RAC filters calibrate the positions of two new fiducials (referred to here as red and magenta fiducials). We started with three known fiducials and two cases were tested. In the *panning* case, new fiducials were placed to the side of the known fiducials, assuming a translated region of interest. In the *zooming* case, new fiducials were placed in the center so that the user could zoom in to a region of interest. Experiment results for both the zoom and pan cases are shown in Figure 10.9, and screen images for the tests are shown in Figures 10.10 and 10.11. The estimates of the filters are compared with analytic solutions, which have the minimum sum of Euclidean distances to the input lines. Figure 10.12 shows a 2D analogy of the analytic solution computation.

A virtual camera view shows the results of the real data experiments graphically (Figure 10.13). The lines are traces of the camera (i.e., the input lines used for the RAC filter). The large spheres indicate positions of known fiducials, which were used for computing camera poses. The dark and bright small spheres represent the estimated positions produced by the RAC and EKF filters, while the black cube represents the analytic solution positions. The results show that the estimates of the filters converge quickly and remain stable after convergence with both real and synthetic data.

Our hardware configuration included a desktop workstation:

- SGI Indy 24-bit graphics system with MIPS4400@200 MHz,
- SONY DXC-151A color video camera with 640×480 resolution, 31.4-degree horizontal and 24.3-degree vertical field of view (FOV), S-video output.

However, this system is similar in computing power to currently available portable laptops or wearable computers.

Our future work entails a solution to the 2D correspondence problem. This is necessary for the system to scale to greater numbers of fiducials. We also hope to improve our ability to select the correct pose solution, and we need to address the error propagation from multiple chained fiducial estimates.

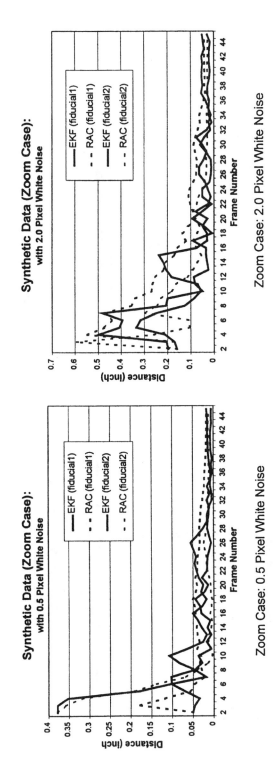

FIG. 10.7. Synthetic data: Camera movement-zooming.

345

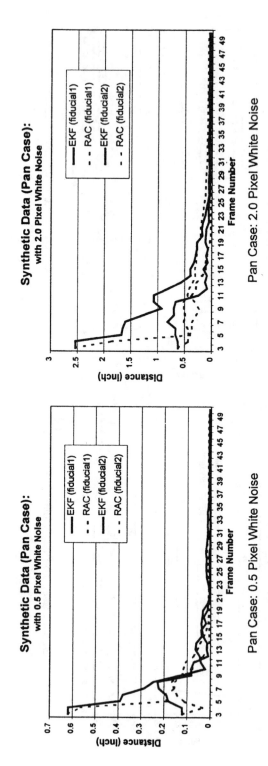

FIG. 10.8. Synthetic data: Camera movement-panning.

346

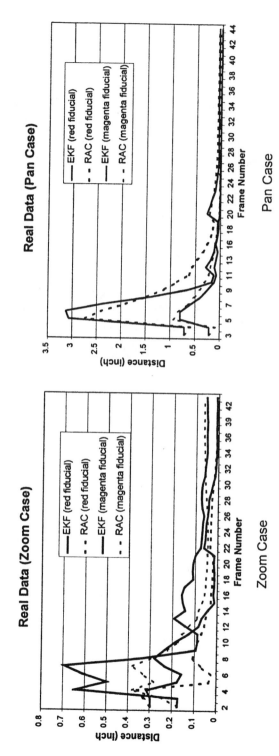

FIG. 10.9. Real data: Zooming and panning cases.

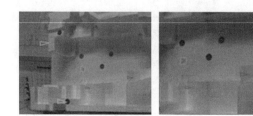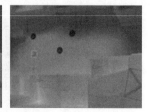

FIG. 10.10. Sequence depicting a zoom into a smaller region of interest. The virtual box around the cross in the lower-right corner of (b) and (c) visually indicates the precision of the alignment. The box corners should be aligned with the ends of the cross.

FIG. 10.11. Sequence depicting a translation of the region of interest. The virtual box around the cross in the lower center of (b) and (c) visually indicates the precision of the alignment. The box corners should be aligned with the ends of the cross.

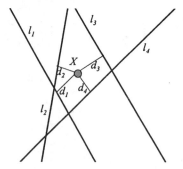

FIG. 10.12. The 2D analytic solution is a point that has the minimum sum of distances to the lines. In this case, the analytic solution is a point X that minimizes $d_1 + d_2 + d_3 + d_4$.

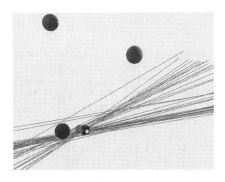 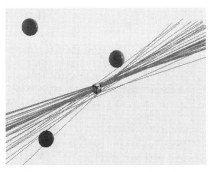

FIG. 10.13. Virtual camera (close-up) views of the results of real data experiments. (Left) Pan case: Magenta fiducial. The EKF result (bright sphere) appears in front of the analytic solution (black cube), and the RAC filter result (dark sphere) is just to their left. (Right) Zoom case: Red fiducial. The EKF result (bright sphere), RAC filter result (dark sphere), and the analytic solution (black cube) overlap and are not separately discernible.

5. SUMMARY

In this chapter, we distinguished between world-centric and object-centric tracking approaches and their qualities that make them appropriate in different application domains. An object-centric AR architecture was presented and an implementation described. This system has the ability to sense and integrate new features into its tracking database, thereby extending the tracking region semiautomatically. We feel strongly that interactive object-centric systems, such as the one described, can provide useful information in an AR metaphor using existing computer technology.

Acknowledgments   This work is supported by NSF CAREER Award CCR-9502830, the Integrated Media Systems Center—a National Science Foundation Engineering Research Center, with additional support from the Annenberg Center for Communication at the University of Southern California, the California Trade and Commerce Agency, and corporate partners Hughes Research Laboratories, Hughes Training Inc., the Boeing Company (Long Beach, CA), and Hewlett Packard. Research collaborators include Jun Park, Youngkwan Cho, Dr. Suya You, and Dr. Anthony Majoros (The Boeing Company).

REFERENCES

[1] Sutherland, I. (1968). "A Head-Mounted Three-Dimensional Display," *Fall Joint Computer Conference*, pp. 757–775.

[2] Foxlin, E. (1996). "Inertial Head-Tracker Sensor Fusion by a Complementary Separate-Bias Kalman Filter," *Proceedings of VRAIS'96*, pp. 184–194.

[3] Ghazisadedy, M., Adamczyk, D., Sandlin, D. J., Kenyon, R. V., and DeFanti, T. A. (1995). "Ultrasonic Calibration of a Magnetic Tracker in a Virtual Reality Space," *Proceedings of VRAIS'95*, pp. 179–188.

[4] Kim, D., Richards, S. W., and Caudell, T. P. (1997). "An Optical Tracker for Augmented Reality and Wearable Computers," *Proceedings of VRAIS'97*, pp. 146–150.

[5] Meyer, K., Applewhite, H. L., and Biocca F. A. (1992). "A Survey of Position Trackers," *Presence: Teleoperator and Virtual Environments, 1*(2), 173–200.

[6] Sowizral, H., and Barnes, J. (1993). "Tracking Position and Orientation in a Large Volume," *Proceedings of IEEE VRAIS'93*, pp. 132–139.

[7] State, A., Hirota, G., Chen, D. T., Garrett, B., and Livingston M. (1996). "Superior Augmented Reality Registration by Integrating Landmark Tracking and Magnetic Tracking," *Proceedings of Siggraph96, Computer Graphics*, pp. 429–438.

[8] Ward, M., Azuma, R., Bennett, R., Gottschalk, S., and Fuchs, H. (1992). "A Demonstrated Optical Tracker with Scalable Work Area for Head-Mounted Display Systems," *Proceedings of the 1992 Symposium on Interactive 3D Graphics*, pp. 43–52.

[9] Welch, G., and Bishop, G. (1997). "SCAAT: Incremental Tracking with Incomplete Information," *Proceedings of Siggraph97, Computer Graphics*, pp. 333–344.

[10] Klinker, G., Ahlers, K., Breem, D., Chevalier, P., Crampton, C., Greer, D., Koller, D., Kramer, A., Rose, E., Tuceryan, M., and Whitaker, R. (1997). "Confluence of Computer Vision and Interactive Graphics for Augmented Reality," *Presence: Teleoperator and Virtual Environments, 6*(4), 433–451.

[11] Feiner, S., Webster, A., Krueger III, T., MacIntyre, M., and Keller, E. (1995). "Architectural Anatomy," *Presence: Teleoperator and Virtual Environments, 4*(3), 318–325.

[12] Holloway, R. (1997). "Registration Error Analysis for Augmented Reality," *Presence: Teleoperator and Virtual Environments, 6*(4), 413–432.

[13] Bajura, M., and Neumann, U. (1995). "Dynamic Registration Correction in Video-Based Augmented Reality Systems," *IEEE Computer Graphics and Applications, 15*(5), 52–60.

[14] Caudell, T. P., and Mizell, D. M. (1992). "Augmented Reality: An Application of Heads-Up Display Technology to Manual Manufacturing Processes," *Proceedings of the Hawaii International Conference on Systems Sciences*, pp. 659–669.

[15] Feiner, S., MacIntyre, B., and Seligmann, D. (1993). "Knowledge-Based Augmented Reality," *Communications of the ACM, 36*(7), 52–62.

[16] Natonek, E., Zimmerman, Th., and Fluckiger, L. (1995). "Model Based Vision as Feedback for Virtual Reality Robotics Environments," *Proceedings of VRAIS'95*, pp. 110–117.

[17] Neumann, U., and Cho, Y. (1996). "A Self-Tracking Augmented Reality System," *Proceedings of ACM Virtual Reality Software and Technology '96*, pp. 109–115.

[18] Rekimoto, J. (1997). "NaviCam: A Magnifying Glass Approach to Augmented Reality," *Presence: Teleoperator and Virtual Environments, 6*(4), 399–412.

[19] Sharma, R., and Molineros, J. (1997). "Computer Vision-Based Augmented Reality for Guiding Manual Assembly," *Presence: Teleoperator and Virtual Environments, 6*(3), 292–317.

[20] Uenohara, M., and Kanade, T. (1995). "Vision-Based Object Registration for Real-Time Image Overlay," *Proceedings of Computer Vision, Virtual Reality, and Robotics in Medicine*, pp. 13–22.

[21] Starner, T., Mann, S., Rhodes, B., Levine, J., Healey, J., Kirsh, D., Picard, R., and Pentland, A. (1997). "Augmented Reality Through Wearable Computing," *Presence: Teleoperator and Virtual Environments, 6*(4), 386–398.

[22] Kutulakos, K., and Vallino, J. (1996). "Affine Object Representations for Calibration-Free Augmented Reality," *Proceedings of VRAIS'96*, pp. 25–36.

[23] Mellor, J. P. (1995). "Enhanced Reality Visualization in a Surgical Environment," Master's Thesis, Dept. of Electrical Engineering, MIT.

[24] Huang, T. S., and Netravali, A. N. (1994). "Motion and Structure from Feature Correspondences: A Review," *Proceedings of the IEEE, 82*(2), 252–268.

[25] Fischler, M. A., and Bolles, R.C. (1981). "Random Sample Consensus: A Paradigm for Model Fitting with Applications to Image Analysis and Automated Cartography," *Graphics and Image Processing, 24*(6), 381–395.

[26] Wells, W., Kikinis, R., Altobelli, D., Ettinger, G., Lorensen, W., Cline, H., Gleason, P. L., and Jolesz, F. (1993). "Video Registration Using Fiducials for Surgical Enhanced Reality," Engineering in Medicine and Biology, IEEE.

[27] Bajura, M., Fuchs, H., and Ohbuchi, R. (1992). "Merging Virtual Reality with the Real World: Seeing Ultrasound Imagery within the Patient," *Computer Graphics (Proceedings of Siggraph 1992)*, pp. 203–210.

[28] Ganapathy, S. (1984). "Real-Time Motion Tracking Using a Single Camera," AT&T Bell Labs Tech Report 11358-841105-21-TM, November.

[29] Horaud, R., Conio, B., and Leboulleux, O. (1989). "An Analytic Solution for the Perspective 4-Point Problem," *Computer Vision, Graphics, and Image Processing, 47*(1), 33–43.

[30] Sharma, R., and Molineros, J. (1997). "Computer Vision-Based Augmented Reality for Guiding Manual Assembly," *Presence: Teleoperator and Virtual Environments, 6*(3), 292–317.

[31] Azarbayejani, A., and Pentland, A. (1995). "Recursive Estimation of Motion, Structure, and Focal Length," *IEEE Transactions on Pattern Analysis and Machine Intelligence, 17*(6).

[32] Broida, T. J., Chandrashekhar, S., and Chellappa, R. (1990). "Recursive Estimation from a Monocular Image Sequence," *IEEE Transactions on Aerospace and Electronic Systems, 26*(4), 639–655.

[33] Soatto, S., and Perona, P. (1994). "Recursive 3D Visual Motion Estimation Using Subspace Constraints," CIT-CDS 94-005, California Institute of Technology, Tech Report, February.

[34] Iu, S-L., and Rogovin, K. W. (1996). "Registering Perspective Contours with 3-D Objects without Correspondence, Using Orthogonal Polynomials," *Proceedings of VRAIS'96*, pp. 37–44.

[35] Cho, Y., Park, J., and Neumann, U. "Fast Color Fiducial Detection and Dynamic Workspace Extension in Video See-Through Self-Tracking Augmented Reality," appeared in *proceedings of Pacific Graphics 98*.

[36] Tsai, R. (1987). "A Versatile Camera Calibration Technique for High Accuracy Three Dimensional Machine Vision Metrology Using Off-the-Shelf TV Cameras and Lenses," *IEEE Journal of Robotics and Automation*, RA-3(4), August, 323–344.

[37] Weng, J., Cohen, P., and Herniou, M. (1992). "Camera Calibration with Distortion Models and Accuracy Evaluation," *IEEE Transactions on Pattern Analysis and Machine Intelligence, 14*(10), 965–980.

[38] Fan, H., and Yuan, B. "High Performance Camera Calibration Algorithm," SPIE Vol. 2067 *Videometrics II*, pp. 2–13.

[39] Haralick, R., Lee, C., Ottenberg, K., and Nolle, M. (1994). "Review and Analysis of Solutions of the Three Point Perspective Pose Estimation Problem," *International Journal of Computer Vision, 13*(3), 331–356.

[40] Tremeau, A., and Borel, N. (1997). "A Region Growing and Merging Algorithm to Color Segmentation," *Pattern Recognition, 30*(7), 1191–1203.

[41] Uenohara, M., and Kanade, T. (1995). "Vision-Based Object Registration for Real-Time Image Overlay," *Proceedings of Computer Vision, Virtual Reality, and Robotics in Medicine*, pp. 13–22.

[42] Grimson, W. E. L., and Huttenlocher, D. P. (1991). "On the Verification of Hypothesized Matches in Model-Based Recognition," *IEEE Transactions on Pattern Analysis and Machine Intelligence, 13*(12), 1201–1213.

[43] Grimson, W. E. L. (1990). *Object Recognition by Computer: The Role of Geometric Constraints*, MIT Press.

[44] Mundy, J., and Zisserman, A. (1992). *Geometric Invariance in Computer Vision,* MIT Press.

[45] Bowyer, K., and Dyer, C. R. (1991). "Aspect Graphs: An Introduction and Survey of Recent Results," *International Journal of Imaging Systems and Technology, 2,* 315–328.

[46] Neumann., U., and Park, J. (1998). "Extendible Object-Centric Tracking for Augmented Reality," *Proceedings of IEEE VRAIS '98,* pp. 148–155.

[47] Mendel, J. M. (1995). *Lessons in Estimation Theory for Signal Processing, Communications, and Control,* Prentice-Hall PTR.

11

NaviCam: A Palmtop Device Approach to Augmented Reality

Jun Rekimoto

Sony Computer Science Laboratory Inc.

1. INTRODUCTION

Current user interface techniques such as WIMP or the desk-top metaphor do not support real-world tasks, because the focus of these user interfaces is only on human–computer interactions, not on human–real world interactions. In this paper, we propose a concept of building computer augmented environments using a situation-aware portable device. With this technology, the user will be able to augment the real world with the computer's synthetic information. The user's situation is automatically recognized by the computer through various recognition methods, and the computer can assist the user without explicit commands from the user. We call this new interaction style *Augmented Interaction*, because this technology enhances the ability of the user to interact with the real-world environment.

Based on the proposed concept, we have developed a portable interaction device called *NaviCam*, which has the ability to recognize the user's situation by detecting tags in real-world environments. It displays situation sensitive information by superimposing messages on its video see-through screen. The combination of ID-awareness and portable video-see-through display solves several problems with current ubiquitous computers systems and augmented reality systems.

1.1   Limitations of Current User Interfaces

Computers are becoming increasingly portable and ubiquitous, as recent progress in hardware technology has produced computers that are small enough to carry easily or even to wear. However, these computers, often referred to as PDAs (Personal Digital Assistant) or palmtops, are not suitable for traditional user-interface techniques such as the desk-top metaphor or the WIMP (window, icon, mouse, and a pointing device) interface. The fundamental limitations of graphical user interfaces (GUIs) can be summarized as follows:

1.   **Explicit operations** GUIs can reduce the cognitive overload of computer operations, but they do not reduce the volume of operations themselves. This is an upcoming problem for portable computers. As users integrate their computers into their daily lives, they tend to pay less attention to them. Instead, they prefer interacting with each other, and with objects in the real world. The user's focus of interest is not the human–computer interactions, but the human–real world interactions. People will not wish to be bothered by tedious computer operations while they are doing a real-world task. Consequently, the reduction of the amount of computer manipulation will become an issue rather than simply how to make existing manipulations easier and more understandable.

2.   **Unaware of the real-world situations** Portability implies that computers will be used in a variety of situations in the real world. Thus, dynamical change of functionalities will be required for mobile computers. Traditional GUIs are not designed for such a dynamic environment. Although some context sensitive interaction is available on GUIs, such as *context sensitive help*, GUIs cannot deal with real-world contexts. GUIs assume an environment composed of desk-top computers and users at a desk, where the real-world situation is less important.

3.   **Gaps between the computer world and the real world** Objects within a database, which is a computer-generated world, can be easily related, but it is hard to make relations among real-world objects, or between a real object and a computer-based object. Consider a system that maintains a document database. Users of this system can store and retrieve documents.

However, once a document has been printed out, the system can no longer maintain such an output. It is up to the user to relate these outputs to objects still maintained in the computer. This is at the user's cost. We thus need computers that can understand real-world events, in addition to events within the computer.

Recently, a research field called *computer augmented environments* has emerged to address these problems [25]. In this chapter, we propose a method to build a computer augmented environment using a portable device that has an ability to recognize a user's situation in the real world. A user can see the world through this device with computer augmented information regarding that situation. We call this new interaction style *Augmented Interaction*, because this device enhances the ability of the user to interact with the real-world environment.

2. SITUATION AWARENESS AND AUGMENTED INTERACTION

Augmented Interaction is a new concept of human–computer interaction that aims to reduce computer manipulations by using environmental information (situations) as implicit input. With this style, the user will be able to interact with a real world augmented by the computer's synthetic information. The user's situation will be automatically recognized by using a range of recognition methods that will allow the computer to assist the user without having to be directly instructed to do so. The user's focus will thus not be on the computer, but on the real world. The computer's role is to assist and enhance interactions between humans and the real world. Many recognition methods and sensing technologies can be integrated with this concept. Time, location, and object recognition using computer vision are possible candidates. Also, we can make the real world more understandable to computers, by putting some marks or tags (barcodes, for example) on the environment.

Figure 11.1 shows the overall architecture of Augmented Interaction. Augmented interaction regards various kind of situational information as implicit inputs from the environment, as well as the explicit inputs from the user. By integrating these two information sources, and accessing a database and other outer information services (such as the WWW), the system generates context-sensitive information.

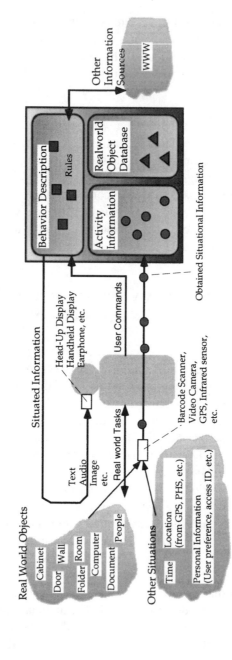

FIG. 11.1. The Augmented Interaction Concept: The user's physical environment is augmented by context-sensitive information.

Augmented interaction brings us many kinds of applications. For example, a user who is new to a town should be able to get location-dependent information from his/her mobile computer. The computer recognizes the user's current location from a GPS sensor and retrieves information from the Internet based on the location information. The user could ask the computer about the finest Italian restaurant, and the computer could show the map of the nearby area and indicate the positions of recommended restaurants. In a shopping mall, suppose that each shop installs its associated ID, and every shopping item has its own barcode. By reading these tags, the computer will be able to assist the user by informing "You can buy the same sweater in the other shop in this mall, at a much cheaper price." Using the augmented interaction technology, our everyday lives will be continuously assisted by highly context-sensitive and personalized information. This "situated assistance" is the crucial part of augmented interaction.

It is, of course, possible to combine augmented interaction with several currently used user interface techniques including WIMP-UIs, pen or touch panel interfaces, button or dial interfaces, and voice inputs. Augmented interaction is not a denial of current interface technologies but is rather an orthogonal concept.

Figure 11.2 shows a comparison of interaction styles involving human–computer interaction and human–real world interaction. In a desk-top computer (with a GUI as its interaction style), interaction between the user

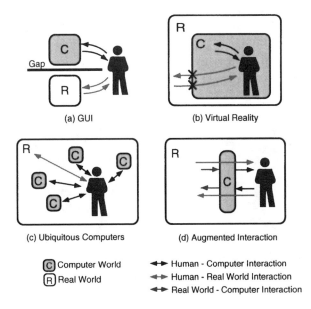

FIG. 11.2. A comparison of HCI styles.

and the computer is isolated from the interaction between the user and the real world. There is a gap between the two interactions. Some researchers are trying to bridge this gap by merging a real desk-top with a desk-top in the computer [14, 24]. In a virtual reality system (Figure 11.2(b)), the computer surrounds the user completely and interaction between the user and the real world vanishes. In the ubiquitous computers environment (Figure 11.2(c)), the user interacts with the real world but can also interact with computers embodied in the real world. Augmented Interaction (Figure 11.2(d)) supports the user's interaction with the real world, using computer augmented information. The main difference between (c) and (d) is the number of computers. The comparison of these two approaches will be discussed later in Section 6.

3. NaviCam: A CONTEXT-SENSITIVE INFORMATION ASSISTANT

As an initial attempt to realize the idea of Augmented Interaction, we have developed a prototype interaction device called *NaviCam* (NAVIgation CAMera). NaviCam is a portable computer with a small video camera to detect real-world situations. This system allows the user to view the real world together with context-sensitive information generated by the computer.

NaviCam has two hardware configurations. One is a palmtop computer with a small CCD camera, and the other is a head-up display with a head-mounted camera (Figure 11.3). Both configurations use the same software. The palmtop configuration extends the idea of position sensitive PDAs proposed by Fitzmaurice [11]. The head-up configuration is a kind of *video*

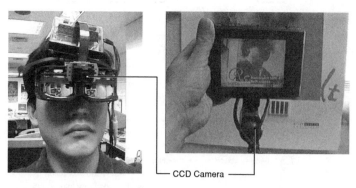

CCD Camera

FIG. 11.3. Palmtop configuration and head-up configuration.

see-through HMD [3], but it does not shield the user's real sight. Both configurations allow the user to interact directly with the real world and also to view the computer augmented view of the real world.

The system uses attached tags to recognize real-world situations. The tag is a 2D cell matrix (black and white) printed on the paper that encodes an ID of a real-world object. For example, the tag on the door of the office identifies the owner of the office. By detecting a specific ID, NaviCam can recognize where the user is located in the real world, and what kind of object the user is looking at.

In addition to the printed 2D markers, we also use infrared (IR) signal beacons for room-level location IDs (Figure 11.5). An IR beacon is a hardware module that periodically transmits its unique ID as an infrared signal. We installed these modules in each room and corridors in a building. When a mobile system detects an IR signal, it is possible to locate the current position by decoding it. This is the reverse idea of the active badge system [22], where a user wears an IR transmitter and receivers installed in the room detect the signal.

Figure 11.4 shows the information flow of this system. First, the system recognizes a 2D code through the camera. Image processing is performed using software at a rate of 10 frames per second. Next, NaviCam generates a message based on that real-world situation. Currently, this is done simply by retrieving the database record matching the ID. Finally, the system superimposes a message on the captured video image.

Using a CCD camera and an LCD display, the palmtop NaviCam presents the view at which the user is looking as if it is a transparent board. We coined the term *magnifying glass metaphor* to describe this configuration

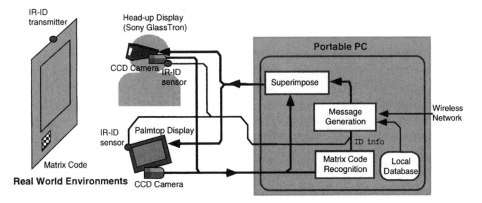

FIG. 11.4. The system architecture of NaviCam.

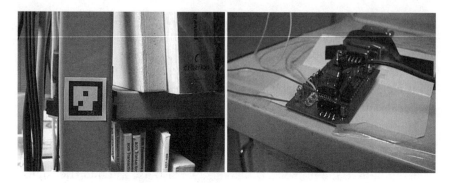

FIG. 11.5.   Two kinds of IDs: printed 2D matrix code and infrared beacon.

FIG. 11.6.   The magnifying glass metaphor.

(Figure 11.6). While a real magnifying glass optically enlarges the real world, our system enlarges it in terms of *information*. Just as with a real magnifying glass, it is easy to move NaviCam around in the environment, to move it toward an object, and to compare the real image and the information-enhanced image.

4.   APPLICATIONS

We are currently investigating the potential of augmented interaction using NaviCam. What follows are some experimental applications that we have identified.

FIG. 11.7. Augmented Museum: NaviCam explains a biography of Rembrandt.

4.1   Information Augmented Physical Spaces

Figure 11.7 shows a sample snapshot of a NaviCam display.[1] The system detects the ID of a picture and generates a description of it. Suppose that a user with a NaviCam is in a museum and looking at a picture. NaviCam identifies which picture the user is looking at and displays relevant information on the screen. This approach has advantages over putting an explanation card beside a picture. NaviCam can provide personalized information depending on the user's age, knowledge level, or preferred language. The explanation cards in today's museums are often too basic for experts, or too difficult for children or overseas visitors. NaviCam overcomes this problem by displaying information appropriate to the viewer.

Since 1996, NaviCam has been used several times as a navigation system for exhibitions and artistic installations. Figure 11.8 is a snapshot from the exhibition of architect Neil Denari, in Tokyo (September 1996) [6]. In this exhibition, works of the architect (writings, design sketches, and computer graphics by Neil Denari) are virtually installed in the physical gallery space, by attaching icons on the surfaces of the room. Visitors walk around the space with the NaviCam device, which serves as an portal to the information space from the physical space.

---

[1]Photographs in this paper with a "NaviCam" logo at the bottom left are snapshots from the NaviCam screen.

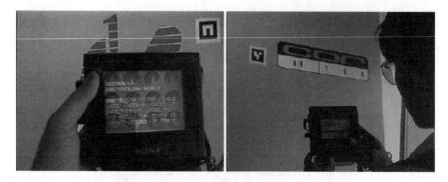

FIG. 11.8. Browsing information space from the physical environment.

FIG. 11.9. Merging real and cyber information: A paper calendar augmented by digital information.

4.2   Active Paper Calendar

Figure 11.9 shows another usage of NaviCam. By viewing a calendar through NaviCam, you can see your own personal schedule on it. This is another example of getting situation-specific and personalized information while walking around in real-world environments. NaviCam can also display information shared among multiple users. For example, you could put your electronic annotation or voice notes on a (real) bulletin board via NaviCam. This annotation can then be read by other NaviCam-equipped colleagues.

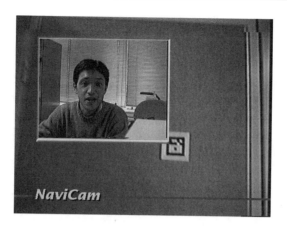

FIG. 11.10.   A pseudo-active office door greets a visitor.

4.3   Active Door

The third example is a NaviCam version of the active door (Figure 11.10). This office door can tell a visitor where the occupier of the office is currently, and when he/she will come back. The system also allows the office occupier to leave a video message to be displayed on arrival by a visitor (through the visitor's NaviCam screen). Therefore, it is in fact a passive door that can behave as an active door.

An important point here is that the door ID does not contain video message data. It only identifies the door. The video message is stored in the computer system, and the ID is used as a retrieval key. Thus, there is no need to embed any computer or storage in the door itself.

4.4   Attaching Information to Movable Objects

ID awareness has some advantages over the position-based approach used in traditional augmented reality systems. By using ID detection, the system can augment information related to an object that might move. For example, the system can display information about a videotape (Figure 11.11). Movable objects include people. Figure 11.12 is an example of people annotation; personal information is obtained from the worn ID badge. Using only position information, these capabilities are very difficult to achieve because it is almost impossible to track the location of all movable objects.

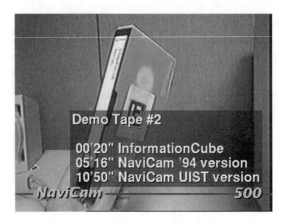

FIG. 11.11.   Annotation of a movable object.

FIG. 11.12.   Showing information about people.

4.5   Spatially Correct Augmenting Information

Using 2D matrix code detection, it is also possible to calculate the camera position as well as its ID number. This information can be obtained from positions of four corners of a 2D matrix pattern on the captured image. Figure 11.13 shows examples of spatially correct annotation using this technique. Unlike other vision-based AR systems [4, 21, 19], the number of distinguishable objects is almost limitless thanks to the ID detection capability. Thus, the annotation information can automatically be switched from object to object, without requiring any explicit commands from the user.

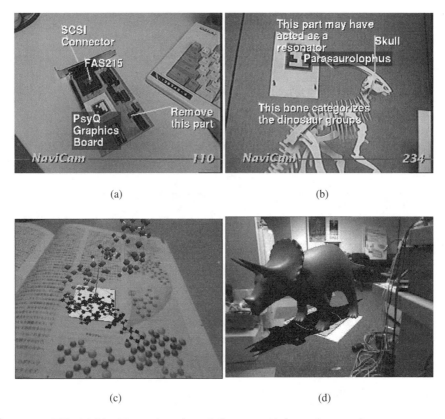

(a) (b)

(c) (d)

FIG. 11.13. Examples of spatially correct information overlay using a matrix code as a visual marker (a) An annotated circuit board. (b) A dinosaur skeleton with 3D annotations. (c) A 3D molecular model pops up from the book. (d) Triceratops in the Laboratory. (The matrix code is spread on the floor.)

The feature can also be applied to more entertainment-oriented domains. Figure 11.13(c) is a virtual pop-up book, and (d) is a "virtual dinosaur room," in which a 3D dinosaur model appears on a floor and the user (having a palmtop NaviCam or wearing a head-mounted NaviCam) can walk around that model for inspection.

4.6 Building Navigation

ID marking has proven to be a simple and useful method to realize several kinds of augmented reality systems. It can also be enhanced by combining other sensing technologies. Figure 11.14 is an experimental navigation

FIG. 11.14. A gyro-ehnahced NaviCam and a building navigation application.

system used in a building, based on a gyro-enhanced NaviCam system. It demonstrates a typical usage of the combination of ID tags and spatial awareness. In this application, a user first puts the NaviCam display in front of a nearby ID on the wall. The system detects the global location including orientation information in the building based on the recognized ID and its shape. Then the user can freely look around the corridor. Even when the ID marker becomes out of the sight of the camera, the system continues to track the relative motion of the device using the gyro sensor and displays proper directional information.[2]

4.7   NaviCam as a Collaboration Tool

In the above three examples, NaviCam users are individually assisted by a computer. NaviCam can also function as a collaboration tool. In this case, a NaviCam user (an operator) is supported by another user (an instructor) looking at the same screen image from probably a remote location. Unlike other video collaboration tools, the relationship between the two users is not symmetric, but asymmetric. Figure 11.15 shows an example of a collaborative task (video console operation). The instructor is demonstrating which button should be pressed by using a mouse cursor and a circle drawn on the screen. The instructor augments the operator's skill using NaviCam.

---

[2]The gyro (JAE MAX3) used with this system is a solid state inertia-based position tracker consisting of three acceleration sensors and three orthogonal angular rate sensors. It is a 6DOF tracker so that it can report x, y, z positions as well as orientations (yaw, pitch, and roll).

FIG. 11.15.　NaviCam can be used as a collaboration tool.

4.8　Situated Conversation with NaviCam

The augmented interaction concept might also be an important contribution to artificial intelligence (AI) research. Recognizing dialogue contexts remains one of the most difficult areas in natural language understanding. Real-world awareness allows a solution to this problem. For example, a system that has an ability to detect near object IDs can respond to a question such as "Where is the book entitled Multimedia Applications?" by answering "It is on the bookshelf *behind* you." This is because the system is aware of which bookshelf the user is looking at. It is almost impossible to generate such a response without using real-world information. The system also allows a user to use deictic expressions such as "Tell me about the author of *this* book," because the situation can resolve ambiguity. This feature is similar to multimodal interfaces such as Bolt's *Put-That-There* system [5]. The unique point in the augmented interaction approach is that it uses real-world situations, other than commands from the user, as a new modality in the human–computer interaction.

Following this idea, Katashi Nagao and the author have developed an extended version of NaviCam that allows a user to operate the system with voice commands [15]. This system consists of the original NaviCam and a speech dialogue subsystem. The speech subsystem has speech recognition and voice synthesis capabilities. The NaviCam subsystem sends the detected ID to the speech subsystem. The speech subsystem generates a response (either voice or text) based on these IDs and spoken commands from the user. The two subsystems communicate with each other through Unix sockets.

An experimental application developed using this prototype is the *augmented library*. In this scenario, the system acts as a personalized library catalogue. The user carries the NaviCam unit around the library and the system assists the user to find a book, or answers questions about the books in the library.

5. IMPLEMENTATION

5.1 ID Recognition

The 2D matrix code used with NaviCam is a 5 black-and-white cell matrix surrounded by a black frame (Figure 11.16). Its 25-bit cell information consists of 16-bit data area and 9-bit error check area. The matrix image can be printed on normal white paper by using black-and-white laser printers.

The system seeks out 2D matrix codes on incoming video images. The image processing is done by software. No special hardware is required apart from video capturing. The code detection algorithm takes the following four steps.

- **Binarization:** First, a captured video image is binarized by using the predefined threshold.
- **Connected Component Analysis:** Then, connected components analysis of binary-1 (black) pixels is performed. For each found connected region, a test based on the size and the aspect ratio of the region bounding the rectangle is applied to select code candidate areas.
- **Code Frame Fitting:** For each selected region, a rectangle is fitted on the contour of the region using the least-square method. Then, distortion-compensation matrix is calculated based on the four corners of the quad-tangle. This matrix cancels the effect of rotation

FIG. 11.16.   2D matrix code examples (left: original, middle: captured, right: restored).

and perspective transformation and maps distorted code frame to the normalized space.

- **Decoding and Error Check:** Using the obtained matrix, project the binary image to the code space (Figure 11.16(right)). Based on the numbers of black and white pixels that are projected within the area of each cell, the cell's bit is determined. Finally, the CRC-error check is applied on the decoded bit pattern, and the certificated cell value is regarded as a recognized code ID.

- **Camera Position and Pose Estimation:** The last step is an optional phase to calculate position and orientation of the camera in relation to the matrix pattern. This information is used to spatially align annotation information. From four positions of known coplanar (real-world) points on the image plane, it is possible to calculate a matrix representing translation and rotation of the camera in the real-world coordinate. We use four corners of the 2D-code frame as such reference points. Our method also tries to minimize the following constraint during estimation, to ensure an estimated coordinate system that is orthogonal:

$$(\vec{v}_0 \cdot \vec{v}_1)^2 + (\vec{v}_1 \cdot \vec{v}_2)^2 + (\vec{v}_2 \cdot \vec{v}_3)^2 + (\vec{v}_3 \cdot \vec{v}_0)^2 + (\vec{v}_4 \cdot \vec{v}_5)^2,$$

where $\vec{v}_0 \dots \vec{v}_3$ are orientation vectors of four edges, and $\vec{v}_4 \dots \vec{v}_5$ are two diagonals of the code frame.

Using the above algorithm, the system can recognize the code (3 cm × 3 cm in size) at a distance of 30–50 cm using a consumer-based small CCD camera (Sony CCD-MC1). IDs are placed in various environments (e.g., offices, libraries, video studios) and so the lighting condition also depends on the place and the time. Even under such conditions, the code detection algorithm was reasonably robust and stable.

5.2 Superimposing Information on a Video Image

The system superimposes a generated message on the existing video image. This image processing is also achieved using software. We could also use chromakey hardware, but the performance of the software-based superimposition is satisfactory for our purposes, even though it cannot achieve a video frame rate. The message appears near the detected code on the screen, to emphasize the relation between cause and effect.

We use a 4-inch LCD screen with pixel resolution of 640×480. The system can display any graphic elements and characters as the X-Window does. However, it was very hard, if not impossible, to read small fonts through this LCD screen. Currently, we use a 24 or 32 point font to increase readability. The system also displays a semitransparent rectangle as a background of a text item. It retains readability even when the background video image (real scene) is complicated.

5.3 Database Registration

For the applications explained in Section 4, the system first recognizes IDs in the real-world environment and then determines what kind of information should be displayed. Thus, the database supporting the NaviCam is essential to the generation of adequate information. The current implementation of the system adopts a very simplified approach to this. The system contains a group of command script files with IDs. On receipt of a valid ID, the system invokes a script having the same ID. The invoked script generates a string that appears on the screen. This mechanism works well enough, especially at the prototype stage. However, we obviously need to enhance this element, before realizing more complicated and practical applications.

6. RELATED WORK

In this section, we discuss our Augmented Interaction approach in relation to other efforts in this area.

6.1 Ubiquitous Computers

Augmented Interaction has similarities to Sakamura's *highly functionally distributed system* (HFDS) concept [18], his TRON house project, and *ubiquitous computers* proposed by Weiser [23]. These approaches all aim to create a computer augmented *real* environment rather than building a *virtual* environment in a computer. The main difference between ubiquitous computing and Augmented Interaction is in the approach. Augmented Interaction tries to achieve its goal by introducing a portable or wearable computer that uses real-world situations as implicit commands. Ubiquitous computing realizes the same goal by spreading a large number of computers around the environment.

These two approaches are complementary and can support each other. We believe that in future, human existence will be enhanced by a mixture of the two: ubiquitous computers embodied everywhere, and a portable computer acting as an intimate assistant.

One problem with using ubiquitous computers is reliability. In a ubiquitous computer world, each computer has a different functionality and requires different software. It is essential that they collaborate with each other. However, if our everyday life is filled with a massive number of computers, we must anticipate that some of them will not work correctly, because of hardware or software troubles, or simply because of their drained batteries. It can be very difficult to detect each problem among so many computers and then fix them. Another problem is cost. Although the price of computers is decreasing rapidly, it is still costly to embed a computer in every document in an office, for example.

In contrast to ubiquitous computers, NaviCam's situation aware approach is a low cost and potentially more reliable alternative to embedding a computer everywhere. Suppose that every page in a book had a unique ID (e.g., barcode). When the user opens a page, the ID of that page is detected by the computer, and the system can supply specific information relating to that page. If the user has some comments or ideas while reading that page, they can simply read them out. The system will record the voice information tagged with the page ID for later retrieval. This scenario is almost equivalent to having a computer in every page of a book but with very little cost. ID-awareness is better than ubiquitous computers from the viewpoint of reliability, because it does not require batteries, does not consume energy, and does not break down.

Another advantage of an ID-awareness approach is the possibility of incorporating existing ID systems. Today, barcode systems are in use everywhere. Many products have barcodes for point of sales (POS) use, while many libraries use a barcode system to manage their books. If NaviCam can detect such commonly used IDs, we should be able to take advantage of computer augmented environments long before embodied computers are commonplace.

6.2 Augmented Reality

Augmented reality (AR) is a variant of virtual reality that uses see-through head-mounted displays to overlay computer-generated images on the user's real sight [20, 10, 8, 3, 9, 7].

AR systems currently developed use only locational information to generate images. This is because the research focus of AR is currently on implementing correct registration of 3D images on a real scene [2][4]. However, by incorporating other external factors such as real-world IDs, the usefulness of AR should be much improved.

We have built NaviCam in both head-up and palmtop configurations. The head-up configuration is quite similar to other AR systems. We thus have experience of both head-up and palmtop type of augmented reality systems and have learned some of the advantages and disadvantages of both.

The major disadvantage of a palmtop configuration is that it always requires one hand to hold the device. Head-up NaviCam allows for hands-free operation. Palmtop NaviCam is thus not suitable for some applications requiring two-handed operation (e.g., surgery). On the other hand, putting on head-up gear is, of course, rather cumbersome and under some circumstances might be socially unacceptable. This situation will not change until head-up gear becomes as small and light as bifocal spectacles are today.

For the ID detection purpose, head-up NaviCam has some difficulties because it forces the user to place his or her head close to the object. Since hand mobility is much quicker and easier than head mobility, palmtop NaviCam appears more suitable for browsing through a real-world environment.

Another potential advantage of the palmtop configuration is that it still allows traditional interaction techniques through its screen. For example, you could annotate the real world with letters or graphics directly on the NaviCam screen with your finger or a pen. You could also operate NaviCam by touching a menu on the screen. This is quite plausible because most existing palmtop computers have a touch-sensitive, pen-aware LCD screen. On the other hand, a head-up configuration would require other interaction techniques with which users would be unfamiliar.

For example, Figure 11.17 shows a variation of NaviCam developed by Yuji Ayatsuka and the author, based on the palmtop-PC and Java. This system allows a user to manipulate the computer using pen inputs, as well as location and object ID information from the camera.

Returning to the magnifying glass analogy, we can identify uses for head-up magnifying glasses for some special purposes (e.g., watch repair). The head-up configuration therefore has advantages in some areas; however, even in these fields hand-held magnifying lenses are still dominant and most prefer them.

FIG. 11.17. A pen-enabled NaviCam variation.

6.3 Chameleon—A Spatially Aware Palmtop

Fitzmaurice's *Chameleon* [11] is a spatially aware palmtop computer. Using locational information, Chameleon allows a user to navigate through a virtual 3D space by changing the location and orientation of the palmtop in his hand. Locational information is also used to display context-sensitive information in the real world. For example, by moving Chameleon toward a specific area on a wall map, information regarding that area appears on the screen. Using locational information to detect the user's circumstances, although a very good idea, has some limitations. First, location is not always enough to identify situations. When real-world objects (e.g., books) move, the system can no longer accurately locate them. Secondly, detecting the palmtop's own position is a difficult problem. The Polhemus sensor used with Chameleon has a very limited sensing range (typically 1–2 meters) and is sensitive to interference from other magnetic devices. Relying on this technology limits the user's activity to very restricted areas.

7. ISSUES IN DESIGNING AUGMENTED INTERACTION

7.1 Situation Sensing Technologies

We are currently using a 2D matrix code and a CCD camera to read the code, to investigate the potential of augmented interaction. Obviously, situation sensing methods are not limited to barcode systems. We should be able to apply a wide range of techniques to enhance the usefulness of the system.

Several, so-called next generation barcode systems have already been developed. Among them, the most appealing technology for our purposes would seem to be the *Supertag* technology invented by CSIR in South Africa [13]. Supertag is a wireless electronic label system that uses a batteryless passive IC chip as an ID tag. The ID sensor is comprised of a radio frequency transmitter and a receiver. It scans hundreds of nearby tags simultaneously without contact. Such wireless ID technologies should greatly improve the usefulness of augmented interaction.

For location detection, we could employ the global positioning system (GPS), which is already in wide use as a key component of car navigation systems. The personal handy phone system (PHS) is another possibility. PHS is a cellular telephone system that came into operation in Japan in 1995. Since this system uses relatively small size cells (typically 100 m in diameter), it is possible to know where the user is located by sensing which cell the user is in.

A more long-range vision would be to incorporate various kinds of computer vision techniques into the system. For example, if a user tapped a finger on an object appearing on the display, the system would try to detect what the user is pointing to by applying pattern matching techniques.

7.2   Combining Information Sources

Obviously, combining several information sources (such as location, real-world IDs, time, and vision) should increase the reliability and accuracy of situation detection. We are currently exploring the several combinations of sensing technologies including wireless tags, gyro, and GPS [1, 17, 16] as well as printed 2D codes and infrared IDs. For example, when a user is outside the building, a GPS receiver could be used to locate the user's current position. When the user enters a building, infrared IDs could be used to detect room-level location. The user's orientation can also be obtained by the combination of ID and gyro sensors. Finally, when the user approaches a specific object such as a book, the attached tag could be used to detect it.

A sequence of sensed information can also be used to determine the system's behavior. Suppose that the user enters the library room (detected by the infrared ID near the entrance), picks up a book (detected by barcode scanning), and leaves the room (detected again by the infrared ID). The sequence of these events is usable to detect the user's activity (borrowing a book, in this case). In an augmented museum, the system would give more detailed explanation when the user visits the same picture again. The system can also explain a picture in relation to another picture according to the user's walk path.

7.3 Inferring the User's Intention from Situations

Recognized situations are still only a clue of the user's intentions. Even when the system knows where the user is and at which object the user is looking, it is not a trivial problem to infer what the user wants to know. This issue is closely related to the design of agent-based user interfaces. How do we design an agent that behaves as we would want? This is a very large open question and we do not have an immediate answer to this. It may be possible to employ various kinds of intelligent user interface technologies such as those discussed in Ref. [12].

One point we would like to argue is that we should not rely too much on *intelligence* in the system because it might make the system unpredictable and thus unreliable. Just like an automatic door can behave exactly as people expect, actions triggered by real-world situations are often smarter than intelligence in the computer. We argue that the combination of rich situational information, with rather simple behavior descriptions, should react more comfortably than its counterpart.

8. CONCLUSIONS

In this chapter, we proposed a new method to realize computer augmented environments. The proposed augmented interaction style focuses on human–real world interaction and not just human–computer interaction. It is designed for the highly portable and personal computers of the future and concentrates on reducing the complexity of computer operation by accepting real-world situations as implicit input. We also reported on our prototype system called NaviCam, which is an ID-aware palmtop interaction device, and we described several applications to show the effectiveness of the proposed interaction style.

Acknowledgments   We would like to thank Mario Tokoro and Toshi Doi for supporting this work. We would also like to thank Katashi Nagao, Satoshi Matsuoka, Yuuji Ayatsuka, Hiroaki Kitano, and members

of Sony Computer Science Laboratory for their encouragement and helpful discussions. Special thanks also go to Masaaki Oka and Terutoshi
Nakagawa for organizing NaviCam-based exhibitions.

REFERENCES

[1] Yuji Ayatsuka, Jun Rekimoto, and Satoshi Matsuoka. Ubiquitous Links: Hypermedia links embe-
ded in the real world. In *IPSJ SIGHI Notes*, number 67-4, pp. 23–30. Japan Information Processing
Society, July 1996 (in Japanese).

[2] Ronald Azuma and Gary Bishop. Improving static and dynamic registration in an optical see-
through HMD. In *Proceedings of SIGGRAPH '94*, pp. 197–204, July 1994.

[3] Michael Bajura, Henry Fuchs, and Ryutarou Ohbuchi. Merging virtual objects with the real world:
Seeing ultrasound imagery within the patient. *Computer Graphics*, Vol. 26, No. 2, pp. 203–210,
1992.

[4] Michael Bajura and Ulrich Neumann. Dynamic registration correction in augmented-reality systems. In *Virtual Reality Annual International Symposium (VRAIS) '95*, pp. 189–196, 1995.

[5] R. A. Bolt. Put-That-There: Voice and gesture at the graphics interface. *ACM SIGGRAPH Comput.
Graph.*, Vol. 14, No. 3, pp. 262–270, 1980.

[6] Neil Denari. *Interrupted Projections*. Toto Publishing, 1996 (in Japanese).

[7] Steven Feiner, Blair MacIntyre, Marcus Haupt, and Eliot Solomon. Windows on the world:
2D windows for 3D augmented reality. In *Proceedings of UIST'93, ACM Symposium on User
Interface Software and Technology*, pp. 145–155, November 1993.

[8] Steven Feiner, Blair MacIntyre, and Doree Seligmann. Annotating the real world with knowledge-
based graphics on a see-through head-mounted display. In *Proceedings of Graphics Interface '92*,
pp. 78–85, May 1992.

[9] Steven Feiner, Blair MacIntyre, and Doree Seligmann. Knowledge-based augmented reality.
Communications of the ACM, Vol. 36, No. 7, pp. 52–62, August 1993.

[10] Steven Feiner and A. Shamash. Hybrid user interfaces: Breeding virtually bigger interfaces for
physically smaller computers. In *Proceedings of UIST'91, ACM Symposium on User Interface
Software and Technology*, pp. 9–17, November 1991.

[11] George W. Fitzmaurice. Situated information spaces and spatially aware palmtop computers.
Communications of the ACM, Vol. 36, No. 7, pp. 38–49, July 1993.

[12] Wayne D. Gray, William E. Hefley, and Dianne Murray, editors. *Proceedings of the 1993 International Workshop on Intelligent User Interfaces*. ACM press, 1993.

[13] Peter Hawkes. Supertag—reading multiple devices in a field using a packet data communications
protocol. In *CardTech/SecurTech '95*, April 1995.

[14] Hiroshi Ishii. TeamWorkStation: Towards a seamless shared workspace. In *Proceedings of CSCW
'90*, pp. 13–26, 1990.

[15] Katashi Nagao and Jun Rekimoto. Ubiquitous Talker: Spoken language interaction with real world
objects. In *Proc. of IJCAI-95*, pp. 1284–1290, 1995.

[16] Katashi Nagao and Jun Rekimoto. Agent augmented reality: A software agent meets the real world.
In *Proceedings of the Second International Conference on Multi-Agent Systems*, pp. 228–235,
1996.

[17] Jun Rekimoto. The magnifying glass approach to augmented reality systems. In *International
Conference on Artificial Reality and Tele-Existence '95 / Conference on Virtual Reality Software
and Technology '95 (ICAT/VRST'95) Proceedings*, pp. 123–132, November 1995.

[18] Ken Sakamura. The objectives of the TRON project. In *TRON Project 1987: Open-Architecture
Computer Systems*, pp. 3–16, Tokyo, Japan, 1987.

[19] Andrei State, Gentaro Hirota, David T. Chen, William F. Garrett, and Mark A. Livingston. Superior augmented reality registration by integrating landmark tracking and magnetic tracking. In *SIGGRAPH'96 Proceedings*, 1996.

[20] Ivan Sutherland. A head-mounted three dimensional display, *Proc. of FJCC 1968*, pp. 757–764, 1968.

[21] M. Uenohara and T. Kanade. Real-time vision based object registration for image overlay. *Journal of the Computers in Biology and Medicine*, pp. 249–260, 1995.

[22] R. Want, A. Hopper, V. Falcao, and J. Gibbons. The active badge location system. *ACM Trans. Inf. Syst.*, January 1992.

[23] Mark Weiser. The computer for the twenty-first century. *Scientific American*, pp. 94–104, September 1991.

[24] Pierre Wellner. Interacting with paper on the DigitalDesk. *Communications of the ACM*, Vol. 36, No. 7, pp. 87–96, August 1993.

[25] Pierre Wellner, Wendy Mackay, and Rich Gold, editors. *Computer Augmented Environments: Back to the Real World*, volume 36. *Communications of the ACM*, August 1993.

12

Augmented Reality for Exterior Construction Applications

Gudrun Klinker
Technische Universität München

Didier Stricker
Dirk Reiners
Fraunhofer Projektgruppe für Augmented Reality am ZGDV, Rundeturmstr. 6, D-64283 Darmstadt, Germany

ABSTRACT

Augmented reality (AR) constitutes a very promising new user interface concept for many applications. In this chapter, we pay particular attention to developing AR technology for exterior construction applications, augmenting video sequences from construction sites with information stored in models. Such augmentations can tremendously benefit several business processes common to many construction projects.

We are focussing on two approaches to augment the reality of construction sites. The first one augments video sequences of large outdoor sceneries with detailed models of prestigious new architectures, such as TV towers and bridges. Since such video sequences are very complex, we currently prerecord the sequences and employ off-line, interactive techniques. The second approach operates on live video streams. To achieve robust real-time performance, we need to use simplified, "engineered" scenes. In particular, we place highly visible markers at precisely measured locations to aid the tracking process.

1. INTRODUCTION

Augmented reality (AR) constitutes a very promising new user interface concept for many applications. Currently, we pay particular attention to developing AR technology for exterior construction applications. In the context of the European CICC project [10], we develop and evaluate the potential of AR in a series of pilot projects, augmenting video sequences from construction sites with information stored in models. Such augmentations can tremendously benefit several processes common to many construction projects.

- **Design and Marketing**: Creating a design and evaluating it for function and aesthetics, and showing a customer what a new structure will look like in its final setting. AR provides the unique opportunity to integrate the design into the real-world context.
- **Construction**: Visualization whether an actual structure is built in accordance with the design; quick update of work plans after a design change; visualization of consequences of potential design changes before they are agreed upon.
- **Maintenance and Renovation**: Visualization of hidden information (wires, pipes, beams in a wall); visualization of nongraphical information (heat and pressure of pipes, maintainance schedules and records); visualization of potential redesigns (interior, exterior) to evaluate their compatibility with existing structures, and placement of new structures onto/into preexisting buildings.

Some of these benefits can also be partially achieved with other graphical approaches, such as Virtual Reality presentations. The level of realism, however, that can potentially be achieved with AR systems far surpasses VR, which seems to asymptotically narrow the gap between synthetic models and the real world (see Figure 12.1). AR, on the other hand, starts with the real world, augmenting it as little or much as is deemed suitable for the task at hand [35].[1]

This gain in realism is coupled with a potential gain in speed since the real environment doesn't have to be rendered but merely mixed with a (much smaller) virtual model. The price to pay is the effort to strive for perfect alignment between the real and the virtual world. Assuming that this can be achieved satisfactorily in real time in the foreseeable future, the overall speed gain by not having to synthesize the real world will be considerable.

---

[1]Sources of all graphical material are listed at the end of this chapter.

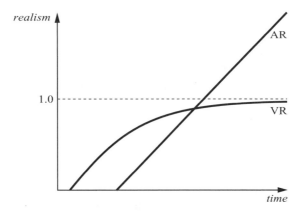

FIG. 12.1. Potential realism of AR vs. VR approaches.

1.1 AR Challenges and Chances in Exterior Construction Applications

Exterior construction applications impose very demanding challenges on the robustness and usability of evolving AR technologies.

- First of all, the mere size of a large construction project (e.g., a bridge, a tower, a shopping mall, or an airport) is overwhelming. The amount of synthetic data is huge and needs special processing technology. To present such a wealth of information in real time, the data need to be reduced and simplified. Concepts such as level of detail and relevance need to be developed with respect to the task at hand. Furthermore, much information is currently only represented in two dimensions. Tools to translate it into a three-dimensional context are necessary. To access all data when and wherever necessary, the system depends on a very good computer infrastructure, including fast and mobile networks, computers, and data repositories.
- Second, the size of not just the synthetic data model but also of the real site impose problems. Users maneuver in a very large space. Some AR devices, such as magnetic trackers or overhead surveillance cameras are rather geared toward indoor applications and unlikely to operate well under such conditions. On the other hand, GPS, optical tracking techniques, and inertial sensors have the potential to fare well. But they must be able to cope with situations when only partial information or only a local view of the entire construction site is available. Exterior construction scenarios thus require more tracking skills than what is currently shown in table-top demonstrations [55, 60, 65, 68].

- Third, AR applications require a very accurate model of the current site (a *reality model*) both to determine the current camera position and to augment the current view realistically with synthetic information (the *virtual model*). Realistic immersion of virtual objects into a real scene requires that the virtual objects behave in physically plausible manners. They occlude or are occluded by real objects, they are not able to move through other objects, and they cast shadows on other objects. To this end, AR systems need geometrically precise descriptions of the real environment. Yet, construction environments are not well structured. Natural objects such as rivers, hills, trees, and also heaps of earth or construction supplies are scattered around the site. Typically, no exact detailed 3D information of such objects exists, making it difficult to generate a precise model of the site. Even worse, construction sites are in a permanent state of change. Buildings and landscapes are demolished; new ones are constructed. People and construction equipment move about, and the overall conditions depend on the weather and seasons. AR applications thus need to identify suitable approaches for generating and dynamically maintaining appropriate models of the real environment. It is also important to decide upon the appropriate level of realism with which virtual objects are rendered into the real world. For safety reasons, construction workers need to have and maintain a very clear understanding of the real objects and safety hazards around them. Virtual objects must not decrease people's awareness of danger (e.g., by perfectly adding virtual floors and walls to the bare wireframe of beams of the next floor being built in a high-rise).

  In other situations, however, the highest level of realism is highly desirable (e.g., when visualizing whether a designed object will integrate well into an existing landscape).
- In addition to augmenting reality, exterior construction scenarios also need tools to diminish reality, since in most cases, objects and landscapes are removed or changed before new ones are built. Thus, techniques for synthetically removing real objects from the incoming video input stream need to be developed.

Despite such challenges, exterior construction is a very suitable application area for AR. Construction, in its very nature, is very much a three-dimensional activity. Business practices and work habits are all oriented toward the design, comprehension, visualization and realization of 3D plans. Workers are used to graphical descriptions such as 2D plots.

Much information is already represented and communicated in graphical form. Thus, new graphical user interfaces like AR fit very naturally into current work practices.

Furthermore, gathering high-precision geodesic measurements of selected points on a construction site and marking them in suitable ways is a well-established practice. Large construction sites use a wealth of high precision equipment, such as theodolites, differential GPS, and laser pointers, that AR can build upon. *Engineering the environment* to suit the current capabilities of the technology is acceptable within limits. Thus, AR can begin by building applications that simplify many of the general challenges, adapting the construction site to suit their skills. Over time more sophisticated and general approaches can be developed.

1.2 Our Approach

Figure 12.2 illustrates our current framework for augmenting images of the real world with virtual objects. The AR viewer takes four kinds of input (shown in the darker, rounded rectangles): virtual object models to be visualized or rendered, a photo or an image sequence to which the virtual objects are added, camera positions to facilitate seamless integration, and

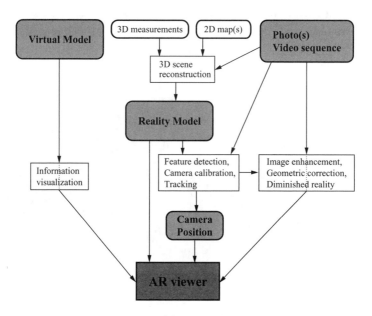

FIG. 12.2. Conceptual framework of an AR system.

a reality model to enable physically correct coexistance of virtual and real objects.

The white, sharp rectangles show our information processing tools: We use a wealth of both commonly available and special purpose visualization and graphical rendering schemes to present the synthetic (virtual) information (Section 4). We have also developed various interactive 3D scene reconstruction techniques to generate reality models from available information, such as photos, maps, or 3D measurements (Section 2). Third, we use various interactive or automatic techniques to calibrate and track cameras for live or prerecorded image sequences, using features that are specified in the reality model (Section 3). Fourth, we are aware of the principle need to synthetically alter the image prior to its display (e.g., to correct for lens distortions or to remove objects from the scene). Yet, we haven't approached the subject in much depth yet (Section 5). All original or processed information flows into the AR viewer where the final three-dimensional integration of real and virtual objects is generated (Section 4), ready for the user to interact with in various ways (Section 6).

Within this framework, we are focusing on two approaches to augment the reality of construction sites.

- The first approach augments video sequences of large outdoor sceneries with detailed models of prestigious new architectures, such as TV towers and bridges that will be built to ring in the new milleneum (see Figure 12.3a). Since such video sequences are very complex,

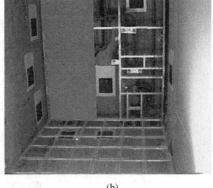

(a)   (b)

FIG. 12.3. Interactive vs. automatic video augmentation. (a) Virtual bridge across a real river. (b) Virtual wall and grid in a real room.

we currently prerecord the sequences and employ off-line, interactive calibration techniques to determine camera positions. Given all calibrations, the augmentation of the images with the virtual object is performed live (i.e., the virtual model can be altered and transformed while it is being seen in the video sequence).

- The second approach operates on live video streams, calibrating and augmenting images as they come in. To achieve robust real-time performance, we need to use simplified, "engineered" scenes. In particular, we place highly visible markers at precisely measured locations to aid the tracking process (Figure 12.3b).

As indicated by these figures and example applications, we focus on only a subset of the challenges posed by exterior construction scenarios. We currently use rather pragmatic simplified or semi-interactive approaches, expecting that future developments will provide more automatic and general solutions.

Furthermore, we focus on the graphical aspects of AR. A full AR system also requires sound and other multi-media interfaces, as well as a complex computer and network infrastructure, such as distributed, mobile, wireless, and wearable computing, to make relevant information and processing power available where the user happens to go [43, 59]. Such aspects of AR are discussed in other chapters of this book.

1.3 Related Work and Current State of the Art

The first chapters of this book and recent surveys have provided excellent general overviews of AR and its young history [3, 6, 44]. We will focus here on topics closely related to our approach.

For the interactive augmentation of landscapes (Figure 12.3a), our approach is most closely related to photomontaging jobs performed by professional photo labs. Selected still photos are augmented to illustrate to the public how new construction projects, such as placing the Munich railroad station underground, will improve the city. The process is currently very tedious and time consuming; the results are individual, unalterable augmentations of individual photos. Using our approach, the understanding (calibration) of the image is decoupled from its augmentation. Thus, augmentations can be altered at interactive speed; innumerous different versions of the virtual objects can be integrated into the photo. Furthermore, the approach works for entire video loops, not just for still photos.

For the automatic augmentation of live video streams (Figure 12.3b), several research groups have begun exploiting the use of special targets in the scene for optical tracking [4, 31, 33, 41, 42, 48, 59, 60, 66]. Experiments indicate that optical approaches provide higher precision than nonoptical ones. Ideally, they are combined in hybrid approaches with other, nonoptical techniques [60]. Yet, since each group uses different targets and equipment, it is unclear which approach works best. So far, no standardized test scene has been shared between several groups.

Little research has focused so far on architectural applications. Feiner's group is exploring approaches to improve the construction, inspection, and renovation of architectural structures with AR [68], focusing on space frame constructions. Bajura and Neumann augment a toy house with a virtual antenna and an annotation arrow [4]. Debevec et al., as well as Faugeras et al., have proposed technologies to semi automatically generate architectural models from images [12, 14].

2. REALITY MODELS

In order to perfectly mix virtual and real objects, AR systems need to calibrate cameras and other sensing and display equipment so that the virtual objects are rendered from the same vantage point as the real objects. Realistic immersion of virtual objects into a real scene further requires that the virtual objects behave in physically plausible manners (i.e., they occlude or are occluded by real objects, they are not able to move through other objects, and they are shadowed or indirectly illuminated by other objects while also casting shadows and mirror images themselves).

For optical camera calibration and to enforce physical interaction constraints between real and virtual objects, augmented reality systems need to have a precise description of the physical scene: a *reality model*.

2.1 Required Complexity of Reality Models

AR reality models don't need to be as complex as, for example, VR models. VR models are expected to synthetically provide a realistic immersive impression of reality. Thus, the description of photometric reflection properties and material textures is crucial. AR, on the other hand, can rely on live optical input to provide a very high sense of realism. The reality model only needs to indicate geometric properties, such as easily identifiable landmarks in the scene for camera calibration and surface shapes for occlusion handling and shadowing between real and virtual objects.

However, AR reality models have to be much more precise than VR models. Since an immersive VR system cuts users off from reality, users can only gain a qualitative impression whether or not the objects are modeled correctly. In AR, on the other hand, users have an immediate quantitative appreciation of the extent of mismatches between the reality model and the live video input from the real scene.

A reality model has to track and adapt to changes in the real world. The need and frequency of model updates depends on the application. Many durable large-scale structures will remain in place during the entire construction work. Thus, such components need to be modeled only once. Using existing CAD models or semi-automatic modeling techniques to generate such models may be sufficient. Other aspects of construction sites are more variable: trucks and cranes move, trees lose their leaves in winter time, and buildings under construction slowly develop. Such gradual changes need to be mirrored in the reality model at an appropriate pace. In some applications, daily or real-time scence changes may also have to be modeled (e.g., when virtual objects have to be integrated into scenes with moving people, material, or equipment).

Reality modeling and *reality tracking* are very complex and demanding tasks. Currently they cannot be achieved in real time in a general way. The subsequent sections present and discuss several approaches to generate reality models.

2.2 Use of Existing Models

The most straightforward approach to acquiring 3D scene descriptions is to use existing geometric models, such as CAD data, output from GIS systems, and maps (see Figure 12.4). When such models are available, they constitute the easiest approach toward integrating virtual objects into the real world. Yet, this approach is not always pursuable for a number of reasons.

- In many applications, reality models are not commercially available. For example, interior restoration of old buildings typically needs to operate without preexisting CAD data.
- The data points in a commercial model don't necessarily coincide well with visible features in images; quite the opposite is true: Geodesic measurements generally are indicated by small, barely visible marks in the ground.

(a) Laboratory setup   (b) At a real construction site

FIG. 12.4.   Example and use of a manually created reality model.

- Available models are not complete. Real physical objects typically show more detail than is represented in the models. Furthermore, scene models cannot fully anticipate the occurrence of nonstationary objects, such as coffee mugs on tables and cars or cranes on construction sites.
- The system needs to account for the changing appearances of existing objects, such as buildings under construction or engines that are partially disassembled.

When users see objects in the scene, they expect the virtual objects to interact with them correctly, independently of whether they are new to the scene or whether they have been there for a long time (i.e., have already been included in a reality model). Thus, it currently is often necessary to create and update reality models explicitly for the AR application.

2.3   Manual Approach

The manual approach involves obtaining 3D measurements within the real world, using measuring tapes, theodolites, GPS-operated laser pointers, information from GIS systems, etc. Such 3D points are entered into a small model, which in turn can be used to calibrate and track a camera by tracking the corresponding image points. Figure 12.4 shows such a generated reality model of our "tracking laboratory," a room with several carefully measured targets on its walls.

The approach is intuitive and works well for very sparse reality models. Yet, it is prohibitively expensive to measure thousands of points this way. Furthermore, the approach depends upon availability of professionals and special equipment. Thus, models cannot be expected to be obtainable on short notice.

2.4 Interactive Approach

As an alternative, reality models can be generated with interactive graphical tools.

We have developed a system, InCal, that begins with a very sparse, initial reality model of a landscape or cityscape, using externally provided information such as the known position and height of a few buildings, electric power poles, and bridge pillars. More information, such as the course of rivers and streets, is measured from two-dimensional maps and inserted at zero height into the model. From this model, we generate an initial camera calibration for a few site photos, interactively indicating how features in the image relate to the model.

Once an image has been successfully calibrated (3.2), the model is overlaid on the image, showing good alignment of the image features with the model features. Models of new structures in the landscape, such as houses or hills, can then be entered into the reality model, using their two-dimensional position in the city map and estimating their height from their alignment with the image. Figure 12.5 shows the final model, compared with a commercially available model of the same area.

2.5 Toward Automatically Generated Models

Computer vision techniques are designed to automatically acquire three-dimensional scene descriptions from image data. Much research is currently under way, exploring various schemes to optically reconstruct a scene from multiple images, such as structure from motion [1, 14, 63, 67], (extended) stereo vision [12, 26, 28, 29, 45], and photogrammetric techniques [21, 23].

(a) Final reality model   (b) Merged with commercial model

FIG. 12.5. Final interactive model superimposed on the commercial model of the city of London.

In the context of the European Realise project [22] and its successor, Cumuli [11], we explore to what extent automatically generated scene models can support AR and VR applications. In collaboration with INRIA and Lund University, we are developing and testing tools to semiautomatically generate descriptions of complex landscapes and cityscapes, such as parts of London along the Thames. Using epipolar relationships between features seen in several images from different unknown vantage points, geometric constraints on architectural structures, as well as city maps, the tools will help determine a set of progressively more precise projective, affine, and finally Euclidean properties of points in the three-dimensional scene.

Figure 12.6 shows the final result of the Realise project, a reconstructed model of the Arcades of Valbonne. Figure 12.6a shows the reconstructed geometric model. In Figure 12.6b, the model has been enhanced by mapping textures from the original image data onto the surfaces. Figure12.6c illustrates how the original image data can be augmented with synthetic objects, such as a Ferrari, once the images have been analyzed and calibrated with the Realise system.

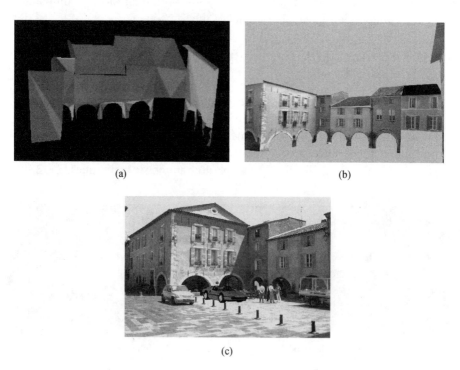

(a)

(b)

(c)

FIG. 12.6. Automatically generated model of the arcades of Valbonne. (a) Geometric model. (b) Enhanced with texture maps. (c) 2D picture of the plaza, augmented with a Ferrari.

2.6   Range Data Models

As an interesting alternative to motion-based scene recognition, approaches using three-dimensional range sensors have been undertaken in efforts such as the RESOLV project [53]. RESOLV is developing a mobile robot (Figure 12.7a) to conduct a 3D survey of a building, including capturing the appearance of the visible surfaces. A portable unit is taken around the environment that is to be captured. The unit includes a spatial camera constituted of a scanning laser range finder for capturing the 3D structure of the surroundings and a video camera for adding the textures.

The environment is scanned from a number of capture positions and reconstructed into a model, unifying measurements from all viewing positions. Surfaces are recognized by processing the range data and are textured from the camera images. By combining what is seen from neighboring capture positions, surfaces that would be occluded from one position are recorded. The spatial camera travels from one capture position to another either on a trolley (Figure 12.7) or an autonomous vehicle. The environment is reconstructed as the robot progresses and each new position is registered with previous ones using key points in the surroundings. The partial reconstruction is used to determine future capture positions.

Figure 12.7b shows a model taken from the interior of the Royal Institute of Chartered Surveyors, London. Of particular note is the fire extinguisher

(a)   (b)

FIG. 12.7.   (a) RESOLV trolley. (b) Automatically generated model of part of the interior of the Royal Institute of Chartered Surveyors, London.

in one corner, and the curved surface of the pillar, showing the usefulness and accuracy of the laser/video combination.

Currently, the robot is optimized for human-scale applications such as indoor refurbishment or maintainance tasks, and not yet designed for capturing external landscapes. Both the trolley shown in the picture and the autonomous vehicle are designed to support capture at two heights—eye level when sitting and standing. The size of the unit is comparable to that of a person to ensure that it can be taken to all the places where people are likely to pause when looking around a building. The data are held in a form suitable for CAD systems and for viewing on a WWW browser—which is also a suitable format for reality models of AR applications.

3. CAMERA CALIBRATION

One of the key issues of AR is the proper alignment of the virtual world with the real world. Such alignment requires that the view onto the real world be accurately determined and simulated for the virtual scene. Since we concentrate in our approach on viewing the real world through a TV camera, proper alignment means calibrating a camera, determining 5 internal parameters (focal length (f_x, f_y), center (c_x, c_y), aspect ratio a) and 6 external parameters (position (x, y, z) and orientation (r_x, r_y, r_z)). We do not yet account for lens distortions.

We are experimenting with two different approaches, depending on various application scenarios. Both approaches proceed in two steps: (a) the determination of suitable feature points in an image and the establishment of proper correspondences between such 2D image features and 3D scene features in the reality model, and (b) the computation of the current camera parameters according to the matches. These two steps, as well as sensor fusion concepts and precision and stability issues, are now discussed.

3.1 Correspondence between the Reality Model and 2D Image Points

3.1.1 Live Automatic Mapping of Specific Targets

To achieve live camera tracking performance, feature detection has to operate fast and automatically. The automatic detection of three-dimensional objects in images is a long-standing research area in computer vision. So far it cannot be achieved in a general way in real time. To provide fast and robust optical tracking performance, it is common practice in AR applications to simplify the computer vision problem by placing special, easily detectable target patterns into the real world [4, 31, 41, 42, 59, 60, 66]. For example, Neumann et al. use circular tracking targets [48]. State et al. use concentric multicolored disks [60]. Bajura and Neumann [4] track bright red LEDs that are significantly brighter than the other objects in the environment and thus can be easily detected. Starner et al. search the environment for visual tags consisting of two red squares bounding a binary pattern of green squares to identify objects of special interest [59]. It is also very common in the car industry to attach special black-and-yellow circular patterns with an internal cross to cars to evaluate car crash tests.

In our approach, we use black squares attached to a planar object with sufficient contrast. In order to uniquely identify each square independently of the current field of view, the black squares contain a labeling region, consisting of 2 rows with 4 positions (bits) for smaller red squares (see Figure 12.8). Using a binary encoding scheme, we can define up to 256 different targets, each of which can be matched against a 3D target in a reality model. In a particular image, any set of two or more targets from the model suffices for our tracking system to work.

At startup time, the optical tracker does not yet have any indication of the viewing direction. Thus, the entire image has to be searched quickly for targets. In a subsampled image, we begin by searching for candidate

(a) Original image

(b) With edge finding results

FIG. 12.8. Processed image of a black target square 13 (binary label 1101).

"blobs," scanning sample lines for strong bright-to-dark and subsequent dark-to-bright transitions. We then follow the contours of each blob. We classify the edge pixels according to their gradient direction as belonging to one of four edge classes, and we fit straight lines to the edges of each class. The intersections of neighboring lines determine the corner points of candidate squares. The algorithm then examines the labeling area within each square, correlating the pixels along sampling lines of the first and second 4-bit row with the ideal binary signal of numbers 0 through 15. The number producing the highest correlation with the image is selected, and the candidate square label is compared against the list of 3D squares in the reality model. If the label exists and is assigned to exactly one image square, a match between the 2D and 3D square is established. To account for camera rolls by more than 90 degrees, we apply the same labeling test along all 4 edges of the square, selecting the labeling with the best match.

Subsequent images do not have to be searched from scratch to find the squares. Rather, tracking algorithms can predict the approximate location of squares from their locations in previous images (see Section 3.4). To determine the exact position of each square, we find strong image gradients in the vicinity of their predicted edge positions. We fit lines to edge pixels and intersect them to determine the corner points of the squares in the new image.

This technique is fast (23 frames per second on an SGI O2, 14 frames on an Indy with an R5000 processor) and robust over quite a range of moderately fast camera motions. Typically, we move the camera on a tripod with wheels. Yet, we can also track an IndyCam in our hand. When the camera moves very fast or is jerked, the predicted square positions are incorrect and/or the image exhibits motion blurring. Under such circumstances, the square redetection algorithm fails, and the square detection system is reinitialized in a third of a second.

3.1.2 Interactive Mapping of Arbitrary Targets

Although automatic target tracking can be demonstrated to operate well in small-scale, engineered environments, it imposes significant restrictions on applications. In many cases, special targets cannot be placed in the scene, or the size of the scene is so large that targets at some distance are barely visible in the images. Thus, more general feature detection approaches need to be investigated. Yet, it will take time for them to mature and to become fast enough for real-time AR applications.

FIG. 12.9. Mapping of selected 3D model features to 2D image features.

On the other hand, quite a few scenarios for exterior construction applications such as project acquisition and design efforts can already benefit from much slower, off-line, photo and video film augmentation. Under such circumstances, we can employ human help and analyze much more general scenes without special targets.

Our interactive calibration system, InCal, provides a user interface to calibrate and track camera positions in images. It superimposes a 3D reality model on an image that the user can move and rotate interactively. Furthermore, the user can interactively indicate correspondences between three-dimensional model features and pixels in the image (see Figure 12.9). Such correspondences are then used to automatically compute the current camera position. When the virtual camera is set to the same position, the reality model "snaps into alignment" with the image (see Figure 12.10).

When calibrating image sequences, InCal exploits interframe coherence to automatically propose feature locations in new images from their locations in previous images. For user-scalable rectangular template areas around each feature in the current image, InCal uses normalized cross-correlation to determine with sub pixel precision the best match between a template area in the current image and pixels within a search window of the

(a)

(b)

FIG. 12.10. Different results of full Tsai calibration, with one of the mapped features having been moved by one pixel. (a) Tsai: Feature at (537, 270). (b) Tsai: Feature at (537, 269).

next image. Such automatic feature tracking substantially helps the user in working through a long image sequence. Many features are well detected. Occasional mismatches can be corrected interactively.

3.2 Calibration

Once correspondences between 3D scene features and 2D image features have been established, they can be used to determine the camera position. Many camera calibration algorithms have been developed in various computer vision research efforts (see Refs. [64, 72] for reviews). The principle problem is well understood: Given a number of matches between 2D and 3D points, compute the camera viewing parameters that minimize the distance between the image points and the projected position of their matched 3D world features. Since the system of equations is not linear, much effort has been spent investigating various approaches for finding stable solutions.

We have worked with two approaches in particular, one developed by Wenig [65, 71] and one by Tsai [64, 72]. Weng's algorithm computes all 11 intrinsic and extrinsic camera parameters. It works only for noncoplanar arrangements of features in the real world. Tsai's algorithm consists of a collection of calibration routines, geared toward computing different subsets of the camera parameters. Thus, different versions of the algorithm compute either all parameters or only the extrinsic ones, assuming that the focal length, center, and aspect ratio are known. The simplified version works also with coplanar 3D feature arrangements. The overall approach begins by adjusting the camera rotation parameters to reduce the misalignment between the world and the image as much as possible. Next, camera translation is explored as a means to further reduce the alignment error. Finally, all six parameters are jointly reconsidered and optimized, using a nonlinear least squares routine to minimize the error.

From Tsai's collection of calibration tools, we have distilled a number of further simplified approaches, assuming that in real applications even some of the extrinsic camera parameters can often be approximated by other means. In our approach, either of the first two steps of Tsai's algorithm (estimation of rotation and translation) can be skipped assuming externally provided data to initialize the nonlinear least squares optimization. The result is a much more stable system that can be adapted quickly to incorporate various external sources of information.

3.3   Precision and Repeatability
of Calibration Results

Calibration precision is a key issue in AR since it determines the quality and credibility of augmented video [27].

We have been able to successfully calibrate many live and prerecorded video sequences. Even for the case of complex landscapes, such as from the river Wear in Sunderland, UK, we have been able to interactively calibrate sequences of hundreds of images nearly automatically within a few hours (e.g., the sequence containing Figure 12.13).

Yet, calibration algorithms are inherently sensitive to noise and to specific properties of the reality model, such as nearly planar or linear groupings of 3D features. We will now report on the more challenging cases. Figures 12.10 and 12.11 illustrate the difficulties we had augmenting photos of the Thames river shore with a new footbridge. For this particular scene, most of the visible houses are nicely aligned along the river. Thus, most targets, such as house corners, lie approximately within a plane. Considering a distance of approximately 400 meters between the camera and the houses, the depth provided within house facades and even the depth of individual houses cannot supply good three-dimensional depth cues for any calibration algorithm.

Figure 12.10 shows that at such camera distance and at the given image resolution slightly different 2D-to-3D mappings—as the result of imprecise user input—have a dramatic effect on the calibration result. In particular, there is a trade-off between the focal length of the camera and the distance of objects from the camera. Small mismatches of features along the line of sight can dramatically change the inferred focal length, altering the perspective appearance (vanishing points) of virtual objects without greatly misaligning their silhouette in the image. Between Figures 12.10a and b, one image feature at the back side of one of the houses is moved up by one pixel. The result is a significant calibration change: The near side of the river shore moves dramatically inward into the river when the feature is moved down.

The calibration results also depend on the particular algorithm that was selected. Since all algorithms use different heuristic assumptions for prioritizing the nonlinear optimization scheme, they fail in different ways when confronted with poor mapping data. Figure 12.11 shows that—in this particular case—the Weng algorithm performed worse than the Tsai algorithm (compare with Figure 12.10a). In other cases, the opposite was true.

(a) Weng: Feature at [537,270]

(b) Weng: After ten 1-pixel corrections

FIG. 12.11. Results of Weng calibration after automatic correction of feature mappings (ten iterations).

These examples demonstrate that real exterior construction applications impose very hard requirements on AR and, in particular, on optical camera calibration. In contrast to artificially created demonstrations shown in laboratory settings to exhibit the general concepts of an approach, real applications provide challenging side constraints. For example, due to the fact that we are visualizing a new bridge, a river keeps us from getting closer to the buildings we use for camera calibration. In our work, we are thus emphasizing pragmatic concepts to cope with such real problems.

- Considering current calibration instability, the need for good reality models becomes evident. They need to be very precise. Furthermore, they should cover targets in a widely spread three-dimensional volume. For example, the inclusion of distant high rises and power poles or some targets on the near side of the river would greatly stabilize the results. Finally, the targets need to be easily detectable and precisely locatable in image data. Tips of power poles have proven very suitable for this purpose.

- To help users correctly identify image features, InCal automatically investigates which image feature currently has the largest influence on a calibration misalignment. By moving that feature by one pixel up, down, left, or right, a new calibration generates a much smaller mismatch between image features and projected scene features. Figure 12.11 shows the results of Weng's algorithm after ten iterations.

- Another pragmatically useful concept suggests exploiting as much externally available information as possible. Thus, we exploit the flexibility of Tsai's calibration system. Using interactively provided data on the camera's focal length, center, and aspect ratio, Tsai's algorithm computes only the external camera position and orientation parameters, as shown in Figure 12.12. The algorithm can be constrained even further by providing approximate camera location and orientation information, as is done in the tracking systems being discussed next.

3.4 Tracking

When live video streams (image sequences) rather than individual images are augmented, camera tracking becomes an issue. Due to the numerical instability of calibration routines, it is not advisable to recalibrate each image from scratch. The result would be a rather bumpy camera path.

FIG. 12.12. Tsai calibration using a fixed focal angle of 42 degrees.

Rather, camera motion must be modeled as part of the parameter estimation process, influencing and stabilizing the system.

Kalman filtering is a well-established technology to stabilize the estimation of camera motion [5, 13] and user motion [2]. With D. Koller, we have developed a three-dimensional camera motion model, which accounts for camera velocity and acceleration [29, 31, 30]. With this motion model, physical camera motion can be calculated and tracked after the first two images have been taken, predicting the real-world camera trajectory for subsequent images. According to the predicted next camera position, the system determines local search areas in the image where the black squares should be. From their actually determined locations in the image a corrective term is calculated to influence the camera motion model. We have used this Kalman filtering approach in live tracking demonstrations, operating at about 10 frames per second on an SGI Indy workstation (with an R5000 processor). Currently, the system tends to adjust only slowly to changing camera motion—due to its built-in technology to assume smooth camera motion. When faced with abrupt camera jerks, the algorithm tends to continue moving the virtual camera in a steady direction for a while before reversing to account for the jerk. As a result, virtual objects tend to have a swinging behavior in the scene and the black square targets are frequently lost such that the tracker needs to be reinitialized. Similar observations have been reported by Lowe and by Ravela et al. [38, 51].

As an alternative, we have begun exploring simpler, more direct tracking schemes. Using the extended tool box of Tsai's calibration routines, we use calibration parameters from the previous image to initialize the next calibration. This works particularly well under circumstances when we know that the camera motion is constrained. When the camera is known to be on a tripod, we can use the same camera position throughout the entire image sequence, recalculating only the rotational components. But even when the camera is not stationary, we have been able to obtain very good and stable calibrations by assuming that—at operating speeds of about 23 frames per second—the camera hasn't moved much in between consecutive frames. We thus initialize the nonlinear least squares motion estimation routine with the previous motion parameters, allowing them to resettle according to the updated matching data. When the camera is rotated too fast, a complete recalibration is achieved in a third of a second.

We use the same technology both for the live demonstrations and for the slower, interactive calibration of complex cityscapes. The system works very well for medium-speed camera motions. We are able demonstrate the system with a hand-held IndyCam. Virtual objects are much more stable within the real scene; the characteristic swinging of the Kalman filtering approach is not observed.

3.5 Tracking Stability

Tracking stability is a key issue in AR. Virtual objects must be precisely positioned in the picture and keep their position over time despite camera motion and noise.

In our demonstrations generating high-quality presentations of new virtual buildings within a given environment, we have observed that appearant stability within the scene is much more important than the precise calibration of individual images by themselves. Thus, it is very important to make suitable stabilizing assumptions. In particular, assuming that the internal camera parameters (focal length) remain constant throughout the demonstration provides significant overall improvements—even though such assumption might be violated. Using schemes that avoid computing all six external parameters together have further stabilized the augmentations.

To this end, we currently determine camera rotation and translation in two discrete steps, optimizing the nonlinear equations for only three unknowns at a time.

3.6 Sensor Fusion:
GPS and Optical Tracking

Many different technologies can be used to determine the current camera location, such as optical, magnetic, mechanical, and inertial trackers. Each technology has advantages and drawbacks when compared to the others. Inertial or magnetic sensors, for example, are fast and can track abrupt motions. Yet, they are not precise enough to allow for an accurate alignment of the graphical object in the pictures. Mechanical trackers are more precise, yet they—as well as magnetic trackers—severely limit camera motions, essentially requiring controlled, indoor application scenarios, such as virtual studios. Optical tracking can deliver precise calibrations both indoors and outdoors. Yet, it has trouble coping with fast camera motion and with scenes that are optically too complex.

GPS [37] is another promising solution—especially for outdoor applications. It is based on a set of 24 satellites orbiting Earth at about 20 km height. Each satellite has a high-accuracy atomic clock and transmits its time signal in a regular interval. A receiver on Earth receives the time signal from at least 4 of these satellites and can calculate its position from the known orbits of the satellites via triangulation. The typical accuracy for standard GPS due to different kinds of error is 98 m in 3D. Using differential GPS (which needs information that comes from a private or public base station or which might be sent from official sources in the near future) brings the accuracy down to the meter level, with expensive equipment available to reach the centimeter level. The problem with GPS is that it only works when the receiver can see the satellites. The signal is too weak to penetrate buildings or other cover. Thus within the canyons of a city's high rises or just standing close to a wall, getting enough satellites can be a problem. Another problem with using GPS for real-time AR tasks is its perfectionism. Rather than giving false data it will give no data. Thus, several seconds can pass between successful measurements. The very high end GPS systems are even slower, requiring a minute or more to initialize under imperfect circumstances. After initialization they are faster, yet when they lose track of some satellites, they require another initialization. In general most GPS receivers are not built for real-time measurements;

a typical update rate is once per second. More expensive systems go up to five per second. Thus GPS alone is not enough for real-time tracking, even besides the fact that GPS only gives positional information, not orientation.

A robustly operating system can be expected to benefit greatly from a well-designed fusion of sensor information provided by several different devices. For example, Jancene et al. combine optical camera calibration with mechanical tracking technology [26]. State et al. combine optical and magnetic trackers to augment table-top scenarios in real time. While several multicolored circular targets are visible, the system relies on optical tracking results. When the camera moves off target, the magnetic trackers ensure that the system maintains an overall sense of orientation and position, reinitializing the optical tracker when the targets come back into sight [60].

In our system we are exploring approaches toward injecting GPS data into optical trackers, since GPS is becoming increasingly available on today's construction sites, and since it is much better suited to outdoor scenarios than magnetic trackers. Attaching a GPS sensor to the camera, we can complement the optical tracking system with camera position data provided by GPS. Since the GPS signal is produced at an unpredictable, asynchronous rate, optical tracking using Tsai's calibration still constitutes the core of the system, integrating GPS information whenever provided and relying on purely optical techniques in the mean time.

4. AUGMENTING REALITY

Once appropriate reality models and camera calibrations have been obtained, they form the basis for mixing real and virtual worlds. The subsequent sections describe the different steps that need to be taken to achieve realistic and fast inclusion of virtual information into a real world.

4.1 Geometric Data

Since exterior construction is a very physical, three-dimensional business, much synthetic data relates directly to the 3D objects being designed and built. Such information is typically represented in 2D or 3D geometric primitives.

With AR, such virtual geometric objects can be integrated into the real environment during all phases of the life cycle of a building. Before the

(a)   (b)

FIG. 12.13. Side view of a new footbridge, planned to be built across the river Wear in Sunderland, UK. (a) Original scene. (b) Augmented with planned footbridge.

(a)   (b)

FIG. 12.14. A virtual wall at a real construction site.

construction project is started, AR can support marketing and design activities to help the customer visualize the new object in the environment (Figure 12.13). During construction, AR can help evaluate whether the building is constructed according to its design (Figure 12.14). After the construction is completed, maintainance and repair tasks benefit from seeing hidden structures in or behind walls (Figure 12.15).

AR thrives on fast, real-time augmentations of the real world. All virtual information thus has to be rendered very quickly. To this end, we

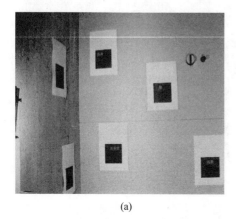
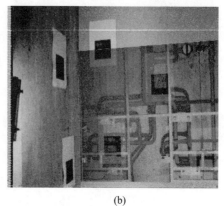

(a)   (b)

FIG. 12.15. Seeing the piping in the wall. (a) Original image. (b) X-ray view into the wall.

carefully tune and prune geometric models to achieve maximal rendering performance while maintaining an acceptable level of realism.

4.1.1   3D Models

D models of medium-sized and large building projects are usually very complex. This is due to the inherent complexity of buildings, as a complete building has thousands of parts. Even if only the outside of the building is of interest, typical models are comprised of several hundred thousand polygons. For off-line augmentation (e.g., for video sequences) this is not a big problem; rendering just takes longer. For on-line interactive or head-mounted augmentations these models are not useful, as even high-powered graphics supercomputers cannot render them at an acceptable frame rate (i.e., more than 10 Hz). Standard geometry optimizations for rendering, such as conversion of individual polygons into triangle strips, increases performance, but not enough in most cases. They thus have to be simplified. We employ an interactive in-house tool [57] building on standard algorithms [56, 58]. Except for closeups, the resulting models are virtually indistinguishable from the originals, albeit at a fraction of the cost (Figure 12.16a).

Another problem stems from the fact that architectural models are usually not created for presentation but rather for building purposes. They are typically generated using standard CAD tools, which work in wireframe mode and do not pay attention to consistent orientation of polygonal

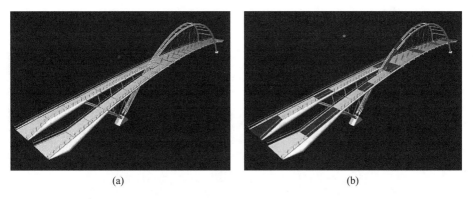

(a)   (b)

FIG. 12.16. Model of Sunderland footbridge (decimated from 45000 down to 25000 triangles). (a) Decimated model. (b) Model without surface normal correction.

faces. When such orientations are used to compute the surface normal for lighting purposes, many of them point the wrong way and thus are not shaded properly. The resulting models then typically have a checkered look (Figure 12.16b). There are some tools to help integrate the faces consistently in the model decimation system, but in general there is no automatic solution and some manual work is required until the modeling tools used and the resulting models get better.

A related problem is the optical appearance of the models. Material parameters such as reflectivity and colors are not included in the standard modeling tools; thus this information has to be generated. The position and strength of light sources and other global lighting parameters have to be added to the model [47], unless they are extracted from the real images [18] or provided by other global information such as date, time, and place [46].

Even if this information were available in the modeling system there is no a standardized way yet to extract it. The most common export format for most CAD systems is DXF, which was designed for the exchange of 2D drawings. It has been extended over the years to handle 3D data as well but is nowhere near being an adequate exchange format for high-quality models for rendering. For these purposes the SGI Inventor format [62] and the VRML1 or VRML2 formats are becoming popular, but they are still not universally supported. Public-domain [24] and commercial [49] converters are available, but most of them do not handle all possible variants of DXF reliably, so that sometimes specialized converters need to be written.

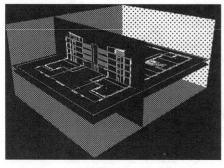

(a) Overview   (b) Inside view

FIG. 12.17. Arrangement of 2D plots in a 3D model.

4.1.2 2D Models

Much information is currently represented as two-dimensional plots rather than three-dimensional models. It currently seems more suitable for individual contractors to maintain their own 2D systems of information relevant to them than for everybody to access an all-inclusive, huge, 3D project model—even at the risk and cost of having to ensure that changes are quickly propagated between the systems of all relevant subcontractors.

For AR applications however, data need to be coordinated within a three-dimensional coordinate system. As one simple approach toward integrating 2D plots into a real-world framework, we have arranged the plots as a system of appropriately stacked panes (see Figure 12.17).

4.2 Fast Rendering and Tracking

Besides fast and robust camera tracking, high-quality, real-time rendering [52, 54] is the essential ingredient to an AR system. To both track and render at high quality and with high speed, a distributed solution is appropriate. This can either mean distributing across several machines or across tasks.

Distribution across several machines allows the use of cheaper machines and more importantly allows the use of vastly different machines. For example the tracking from simple sensors could be done by a small wearable machine while the high-quality rendering of a complex model is done on a stationary graphics supercomputer [59]. Or the situation can be the other way around, in which an expensive optical tracking algorithm using very little information about the scene runs on a supercomputer while simple models or textual information are rendered by the wearable machine. The problem of this multimachine distribution is the neccessary communication. The amount of data that has to be communicated can be quite high (e.g., video images for tracking or from rendering). But in any case this distribution adds lag to the entire system, which is quite detrimental to the effect of immersion.

The alternative is to use multiple threads on a single machine. For this to be useful the machine has to have multiple processors. A short time ago this meant an expensive workstation, but multiprocessor Pentium boards are becoming commonplace. Running both tasks on one machine shortens the communication pathways dramatically, as in the worst case a memory to memory copy has to be done and in the best case of a shared memory system just pointers have to be moved. This is the model that we use for our high-quality real-time applications, running the optical tracking on one processor while the other is feeding the graphics pipeline.

Thus the natural separation of an augmented reality system into the two tasks of tracking and rendering allows parallel processing for high throughput and also asymmetric distributed processing to comply with constraints of specialized machinery such as wearable computers.

4.3 Interactions between Virtual and Real Objects

Realistic immersion of virtual objects into a real scene requires that the virtual objects behave in physically plausible manners (i.e., they occlude or are occluded by real objects, they are not able to move through other objects, and they are shadowed or indirectly illuminated by other objects while also casting shadows and mirror images themselves).

4.3.1 Occlusion

Occlusions between real and virtual objects can be computed quite efficiently by the geometric rendering hardware of high-quality graphics workstations, when provided with a list of the geometric descriptions of all real and virtual models. By drawing real objects in black, the luminance keying feature of video mixing devices can be activated to substitute the respective image area with live video data. As a result, the user sees a picture on the monitor that blends virtual objects with live video, while respecting 3D occlusion relationships between real and virtual objects (Figure 12.18).

Other mixing approaches use depth maps of the real world obtained with a laser scanner [53] or vision-based scene recognition approaches

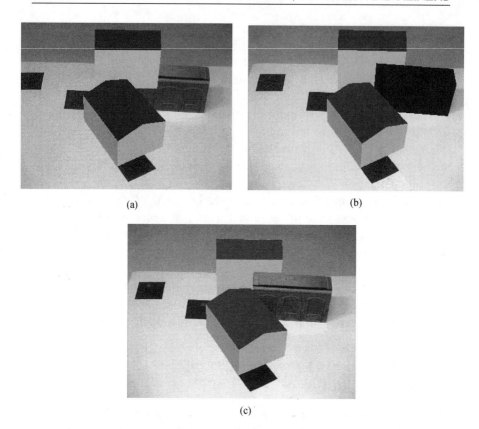

(a)   (b)

(c)

FIG. 12.18. A toy house occluding parts of a virtual pink house
while being partially occluded by a virtual green house. (a) Without occlusion handling. (b) Reality model shown in black. (c) With
occlusion handling.

[11, 12, 28]. The depth maps can be used to initialize the Z-buffer of the
graphics hardware [29, 73]. Occlusion of virtual objects is then performed
automatically. When the virtual object is rendered, pixels that are further
away from the camera than the Z values in the depth map are not drawn.
By setting the background color to black, the real objects present in the
original video are displayed in these unmodified pixels.

   Both approaches have advantages and disadvantages, depending on the
application. Full 3D geometric models are best for real-time movement
of cameras. Polygonal approximations to depth maps can be used over
a certain range of camera positions since the synthesized scene model is
rerendered when the camera moves. Copying the depth maps directly into
the Z-buffer is the hardest approach: The map needs to be recomputed after

each camera motion because the new projective transformation "shifts" all depth values in the depth map. Thus, this approach only works with stationary cameras or with shape extraction algorithms that perform at interactive speeds.

On the other hand, the geometric modeling approach suffers from an inherent dependence on scene complexity. If the scene needs to be represented by a very large polygonal model, the rendering technology may not be able to process it in real time. In contrast, the size of a depth map does not depend on scene complexity. Which approach to use in an application depends on the overall requirements and the system design.

4.3.2   Shadows and Reflections

Objects in the real world not only determine their own shading, they also have an influence on the appearance of other, distant objects by means of shadows and reflections [18, 19].

With the availability of reality models, standard computer graphics algorithms [17] can be used to compute the geometry of shadows cast by virtual objects onto real ones (see Figures 8 and 9 in Ref. [60]). Given the right hardware, this can even be done in real time. Depending on the amount of ambient light in the scene, shadows should not completely replace the object they are falling on, but should rather be blended with the image of the underlying object [26].

Reflections are a more difficult topic that can be solved for many useful special cases, but not in general. Reflections of virtual objects in planar real mirrors can be resolved using standard computer graphics techniques [17]. Similar to shadows, perfect reflections are rare in the real world; the reflected object should be blended with the mirror image rather than replacing it. This also allows the simulation of essentially planar reflective surfaces such as water (see Figure 12.19).

Difficult to handle are reflective virtual objects, as in general they would have to reflect things from the surrounding environment that are not visible in the image. For special cases this can be circumvented by placing real reflecting objects, such as a silver sphere, in carefully chosen locations in the scene and using their reflections as an environment map to determine the light reflections for the virtual object [61].

An alternative approach would be to use a high-quality reality model to render the reflection onto the virtual object, but that would defeat the idea of augmented reality not having to build such a complex, photometrically precise model.

FIG. 12.19.   Virtual London bridge reflecting in the real water.

4.3.3   Physical Constraints:
Solid Virtual Objects, Gravity

For an augmented world to be realistic the virtual objects not only have to interact optically with the real world but also physically. This applies to virtual objects when animated or manipulated by the user. For example, a virtual chair shouldn't go through walls when it is moved, and it should exhibit gravitational forces [7].

According to the law of nonpenetration, two solid objects cannot be in the same place at the same time. Thus, virtual objects should prevent themselves from moving into or through other objects. Given a reality model, this behavior can be achieved using the same collision detection and avoidance systems that are used for virtual reality systems [75].

Another important physical law concerns gravity: When not supported by anything virtual objects should move downwards while obeying the law of nonpenetration until they reach the lowest possible position.

These two laws make up the most important physical constraints. A full physical simulation including more aspects of the interaction between real and virtual objects, such as elastic behavior and friction, would be desirable. For off-line applications this is possible if enough information about the virtual objects and a complete enough reality model are available. For real-time applications most simulation systems are not fast enough. Yet, even simple implementations of the above rules will make the system much more realistic.

4.4 Leaving Reality Behind

Augmented reality and virtual reality are not two discrete alternatives but rather part of a spectrum of mixed realities [44] with full virtual reality on one end and full physical reality on the other. Augmented reality is in the middle, combining the best of both worlds. But sometimes it might be desirable to lean more in one direction or the other.

Because registered augmented reality by concept needs real images, its freedom of movement is limited to the places where an image recording device (possibly a human eye) can go. Unless employing rather exotic hardware such camera-carrying blimps [50] this limits the possible positions of the viewer. Virtual reality on the other hand allows complete freedom of movement, as computer generated images can be generated for every possible viewpoint. Thus it is sometimes desirable to leave the augmented reality behind and switch into the virtual reality to take a look from a point where it's physically impossible to get (e.g., from above).

The disadvantage of leaving the augmented reality behind is that the view now has to be constructed entirely from synthetic information (i.e., the virtual objects and the reality model) and from previous image data. A very promising area of current computer graphics research to circumvent this shortcoming is image-based rendering [9, 20, 36, 40], which strives toward generating images from new viewpoints given some images from other viewpoints. A future system might employ a camera to take images while viewing the augmented scene and later using these images to give the freedom of movement to the user while incorporating images taken on site just a short time ago, thus being as current as possible.

4.5 Nongeometric Data

Virtual information doesn't have to be exclusively three dimensional and geometric (polygonal) [59]. In large construction projects, many kinds of information are gathered, stored, maintained, and shared digitally in many formats [34]. When suitable information visualization schemes are used, AR can bring any such information to users roaming the real world. Examples of such sources of information include:

- **Business data**, such as project schedules and time lines, building codes and tolerances, customer preferences, as well as texts describing

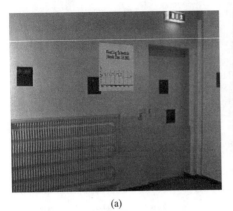 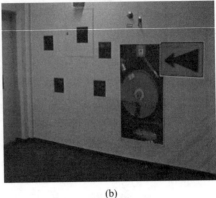

(a)   (b)

FIG. 12.20. (a) Heating schedule and visualized temperatures within heating pipes. (b) Picture of a fire hose inside a cabinet and an instruction sheet, superimposed on (next to) the cabinet door.

subcontractors (e.g., their WWW entries). When available "on location," such information can provide the basis for on-site evaluations analyzing whether the construction is progressing according to schedule, which firm is to come in next, etc. Furthermore, contractors can be contacted immediately to discuss discrepancies between the specification and what was really built [10]. Business information can be made available as 2D windows on the world [15] or on virtual sheets of paper or panels attached to real walls or floating next to objects under discussion (see the virtual panel showing a heating schedule in Figure 12.20a and the fire hose instruction sheet in Figure 12.20b).

- **Image/video data.** Using on-line brochures of contractors and catalogs of material suppliers, customers can choose among various options (e.g., different wood grains for doors), within the real life context. Like business data, brochures and catalogs can be presented (texture-mapped) on virtual panels. Even entire videos can be presented, advertising how novel window or door designs will much improve their ease of use, how they will reduce the evaporation of heat, or other things. Multimedia information will be at the architect's and customer's fingertips.

Images and videos can also show equipment such as pipes or electric wiring inside walls from photos taken before the equipment was covered with plaster. When such images are shown well aligned with the walls they provide the illusion of X-ray vision skills (see Figure 12.15b and Figure 12.20b).

- **Process data**, indicating the operating conditions of machinery in use. Such information is essential to building maintainance and renovation tasks. It can help find leaks in pipes etc. AR can show the information right on the device that is being inspected. In Figure 12.20a, we use a red-to-blue coloring scheme to symbolically represent warm-to-cold temperature variations within heating pipes.
- **Instructions**, such as what to build next. AR can remind people of the correct scheduling of tasks, showing them one step at a time what to do next (see Figure 12.25). Much time and material is wasted when a wall is erected too early and has to be removed again so that large equipment (e.g., elevator equipment or a water tank) can be put into its correct place.

  Further instructions can serve as navigation aides. Large constructions sites such as a new airport or shopping mall are ever-changing mazes of roads. AR can help people navigate within the area (e.g., to find the currently shortest or safest path from one place to another) (see Figure 12.21).
- **Simulated data**, such as the expected circulation of air within a building and lighting simulations to verify and optimize the placement of windows and artificial light sources inside buildings (Figure 12.22).

Various exterior construction applications benefit from granting workers and engineers on-site access to all of these sources of information, augmenting reality in suitable, nondisturbing ways.

FIG. 12.21. Navigation aide.

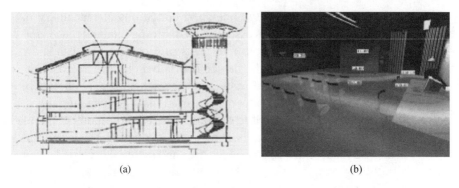

(a)   (b)

FIG. 12.22. Simulated illuminance values on a working plane.

5. DIMINISHING REALITY

Many construction projects require that existing structures be removed before new ones are built. Thus, just as important as augmenting reality is technology to diminish it.

Figure 12.23a shows one of several pictures of TV towers on Monte Pedroso near Santiago de Compostela, Spain. The project to build a new communications tower on Monte Pedroso for completion by 1999 was conceived by the Concello de Santiago as part of an overall plan for the city and for the mountain. The new tower, designed by the architects Sir Norman Foster and Parners and consulting engineers at Ove Arup and Partners, UK, is primiarily intended for telecommunications users, but because of the drawing power of unusual buildings and its location, the tower is also intended to appeal to public visitors.

Prior to augmenting the image with a model of the new tower, the existing towers need to be removed (Figure 12.23b). To this end, a part of the sky has to be extrapolated into the area showing the TV towers and the barracks. Then the new tower can be put into place (Figure 12.23c).

In principle, the problem of diminshing reality consists of two phases. First expiring buildings have to be identified in an image. When such structures are well represented in a reality model, they can be located by projecting the model into the image according to the current camera calibration.

Next, the outdated image pixels need to be replaced with new pixels. There is no general solution to this problem since we cannot know what a dynamically changing world looks like behind an object at any specific instant in time—unless another camera can see the occluded area and provides us with the information. Yet, some heuristics can be used to solve

(a)   (b)

(c)

FIG. 12.23.   Monte Pedroso near Santiago de Compostela, Spain.
(a) Original image. (b) Diminished reality. (c) Augmented diminished reality.

the problem for various realistic scenarios:

- The most simple approach might just deemphasize outdated areas,
  graying them out or smoothing across them with large convolution
  windows.
- More sophisticated morphological approaches might extrapolate properties of surrounding "intact" areas (e.g., a cloudy sky) across outdated areas.
- When a building is to be removed from a densely populated area in a
  city, particular static snapshots of the buildings behind it can be taken
  and integrated into the reality model. Computer graphics technology
  can then map those textures into the appropriate spaces of the current
  image.

    First results of such "X-ray vision" capabilities are shown in Fig-
  ure 12.15b. In the case of diminished reality, the supplemented video
  image (piping) must be displayed a full alpha-level, thus completely
  hiding the current video (wall).

- For video loops of a dynamically changing world, computer vision techniques can be used to suitably merge older image data with the new image. Faugeras et al. have shown that soccer players can be erased from video footage when they occlude advertisement banners: For a static camera, changes of individual pixels can be analyzed over time, determining their statistical dependence on camera noise. When significant changes (due to a mobile person occluding the static background) are detected, "historic" pixel data can replace the current values [76].

  In more general schemes using mobile cameras, such techniques can lead toward incremental techniques to diminish reality. While moving about in the scene, users and cameras see parts of the background objects. When properly remembered and integrated into a three-dimensional model of the scene, such "old" image data can be reused to diminish newer images, thus increasingly effacing outdated objects from the scene as the user moves about.

We currently use interactive 2D tools to erase old structures from images. This approach can only be used for static, individual photos, but not for video sequences from a live, dynamically moving camera.

6. USER INTERACTION IN A THREE-DIMENSIONAL AUGMENTED WORLD

Augmented reality is a technology by which a user's view of the real world is augmented with additional information from a computer model. Users can work with and examine real 3D objects while receiving additional information about those objects or the task at hand. Exploiting people's visual and spatial skills, AR thus brings information into the user's real world rather than pulling the user into the computer's virtual world.

6.1 A Real-World Interface to Virtual Worlds

The power of AR as a real-world interface to virtual worlds becomes evident when we present a virtual building on a real table top (Figure 12.24a). Rather than using complex 2D or 3D interaction metaphors, users can walk around the table to inspect the building from all sides, rotating a reference pattern

(a)   (b)

FIG. 12.24. (a) Virtual buildings viewed on a real table. (b) Inter-active layout of a city scape.

on the table to turn the building. Using several reference patterns, several virtual objects can be moved independently of each other (Figure 12.24b). At the same time, users can reference other material on the table, such as maps, and discuss the model with colleagues.

In these examples, our approach incorporates the concepts of the DigitalDesk [69, 70], the metaDESK [25], and the Responsive Workbench [32], extending them toward augmenting more complex realities than planar desktops. It leads to hybrid digital/real mock-ups (e.g., of a complex construction site for which a scaled-down physical model of the environment might already exist). AR can augment it with digital prototypes of new buildings or objects until their design and layout matures. Similar concepts apply when extensions or renovations of large manufacturing facilities are planned.

6.2 Computer-Provided Guidance to Real-World Tasks

Computer augmentations of the real world can provide the user with dynamic, up-to-date instructions on how to perform a task [8]. We explore and demonstrate the potential of this new paradigm in our laboratory with the example of a Tangram game (Figure 12.25).

In contrast to 2D games played on a monitor screen, the Tangram game takes place on a real table using real Tangram pieces. Figure 12.25a shows our setup. The user sits in front of a small cubicle in which the game takes place. A camera behind his shoulder records the scene. Our live AR

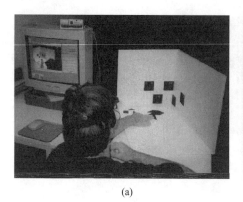
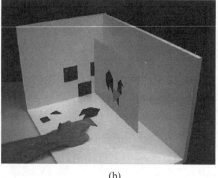

(a)   (b)

FIG. 12.25. Computer-guided assembly of a Tangram shape.
(a) Real world. (b) Augmented world.

system (See Sections 1.2 , 3.1.1, and 3.4) running on the Indy in the corner
is capable of tracking camera or cubicle motion, while maintaining the
three-dimensional augmentations of the scene.

The computer three-dimensionally augments the real-world view with
instructions on how to assemble a complex Tangram shape. A virtual sheet
shows the entire shape (Figure 12.25b). Both on the sheet and superimposed
on the table the next piece to align is highlighted. Either on the monitor or
in a "feed-through" head-mounted display, the user sees the augmentations
while working in the scene. The user can now proceed to assemble the
entire Tangram shape.

Similar concepts also apply to exterior construction tasks, such as the
computer-guided installation of an elevator in its shaft where extreme care
has to be taken to plan the path along which to insert the elevator. Other
applications include the assembly or repair of machines [15] and the installation of alluminum struts in a diamond shaped spaceframe [68].

6.3   Two-Way Human–Computer
Interaction in the Real World:
Reality Tracking

To fully exploit the AR paradigm, the computer must not only augment
the real world but also receive input (feedback) from it. In truely three-
dimensional human–computer interaction, actions or instructions issued
by the computer cause the user to change the real world—which, in turn,
prompts the computer to perform further actions, as demonstrated in the

ALIVE project [39, 74] in which a person can interact with a virtual dog, gesturing it to sit down, etc.

The use of magnetically tracked devices, such as data gloves and body suits, and GPS-tracked laser pointers provides three-dimensional, interactive schemes to communicate with the computer. Speech or sound input, as well as gesture recognition, provide further interaction means. In their spaceframe construction demonstration, Webster et al. equip users with barcode readers to enable them to inform the computer of newly selected struts [68].

Yet, not all changes in the real world can be tracked by a discrete set of (more or less bulky) physical tracking devices, and the user also cannot describe all changes verbally or with a barcode reader. To this end, AR systems must be capable of automatically detecting and tracking changes in the real world. We explore *optical reality tracking* approaches at the example of a tic-tac-toe game (Figure 12.26). The physical setup of the game is similar to the one used in the Tangram scenario (Figure 12.25a). In this case, a real tic-tac-toe board is sketched out on the surface of the cubicle. After the user has placed a stone on the board and hit the virtual "GO" button by moving his or her hand across it, the computer detects the stone in the image and plans a counter move, indicating its decision by a virtual cross on the board and instructing the user on the virtual panel to continue. In this example, the user does not need to touch a keyboard or mouse once the game has started. All interactions occur directly on the game board, embedded in the real world.

FIG. 12.26. Real-world-based tic-tac-toe.

These demonstrations are a first, essential step toward enabling truely real-world-based interactive AR applications. They form the basis for real construction, maintenance, and repair tasks.

SOURCES OF GRAPHICAL MATERIAL

- Figure 12.1: Private communication with David Leevers, BICC (CICC Project).
- Sunderland Newcastle, UK: Work conducted in collaboration with Ove Arup and Partners (CICC Project).

  Figure 12.3a, Figures 12.13a,b, Figures 12.16a,b: Picture of the river Wear, Sunderland Newcastle, UK. The bridge model was provided by Sir Norman Foster and Partners.

- Thames river, London, UK: Work conducted in collaboration with Ove Arup and Partners (CICC Project).

  Figures 12.9–12.12, Figure 12.19: Pictures of the river Thames, London near St. Paul's Cathedral. The 3D model of a designed millennium footbridge was provided Sir Norman Foster and Partners.

  Figure 12.5: Three-dimensional model of selected areas (four tiles) of the city of London. Acquired for the CICC project by Ove Arup and Partners.

- Bluewater Kent, UK: Work conducted in collaboration with Bovis and Trimble Navigation Limited (CICC Project).

  Figure 12.14: Picture from a video sequence.

- Santiago de Compostela, Spain: Work conducted in collaboration with Ove Arup and Partners (CICC Project).

  Figures 12.23a,b,c: Picture of Monte Pedroso near Santiago de Compostela, Spain. The model of the TV tower was provided by Sir Norman Foster and Partners.

- Gmunder Straße, Munich, Germany: Work conducted in collaboration with Philipp Holzmann AG, Germany.

  Figure 12.3b, Figures 12.15a,b: Indoor pictures of a bathroom under construction.

  Figure 12.17: View of the CAD model.

- Valbonne, France: Work conducted in collaboration with INRIA Sophia-Antipolis (Realise and Cumuli Projects).

Figures 12.6a,b,c: Picture and reconstructed model of the Arcades in Valbonne, France.

- Royal Institute of Charted Surveyors, London, UK: Work conducted by U. Leeds, JRC, and BICC (RESOLV Project).

  Figure 12.7: Pictures of the RESOLV trolley and the reconstructed model of the Royal Institue of Charted Surveyors. Courtesy of the RESOLV project.

Acknowledgments   The research was conducted during 1995–1998 and was financially supported by the CICC project (ACTS-017) in the framework of the European ACTS programme and by the Cumuli project (LTR-21914) in the ESPRIT programme. The laboratory space and the equipment were provided by the European Computer-industry Research Center (ECRC).

We are grateful to the current and former colleagues at Fraunhofer IGD, ZGDV, and ECRC for many useful comments and insights which helped us develop and refine the work. Particular thanks go to Dieter Koller and Eric Rose. Both the CICC and the Cumuli consortium have deeply influenced our approach.

REFERENCES

[1] A. Azarbayejani and A. P. Pentland. Recursive Estimation of Motion, Structure, and Focal Length. *IEEE Trans. on Pattern Analysis and Machine Intelligence (PAMI)*, 17(6):562–575, June 1995.

[2] R. Azuma and G. Bishop. Improving Static and Dynamic Registration in an Optical See-Through HMD. *Proc. Siggraph'94,* Orlando, FL, July 1994, pp. 194–204.

[3] R. T. Azuma. A Survey of Augmented Reality. *Presence, Special Issue on Augmented Reality*, 6(4):355–385, August 1997.

[4] M. Bajura and U. Neumann. Dynamic Registration Correction in Video-Based Augmented Reality Systems. *IEEE Computer Graphics and Applications*, 15(5):52–60, 1995.

[5] Y. Bar-Shalom and T.E. Fortmann. *Tracking and Data Association*. Academic Press, New York, 1988.

[6] J. Bowskill and J. Downie. Extending the Capabilities of the Human Visual System: An introduction to Enhanced Reality. *Computer Graphics*, 29(2):61–65, 1995.

[7] D. E. Breen, E. Rose, and R. T. Whitaker. Interactive Occlusion and Collision of Real and Virtual Objects in Augmented Reality. Technical Report ECRC-95-02, ECRC, Arabellastr. 17, D-81925 Munich, 1995.

[8] T. Caudell and D. Mizell. Augmented Reality: An Application of Heads-Up Display Technology to Manual Manufacturing Processes. *Proc. Hawaiian International Conference on System Sciences (HICSS'92)*, pp. 659–669, 1992.

[9] S. E. Chen and L. Williams. View Interpolation for Image Synthesis. *Computer Graphics (Proc. Siggraph'93)*, 27:279–288, August 1993.

[10] CICC: Collaborative Integrated Communications for Construction. ACTS AC-0017, 1995–1998, *http://www.hhdc.bicc.com/cicc/*, 1995.

[11] CUMULI: Computational Understanding of Multiple Images. Esprit LTR-21914, 1996–1999, *http://www.inrialpes.fr./CUMULI/*, 1996.

[12] P. E. Debevec, C. J. Taylor, and J. Malik. Modelling and Rendering Architecture from Photographs: A Hybrid Geometry- and Image-Based Approach. *Proc. Siggraph'96*, New Orleans, Aug. 4–9, 1996, pp. 11–20.

[13] A. Gelb (ed.). *Applied Optimal Estimation*. MIT Press, Cambridge, MA, 1974.

[14] O. Faugeras, S. Laveau, L. Robert, G. Csurka, and C. Zeller. 3D Reconstruction of Urban Scenes from Sequences of Images. In A. Gruen, O. Kuebler, and P. Agouris (eds.), *Automatic Extraction of Man-Made Objects from Aerial and Space Images*. Birkhauser, 1995.

[15] S. Feiner, B. MacIntyre, M. Haupt, and E. Solomon. Windows on the World: 2D Windows for 3D Augmented Reality. *Proc. UIST'93*, Atlanta, GA, 1993, pp. 145–155.

[16] S. Feiner, B. MacIntyre, and D. Seligmann. Knowledge-Based Augmented Reality. *Communications of the ACM (CACM)*, 36(7):53–62, July 1993.

[17] J. D. Foley, A. Van Dam, S. K. Feiner, and J. F. Hughes. *Computer Graphics, Principles and Practice*, 2nd Edition. Addison-Wesley, 1989.

[18] A. Fournier. Illumination Problems in Computer Augmented Reality. *Journee INRIA, Analyse/Synthese d'Images (JASI)*, January 1994, pp. 1–21.

[19] A. Fournier. Computer Augmented Reality and Illumination. *Proc. International Workshop MVD'95: Modeling—Virtual Worlds—Distributed Graphics*, St. Augustin, Germany, Nov. 1995.

[20] S. J. Gortler, R. Grzeszczuk, R. Szeliski, and M. F. Cohen. The Lumigraph. *Proc. Siggraph'96*, New Orleans, Louisiana, Aug. 4–9, 1996, pp. 43–54.

[21] D. S. Greer and M. Tuceryan. Computing the Hessian of Object Shape from Shading. Technical Report ECRC-95-30, *http://www.ecrc.de*, 1995.

[22] A. Hildebrand, S. Müller, and R. Ziegler. REALISE—Computer Vision Basierte Modellierung für Virtual Reality. *Proc. International Workshop MVD'95: Modeling—Virtual Worlds—Distributed Graphics*, Sankt Augustin, Germany, Nov. 1995, pp. 159–168.

[23] B. K. P. Horn and M. J. Brooks. *Shape from Shading*. MIT Press, Cambridge, MA, 1989.

[24] Silicon Graphics Inc. and Abaco Systems Inc. *DxfToIV*. Tool Converting Autodesk DXF R12 Format into Open Inventor 2.0 Files, 1995.

[25] H. Ishii and B. Ullmer. Tangible Bits: Towards Seamless Interfaces between People, Bits and Atoms. *In Proc. CHI 97*, Atalanta, GA, March 1997.

[26] P. Jancene, F. Neuret, X. Provot, J.-P. Tarel, J.-M. Vezien, C. Meilhac and A. Verroust. RES: Computing the Interactions between Real and Virtual Objects in Video Sequences. *Proc. 2nd IEEE Workshop on Networked Realities*, Boston, MA, Oct. 1995, pp. 27–40.

[27] A. Janin, D. Mizell, and T. Caudell. Calibration of Head-Mounted Displays for Augmented Reality Applications. *Proc. of the Virtual Reality Annual International Symposium (VRAIS'93)*, pp. 246–255, 1993.

[28] T. Kanade, A. Yoshida, K. Oda, H. Kano, and M. Tanaka. A Stereo Machine for Video-Rate Dense Depth Mapping and Its New Applications. *Proc. 15th IEEE Computer Vision and Pattern Recognition Conference (CVPR)*, 1996.

[29] G. J. Klinker, K. H. Ahlers, D. E. Breen, P.-Y. Chevalier, C. Crampton, D. S. Greer, D. Koller, A. Kramer, E. Rose, M. Tuceryan, and R. T. Whitaker. Confluence of Computer Vision and Interactive Graphics for Augmented Reality. *Presence, Special Issue on Augmented Reality*, 6(4):433–451, August 1997.

[30] D. Koller, G. Klinker, E. Rose, D. Breen, R. Whitaker, and M. Tuceryan. Automated Camera Calibration and 3D Egomotion Estimation for Augmented Reality Applications. *7th Int. Conference on Computer Analysis of Images and Patterns (CAIP'97)*, Kiel, 1997.

[31] D. Koller, G. Klinker, E. Rose, D. Breen, R. Whitaker, and M. Tuceryan. Real-time Vision-Based Camera Tracking for Augmented Reality Applications. *Proc. ACM Symposium on Virtual Reality Software and Technology (VRST'97)*, Lausanne, Switzerland, Sept. 15–17, 1997.

[32] W. Krueger, C.-A. Bohn, B. Froehlich, H. Schueth, W. Strauss, and G. Wesche. The Responsive Workbench: A Virtual Work Environment. *IEEE Computer*, pp. 42–48, 1995.

[33] K. N. Kutulakos and J. Vallino. Affine Object Representations for Calibration-Free Augmented Reality. *Proc. Virtual Reality Ann. International Symposium (VRAIS'96)*, 1996, pp. 25–36.

[34] D. Leevers. Inner Space, the Final Frontier. *Proc. Int. Conference From Desktop to Web-Top: Virtual Environments on the Internet, WWW and Networks*. Picturevill, Bradford UK, April 1997.

[35] D. Leevers. Private Communication. May 1997.

[36] M. Levoy and P. Hanrahan. Light Field Rendering. *Proc. Siggraph'96*, New Orleans Louisiana, Aug. 4–9, 1996, pp. 31–42.

[37] Trimble Navigation Ltd. *All about GPS. http://www.trimble.com/gps/index.htm*. 1997.

[38] D. G. Lowe. Robust Model-Based Motion Tracking Through the Integration of Search and Estimation. *International Journal of Computer Vision (IJCV)*, 8(2):113–122, 1992.

[39] P. Maes, T. Darrell, B. Blumberg, and A. Pentland. The ALIVE System: Full-body Interaction with Autonomous Agents. *Proc. Computer Animation'95*, 1995.

[40] L. McMillan and G. Bishop. Plenoptic Modeling: An Image-Based Rendering System. *Proc. Siggraph'95*, Los Angeles, CA, Aug. 6–11, 1995, pp. 39–46.

[41] J. P. Mellor. *Enhanced reality visualization in a surgical environment*. Master's Thesis. Technical report 1544, MIT AI-Lab, 1995.

[42] J. P. Mellor. Realtime Camera Calibration for Enhanced Reality Visualization. *Proc. IEEE Conference on Computer Vision, Virtual Reality and Robotics in Medicine (CVRMed'95)*, 1995, pp. 471–475.

[43] MICC: Mobile Integrated Communications for Construction. ACTS AC-0088, 1995–1998, *http://www.uk.infowin.org/ACTS/RUS/PROJECTS/ac088.htm*, 1995.

[44] P. Milgram and F. Kishino. A Taxonomy of Mixed Reality Visual Displays. *IEICE Transactions on Information Systems*, E77-D(12), December 1994.

[45] P. Milgram, S. Zhai, D. Drascic, and J. J. Grodski. Applications of Augmented Reality for Human-Robot Communication. *Proc. International Conference on Intelligent Robots and Systems (IROS'93)*, 1994, pp. 1467–1472.

[46] S. Müller, W. Kresse, N. Gatenby and F. Schöffel. Approach for the Simulation of Daylight. *Proc. 6th Eurographics Workshop on Rendering*, Springer-Verlag, 1995, pp. 137–146.

[47] S. Müller, M. Unbescheiden, and M. Göbel. GENESIS—Eine interaktive Forschungsumgebung zur Entwicklung parallelisierter Algorithmen für VR-Anwendungen. In *Virtual Reality—Anwendungen und Trends* (H. J. Warnecke and J.-J. Bullinger, eds.), Springer-Verlag, Reihe: Forschung und Praxis, Bd. T35, pp. 321–341, 1993. (In German.)

[48] U. Neumann and Y. Cho. A Self-Tracking Augmented Reality System (USC). *IEEE Virtual Reality Annual International Symposium (VRAIS'96)*, Hong Kong, 1996.

[49] Okino. Model Translation Software. *http://www.okino.com/*, 1997.

[50] E. Paulos and J. Canny. Ubiquitous Tele-Embodiment: Applications and Implications. *International Journal of Human–Computer Studies, Special Issue on Innovative Applications of the World Wide Web*, 1997.

[51] S. Ravela, B. Draper, J. Lim, and R. Weiss. Adaptive Tracking and Model Registration Across Distinct Aspects. *Proc. International IEEE Conference on Intelligent Robots and Systems*, Pittsburgh, PA, 1995.

[52] D. Reiners. *High-Quality Realtime Rendering for Virtual Environments*. Diplomarbeit, TU Darmstadt, 1994.

[53] RESOLV: Reconstruction using Scanned Optical Laser Video. ACTS AC-0021, 1995–1998, *http://www.hhdc.bicc.com/resolv/*, 1995.

[54] J. Rohlf and J. Helman. IRIS performer: A High Performance Multiprocessing Toolkit for Real-Time 3D Graphics. *Proc. ACM Siggraph'94*, Orlando, FL, July 1994, pp. 381–395.

[55] E. Rose, D. Breen, K.H. Ahlers, C. Crampton, M. Tuceryan, R. Whitaker, and D. Greer. Annotating Real-World Objects Using Augmented Reality. *Computer Graphics: Developments in Virtual Environments*. Academic Press, 1995.

[56] J. Rossignac and P. Borrel. Multi-Resolution 3D Approximation for Rendering Complex Scenes. *Proc. 2nd Conference on Geometric Modelling in Computer Graphics*, Genova, Italy, June 1993, pp. 453–465.

[57] J. Schiefele. *Methoden der automatischen Komplexitätsreduktion zur effizienten Darstellung von CAD- Modellen*. Diplomarbeit, TU Darmstadt, 1996.

[58] W. J. Schröder, J. A. Zarge, and W. E. Lorensen. Decimation of Triangle Meshes. *Computer Graphics (Proc. Siggraph'92)*, 26:65–70, July 1992.

[59] T. Starner, S. Mann, B. Rhodes, J. Levine, J. Healey, D. Kirsch, R. W. Picard, and A. Pentland. Augmented Reality Through Wearable Computing. *Presence, Special Issue on Augmented Reality*, 6(4): 386–398, August 1997.

[60] A. State, G. Hirota, D. T. Cheng, W. F. Garrett, and M. A. Livingston. Superior Augmented Reality Registration by Integrating Landmark Tracking and Magnetic Tracking. *Proc. Siggraph'96*, New Orleans, Aug. 4–9. 1996, pp. 429–438.

[61] A. State, G. Hirota, D. T. Cheng, W. F. Garrett, and M. A. Livingston. Superior Augmented Reality Registration by Integrating Landmark Tracking and Magnetic Tracking. First picture on *http://www.cs.unc.edu/us/hybrid.html*; unfortunately, this figure is not included in the Siggraph paper, 1996.

[62] P. S. Strauss and R. Carey. An Object-Oriented 3D Graphics Toolkit. *Computer Graphics (Proc. Siggraph'92)*, 26:341–349, 1992.

[63] R. Szeliski and S. B. Kang. *Recovering 3D Shape and Motion from Image Streams Using Non-linear Least Squares*. Technical Report CRL 93/3, Cambridge Research Lab, Digital Equipment Corporation, One Kendall Square, Bldg. 700, March 1993.

[64] R. Y. Tsai. An Efficient and Accurate Camera Calibration Technique for 3D Machine Vision. *Proc. CVPR*, pp. 364–374, 1986. See also *http://www.cs.cmu.edu/rgw/TsaiCode.html*.

[65] M. Tuceryan, D. Greer, R. Whitaker, D. Breen, C. Crampton, E. Rose, and K. Ahlers. Calibration Requirements and Procedures for a Monitor-Based Augmented Reality System. *IEEE Transactions on Visualization and Computer Graphics*, 1:255–273, Sep. 1995.

[66] M. Uenohara and T. Kanade. Vision-Based Object Registration for Real-Time Image Overlay. *Proc. IEEE Conference on Computer Vision, Virtual Reality and Robotics in Medicine (CVRMed'95)*, 1995, pp. 13–22.

[67] Vanguard: Visualization Across Networks using Graphics and Uncalibrated Acquisition of Real Data. ACTS AC-0074, 1995–1998, *http://www.esat.kuleuven.ac.be/konijn/vanguard.html*, 1995.

[68] A. Webster, S. Feiner, B. MacIntyre, W. Massie, and T. Krueger. Augmented Reality in Architectural Construction, Inspection, and Renovation. *Proc. ASCE Third Congress on Computing in Civil Engineering*, Anaheim, CA, June 17–19, 1996, pp. 913–919.

[69] P. Wellner. Interacting with paper on the digital desk. *Communications of the ACM (CACM)*, 36(7):87–96, July 1993.

[70] P. Wellner, W. Mackay, and R. Gold. Computer Augmented Environments: Back to the Real World. *Communications of the ACM (CACM)*, 36(7):87–96, July 1993.

[71] J. Weng, T. S. Huang, and N. Ahuja. Motion and Structure from Two Perspective Views: Algorithms, Error Analysis, and Error Estimation. *IEEE Transactions on Pattern Analysis and Machine Intelligence*, 11(5):451–476, 1989.

[72] R. G. Willson. Modeling and Calibration of Automated Zoom Lenses. Ph. D. thesis, Robotics Institute, Carnegie Mellon, Pittsburgh, PA. Jan. 1994. (*http://www-cgi.cs.cmu.edu/ afs/cs/usr/rgw/www/thesis.html*.)

[73] M. Wloka and B. Anderson. Resolving Occlusion in Augmented Reality. *Proc. of the ACM Symposium on Interactive 3D Graphics*, 1995, pp. 5–12.

[74] C. Wren, A. Azarbayejani, T. Darrell, and A. Pentland. Pfinder: Real-Time Tracking of the Human Body. *IEEE Trans. on Pattern Analysis and Machine Intelligence (PAMI)*, 19(7): 780–785, July 1997.

[75] G. Zachmann. Real-time and Exact Collision Detection for Interactive Virtual Prototyping. *Proc. ASME Design Engineering Technical Conference*, CIE-4306, 1997.

[76] I. Zoghlami, O. Faugeras, and R. Deriche. Traitement des occlusions pour la modification d'objet plan dans une sequence d'image. *http://www.inria.fr/robotvis/personnel/zimad/ Orasis6/Orasis6/html*, 1996.

13

GPS-Based Navigation Systems for the Visually Impaired

Jack M. Loomis, Reginald G. Golledge
University of California Santa Barbara

Roberta L. Klatzky
Carnegie Mellon University

INTRODUCTION

According to the 1990 U.S. Census of Population, there are approximately 1.1 million individuals registered as legally blind and up to 3 million reporting severe vision impairment. Yet another 3 to 4 million are visually impaired to the degree that they cannot drive and/or have difficulty reading signs or printed material. The most fundamental needs of visually impaired or blind populations include access to information (particularly that presented in written format), accessibility to the environment, and independence of movement. Our focus here is on the latter two needs.

Accessibility to the environment is important for all individuals. Access includes not only physical mobility, such as making a trip to a store by a selected transportation mode, but also being able to recognize key choice points or decision points in the environment (e.g., landmarks, streets, or neighborhoods). Accessibility therefore involves the ability to interpret, recognize, and understand the layout of features in the environment as well as being able to travel in as obstacle-free a manner as possible.

For many blind people the loss of sight is paralleled by a loss of independence. Of the 1.1 million legally blind persons in the United States, approximately 10,000 use guide dogs and 100,000 are able to travel somewhat independently using a long cane. This leaves approximately 1 million people who are dependent on other humans for movement, information processing, and environmental interpretation and use. Loss of independence is probably the most humbling of all the disadvantages associated with the loss of sight. A wearable device that can reduce dependence in all manners of interaction with the local environment is of the utmost importance to increasing the quality of life for the blind or visually impaired individual.

This chapter details the current state of research and development on GPS-based navigation systems for the visually impaired, most of which are portable, verging on wearable. It begins with a consideration of the need for such systems by the visually impaired, distinguishing between two different aspects of wayfinding—obstacle avoidance and navigation. It then reviews a number of efforts aimed at developing navigation aids for the visually impaired and then focuses first on other projects dealing with GPS-based navigation systems and then on our own project. It then briefly describes some research we have done on the display of information for route guidance and for conveying the spatial layout of important off-route entities (e.g., landmarks) and ends with a consideration of the obstacles to be overcome in implementing effective practical systems.

NAVIGATION VERSUS OBSTACLE AVOIDANCE

Human wayfinding consists of two very different functions: sensing of the immediate environment for obstacles and hazards and navigating to remote destinations beyond the immediately perceptible environment (Golledge, Loomis, Klatzky, Flury, & Yang, 1991; Golledge, Klatzky, & Loomis, 1996; Rieser, Guth, & Hill, 1982; Strelow, 1985; Welsh & Blasch, 1980). Navigation, in turn, involves updating one's position and orientation during travel with respect to the desired destination, usually along some intended route. Methods of updating position and orientation can be classified on the basis of kinematic order. Position-based navigation relies on external signals indicating the traveler's position and orientation (generally in conjunction with an external or internal map). Velocity-based navigation (usually referred to as "dead reckoning") relies on external or internal signals indicating the traveler's velocity vector; displacement and heading change from the origin of travel are obtained by integrating the velocity vector. Acceleration-based navigation (usually referred to as "inertial navigation") involves double integration of the traveler's linear and rotary accelerations to obtain displacement and heading change from the origin; external signals are not required. Most navigation, whether by animal, human, or machine, involves a combination of two or more of these methods.

The visually impaired are at a huge disadvantage, especially in unfamiliar environments, for they lack much of the information needed for planning detours around obstacles and hazards and have little information about distant landmarks, heading, and self-velocity, information that is essential when traveling through unfamiliar environments on the basis of maps and verbal directions.

ELECTRONIC TRAVEL AIDS

Following adoption of the long cane by the blind community as the primary mobility aid in the 1940s (Farmer, 1980), considerable effort has been expended to supplement or replace the long cane with electronic travel aids (ETAs) to assist with obstacle avoidance and navigation. Electronic obstacle avoiders, such as the Laser Cane and ultrasonic-based Binaural Sonic Aid, inform the traveler of nearby barriers to free and safe movement and are used to find paths that circumvent such obstacles (Brabyn, 1985; Farmer, 1980). Even with these devices, however, the visually impaired traveler has lacked the freedom to travel without assistance, for efficient navigation through unfamiliar environments relies on information beyond the limited sensing range of these devices.

Within the past decade, development of ETAs has been directed much more to the navigation function. One approach has been to put location identifiers throughout the environments traveled by visually impaired persons. Tactual identifiers are not effective, for the visually impaired traveler needs to tactually scan the environment just to know of their existence. Instead, developers have come up with identifiers that can be remotely sensed by the visually impaired traveler using special equipment. Two such systems of remote signage are Talking Signs and Verbal Landmark (Bentzen & Mitchell, 1995; Crandall, Gerry, & Alden, 1993; Loughborough, 1979). In the Talking Signs system, currently being deployed around the world, infrared transmitters are installed throughout the

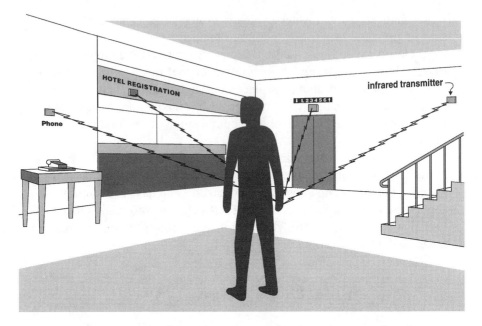

FIG. 13.1.   The Talking Signs system of remote signage. Infrared
transmitters are installed near important environmental entities.
Each transmitter continuously sends out a digitally encoded utterance about the nearby entity. A person holding a Talking Signs
receiver hears the utterance when pointing the receiver in the direction of the transmitter. (Adapted from Figure 1 in: Crandall, W.,
Bentzen, B. L., Myers, L., & Mitchell, P (1995). *Transit accessibility improvement through Talking Signs remote infrared signage: A
demonstration and evaluation.* Document #95-0050, Project AC-
TION, 700 Thirteenth St. NW, Suite 200, Washington, DC.)

travel environment, such as in airports, shopping centers, and hotels (see
Figure 13.1). These highly directional transmitters continuously transmit
digital speech indicating their location; within a range of 20 m, a visually
impaired traveler with an infrared receiver can pick up the signal from the
transmitter and hear the digitally encoded utterance; directional localization of the transmitter is sensed by aiming the hand-held receiver to obtain
maximum signal strength. The Verbal Landmark system is of similar design, but differs in significant ways (Bentzen & Mitchell, 1995). The RF
transmitter has a more limited range (2 m) and is omnidirectional, making
precise localization impossible. A visually impaired traveler within reception range receives verbal instructions for travel to other locations within the
environment.

The obvious disadvantage of placing a network of location identifiers within the environment is the cost of installing and maintaining the network relative to the coverage achieved. An alternative is to use computer technology to locate the traveler, there being a multitude of methods. These vary in the extent to which they require sensing of the environment, reception of signals provided by external positioning systems, and stored information about the environment. Inertial navigation is attractive, for it requires neither external sensing nor stored information about the environment; unfortunately, the unavailability of accelerometers of sufficiently high sensitivity and low noise rules out inertial navigation as the primary basis for pedestrian travel. A very different approach is to use map correlation, whereby video images of the environment are matched to 3-D models of the environment stored in memory. More common, however, are methods that rely on positional signals received by the traveler and processed by computer. Ogata, Makino, Ishii, and Nakashizuka (1997) have experimented with infrared barcode labels that appear as uniformly colored strips and can be attached to objects and wall surfaces within a building; infrared video sensing and computer processing provide the traveler with locational speech information encoded on the labels. Local positioning systems based on low-power high-frequency transmitters are another approach. However, the method that is most frequently considered because of its high positional accuracy, wide signal coverage, and ready availability is that based on satellite signals. The Global Positioning System (GPS) and its Russian equivalent (GLONASS) are now widely used for many positioning applications, including navigation (for details on GPS, see Parkinson & Spilker, 1996).

Pedestrian use of GPS for positioning has three shortcomings. The first is that the accuracy of stand-alone commercially available GPS receivers is limited to 20 m or so. Much higher accuracy is afforded by differential correction (DGPS), in which correction signals from a GPS receiver at a known fixed location are transmitted by radio link to the mobile receiver, allowing the latter to determine its position with an absolute positional accuracy on the order of 1 m or better; however, differential correction requires a separate receiver, and service is not available in many locations. The second shortcoming is the possible loss of satellite visibility when nearby buildings or dense foliage block a substantial part of the sky. Signal loss is most noticeable in the downtown areas of moderate to large cities where a concentration of tall buildings at times occludes much of the sky from a street-level location. The third shortcoming involves multipath distortion resulting from reflections of the GPS signal from nearby structures. Because distance to a satellite is computed from the time delay of signal transmission and signal reception, positions derived from reflected signals are in error.

For environments in which GPS signals are only intermittently available or degraded by multipath distortion, GPS needs to be supplemented by dead reckoning (based on measurements of travel velocity) or inertial navigation (based on measurements of travel acceleration). When GPS signals are unavailable (e.g., indoor environments), either some form of local positioning system or a network of location identifiers, such as Talking Signs, will be needed in assisting visually impaired persons with navigation.

GPS-BASED NAVIGATION AIDS
FOR THE VISUALLY IMPAIRED

The idea of using GPS to assist with navigation by the visually impaired goes back over a decade (Collins, 1985; Loomis, 1985). The first evaluation of GPS for this purpose was carried out by Strauss and his colleagues (Brusnighan, Strauss, Floyd, & Wheeler, 1989). Because their research was conducted during early deployment of GPS, the poor positioning accuracy available to them precluded practical studies with visually impaired subjects.

There are now a number of research and commercial endeavors around the world utilizing GPS or DGPS for determining the position of a visually impaired traveler. One commercial GPS-based system has been under development by Arkenstone of Sunnyvale, California for several years (Fruchterman, 1996). It makes use of detailed digital street maps covering the United States as well as selected locations in other countries. A synthetic speech display provides both information about the locations of nearby streets and points of interest and instructions for traveling to desired destinations. The system uses neither a compass nor differential correction for GPS localization and, thus, affords only approximate information for orientation and route guidance.

A similar research and development effort is the Mobility of Blind and Elderly People Interacting with Computers (MoBIC) project that has been conducted by a UK/Swedish/German consortium (Petrie, Johnson, Strothotte, Raab, Fritz, & Michel, 1996). The MoBIC Outdoor System (MoODS) is similar to the Arkenstone system but includes differential correction and a compass worn on the body for heading information.

Another GPS-based system is that being developed as part of a research project in Japan by Makino and his colleagues (Makino, Ishii, and

Nakashizuka, 1996; Makino, Ogata, & Ishii, 1992). A distinguishing feature of this system is its use of a digital mobile phone for communication between the traveler and the computer that contains the spatial database. The mobile phone transmits the traveler's GPS coordinates to the computer at a central facility, which in turn outputs synthetic speech, which is transmitted back to the traveler, providing information on his/her position. Use of a mobile phone link has the advantages of minimizing the cost, weight, and computing power of the unit carried by the traveler and of simplifying the updating of the spatial database. A related system is the "Electronic Guide Dog" project in Europe (Talkenberg, 1996). It too uses a mobile phone link between the traveler and a central facility, but, in contrast to the Makino design, uses a human agent at the central facility, who communicates by voice to give the traveler positional and other information.

THE PERSONAL GUIDANCE SYSTEM

The system our group has developed, the Personal Guidance System, is being used as a research test bed (Golledge, Klatzky, Loomis, Speigle, & Tietz, 1998; Loomis, Golledge, Klatzky, Speigle, & Tietz, 1994; Loomis, Golledge, & Klatzky, 1998). Our long-term goal is to contribute to the development of a portable, self-contained system that will allow visually impaired individuals to travel through familiar and unfamiliar environments without the assistance of guides. The basic conception that has guided us from the beginning (Loomis, 1985) relies on a virtual acoustic display as part of the user interface. A virtual acoustic display takes a monaural audio signal (e.g., speech or environmental sound) and transforms it into a binaural signal delivered by earphones, the result being a sound that appears to emanate from a given environmental location (Begault, 1994; Carlile, 1996; Gilkey & Anderson, 1997; Loomis, Hebert, & Cicinelli, 1990; Wenzel, 1992; Wightman & Kistler, 1989). In our conception, as the visually impaired person moves through the environment, he/she would hear the names of buildings, street intersections, etc. spoken by a speech synthesizer, coming from the appropriate locations in auditory space (Figure 13.2), as if they were emanating from loudspeakers at those locations, in analogy with the Talking Signs system. Besides leading the visually impaired person along a desired route, the system would hopefully allow the person to develop a much better representation of the environment than has been the case so far. Our system is not intended to provide the visually impaired person with detailed information

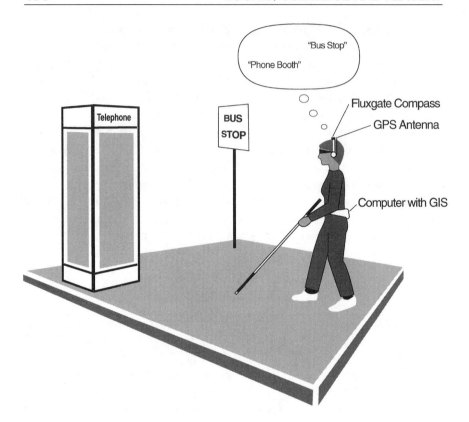

FIG. 13.2. Artist's conception of a future implementation of the Personal Guidance System. The GPS receiver locates the traveler with respect to the surrounding environment, represented in a spatial database that is part of the Geographic Information System within the computer. The fluxgate compass provides the computer with orientation of the traveler's head. Wearing earphones, the traveler hears spatialized virtual sound spoken by a speech synthesizer, with the spoken labels of entities appearing to come from their locations in the environment. (Adapted from Figure 1 from Loomis, Golledge, Klatzky, Speigle, and Tietz [1994]. © 1994 Association for Computing Machinery, Inc. Reprinted by permission.)

about the most immediate environment (e.g., obstacles); thus, the blind traveler will still have to rely on the long cane, seeing-eye dog, or ultrasonic sensing devices for this information. The current implementation of our system weighs 11 kg and is carried in a backpack worn by the user (Figure 13.3), but the version being developed will be truly wearable.

FIG. 13.3. Current implementation of the Personal Guidance System being worn by author Reginald Golledge. Miniaturization will eventually result in a wearable system like that depicted in Figure 13.2.

The first module of our system (Figure 13.4) determines the position and orientation of the traveler. For positioning, we have used a number of DGPS configurations. The configuration used in the experiments mentioned below consisted of a Trimble 12-channel GPS receiver with differential correction from a base station located 20 km away (Accqpoint Wide Area DGPS service). DGPS fixes, with absolute errors of about 1 m and much better relative accuracy, were provided at the rate of 0.67 Hz. Although GPS can indicate the traveler's course (direction of travel over the ground) on the basis of successive position fixes, navigation systems are more effective when heading is independently available, for travel instructions are usually expressed relative to the traveler's heading rather than course, and course is not defined for a stationary traveler. For the sensing of heading (of either the head or of the body), we have used a fluxgate compass attached either

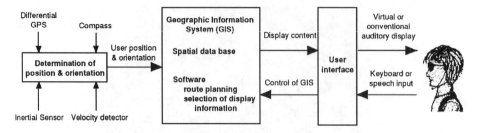

FIG. 13.4. Functional components of a GPS-based navigation system for the visually impaired. (From Loomis, Golledge, Klatzky, Speigle, and Tietz [1994]. ©1994 Association for Computing Machinery, Inc. Reprinted by permission.)

to the strap of the earphones worn on the head or to the backpack carrying the rest of the equipment.

The second module of our system (Figure 13.4) is the subnotebook computer containing the Geographic Information System (GIS), which comprises the environmental database and system software. Our test site is the University of California, Santa Barbara campus, for which we have developed a spatial database containing all buildings, roads, walkways, bikeways, trees, and other details. Our main development efforts have gone into creating a reliable system, developing the database of campus and developing the GIS software that provides the traveler with the desired functionality (Golledge, Loomis, Klatzky, Flury, & Yang, 1991; Golledge, Klatzky, Speigle, Loomis, & Tietz, 1998).

The third module of our system (Figure 13.4) is the user interface. For user input, we currently use a 24-button keypad, but in the version being developed we will use a lapel microphone and speech-recognition software. Almost all of our development effort has gone into the display component. Our approach of using virtual sound contrasts with all other projects on GPS-based navigation systems for the visually impaired, for these others use conventional synthesized speech to convey information to the traveler. Research and user preferences will ultimately determine which approach is better.

The system provides information for route guidance as well as about the spatial disposition of important off-route entities (e.g., buildings, well-known landmarks). Route guidance information is provided by a succession of virtual beacons placed at significant choice points (waypoints) along that path. The virtual beacons are activated in sequence as a path is followed, and each beacon in turn becomes more intense as the traveler approaches it. Homing occurs by turning one's head until the sound source appears directly in front of the head and then orienting the body in that direction and walking toward the source. Additional information can be provided about off-route entities, also using virtual sound. Here, the traveler is informed about some subset of the surrounding environmental entities by having their names spoken by a speech synthesizer and then rendered as spatialized sound by the virtual display. The traveler can activate this additional layer of information whenever it is desired.

We have conducted a number of informal demonstrations of the system at our test site, the UCSB campus. With these demonstrations, we have shown the capability of the system to guide an unsighted person with normal binaural hearing to some specified destination using a sequence of virtual beacons; under conditions of good satellite availability, the DGPS component functions well enough to keep the traveler within sidewalks about 5 m in width.

RESEARCH ON AUDITORY DISPLAY MODES

The formal research we have done with the current system has been concerned with comparing the effectiveness of spatialized speech from a virtual acoustic display with nonspatialized speech that conveyed the spatial information in words. The first of two experiments we have conducted was concerned with route guidance (Loomis, Golledge, & Klatzky, 1998). Our primary interest was in determining whether spatialized speech resulted in better or worse route following performance than verbal guidance commands provided by a synthetic speech display. Of secondary interest was a comparison of guidance with and without heading information, as provided by the fluxgate compass.

In the experiment, the subject was led along one of four paths, each comprising 9 linear segments defined by 10 waypoints (specified by their DGPS coordinates). These were situated within a large open grassy field on campus. Each path was 71 m long. We evaluated four display modes in the experiment, three involving a conventional speech display and the fourth involving spatialized (virtual) sound (Figure 13.5). Auditory guidance information was given at two intermittency rates: fast (once every 1.5 s) or slow (once every 5.0 s).

In the Virtual mode, the fluxgate compass was mounted on the earphone strap and thus provided the heading of the person's head. The navigation system computer constantly updated the distance and relative bearing of

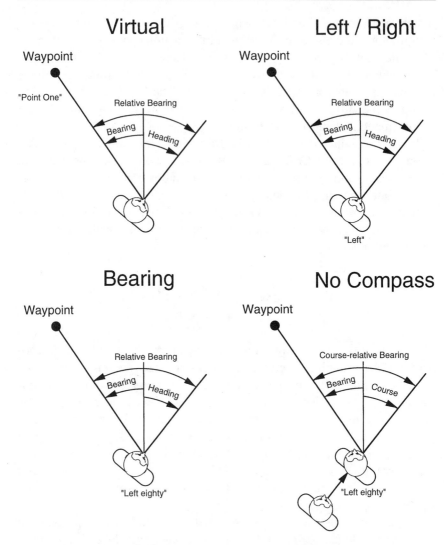

FIG. 13.5. Auditory display modes used in the experiment by Loomis, Golledge, and Klatzky (1998). See text for details. (© MIT Press. Reprinted by permission.)

the waypoint (here, Point 1) with respect to the subject's head. By turning his/her head in its direction, the subject could center the perceived sound within the median plane and then walk in that direction. As the subject approached the computer-defined waypoint, the sound level of the utterance increased appropriately. When the subject approached within 1.5 m of the waypoint, the computer took this as being at the waypoint and then activated the next waypoint in sequence. In the Left/Right mode, the fluxgate compass was mounted on the backpack worn by the subject and indicated heading of the subject's torso. The speech synthesizer provided information about the bearing of the next waypoint ("left"/"straight"/"right") relative to the subject's heading. The Bearing mode was like the Left/Right mode except that more information about relative bearing was provided. Here, the relative bearing between the body and the next waypoint, rounded to the nearest 10 deg, was spoken (e.g., "left 80"). Finally, the No Compass mode was like the Bearing mode, in that the subject received the same type of verbal command from the computer (e.g., "left 80"). However, the bearing of the next waypoint relative to the subject's course (based on two successive DGPS fixes) was spoken by the synthesizer. If the subject stopped moving, however, course was not defined, and the computer stopped issuing commands.

The two performance measures (time to complete the route and travel distance) as well as the subjective ratings by the ten blind subjects showed a slight superiority of the Virtual mode over the next best mode, Bearing (time to completion is shown in Figure 13.6). The No Compass mode was decidedly worst, both in terms of performance and ratings. The experimental findings show the importance of using a compass to provide heading information for route guidance and the potential benefit of using spatialized virtual sound over conventional speech information.

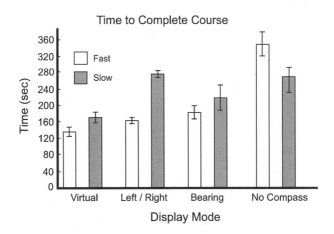

FIG. 13.6. Results of the experiment on guidance using different auditory display modes. Time to finish walking the 71 m path is given as a function of display mode and rate at which information was given to the subject (once every 1.5 s or every 5.0 s). (From Loomis, Golledge, and Klatzky [1998]. © MIT Press. Reprinted by permission.)

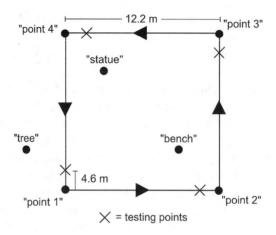

FIG. 13.7. One of the spatial configurations of landmarks along with the walking route used in the experiment on spatial learning. Subjects walked around the square in a counterclockwise direction. They were led to each waypoint (vertex) by means of spatialized speech indicating the next waypoint. Information about the locations of three off-route landmarks using either spatialized virtual speech (e.g., "statue") or nonspatialized speech (e.g., "statue, 4 o'clock") was presented on each leg of the square. After four traverses of the square during the learning phase, the subject was tested at each of the four marked locations.

The second experiment we have conducted on auditory display modes was concerned with the learning of spatial layout. In the training phase, the subject was guided five times around a square (12.2 m on a side) using virtual beacons located at the four vertices (Figure 13.7). Along each side of the square, the subject received information about the location of each of three off-route landmarks, identified by names (e.g., "statue") spoken by a synthesizer. In the Virtual mode, subjects heard the name as spatialized speech from the virtual display, as in the preceding experiment. In the Bearing mode, the subjects heard nonspatialized speech giving the approximate relative bearing to each landmark in terms of a clockface (e.g., "statue, 3 o'clock"). Two different spatial configurations of landmarks were used, with proper counterbalancing of their assignment to the two conditions across subjects. Spatial learning was assessed using both tactual sketch maps and direction estimates. The latter were obtained during a sixth traverse of the square following the training phase. No information about the landmarks was provided during this sixth traverse. Along each leg, the subject was instructed to stop at a testing location (indicated by the "X" in Figure 13.7) and there the subject was given the name of each landmark

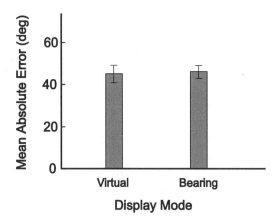

FIG. 13.8. Results of the spatial learning experiment.

(without directional information) and responded with its relative bearing (stated in terms of a clockface, such as "ten o'clock" or "eleven thirty"). Absolute error was the primary performance measure; it is the absolute value of the difference between the relative bearing of the target and the subject's estimate.

Nine blind subjects performed in both conditions of the experiment (with counterbalancing of order and landmark configurations). Even with outlying data excluded, mean absolute error for the direction estimates was large but considerably better than chance performance (90 deg), as shown in Figure 13.8. There were no statistically reliable differences between the two modes. Sketch map performance was similarly unimpressive, with there being no differences between modes. Although performance was disappointing, a brief spatial learning experiment such as this surely does not indicate how well visually impaired travelers could perform following extensive practice with such a system. A visually impaired person using such a system every day while walking to and from work could eventually learn a great deal about the surrounding environment, even if only a small amount of landmark information were acquired each day.

The two experiments together indicate that spatialized speech from a virtual acoustic display has promise as part of the user interface of a blind navigation system. Performance was at least as good as with the other display modes investigated, and spatialized speech has the additional advantage of consuming less time than conventional speech, for the latter must include the spatial information as part of the utterance. On the other hand, there are two difficulties currently associated with the use of spatialized sound. First, present earphone designs attenuate or distort some of the environmental information that is important to visually impaired travelers. This difficulty might be ameliorated, however, by using small transducers mounted a few centimeters from the ears. Second, realistic virtual sound that appears to come from moderate to large distances has been difficult to achieve so far. However, given that the difficulty lies more with current implementations of virtual sound than with the use of earphones per se (Loomis, Klatzky, & Golledge, 1999), it is probably just a matter of time before more effective algorithms for realistic virtual sound are developed.

We should mention, of course, that spatialized speech and nonspatialized speech do not exhaust the possibilities for the display of spatial information. Another promising display technique is based on the Talking Signs system for remote signage. As mentioned earlier, directional localization of the infrared transmitter is obtained by aiming the hand-held receiver to obtain maximum signal strength. The haptic information from the hand, wrist, and arm when the signal is maximal is apparently quite effective in allowing the user to perceive the direction of the transmitter (Bentzen & Mitchell, 1995). Thus, a navigation system interface that uses haptic information in a similar fashion ought to be similarly effective. Such an interface could be readily implemented in the form of a hand-held unit comprising a loudspeaker and an electronic compass. Rotation of the hand would result in a change in the compass signal. Software could then mimic the operation of the Talking Sign receiver so that audible speech from the speech synthesizer is produced only when the hand is pointed in the approximate direction of the virtual beacon or off-route landmark.

PROSPECTS FOR THE FUTURE

In view of the ever improving accuracy of GPS receivers, increasing coverage of differential correction, decreasing size and cost of electronics, increasing sophistication of GIS software, and growing availability of digital maps suitable for pedestrian travel, the prospects are excellent that truly wearable GPS-based navigation systems will someday be used by both the visually impaired and sighted populations (in connection with the latter, see Feiner, MacIntyre, Hollerer, & Webster, 1997). Surely, obstacles remain, such as the development of low-cost alternatives to GPS when GPS coverage is lacking, creation and maintenance of digital maps appropriate to blind travel, fabrication of reliable, affordable, and lightweight systems for all-weather operation, and coping with the inevitable liability issues.

However, because these are not insurmountable obstacles, we are confident that it is just a matter of time before navigation systems using GPS and local positioning technology (e.g., Talking Signs) will be routinely guiding visually impaired travelers through outdoor and indoor environments. Hopefully these navigation systems will provide the visually impaired with much more functionality than simple route guidance. As rich databases for town and cities are developed for the larger population, databases that inform the traveler about nearby restaurants, businesses, etc., there is every reason to expect that the visually impaired population will eventually have access as well. Moreover, we are hopeful that navigation systems with this information will allow the visually impaired to gradually develop more extensive and more coherent mental representations of the environment than they currently have.

Acknowledgments   Our reseach and development efforts have been supported by National Eye Institute grants EY07022 and EY09740. The authors thank Jerome Tietz, Jon Speigle, and Chick Hebert for technical support and Mike Provenza for his assistance in conducting the two experiments described.

REFERENCES

Begault, D. R. (1994). *3-D Sound for Virtual Reality and Multimedia*. New York: AP Professional.

Bentzen, B. L. & Mitchell, P. A. (1995). Audible signage as a wayfinding aid: Verbal Landmark versus Talking Signs. *Journal of Visual Impairment and Blindness*, 89, 494–505.

Brabyn, J. A. (1985). A review of mobility aids and means of assessment. In D. H. Warren & E. R. Strelow (Eds.), *Electronic Spatial Sensing for the Blind* (pp. 13–27). Boston: Martinus Nijhoff.

Brusnighan, D. A., Strauss, M. G., Floyd, J. M., & Wheeler, B. C. (1989). Orientation aid implementing the Global Positioning System. In S. Buus (Ed.), *Proceedings of the Fifteenth Annual Northeast Bioengineering Conference* (pp. 33–34). Boston: IEEE.

Carlile, S. (1996). *Virtual Auditory Space: Generation and Applications*. New York: Chapman & Hall.

Collins, C. C. (1985). On mobility aids for the blind. In D. H. Warren & E. R. Strelow (Eds.), *Electronic Spatial Sensing for the Blind* (pp. 35–64). Dordrecht: Martinus Nijhoff.

Crandall, W., Gerrey, W., & Alden, A. (1993). Remote signage and its implications to print-handicapped travelers. *Proceedings: Rehabilitation Engineering Society of North America (RESNA) Annual Conference*, Las Vegas, June 12–17, 1993, pp. 251–253.

Farmer, L. W. (1980). Mobility devices. In Welsh, R. & Blasch, R. (Eds), *Foundations of Orientation and Mobility*. New York: American Foundation for the Blind.

Feiner, S., MacIntyre, B., Hollerer, T. & Webster, A. (1997). A touring machine: Prototyping 3D mobile augmented reality systems for exploring the urban environment. *Proceedings of the International Symposium on Wearable Computing* (pp. 74–81), Cambridge, MA, October 13–14, 1997.

Fruchterman, J. (1996). Talking maps and GPS systems. Paper presented at The Rank Prize Funds Symposium on Technology to Assist the Blind and Visually Impaired, Grasmere, Cumbria, England, March 25–28, 1996.

Gilkey, R. & Anderson, T. R. (1997). *Binaural and Spatial Hearing in Real and Virtual Environments.* Hillsdale, NJ: Lawrence Erlbaum Associates.

Golledge, R. G., Klatzky, R. L., & Loomis, J. M. (1996). Cognitive mapping and wayfinding by adults without vision. In J. Portugali (Ed.), *The Construction of Cognitive Maps* (pp. 215–246). Dordrecht, The Netherlands: Kluwer Academic.

Golledge, R. G., Klatzky, R. L., Loomis, J. M., Speigle, J., & Tietz, J. (1998). A Geographical Information System for a GPS based Personal Guidance System. *International Journal of Geographical Information Science, 12,* 727–749.

Golledge, R. G., Loomis, J. M., Klatzky, R. L., Flury, A., & Yang, X. L. (1991). Designing a personal guidance system to aid navigation without sight: Progress on the GIS component. *International Journal of Geographic Information Systems, 5,* 373–395.

Loomis, J. M. (1985). Digital map and navigation system for the visually impaired. Unpublished manuscript, Department of Psychology, University of California, Santa Barbara.

Loomis, J. M., Golledge, R. G., & Klatzky, R. L. (1998). Navigation system for the blind: Auditory display modes and guidance. *Presence: Teleoperators and Virtual Environments, 7,* 193–203.

Loomis, J. M., Golledge, R. G., Klatzky, R. L., Speigle, J., & Tietz, J. (1994). Personal guidance system for the visually impaired. *Proceedings of the First Annual International ACM/SIGCAPH Conference on Assistive Technologies* (pp. 85–90), Marina Del Rey, California, October 31–November 1, 1994. New York: Association for Computer Machinery.

Loomis, J. M., Hebert, C., & Cicinelli, J.G. (1990). Active localization of virtual sounds. *Journal of the Acoustical Society of America, 88,* 1757–1764.

Loomis, J. M., Klatzky, R. L., & Golledge, R. G. (1999). Auditory distance perception in real, virtual, and mixed environments. In Y. Ohta & H. Tamura (Eds.), Mixed Reality: Merging Real and Virtual Worlds (pp. 201–214). Tokyko: Ohmsha.

Loughborough, W. (1979). Talking lights. *Journal of Visual Impairment and Blindness, 73,* 243.

Makino, H., Ishii, I., & Nakashizuka, M. (1996). Development of navigation system for the blind using GPS and mobile phone connection. *Proceedings of the 18th Annual Meeting of the IEEE EMBS,* Amsterdam, The Netherlands, October 31–November 3, 1996.

Makino, H., Ogata, M., & Ishii, I. (1992). Basic study for a portable location information system for the blind using a Global Positioning System. *MBE92-7, Technical report of the IEICE,* 41–46 (in Japanese).

Ogata T., Makino H., Ishii I., & Nakashizuka M. (1997). Location guidance system for the visually impaired using an invisible bar code. *Transactions of IEICE,* J80/D-2, 3101–3107 (in Japanese).

Parkinson, B. W., & Spilker, J. J., Jr (1996). *The Global Positioning System : Theory and Applications.* Washington, DC : American Institute of Aeronautics and Astronautics.

Petrie, H., Johnson, V., Strothotte, T., Raab, A., Fritz, S., & Michel, R. (1996). MoBIC: designing a travel aid for blind and elderly people. *Journal of Navigation, 49,* 45–52.

Rieser, J. J., Guth, D. A., & Hill, E. W. (1982). Mental processes mediating independent travel: Implications for orientation and mobility. *Journal of Visual Impairment and Blindness, 76,* 213–218.

Strelow, E. R. (1985). What is needed for a theory of mobility: Direct perception and cognitive maps—lessons from the blind. *Psychological Review, 92,* 226–248.

Talkenberg, H. (1996). Electronic Guide Dog—A technical approach on in-town navigation. Paper presented at The Rank Prize Funds Symposium on Technology to Assist the Blind and Visually Impaired, Grasmere, Cumbria, England, March 25–28, 1996.

Welsh, R. & Blasch, R. (1980). *Foundations of Orientation and Mobility.* New York: American Foundation for the Blind.

Wenzel, E. M. (1992). Localization in virtual acoustic displays. *Presence: Teleoperators and Virtual Environments, 1,* 80–107.

Wightman, F. L. & Kistler, D. J. (1989). Headphone simulation of free-field listening. II: Psychophysical validation. *Journal of the Acoustical Society of America, 85,* 868–878.

14

Boeing's Wire Bundle Assembly Project

David Mizell
The Boeing Company

INTRODUCTION

This chapter is a history, not a standard scientific report. It tells the story of an industrial research and development project in Augmented Reality (AR)—the Boeing project aimed at using AR to guide the assembly of electrical wire bundles. It describes the goal of the project, the technical approaches we chose, some of the problems and surprises we encountered along the way, and some of the lessons we learned in trying to bring a new computing technology into the world's largest factory.

INCEPTION

Boeing's project on using Augmented Reality for wire bundle assembly was started on January 24, 1990. That was the day that Tom Caudell and I visited Boeing's engineering and manufacturing facility in Everett, Washington. Our objective was to learn what applications there might be for virtual reality technology in commercial aircraft design and manufacturing. We had both only been with the company a few months, we were both interested in VR, and since neither of us knew a lot about aircraft design and manufacturing, we wanted to hear from design and manufacturing engineers how this technology might help them. We met Karl Embacher, who had just finished a stint as manager of Everett's wire shop. His question to us was, "What can you do about 'formboard?'"

Karl explained what he meant. The electrical wiring that goes on board an aircraft is assembled into bundles, or harnesses, groups of wires bound and sleeved together. The wire bundles are assembled, before installation on the aircraft, on one or more 3′ × 8′ easel-like boards, called formboards. Each board has a sheet of plotter paper glued to its front surface, which contains a full-scale schematic diagram of the wire bundle that is assembled on the board (see Figure 14.1). Pegs are mounted on the board to hold the bundle in place as it is assembled. Workers refer to the diagram on the board, as

FIG. 14.1.   A typical wire bundle assembly formboard in Boeing's Everett, WA factory.

well as to a separate set of $8.5'' \times 11''$ printout, as they route, sleeve, and tie off the bundle. This approach makes every formboard unique to the bundle that is assembled on it, implying significant construction, storage and other costs for the wire shop.

We told Karl we'd think about it, and we started to leave. I could not imagine how virtual reality (VR) could help with an assembly job in the factory. Before I could say so to Tom, he said, "Hey, you know what we could do—we could use a see-through head-mounted display with a head tracker, and mathematically project the bundle diagram on a blank formboard." We turned around and told Karl the idea. He immediately saw potential savings in this approach and thought that the technology would have many other manufacturing applications, as well. Many steps of aircraft manufacturing and assembly are done by hand. They may be too complex to automate, as is the case with wire bundle assembly, or may not be done often enough to cost-justify automation. Tom's idea had the potential of giving aerospace workers who had to perform these manual tasks better information than they had ever had before, superimposing diagrams or textual information on the surface of the workpiece exactly where it was needed. Over the next couple of weeks, Tom and I fleshed out his concept. We perceived that a system that could be applied to wire bundle assembly needed three components:

- *A see-through, head-mounted display.* Front projection, rear projection or large, fixed beam splitter panels wouldn't work. The scale is too large; some of the bundles are assembled on eight or more formboards butted together.
- *A head position/orientation tracker.* Commercially available systems weren't adequate: They didn't have the range, the accuracy, or the ability to operate untethered, which the long bundles would require.
- *A wearable computer.* Again, at least in the case of the longer, multiple-board bundles, tethering the worker to a fixed base was not feasible.

A new R&D idea that is hatched inside a large corporation faces the same problem as a new startup company. The first thing you have to do is get the money. Unless you carefully time your new inspiration to arrive just before the annual budget planning cycle, all the R&D budget is already allocated. Tom and I spent the next six months briefing potential Boeing customers on the "see-through virtual reality" concept (as we were calling it back then), who would invariably show curiousity, interest, and empty pockets, in that order. We learned a lot about our company during the process but didn't find any budget.

Finally, some reorganizations left us with a third-level manager who was sympathetic. He gave us enough budget to develop a "proof-of-concept" demo. We started out in the now-traditional manner of AR and wearable computer researchers without adequate budget: We bought a $500 Private Eye from Reflection Technologies, Inc. to use as a head-mounted display. The Private Eye is not a see-through display, so our proof-of-concept demo required the user to look at the workpiece with his/her left eye, the display with the right eye, and do the superimposition in his/her own brain. We borrowed a Polhemus from a VR project and scrounged an old 386 PC somewhere. Tom built a miniature formboard and mounted the Polhemus transmitter on it. He made some wooden pegs that fit into the holes in the pegboard that he used to build the little formboard. Mike McAlister and I wrote the software for the demonstration system. The

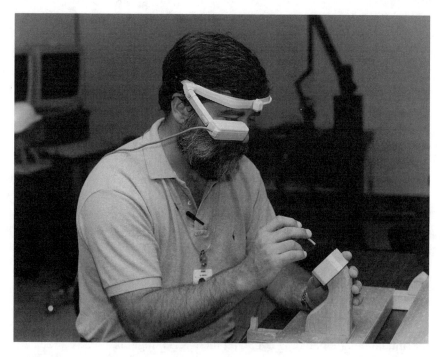

FIG. 14.2. David Mizell using the 1990 Augmented Reality proof-of-concept demonstrator.

user could route three or four wires on the board, one wire at a time, following a line that the system displayed. Our demonstrator is shown in Figure 14.2.

THE EARLY YEARS

The demo convinced enough people of the potential of the technology that funding over the next couple of years was fairly ample. In that time, we accomplished several things:

- Tom built a new head-mounted display that was truly see-through. In it, the Private Eye was mounted above the eye and faced downwards toward a piece of dichroic glass, which served as a beamsplitter. (See Figure 14.3.)
- Jeff Heisserman built another HMD with the same optical approach, mounted on a standard construction "hard hat." (See Figure 14.4.)

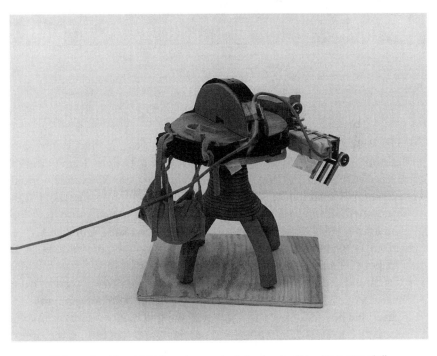

FIG. 14.3. The 1991 head-mounted display built by Tom Caudell, using a Private Eye and a dichroic glass beamsplitter.

FIG. 14.4. Henry Sowizral using the miniature, demonstration wire bundle formboard with Jeff Heisserman's design for an AR head-mounted display built into a hard hat. This design also used a Private Eye image source and a dichroic glass beamsplitter.

- In 1992, we contracted out the prototyping of our first stereo see-through display. The optics were designed and built by John Ferrer of Vis-O-Displays and the driver electronics by Dean Kocian of Virtual Systems Consultants. This system used two high-brightness CRTs mounted on top of a bicycle helmet, with relay optics directing the images toward a polycarbonate beam-splitter. (See Figure 14.5.)
- Tom coined the term "Augmented Reality," for our 1992 HICCS paper (Caudell and Mizell, 1992).
- Tom and Adam Janin worked out an approach for registering the user's eyes into the coordinate system of the HMD by having the user rest his/her chin on a wooden frame that had two video cameras attached to it. Adam's image processing code would find the user's pupils in each camera image and triangulate to compute their exact positions relative to the HMD. (See Figure 14.6.)

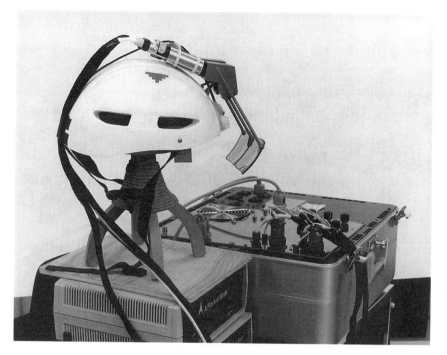

FIG. 14.5. The stereo, monochrome VGA head-mounted display delivered to Boeing in 1993. This laboratory prototype supported much of our early software development.

- Adam, Karel Zikan, Dan Curtis, and Henry Sowizral (1995) worked out a videometric tracking approach. A video camera mounted on the HMD looked at an array of LEDs mounted above the formboard. A fast transportation algorithm code matched the imaged LEDs with their previous position, in order to compute the current position of the HMD.
- Adam designed and implemented an optimization-based approach to registering the user, the tracker, and the display into the coordinate system of the workpiece (Janin, Mizell, and Caudell, 1993). The user was directed by Adam's software to repeatedly superimpose a pair of crosshairs drawn in the display on each of a set of known locations on the workpiece. Collecting enough such data enabled Adam's code to compute a least-squares approximation to the values of each of the 19 calibration parameters of a monocular see-through system.

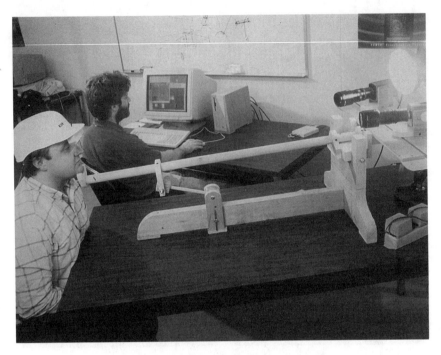

FIG. 14.6. Adam Janin uses his software and the wooden boom with video cameras built by Tom Caudell to calibrate Henry Sowizral. This system could fairly precisely determine the position of the user's pupil relative to the head-mounted display.

THE DARPA TRP PROJECT

As useful as this work was, however, we felt that we were about to "hit a wall." We were not in a position to develop any hardware. Even if we could build hardware prototypes, it was irrelevant to a company of Boeing's size. Several hundred people assemble wire bundles for Boeing. For AR to be deployed on that kind of scale, a well-financed, motivated vendor of this technology had to be found or cultivated. DARPA's Technology Reinvestment Project (TRP) came along at just the right time. Partnered with Honeywell Military Avionics Division of Minneapolis and Virtual Vision, Inc. of Redmond, Washington, two companies in the business of productizing trackers and head-mounted displays, and another research organization, Carnegie Mellon University's Engineering Design Research Center, we proposed to design, implement, and test under realistic conditions, in military and commercial applications, two generations of wearable computer systems and Augmented Reality systems.

FIG. 14.7. Adam Janin assembling a wire bundle using the 1995 Honeywell AR System prototype.

We actually stretched the TRP project out to three years and developed and tested three generations of wearable computers and AR systems. The prototypes we developed and other components we used over this period illustrate how rapidly this technology is developing. In 1995, the first year of the project, for example, our wearable PC was a one-of-a-kind system handbuilt by Honeywell and carried in a backpack (see Figure 14.7). By 1997, powerful, compact, and reliable wearable computers were available off-the-shelf for use in our system prototypes.

Along with this homemade wearable PC, our 1995 AR system prototype used a commercially available magnetometer tracker from Ascension, Inc. and a see-through HMD based on Honeywell's Combat Vehicle Crewman display and optics. We had implemented a videometric tracker based on the Zikan/Curtis/Sowizral/Janin (1995) algorithm, but the video processing hardware we chose to use in the 1995 prototype was too slow, so we were forced to stay within the work volume limits of the tethered magnetometer tracking system. This was adequate for one-board wire bundle assembly tasks, however, and we were able to demonstrate some promise

FIG. 14.8. David Mizell demonstrating the 1996 prototype system from TriSen, Inc. with its proprietary videometric tracker.

for productivity improvements using the AR system compared to the traditional formboard assembly method. It turns out that there are enough people assembling wire bundles in Boeing's wire shops that a fairly small productivity improvement is more significant in terms of financial savings than all the savings incurred by no longer having to build and store the bundle-unique formboards.

In 1996, Honeywell subcontracted the development of a new videometric tracker system (see Figure 14.8) to TriSen, a new Minneapolis company founded by former Honeywell employees. This second generation system's other main components included a Via wearable computer and a Honeywell-designed HMD, which incorporated an electroluminescent flat panel display from Planar. The tracker utilized a video camera mounted above the display optics of the HMD. It detected white dots painted on a black aluminum formboard. The dots were arranged in a special pattern designed by TriSen, so that if the camera saw a small subset of the dots, the tracker code could identify which ones they were, and compute the user's position and orientation relative to them.

In 1997, we had two vendors providing prototype AR systems. TriSen implemented a second generation of the videometric tracker with higher performance and accuracy. Their system used a monocular, monochrome VGA see-through head-mounted display of their own design and a wearable computer from Via, Inc. of Northfield, Minnesota. It had a 133-MHz 586-class processor. Virtual Vision implemented an AR system using their V-Cap 1000 see-through HMD, which was also monocular, monochrome VGA, and an acoustic/inertial hybrid tracker prototyped under subcontract by InterSense, Inc. of Cambridge, Massachusets. The Virtual Vision system (see Figure 14.9) was able to use either a Via wearable computer or the similar-performance Trekker model from Rockwell Electronics.

The InterSense tracker combined an acoustic system with InterSense's VR360 inertial tracker. A row of infrared LEDs on the brim of the V-Cap strobed to synchronize the acoustic system. This was detected by battery-operated speakers mounted on the formboard, each of which contained a photosensor. In response to the LED strobe, each sensor would emit an

FIG. 14.9. Dan Curtis assembles a wire bundle using Virtual Vision's 1997 prototype AR system with InterSense's acoustic/inertial hybrid tracker.

ultrasonic beep at its identifying frequency. Three microphones mounted on the HMD would receive the ultrasonic signals, allowing the tracker software to compute a time-of-flight measurement from each responding speaker to each receiving microphone. The inertial system would keep track of position and orientation between acoustic measurements. This first-generation system was not as accurate as TriSen's second-generation videometric system. We expect further development and tuning to improve the hybrid system's performance.

THE 1997 EVERETT PILOT PROJECT

In the summer of 1997, we conducted a system validation pilot in Boeing's Everett factory. Our goal was to determine whether or not AR had been developed fully enough to be deployed in the wire shop for production use. TriSen supplied us with three AR systems based on their second-generation videometric tracker, and Virtual Vision provided two of their 1997 proto-types equipped with the InterSense acoustic/inertial tracker. Unfortunately, the InterSense tracker was not working accurately enough at that time to use in the pilot project, so only the TriSen systems were used.

Several workers with wire shop experience volunteered for the pilot. We selected two with several years' experience and two with little or no experience assembling bundles. Three bundles were selected to be used in the pilot, one simple, one-board bundle, a one-board bundle of medium complexity, and a complex (several hundred wires) bundle that took up two boards. Each worker assembled each bundle three times, for a total of 36 trials. Half of the trials used AR and half used the traditional formboard method (see Figure 14.10). Bill Fortney, a Boeing statistician, assisted us with the experiment design. Bundles built during the trial were not production bundles. After each was assembled, we took it apart again and re-kitted it to be used in the succeeding trial.

Not only did we need to obtain adequately performing hardware proto-types from the vendors to carry out the pilot project, but there were two significant software developments as well. First, the data to be displayed in the AR system had to be translated into a representation suitable for display in this manner. For the most part, the geometric information drawn on the traditional formboards was stored in CATIA data sets for each bun-dle. The textual information about each wire and each wire group is stored in an ancient, Cobol-based database. Peter Gruenbaum wrote a translator to extract information from both of these systems and combine them in a meaningful way in the AR display.

(a)

(b)

FIG. 14.10. (a) Ripp Brammer of Boeing's Everett factory assembling a wire bundle using the traditional formboard, during the 1997 pilot project. (b) Sam Couch using the 1997 TriSen AR prototype during the 1997 pilot project.

Second, while the paper diagram glued to the traditional formboard shows the worker all the geometric information at once, we weren't going to do that with the AR system. We wanted to show the worker diagrammatic information for only those wires that he/she was currently routing. There were three reasons for doing this:

1. The formboard diagram can be ambiguous. All the worker is given are the beginning and ending points of each wire. It is up to him or her to figure out which possible path along the lines on the board is the correct one for that wire. Labor expenses are critical; thus trial-and-error processes are undesirable.

2. Wire shop management has been pushing for some time for what they call a "preferred process" for each wire bundle. Using the traditional method, each worker is free to assemble the bundle in whatever sequence he or she chooses. This tends to produce bundles that are inconsistent in terms of thickness and bendability. This is considered undesirable, from a maintenance point of view, by our airline customers. Management's goal was to figure out what would be the best assembly sequence for each bundle, and then use some means to require the worker to follow that sequence. To us, AR was the most natural method of making a worker follow a specified sequence. At any step in the process, we would only show the worker the diagrams and text relevant to that step.

3. The goal of stepping the worker through a "preferred process" implied that we had to create this desired sequence of assembly steps. All that previously existed was a set of general guidelines that the wire shop experts had listed. Dan Curtis wrote software that took the output of Peter Gruenbaum's translator and generated a sequence of diagram/text frames to be displayed by the AR system, which corresponded to a reasonable series of assembly steps for the worker to follow, consistent with the general guidelines the wire shop had produced. Ian Angus prototyped an editor that the wire shop people could use to revise this sequence, to manually take care of exceptional cases.

RESULTS OF THE PILOT PROJECT

We learned several things from this pilot project:

1. We proved that we can build real bundles using AR. Several bundles built using AR during the trials were subsequently removed from the

AR formboard and put on the corresponding traditional formboard. They were found to fit extremely well, and to an extent sufficient to pass a QA inspection.

2.   On the negative side, we didn't show a significant productivity improvement. There are hundreds of people forming bundles every day at Boeing, so even small productivity improvements would translate into large dollar savings. Accordingly, it was an important objective of the project to demonstrate productivity improvements, and we believed (and still believe) that AR has an inherent potential to improve the workers' productivity, because they never have to turn away from the bundle. In the traditional approach, the worker must frequently turn away from the bundle to refer to the 8.5″ × 11″ "shop aid" paperwork included in the bundle kits. Our AR application software included all the relevant information from the shop aid paper in the display.

In the 1997 pilot, the traditional formboard method and the AR approach roughly tied in elapsed bundle-forming time. We blame our user interface. AR represents a new computing paradigm, and it's hard to get the user interface right the first time. Our interface had some characteristics that actually slowed the user down. For example, the system, as used in the trials, forced the worker to search repeatedly through a large handful of wires to find particular wires.

The root cause of the user interface–induced inefficiencies, in my opinion, was that the bandwidth from the user to the computer in our pilot system implementation was too narrow. It consisted solely of mouse clicks. One click meant "up" or "yes" or "next," a double click meant "down" or "no" or "back." Because the user interface had such a low bandwidth, our application software had to say to the user, in effect, "Next, I'm going to tell you where to route wire W33-20. Click once after you have found it and routed it." Speech recognition or some automatic ability to read wire ID numbers into the computer, such as barcoded wires, would enable the user to say to the system, in effect, "I've just picked out wire W49-24. Where should I route it?" Rather than repeated searches through the group of wires now being routed, the user could sort the wires in a single pass. We are implementing speech recognition as part of our next-generation user interface.

3.   The TriSen videometric tracking system is very accurate, allowing a high level of precision in the forming process. The Boeing wire shop uses a .25″ accuracy tolerance for the wire bundle assembly process, and the TriSen system was giving us at least that accuracy in the location of the information superimposed on the formboard. The Virtual

Vision/InterSense tracking system is not yet accurate enough for efficient and reliable use in the forming process. It has, however, been significantly improved since original delivery and it is currently not far from being a usable system.

4.   The TriSen system does have problems if the bundle is so dense that the fiducial pattern becomes substantially occluded. We had to eliminate one very complex, 800-wire, single-board bundle from the pilot project because the wires became so thick in a central area of the board that the system would lose track when the former was working in that area (see Figure 14.11). Because it only depended on small speakers mounted above the wiring level on the board, the Virtual Vision/InterSense system was immune to this type of occlusion problem. In fact, Dan Curtis and I, independently of the pilot project, assembled the complex bundle on which

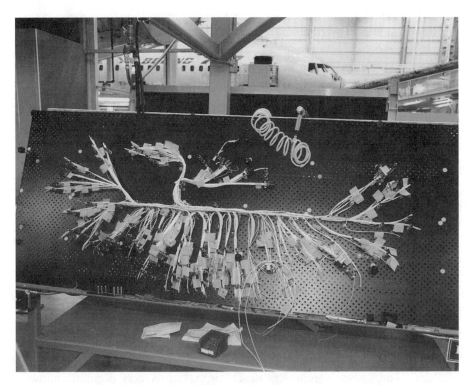

FIG. 14.11. This 800-wire bundle occluded too many fiducial marks in the lower center area of the formboard for the TriSen system to maintain tracking. It was successfully assembled by Dan Curtis and David Mizell using the Virtual Vision/InterSense prototype.

the TriSen tracker had failed, using the Virtual Vision system with the InterSense acoustic/inertial tracker. Bundle density did not cause it to lose track, and we completed the bundle, but the inaccuracy of the tracker forced us to do so much rechecking of our routes that the process took twice as long as the average on the traditional formboard. The InterSense tracker had its own type of occlusion problem, as well, in that the hybrid system could lose track if the user's arm or hand interfered with the line of sight between the microphones on the HMD and the acoustic beacons.

5. The TriSen system is easily calibrated, using software supplied by TriSen. Virtual Vision/InterSense needs to develop a similarly effective calibration method. The currently used method is very tedious and we experienced great difficulty in getting a sufficiently accurate calibration.

6. We found that workers, both experienced and novices, were able to learn to use the AR system very quickly. After an introductory session on the laptop and further training building a simple bundle, they were able to build subsequent bundles on their own. The reason for this is that the AR user interface is designed to lead the worker step-by-step through the build process. In the traditional forming method, the worker must read through paperwork and figure out from the paperwork where the various wires go. Workers go through a two-week class to learn how to read the shop aid paperwork. We could train beginners to assemble wire bundles using the AR system in half an hour.

7. When using the AR system, not only does the need for the shop aid paperwork go away, but the build sequence defined by the system can be enforced, thus providing a means for implementing a preferred build process, as was discussed above.

8. User safety and comfort are always a concern, but they were not a significant problem during the trials. We had a Boeing optometrist perform visual acuity tests on workers participating in the trials. Temporary levels of eyestrain were observed, but there was no significant differences in the results whether the AR system or the traditional method had been used. Reading tiny ID numbers on wires all day seems to be the cause of the problem. Note that implementation of wire barcoding could alleviate this problem considerably.

9. Other comfort issues included weight of the HMD and the weight of, and heat generated by, the belt-worn computer. The vest provided by TriSen, intended to distribute the weight of the system and hide the cables, was found to be warm by some users. All of these were relatively minor problems and will naturally improve as the technology evolves. The next generation of VIA wearable computer will be smaller, lighter, and, at the

same time, more powerful than the present system. Newer cpu technology will give increased compute power with less power consumption and hence less heat generation. In addition, battery technology is improving, promising smaller and lighter batteries with increased storage capacity.

10.   Worker reaction to using the AR system was somewhat bipolar. Some workers involved in the pilot project wore the TriSen system for a full 8-hour work shift, except for their scheduled breaks, reporting to us that they enjoyed it and thought that it was improving their productivity. Others wore it for half a minute and then told us "This thing makes my head hurt/neck hurt/back hurt. I'm going to go tell my supervisor I can't participate in this project." Women almost unanimously refused to wear the system. We concluded that there was probably another type of discomfort, rather than physical discomfort, at work. First, the HMD, especially after being worn for an hour or two, causes a terminal case of "hat hair." Second, although the TriSen HMD was lighter and better balanced than most designs we AR researchers had worn before, it looked boxy and clunky. Users could expect lots of comments from passers-by about having been "assimilated by the Borg." Third, we created ample opportunity for such comments. The place where our pilot project was in operation was right next to the wire shop and mockup shop's break room. Participants could expect lots of their friends and coworkers to pass by on their way to a break or to lunch and see them wearing the "geek gear." As HMD technology continues to be miniaturized, enabling HMD designers to trend toward an eyeglasses-like form factor, we expect this appearance-related problem to greatly diminish.

FUTURE STEPS

Right now we are in the planning stages for the next iteration of development and testing for the wire bundle assembly application of AR. Over the first half of this year, we intend to incorporate speech recognition into the system and to develop a new user interface based on speech input. We will experiment with the sorting-instead-of-searching approach described above. In the second half of the year, we hope to take a new generation of hardware, with faster processors, improved trackers, and possibly improved HMDs, together with our new user interface software, into the factory for another round of testing and evaluation. We will probably test by conducting a project at about the same scale as the 1997 pilot, but we hope to be able to be building production bundles this time instead of throwaway test bundles. Ideally, the outcome of this round of testing would be the conclusion that the technology is ready for deployment on a larger scale.

There are many other possible applications to manual manufacturing and assembly tasks at Boeing for AR systems, but not many would be able to use the TriSen videometric tracker, the head tracker that has worked the best for wire bundle assembly. Factory managers won't be willing to paint the inside of the fuselages with an array of polka dots in order to use AR systems to guide part installations, for example. An important open research question, therefore, is the design of a portable, easily deployed position/orientation measurement system. We want something that we can bring to the place on the aircraft where we want to do the work, quickly attach to the aircraft or set nearby, register into the coordinate system of the aircraft, and start using.

Such a tracker would not only enable other manufacturing applications, but applications in other domains, as well. An important potential application area for the aerospace industry is maintenance and maintenance training. AR has the potential, tracker technology permitting, of being used to guide a maintainer through the detailed steps of a diagnosis and maintenance procedure. Refresher training, sometimes even initial training, could take place on the spot.

CONCLUDING THOUGHTS

There is an experience-based rule of thumb in industrial and military R&D that it takes ten years to bring a technology from the new idea stage to the product stage. AR per se is not a new idea; Sutherland prototyped a see-through display with a mechanical head tracker in the sixties (Sutherland, 1968), and military helicopter pilots have used see-through helmet-mounted displays with in-cockpit head trackers ever since the Vietnam war. What was new about our concept of AR was the context: the manufacturing and maintenance application domains, which the advent of small, low-power, wearable computers and lightweight, low-power head-mounted displays had made feasible. In any case, we seem to be more or less on schedule, given that 1999 is the tenth year of the Boeing AR project, with some months to go before any possible production use of the technology.

Why does it take so long? For AR, it really was not because the technology was not well understood. People knew how to build HMDs. We got the rendering algorithms straight out of Foley and Van Dam (1982). Wearable PCs were simply an issue of appropriately repackaging standard PC electronics. Robust, long-range trackers weren't available off the shelf, but several approaches were known. Certainly, there were many engineering and ergonomics issues involved, which have been solved by some designers in far better ways than others, but there were no fundamental unknowns in the design and implementation of AR systems; no conceptual breakthroughs were required.

It all boils down to money. The people managing industrial R&D never give you enough funding to carry out the development of a new technology the way you'd like to. R&D funding is just too scarce. The end customers of the new technology typically aren't chartered to spend funds on the development of new technologies, period. The organizations that do have a charter to spend R&D funding have large constituencies around which they have to spread the money. So you have to stretch the project out along the time axis to fit it within its budget.

Aerospace companies aren't an easy place in which to develop a new computer technology, either. For companies like Microsoft and HP, developing new computer technologies is central to their continued success, and they support it accordingly. For an aerospace company, computers are perceived as cost items, analogous to the toilet paper rolls in the company restrooms. We just buy them because the employees somehow seem to expect them to be there. New computer technologies are rarely, if ever, perceived as something that would increase profits. "We didn't use computers at all on the 747, and look how successful it is!" To the extent that they do support new technology development, they are expecting very near-term payoffs. Complaints about insufficient funding to make progress fall on deaf ears. The Boeing AR project would have died in 1994 if it had not been for DARPA's selection of our TRP proposal in 1993.

There's another implication of trying to develop new computer technology inside a company that doesn't make its money doing that: You have to team. Boeing was never going to go into the business of manufacturing and marketing Augmented Reality systems. We had to find or develop vendors who were willing to invest in AR themselves because they believed in their own ability to make money selling AR systems.

Other than the very beneficial relationship we had with the Carnegie Mellon University wearable computing research people, the teaming successes we had were with startups. The good side of working with a startup is that the startup's employees are usually talented and always motivated. They respond more quickly to requests and direction changes than large companies do. The negative aspect of working with a startup is that they can't cover all the bases. The one that did the best HMD optics didn't get the HMD ergonomics right. The one that did the best head tracker had the least comfortable wearable computer. If IBM had decided to invest in

AR systems, they would have been capable of assigning world-class hardware designers, optics designers, ergonomicists, battery experts, etc. to the project, and practically every design aspect could have been addressed at least adequately. You rarely get that from a ten-person company.

One of the most fun things about the project was its multidisciplinary nature. We had to deal with computer hardware, computer software, linear algebra, algorithms, ergonomics, human–computer interface design, the physics of head trackers, the design of HMD optics, etc. But the best thing about the project was the people I had the pleasure and privilege of working with:

Tom Caudell, who thought of the idea in the first place;
Adam Janin, who did most of the algorithm and software development; and
Dan Curtis and Peter Gruenbaum, who did all the software design and development for the 1997 factory pilot project.

They, and the prospect of seeing this technology in use in the Boeing factory, have kept me working on the project all these years. That deployment vision hasn't been realized yet, in the sense that AR has not yet been used on production wire bundles at Boeing. We can declare the project a limited success already, though. TriSen has sold several AR systems to smaller companies in the wire bundle manufacturing business. In several places in the United States, Augmented Reality is being used to assemble wire bundles.

REFERENCES

Caudell, T. P. and Mizell, D. W., "Augmented Reality: An Application of Heads-Up Display Technology to Manual Manufacturing Processes," *Proceedings, IEEE Hawaii International Conference on Systems Sciences, 1992*, January 1992, Kauai, HI, IEEE, 0073-1129-1/92, 1992, pp. 659–669.

Foley, J. D., van Dam, A., *Computer Graphics, Principles and Practice*, Addison-Wesley, 1982.

Janin, A. L., Mizell, D. W., and Caudell, T. P., "Calibration of Head-Mounted Displays for Augmented Reality Applications," *Proceedings, Virtual Reality Annual International Symposium, 1993*, Seattle, WA, September 1993, IEEE, pp. 246–255.

Sutherland, I., "A Head-Mounted Three-Dimensional Display," *Proceedings, Fall Joint Computer Conference, 1968*, pp. 757–764.

Zikan, K., Curtis, W. D., Sowizral, H. A., and Janin, A. L., "Fusion of Absolute and Incremental Position and Orientation Sensors," *SPIE Proceedings Vol. 2351: Telemanipulator and Telepresence Technologies*, Boston, pp. 316–327, 1995.

IV

Wearable Computers

15

Computational Clothing and Accessories

Woodrow Barfield
Virginia Tech

Steve Mann
University of Toronto

Kevin Baird
Virginia Tech

Francine Gemperle, Chris Kasabach,
John Stivoric, Malcolm Bauer,
Richard Martin
Carnegie Mellon University

Gilsoo Cho
Yonsei University

1. INTRODUCTION

Wearable computers are fully functional, self-powered, self-contained computers that allow the user to access information anywhere and at any time (Mann, 1996; Mann, 1997a; Barfield and Baird, 1998; Bass, Kasabach, Martin, Siewiorek, Smailagic, and Stivoric, 1997). Until just recently,

FIG. 15.1. Wearable device from the late 1980s—a multimedia computer with a 0.6-inch CRT, invented and worn by Steve Mann (http://wearcomp.org/ieeecomputer/r2025.html).

wearable computers consisted of fairly obtrusive computer displays, CPUs, and input devices, worn on the user's body (Figure 15.1). However, due to advances in microelectronics, developments in networking, and growing interest from the public, designers of wearable computers are beginning to focus on the issue of making these systems look more like clothing and less like computers. For example, in the area of visual displays, a covert eyeglass-based display was built in 1995 (Mann, 1997a,b). Others later built an eyeglass-based display in 1997 (Spitzer, Rensing, McClelland, and Aquilino 1997) that was intended for eventual commercialization. Beginning in 1982, Eleveld and Mann began to experiment with the design of computers built directly into clothing, and the notion of flexible computational garments was developed (Mann, 1997a,b). Others have more recently experimented with the idea of flexible clothing based computing (Post and Orth, 1997).

To emphasize the attempt to integrate computing into clothing, we use the term "computational clothing" to refer to clothing that has the ability to process, store, retrieve, and send information. These capabilities will allow clothing and clothing accessories to function as stand-alone computers, to react to sensors in the environment, or to link to the

World Wide Web and other networked systems. Most importantly, computational clothing will allow users to access the functionality associated with modern computational devices while conforming to traditional fashion trends.

Networked clothing will represent an especially important development in personal information technology. Applications previously implemented with networked clothing include devices that serve the function of turning on the lights of a room and adjusting and controlling the heating and cooling in the room (Mann, 1997c). Future proposed applications also include performing medical tests on the wearer (Lind, Jayaraman, Rajamanickam, Eisler, and McKee, 1997). On the issue of collecting physiological data, the early work of Mann during the 1980s (Mann, 1997a,c) illustrates an attempt to integrate biosensors into wearcomps that detect the user's physiological state. The more recent work of Picard and Healey (1997) attempts to correlate physiological states with emotion. Most importantly, by actually embedding computer technology into clothing, people can have continual access to resources without the inconvenience and obtrusiveness of carrying computational devices around separately. Thus, clothing with computational capability will facilitate a form of human–computer interaction comprising small body-worn computers (e.g., user-programmable devices integrated into clothing) that are always on and always ready and accessible. The "always ready" capability of computational clothing will lead to a new form of synergy between human and computer, characterized by long-term adaptation through constancy of user interface (Mann, 1997a).

As noted, before the general public will accept computers as everyday apparel, the computers will need to behave like and feel like actual clothing. For this reason, it will be important for designers to consider information from a wide variety of sources when designing computational clothing. For example, information about textiles and fabrics, microelectronics, human interface design, networking, power sources, and cultural fashion trends will need to be considered when designing computational clothing. The purpose of this chapter is to present information from these diverse areas in the context of the design and use of computational clothing. The material presented in this chapter represents a starting point for those interested in designing clothing with computational capability—what we present is an overview of several related areas that impact the design of computational clothing. We expect significant advances to be made on this topic in the next five to ten years.

2. SYSTEM DESCRIPTION
AND GENERAL CAPABILITIES

Many of the current research directions in the design and use of computational clothing can be traced to the early 1960s when Ivan Sutherland (1968) from MIT first developed a see-through display that allowed graphics to be superimposed over a real-world scene. In addition, the ideas associated with ubiquitous computing have also contributed to developments in computational clothing. What better way to access computing resources anywhere and at any time than to be wearing them on your body? In this regard, advances in microelectronics and wireless networking are making ubiquitous computing a reality. However, as noted previously, before the general public will be seen wearing a computer, components of the wearable computer (e.g., the CPU housing unit, input and output devices, etc.) will have to look far more like clothing or clothing accessories than the commercial wearable computer systems available now (Barfield and Baird, 1998).

As indicated by Mann (1997c), clothing with computational capability will consist of a computer that is subsumed into the personal space of the user, controlled by the user, and will have both operational and interactional constancy (i.e., will always be on and will always be accessible). Most notably, since computational clothing will be worn by the user as everyday apparel, the user will always be able to enter commands and execute a set of such entered commands, and the user will be able to do so while walking around or doing other activities. The issue of how to design computational clothing so as to both fit the user's apparel and to become an accepted component of the user's apparel is a current and challenging design problem (Figure 15.2).

The most salient aspect of computers, in general (whether computational clothing or not), is their reconfigurability and their generality (e.g., that their functions can be made to vary widely, depending on the instructions provided for program execution). With computational clothing, this is no exception; that is, the wearable computer clothing is more than just a wristwatch or regular eyeglasses. Computational clothing will have the full functionality of a computer system but, in addition, will also be inextricably intertwined with the wearer. This is what will set computational clothing apart from other wearable devices that are more in the area of computational (or digital) accessories, such as wristwatches, regular eyeglasses, wearable radios, etc. Important aspects of computational clothing will be discussed next in terms of basic modes of operation and fundamental attributes.

FIG. 15.2. A recent, nearly undetectable, prototype wearable computer consisting of eyeglasses, a handheld control, and a computer worn in back under the shirt, invented and worn by Steve Mann, http://wearcomp.org/ieeecomputer/r2025.htm.

- Photographic memory: When using a camera as a component of the system, the user will be able to experience perfect recall of previously collected information.
- Shared memory: In a collective sense, two or more individuals may share in their collective information of the other, so that one may have a recall of information that one need not have experienced personally.
- Connected collective intelligence: In a collective sense, two or more individuals may collaborate while one or more of them is doing another primary task.
- Personal safety: In contrast to a centralized surveillance network built into the architecture of the network, with computational clothing the surveillance will be built into the architecture (clothing) of the individual.
- Tetherless operation: Computational clothing will afford mobility, essentially the freedom from the need to be connected by wire to an electrical outlet or communications line.
- Synergy: The goal of clothing with computational capability is to produce a synergistic combination of human and machine, in which the human performs tasks that it is better at, while the computer

performs tasks that it is better at. Over an extended period of time, computational clothing will begin to function as an extension of the mind and body and will no longer be treated as if it is a seperate entity. This intimate and constant bonding is such that the combined capabilities of the resulting synergistic whole will exceed the sum of either. Mann (1998) calls synergy, in which the human being and computer become elements of each other's feedback loop, Humanistic Intelligence (HI).

In addition to the above points, there are three operational modes representing the interaction between human and computer associated with computational clothing:

1.   Constancy: The computer aspect of computational clothing will run continuously and will always be ready to interact with the user. Unlike a hand-held device, laptop computer, or PDA, it does not need to be opened up and turned on prior to use.

2.   Augmentation: Traditional computing paradigms are based on the notion that computing *is* the primary task. In contrast, with computational clothing, computing is *not* the primary task. The assumption behind a user wearing a computer as clothing is that the user will be doing something else at the same time as doing the computing. Thus the computer should serve to augment the human's intellect, or augment the senses.

3.   Mediation: Unlike hand-held devices, laptop computers, and PDAs, computational clothing will encapsulate us. It doesn' t necessarily need to completely enclose us, but the concept allows for a greater degree of encapsulation than traditional portable computers. There are two aspects to this encapsulation:

3a.   Solitude: Clothing with computational capability can function as an information filter, allowing the user to block out material he/she might not wish to experience, whether it be offensive advertising or simply a desire to replace existing media with different media.

3b.   Privacy: Mediation allows us to block or modify information leaving our encapsulated space. In the same way that ordinary clothing prevents others from directly seeing our bodies, computational clothing may, for example, serve as an intermediary for interacting with untrusted systems.

The privacy aspects of computational clothing are especially interesting and germane to design. Although other technologies, such as desktop computers, can help us protect our privacy with programs such as Pretty Good Privacy (PGP), the Achilles heel of these systems is the space between us and them. It is generally far easier for an attacker to compromise the link between us and the computer (perhaps through a so-called Trojan horse or other planted virus) than it is to compromise the link between our computer and other computers. Thus computational clothing can be used to create a new level of personal privacy because it can be made much more personal (e.g., so that it is always worn, except perhaps during showering, and therefore less likely to fall prey to covert attacks upon the hardware itself). Moreover, the close synergy between the human and computers will make it harder to attack directly (e.g., as one might peek over a person's shoulder while they are typing, or hide a video camera in the ceiling above their keyboard). Furthermore, the wearable computer can take the form of undergarments that are encapsulated in an outer covering or outerwear of fine conductive fabric to protect from an attacker looking at radio frequency emissions. The actual communications between the wearer and other computers (and thus other people) can be done by way of outer garments, which contain conformal antennas, or the like, and convey an encrypted bitstream.

3. GENERAL CHARACTERISTICS OF CLOTHING

To better understand how to design clothing with the capabilities that we attribute to computer systems it is relevant to review the basic principles associated with clothing in general. Clothing is often referred to as a "portable environment" or as a "second skin". Furthermore, clothing has three different aspects, which include the physiological, social-psychological, and cultural. Each of these areas should be considered in the design of computational clothing.

Clothing specialists have delineated the main reasons why people wear clothing. These reasons include issues of protection, modesty+privacy, status, identification, and self-adornment and self-expression (Lyle and Brinkley, 1983). In terms of self- expression, witness the current computational devices that change color to reflect the user's mood. People also use clothing as a means to project status, role, and gender, as well as cultural differences. In our view, the fact that future clothing will be programmable will allow the user to change the "cultural" appearance of their clothing as a function of mood or situation.

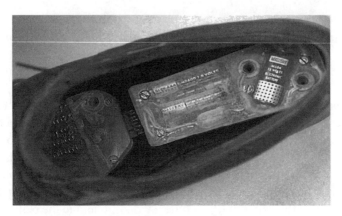

FIG. 15.3.   The Eudaemon shoe (Bass, 1985).

It is also interesting to note that people use the term's "clothes" and "clothing" interchangeably. The term "clothing" has a broader and more comprehensive meaning than the term "clothes", which by itself indicates garments or apparels. The term "clothing" refers not only to clothes but also to the accessories that people place on their body. Noncloth items include hats, shoes, handbags, belts, watches, gloves, accessories, glasses, umbrellas, etc. Figure 15.3 shows an early example of a computer integrated into a traditional noncloth item, the shoe (Bass, 1985). This shoe (Eudaemon shoe) used a short-range inductive system to receive signals from a calculator operated by another person and a vibrotactile display as output to the foot. The application for the Eudaemon shoe was to predict the landing location of a roulette ball. Note that this shoe was not a general-purpose computer in the sense that the user could not change its functionality by writing a new program into it, while walking around. It was more like wearable technology than the wearable computing we know and use today, but it nevertheless serves as a good example of useful wearable technology. A more recent shoe with computational capability was developed by Paradiso and Hu (1997). The purpose of their "electronic shoe" was to transmit data about foot position through sensors that were designed to drive music synthesizers and computer graphics in real time. Their shoe is instrumented with pizoelectric pads that measure differential toe pressure and dynamic pressure at the heel, a bidirectional FSR strip that measures sole deflection, and a micromechanical accelerometer used to detect tilt (pitch) and foot velocity. Other sensors include an electronic compass to measure yaw and a magnetic vector sensor to measure bearing (from the Earth's magnetic field). Finally, Paradiso and Hu (1997) indicate that the translational position of the shoes can be measured by use of a scanning laser range finder or directly using sonar.

Even though various types of clothing can be differently classified according to the wearer's lifestyle or viewpoint, clothing is usually divided into four groups. These include business or dress clothing, casual clothing, sports or leisure clothing, and sleepwear or underwear (Erwin, Kinchen, and Peters, 1979). Dress items include suits, dress shirts, ties, jackets, slacks, shoes, and socks for men, and dresses, pants or skirts, blouses, and shoes for women. Casual jacket, T-shirts, slacks, sweaters, and casual shoes belong to the group of casual clothes. More or less, the casual items are unisexual and genderless in terms of style. Active sportswear, cold-weather jackets, pants, and resort wear belong to the sports or leisure clothing category. Pajamas, undershirt, corsets, etc. are examples of sleepwear and underwear. Protective garments for fire fighters, pesticide applicators, hockey players, the military, and surgeons have their own distinct functions. Protective garments are classified into four groups according to the following categories or characteristics of clothing: thermal, chemical, mechanical, and biological (McBriarity and Henry, 1992). In the area of protective clothing for the military, Lind, Jayaraman, Rajamanickam, Eisler, and McKee (1997) have developed a wearable motherboard or what they term "sensate liner" (Figure 15.4). The sensate liner is a form-fitting garment that consists of sensing devices containing a processor and transmitter. The sensate liner textile consists of a mesh of electronically and optically conductive fibers integrated into the normal structure of fibers and yarns used to generate the garment.

In order to discuss the requirements of clothing, it is necessary to discuss the serviceability of clothing. Serviceability is the measure of the clothing products' ability to meet consumer's needs. Serviceability concepts include: aesthetics, durability, comfort, care, safety, environmental impact, and cost (Hatch, 1993; Kadolph and Langford, 1998). Serviceability concepts for clothing will need to be considered when designing computational clothing. Aesthetics refers to the attractiveness or appearance of clothing. Aesthetic appeal is becoming a more important criterion for clothing than ever before (see Figures 15.18–15.23). In addition, color, texture, and luster of fabrics and even silhouette, style, and coordination of outfits are very critical factors in choosing garments. Finally, durability denotes how the product withstands use.

The meaning of comfort in relation to clothing is as follows: Generally, comfort can be defined as freedom of discomfort and pain. It is a neutral state. When comfort is discussed, the relationships among the type of

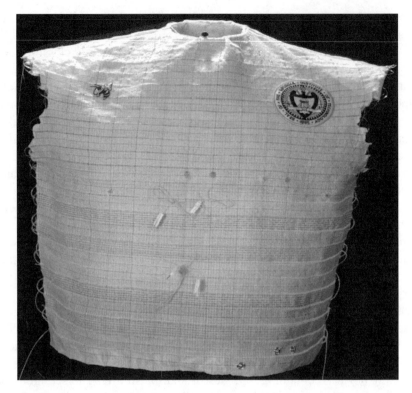

FIG. 15.4. Sensate liner developed by Lind, Jayaraman, Rajamanickam, Eisler, and McKee (1997) at the Georgia Institute of
Technology (http://vishwa.tfe.gatech.edu/gtwm/gtwm.html).

clothing, characteristics of the person, and characteristics of the environment have to be considered. Comfort can be divided into several components as follows: Comfort relates to the way in which clothing affects heat,
moisture, and air transfer as well as the way in which the body interacts
with clothing. This aspect of comfort is referred to as "thermophysiological comfort." In addition, comfort is related to the issue of how consumers
actually feel when clothing comes into contact with the skin. This is referred to as "sensorial or neurophysiological comfort." Finally, "comfort"
is related to the ability of clothing to allow freedom of movement, reduced
burden, and body shaping as required. This is "body-movement comfort."
One of the later sections of this chapter will focus on the issue of wearability as a function of body movement. Each of these aspects of comfort will
be discussed in more detail below given their importance for the design of
computational clothing.

Awareness of clothing usually leads to an expression of discomfort such as too hot, too cold, too cool, and too wet. In general, clothing is considered thermally comfortable when there is no need to take off or put on additional clothing and the fabric is not sensed as wet or humid. Thermophysiological comfort depends on the clothing microclimate developed between the body skin and inner layer of clothing. In order to be comfortable "thermophysiologically," the clothing microclimate should lie in the range of 35 ± 2 deg C temperature, $50 \pm 10\%$ relative humidity, and 25 ± 5 m/s air velocity. Attainment of a comfortable thermal and wetness state involves transport of heat and moisture through a fabric. Fabric has its own insulative ability, water vapor permeability, absorbency, wickability, and other properties related to thermal comfort. It is also interesting to note that some unpleasant sensations, such as prickliness, itchiness, inflammation, roughness, and warm and cool sensations, are produced when clothing irritates the sensory receptors and nerve endings in the skin. It is generally agreed that there are three categories of sensory nerves, that cover haptic sensations: the pain group, the touch group (pressure and vibration), and the thermal group (warmth and coolness). In the context of the haptic modality, static charged fabrics cling to the body, resulting in an uncomfortable feeling. Charged fabrics may lead to shocks when the wearer touches a metal object. These side effects are due to the electrical nature of the textiles and the skin.

People must be able to move around in the apparel items they wear. Discomfort may result when clothing restrains movement, creates a burden, or exerts pressure on the body—this aspect of clothing design is particularly relevant to computational clothing. Textile materials must be flexible and elastic. Also, when people move, their skin stretches and recovers, and so fabric must elongate to accommodate body movements and then must be able to recover. Generally, fabrics with less than 15% elongation values are referred to as rigid fabrics, and fabrics with more than 15% elongation are stretch fabrics. Tailored clothing requires 15 to 25% elongation, whereas sportswear requires about 20 to 35% elongation. Finally, active wear needs 35 to 50% elongation for comfort.

Garment weight also contributes to comfort and discomfort because it determines the burden the wearer must carry. Garment weight mainly depends on the amount of fiber in the garment and with computational clothing the added weight resulting from the devices worn on the body. For example, some commercial wearable computers can weigh between 5 and 9 kilograms and devices worn by soldiers can add another 66 kilograms of weight to the soldiers clothing. In this context, the average weight of men's garments totals up to 3.8 kilograms and that of women's clothing about 2.3 kilograms. The weight of clothing may also result in pressure being applied to the skin. Clothing pressure depends on the garment design and fit and the stretchability of fabric. These variables determine the amount of pressure exerted on the body that results from clothing. Pressures of less than 60 grams per square meter exerted by clothing on the body are considered to be comfortable. Pressures of from 60 to 100 grams per square meter are considered uncomfortable and pressures over 100 grams are not tolerable. As another design consideration for computational clothing, the pressure exerted by clothing on the body becomes greater as the curvature of the body increases. In addition, safety, care, environmental impact, and cost are important factors for serviceability.

Clothing is made of textile fabrics, which are materials characterized as planar structures consisting of yarns or fibers. Using these materials, clothing is constructed into three-dimensional forms. Clothing is formed by cutting appropriately shaped pieces from fabrics and sewing them together. The major components of textile fabrics are fibers, yarns, fabrics, and colorants and chemicals. Fibers are tiny substances, which have a length at least 100 times its diameter. This large ratio between length and diameter enables fibers to be spun into yarns or made into fabrics. Many different types of fibers such as cotton, polyester, nylon, wool, silk, acrylics, olefin, and linen are used as sources of textile fabrics. Whether they are natural or man-made, fibers differ from each other in their chemical nature, in other words, in polymeric substance. For example, cotton consists of a polymeric substance that is a cellulose, whereas wool is a protein consisting of amino acids.

Yarns are continuous strands of textile fibers suitable for weaving, knitting, or intertwining to form textile fabrics. Fabrics have yarns interlaced at right angles, or interlooped horizontally or in zigzag form. Some fabrics do not have yarns and are made from some arrangement of fibers. These are so-called woven, knitted, and nonwoven fabrics. Fabrics are highly porous. Much of the thermophysiological comfort is provided by the porous structure of the fabric. Colorants and chemicals constitute a substantial portion of the finished and dyed textile fabrics. One or more chemicals are used to improve the fabric's properties. Colorants are used to modify the perceived color of fabrics or to impart color to colorless fabrics.

Apparel is made of patterned fabrics sewn together by thread. Appareal sometimes has inner and outer fabrics and at the same time, in the center of these two layers, an interfacing fabric. A variety of materials and techniques are employed in constructing well-made and functional closures.

These depend on the type and design of the garment, the fabric, and the location of the opening in the fabric. Closure utilizes a wide use of zippers. There are three ways to insert zippers: invisible applications, lapped applications, and centered applications. In addition to buttonshooks, fabric loops and velcro fasteners are easily used. Couture techniques such as stitch, knot, covered snap, covered cords, braided belt, tassels and fringes, and embroidery stitches are frequently chosen to enhance the garment and emphasize its beauty, individuality, and quality. Nylon was first produced by Carothers in 1935, and synthetic fibers such as polyester, polypropylenes, and acrylics were discovered during the 1950s. Now new materials of high functionality and high performance are designed and produced according to the nature of their utilization. Some of these materials are discussed in this chapter as possible fabrics suitable for computational clothing. The static charge of synthetic fabrics, due to their low water content, causes many problems, such as fabric cling, electric shock, and dust adsorption. It is therefore necessary to utilize electroconductive fabrics in the design of computational clothing. This is because electroconductive fabrics are good not only for the protection of the digital circuits but also for the safety of the wearer. Recent developments in the design of electroconductive fibers are to use carbon black as the core component and conjugate-spin as a nylon filament. The carbon black is used so that electroconductive static charges are not built up on the clothing surface. The specific resistivity of the fabric made of the conjugate spinning process using the carbon black is 10^3–10^5 $\Omega\cdot$cm, whereas the specific resistivity of ordinary nylon fabric is 10^{11}–10^{13} $\Omega\cdot$cm and that of cotton fabric is 10^8–10^9 $\Omega\cdot$cm, which means that for cotton fabric static is not a problem during daily wear.

There are some fabrics that control the microclimate temperature automatically. If the wearer feels hot, the fabric absorbs the excess heat, and vice versa. This intelligent fabric is made possible by using some phase-change material as a finishing agent. Phase-change material absorbs and preserves the optical energy of the sun; it releases heat when the material is cooled, and it absorbs heat when the material is heated. Materials such as polyethylene glycol (PEG) and zirconium carbide compounds are typical phase-change materials. PEG was first adopted for use in fabrics and used for winter sportswear. Zirconium carbide was used in the form of particles in polyamide and polyester fibers. The particles are enclosed within the core of synthetic fibers. The garment made of this fiber absorbs solar visible radiation, which is released in the clothing. Furthermore, sweat absorbent fabric serves as a functional fabric for sportswear. To be comfortable and functional, the fabric used for sportswear needs to have the capability to absorb moisture and sweat. If athletes wear conventional clothing during sporting events, the effect is that they will feel hot and thus sweat sufficiently to result in wet garments. In this case the fabric will stick to their body, and behaviorally they will try to detach the "sticked fabric" from their body. Nowadays, sportswear is also used as leisure clothing, which considerably extends its scope. Sweat absorbed fabric consists of polyester fiber, which has a hollow center, and with a large number of micropores at the surface of the polyester fiber. The micropores on the surface are homogeneously distributed throughout the surface and some run through into the hollow part. Sweat is immediately absorbed through the pores and diffused into the hollow center; the result is that the fiber surface is kept dry. The hollow center acts like a reservoir for sweat.

4. COMPUTATIONAL CLOTHING DESIGN CHARACTERISTICS

Essentially, for clothing to have computational capability, digital circuits must be integrated into the clothing. There are many ways that circuits may be integrated into clothing. As an example, when polymers are extruded through spinnerets as hollow fiber, wires could be centered in the fiber. In addition, conjugate spinning can also utilize chips or wires as one or two components in the polymer solution. When yarns are manufactured, wrapped yarns can be made such that the wires are the core part of the yarn and the ordinary polymer filaments are wrapped around the core part. Another method to integrate digital devices with clothing is to make metallic yarn, for example, yarns made of silver or gold. Computer chips and polymer solutions may be put together into sheet form and then split plotted into thin yarn. This procedure offers a way to form metallic yarns of materials such as silver, gold, and aluminum. When weaving fabrics, wire itself can be core-spun or wrapped and can be used as warp yarns or filling yarns in constant intervals. This process is similar to the electroconductive fabric process where carbon black yarns are used as warp and as filling yarns in some interval. In addition, wires can be used as embroidery yarns onto conventional fabric surface. Moreover, other couture methods can be adopted as a possible way to include wires containing digital circuits or wireless networking. Computer chips can also be integrated into various forms of closures that exist now with clothing, such as zippers, hooks, or metallic buttons. According to the end-use, various types of clothing like underwear, sportswear, casual wear, or work clothes could be developed into wearable clothing. For formal wear, underlining fabric and interfacing are the possible targets for integrating digital circuits into clothing.

The issue of how to integrate computing capability into clothing has been investigated for over twenty years. Some of what has been learned in past efforts is relevant to current implementations of computational clothing. For example, the wearable signal processing apparatus of the 1970s and early 1980s was quite cumbersome, so an effort was directed by Steve Mann and other researchers toward not only reducing the size and weight but, more importantly, reducing the undesirable and somewhat obtrusive appearance. The designer discovered that the same apparatus could be made much more comfortable by bringing the components closer to the body, which had the effect of reducing both the torque felt bearing the load, as well as the moment of inertia felt in moving around. This effort resulted in version of a wearable computer invented by Steve Mann called the "Underwearable Computer" shown in Figure 15.5.

Typical embodiments of the underwearcomp shown in Figure 15.5 resemble an athletic undershirt (tank top) made of durable mesh fabric, upon which a lattice of webbing is sewn. This facilitates quick reconfiguration in the layout of components and rerouting cabling. Note that in this system, wire ties were not needed to fix cabling, as it was simply run through the webbing, which held it in place. All power and signal connections were standardized, so that devices could be installed or removed without the use

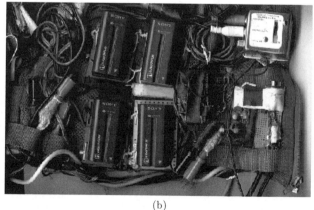

(a)   (b)

FIG. 15.5. The "underwearable" signal processing hardware: (a) as worn by Steve Mann, (b) close-up showing webbing for routing of cabling.

of any tools (such as a soldering iron) by simply removing the garment and spreading it out on a flat surface. Some more recent related work by others (Lind, Jayaraman, Rajamanickam, Eisler, and McKee, 1997), also involves building circuits into clothing, in which a garment is constructed as a monitoring device to determine the location of a bullet entry (see Figure 15.4). Conductive materials have been used in certain kinds of drapery for many years for appearance and stiffness, rather than electrical functionality, but these materials can be used to make signal processing circuits, as depicted in Figure 15.6. Simple circuits like this suggest a future possible direction for research in the design of computational clothing.

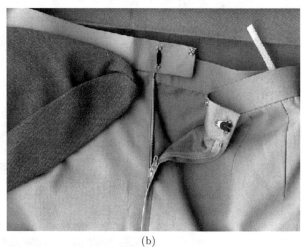

(a)   (b)

FIG. 15.6. Some simple examples of cloth that has been rendered conductive. (a) Cords on early headsets, telephones, etc. often felt more like rope than wire. (b) A recent generation of conductive clothing made from bridged-conductor two-way (BC2) fabric. Although manufactured to address the growing concerns regarding exposure to electromagnetic radiation, such conductive fabric may be used to shield signal processing circuits from interference. Signal processing circuits worn underneath such garments were found to function much better due to this shielding. This outerwear functions as a Faraday cage for the underwearable computing. The notion that cloth be rendered conductive, through the addition of metallic fibers interwoven into it, is one thing that makes possible clothing that serves as an RF shield (Fig 15.5(b)), manufactured to address response to the growing fear of the health effects of long-term exposure to radio-frequency exposure. However, it may also be used to shield signal processing circuits from outside interference or as a ground plane for various forms of conformal antennas sewn into the clothing.

Another characteristic of computational clothing is that the clothing may be "smart." Smart clothing is inspired by the need for comfortable signal processing devices that can be worn for extended periods of time and can provide the wearer a level of intelligence to assist in performance of tasks. Smart clothing is made using either of the following two approaches: additive or subtractive. In the additive approach, the process begins with ordinary cloth by sewing fine wires or conductive threads into the clothing to achieve the desired current-carrying paths. In the subtractive approach, the process begins with conductive cloth, which is cut away in certain places, to leave behind the desired pattern, or with conductive cloth in which the conductors are insulated, and the insulation is removed in only certain locations. Smart clothing has been proposed as a form of existential media (Mann, 1997b) (Figure 15.7). Existential media defines new forms of social interaction through enhanced abilities for self-expression and self-actualization, as well as through self-determination. The aspects of existentialism pertaining to existential media are best understood through a reading of Frankl (1994). Examples of smart clothing include internet-connected shoes that allow one to run with a jogging partner located in some distant place, but connected via the network. Viewpoints might also be shared using the "eye-to-eye" glasses (where a portion of each runner's visual field comes from the other runner (Mann, 1994)).

5. DESIGN FOR WEARABILITY

In the sections that follow, we present material that bridges the obtrusive wearable computers of the present with the fashionably integrated wearables of the future. The information in the following sections was developed based on a wearability study conducted by Interaction Designers at Carnegie Mellon University. The team's thesis is that despite visions of smart textiles and computational clothing in the future, there will still be a need for the solid forms of today's computers to be integrated comfortably with the human body.

Networked and computational clothing may be available in the near future because of our ability to place circuits into fibers and weaves. There are however going to be parts of wearable computers (power supplies for example) that are not going to be easily made of fabric. These parts will always be solid forms, but they need not be plastic bricks. To solve the problems of integrating solid and flexible three-dimensional forms with the human body, the design team at Carnegie Mellon University has developed

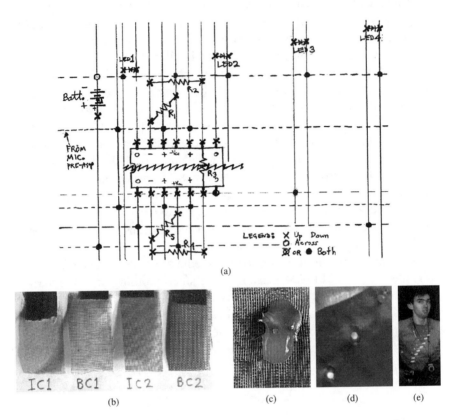

(a)

(b)   (c)   (d)   (e)

FIG. 15.7. Signal processing with "smart clothing". (a) Portion of
a circuit diagram showing the new notation developed to denote
four LED indicators and some comparators. The "X" and "O" notation borrows from the tradition of depicting arrows in and out of
the page (e.g., "X" denotes connection to top layer, which is oriented in the up-down direction, while "O" denotes connection to
bottom "across" layer). The "sawtooth" denotes a cut line where
enough of the fabric is removed so that the loose ends will not
touch. Optional lines were drawn all the way from top to bottom
(and dotted or hidden lines across) to make it easier to read the
diagram. (b) Four kinds of conductive fabric. (c) Back of a recent
article of smart clothing showing a solder joint strengthened with a
blob of glue). Note the absence of wires leading to or from the glue
blob, since the fabric itself carries the electrical current. (d) Three
LEDs on type-BC1 fabric, bottom two lit, top one off. (e) A signal
processing shirt with LEDs as its display medium. This apparatus
was made to pulse to the beat of the wearer's heart as a personal
status monitor, or to music, as an interactive fashion accessory.

"Design for Wearability." Design for Wearability is a tool that helps designers of wearable computers create forms that work well with any size adult human body. The basis for this tool is a set of design guidelines that describe how to make a wearable form.

5.1 Wearable Forms

In the following sections we present a set of three-dimensional forms for the human body that employ design guidelines for wearability. These forms outline the ideal envelope for dynamic wearability. The creation of these forms is based on an iterative process based on extensive field and laboratory experience designing wearable computers from CMU researchers. The general methodology was to make two-dimensional drawings and three-dimensional foam models and then to apply the models to human bodies. In addition, user studies were conducted for two purposes: (1) to better understand the complex curves of the body and (2) to verify that the developed forms were indeed wearable on the dynamic human form.

Each of the above thirteen guidelines listed in Table 15.1 is key to making a wearable computer into something that is really wearable. Given use of these guidleines, the user's comfort and freedom of movement are preserved. In the view of CMU researchers, wearable computing should be a positive and empowering experience, not one of discomfort or with cyborg connotations. The CMU team has employed the first six of those

TABLE 15.1

Guidelines for Wearability

---

1. Placement (where on the body it should go)
2. Form Language (defining the shape)
3. Human Movement (consider the dynamic structure)
4. Proxemics (human perception of space)
5. Sizing (for body size diversity)
6. Attachment (fixing forms to the body)

....

7. Containment (considering what's inside the form)
8. Weight (as its spread across the human body)
9. Accessibility (physical access to the forms)
10. Sensory Interaction (for passive or active input)
11. Thermal (issues of heat next to the body)
12. Aesthetics (perceptual appropriateness)
13. Long-Term Use (effects on the body and mind)

---

guidelines and created a reference set of wearable forms to be used in the development of wearable computers. While all the above guidelines are important, we will only focus on the first six guidelines. The latter seven design guidelines are not easily generalizable since they are much more dependent on the context and constraints of a specific design problem.

Guideline 1   Design for dynamic wearability requires unobtrusive placement on the human body. Placement is determined by editing the extensive human surface area with the use of criteria. Criteria for placement can vary with the needs of functionality and accessibility; however, it is important to work within the appropriate areas (indicated below) for the dynamic human body. The criteria used for determining placement for dynamic wearability are:

- areas that are relatively the same size across adults,
- areas that have low movement/flexibility even when the body is in motion, and
- areas that are larger in surface area.

Applying these criteria results in the most unobtrusive locations for placement of wearable objects on the human body. These are depicted in Figure 15.8.

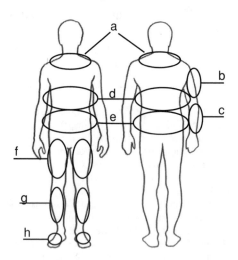

FIG. 15.8.   The general areas found to be the most unobtrusive for wearable objects are: (a) collar area, (b) rear of the upper arm, (c) forearm, (d) rear, side, and front ribcage, (e) waist and hips, (f) thigh, (g) shin, and (h) top of the foot (from Gemperle, Kasabach, Stivoric, Bauer, and Martin, 1998).

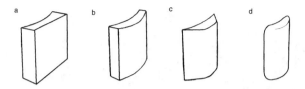

FIG. 15.9. Combining elements of concavity (a) against the body, convexity (b) on the outside surfaces of the form, tapering (c) as the form extends off the body, and (d) radii softening up the edges to create a humanistic form language (from Gemperle, Kasabach, Stivoric, Bauer, and Martin, 1998).

Guideline 2   Design for the human body also requires a humanistic form language. This works with the dynamic human form to ensure a comfortable, stable fit. Humanistic form language includes forming a concavity on the inside surface touching the body, to accept human convexities. On the outside surface, convexity will deflect objects in the environment, thereby avoiding bumps and snags. Tapering of the form' s sides will stabilize the form on the body. Radiusing all edges and corners creates a safe, soft, and wearable form. These steps are illustrated in Figure 15.9, taking a simple block to a wearable form. The humanistic form language not only makes forms wearable, it adds structural ruggedness that is crucial in an active environment.

Guideline 3   Human movement provides both a constraint and a resource in the design of dynamic wearable forms. Human movement is useful in determining a profile or footprint for wearable forms, as well as to shape the surface of forms. Consider the many elements that make up any single movement. Elements include the mechanics of joints, the shifting of flesh, and the flexing and extending of muscle and tendons beneath the skin. The photographs in Figure 15.10 illustrate how much the form of the body changes with simple motion. Allowing freedom for these movements can be accomplished in one of two ways: by designing around the more active areas of the joints or by creating spaces on the wearable form into which the body can move. For example, the torso is a good place to put a wearable, but the arms need to have full freedom to swing around the side and front of the torso. In addition, the torso needs the full ability to twist and bend. These movements can help sculpt the surface of the form.

Guideline 4   Design for human perception of size. The brain perceives an aura around the body that should be considered when determining the distance a wearable form should project from the body. The

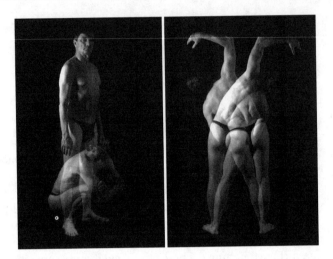

FIG. 15.10. Even through simple motions, our bodies change significantly (from Gemperle, Kasabach, Stivoric, Bauer, and Martin, 1998).

FIG. 15.11. Aura around the human body that the brain will perceive as part of the body (from Gemperle, Kasabach, Stivoric, Bauer, and Martin, 1998).

understanding of these layers of perception around the body is referred to as proxemics (Hall, 1982). Forms should stay within the wearer's intimate space, so that perceptually they become a part of the body. The intimate space is illustrated in Figure 15.11 and can be between 0 and 5 inches off the body. Compromises are often necessary but a general rule of thumb is to minimize thickness as much as possible. This increases safety and comfort, both physical and perceptual. A good example to observe is when a young American football player first dons shoulder pads and immediately starts bumping into people and door ways because of the extra bulk.

Guideline 5   Size variation provides an interesting challenge when designing wearable forms. Both the build of a body and the ways in which it will gain and lose weight and muscle are important. Wearables must be designed to fit as many types of users as possible. Allowing for these size variations is achieved in two ways. The first is the use of static anthropometric data, which detail point to point distances on different sized bodies (Tilley, 1993; McCormick and Sanders, 1982) (Figure 15.12). The second is consideration of human muscle and fat growth in three dimensions. Fitting these changing circumferences can be achieved through the use of solid rigid areas coupled with flexible areas. The flexible areas should either be located between solid forms as joints or extending from the solid forms as wings.

Guideline 6   Comfortable attachment of forms can be created by wrapping the form around the body, rather than using single point fastening systems such as clips or shoulder straps (Figure 15.13). As in guideline 5,

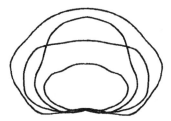

FIG. 15.12.   Torso cross sections of various sized bodies show how sizes vary (from Gemperle, Kasabach, Stivoric, Bauer, and Martin, 1998).

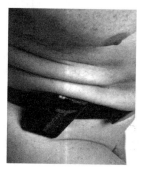

FIG. 15.13.   Single point attachment of a common pager or portable stereo is unstable and perceptually separate from the body (from Gemperle, Kasabach, Stivoric, Bauer, and Martin, 1998).

it is also important to have attachment systems that can accommodate various physical sizes. Design for stable, solid, and comfortable attachment draws on the clothing and outdoor equipment industries. Design for size variations in attachment systems can be obtained in two simple ways. The first is through adjustability (e.g., straps that can be extended as seen on backpacking equipment). The second is through the use of standardized sizing systems from the clothing industry.

Guideline 7   Designing wearable objects generally requires the object to contain materials such as digital technology, water, food, etc. While some of these things are malleable in form, there are many constraints that these "insides" bring to the outer form.

Guideline 8   The weight of a wearable should not hinder the body's movement or balance. The human body bears its own extra weight on the stomach, waist, and hip area. Placing the bulk of the load there, close to the center of gravity, and minimizing as it spreads to the extremities is the rule of thumb.

Guideline 9   For any wearable it is important to consider the sort of accessibility necessary to render the product most usable. Extensive research exists in the areas of visual, tactile, auditory, or kinesthetic access on the human body. Simple testing should be conducted to verify the accessibility of specific wearables.

Guideline 10   Sensory interaction, both passive and active, is a valuable aspect of any product. It is important to be sensitive to how one interacts with a wearable—something that exists on one's body. This interaction should be kept simple and intuitive.

Guideline 11   There are three thermal aspects of designing objects for the body: functional, biological, and perceptual. The body needs to breathe and is very sensitive to products that create, focus, or trap heat.

Guideline 12   An important aspect of the form and function of any wearable object is aesthetics. Culture and context will dictate shapes, materials, textures, and colors that perceptually fit the users and their environment (Craik, 1994). For example, CMU researchers created a wearable computer for an airplane repair situation, depicted in Figure 15.14. Using the heavy leather of the traditional tool belt, it is possible to increase the comfort and acceptance by the repair technicians.

FIG. 15.14. Navigator 2 wearable computer for aircraft maintenance engineers integrates a humanistic form language with attachment guidelines, placement guidelines (small off the back),
and aesthetic, perceptual, and sensory informed use of materials
(from Gemperle, Kasabach, Stivoric, Bauer, and Martin, 1998).

Guideline 13   The long-term use of wearable computers has an unknown physiological effect on the human body. As wearable systems become more and more useful and are used for longer periods of time, it will
be important to test their effect on the wearer's body.

The Design Guidelines alone cannot convey all significant aspects of
designing for wearability. They communicate a means to consider all the
issues involved when creating wearable forms. The design guidelines for
dynamic wearability as presented in conjunction with the development of
a family of wearable forms is presented next.

5.2   Dynamic Wearable Forms

In the following material a set of three-dimensional forms for the human
body that employ design guidelines for wearability are presented. These
forms outline the ideal envelope for dynamic wearability. The first design
object was to generate data defining the complex convex curves of the
determined placement areas. The goal was to not only define but also
to further understand those curves and how they changed with bodies of
different sizes and shapes. For this goal, a set of tools were developed,
depicted in Figure 15.15. With this flexible tool, the CMU team was able to
map the arcs of several parts of the body—the collar area, triceps, forearm,
ribcage, thigh, and shin—and to both compare and measure the arcs from

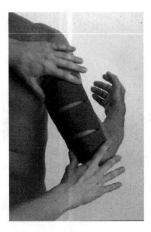

FIG. 15.15. A shapeable tool used to map the arc of the triceps. Different peoples arcs are then mapped together to create an average concavity and width dimension to apply to wearable designs (from Gemperle, Kasabach, Stivoric, Bauer, and Martin, 1998).

various bodies. This allowed the CMU team to determine an appropriate radius/spherical section for the concave inside of the forms, as well as the starting point and length of the flexible areas for each form.

In addition, a usability study was performed to test the comfort level and freedom of movement allowed by the forms that were designed. Thus far, ten people have tested the wearability of these forms. These test subjects were chosen to represent extreme diversity in body shape and size. The subjects were required to perform a series of simple activities, once in their regular clothing and once with the full set of wearable forms on their bodies, over their clothing. The activities included walking, carrying a box, bending, squatting to lift a box, reaching, climbing, and sitting. Subjects were also asked to rate their freedom of movement and comfort levels during each activity and for each area of their bodies. Each of the forms worn by the test subjects was developed by applying design guidelines as outlined previously. Beginning with placement in acceptable areas on the body and the humanistic form language, human movements in each individual area were considered. Each area was unique; thus some study of the muscle and bone structure was required along with common movement. Perception of size was also studied for each individual area. The general principles for size variations were applied and customized for each unique area. A preliminary data analysis indicated that levels of comfort and freedom of movement appeared nearly identical with or without the pods.

FIG. 15.16. Example of spandex pockets placed on the body for use with wearable computers (from Gemperle, Kasabach, Stivoric, Bauer, and Martin, 1998).

The attachment system designed for testing with the forms was minimal spandex that stretched around the body. Spandex pockets held each of the forms close to the body. The image below (Figure 15.16) depicts the attachment system on our model. One additional constraint in developing these pods was that they must be able to house electronic componentry. As a result, all of the forms were between 3/8″ and 1″ thick and we anticipate that flexible circuits could fit comfortably into the 1/4″ thick flex zones.

Descriptions of each of the dynamic wearable forms with photos, charts, and maps of the body and the unique details of the individual areas are too extensive to list here. This chapter will detail the neck area to illustrate the work.

5.3 Dynamic Wearable Forms around the Neck

The three-dimensional forms that were developed are referred to as "pods." A group of pods strung together are "pod sets." Around the neck there is a pod set consisting of four pods. Two pods rest on the front of the body and two on the back. Each individual pod is made up of three parts: two thin solid forms with a flexible material sandwiched between them. The flexible material extends beyond the solid pod structure, serving as a flex zone. On the neck the flex zone creates a collar that encircles the neck and connects all four pods. The two pods on the front of the body sit just below the collar bone, on the pectoral muscle and above the breast.

The two pods on the back of the body sit on the large triangular muscle that connects the shoulders to the neck, the trapezius. Placement of four neck pods allows for all movement of the shoulders, arms, and head. The flex zone connecting these four pods flexes to accommodate the various different torso depths and chest and trapezius arcs. Pods on the chest follow the curves defined by the first and second ribs below the collar bone. The trapezius pods have a top profile determined by the curve where the neck meets the shoulders and a bottom profile determined by the movement space for the shoulder blades and the spine. These pods on the shoulder area are designed to move and float over the movement of the trapezes (shrugged shoulders). The pods on the chest extend 1/2 inch off the body and are 4 by 2 inches. The pods on the shoulder extend 3/4 inch off the body and are 2.3 by 3.4 inches. The neck and chest pods are contained in a collar that encircles the neck and holds them in place. These pods can also be attached by containing them in a minimal vest structure that supports pods on the rest of the torso (Figure 15.17.)

By making dynamic wearability constraints explicit, it is hoped that designers will treat wearability requirements as concretely as technological constraints and match them to users' functional requirements in the early

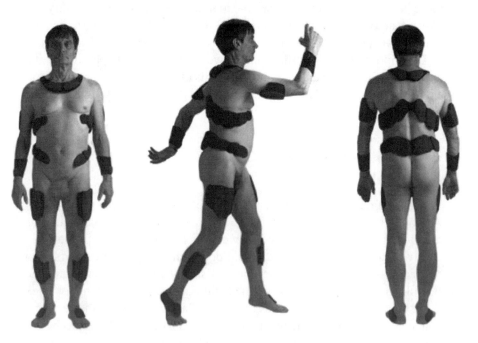

FIG. 15.17. Additional example of spandax pockets placed on the body for use with wearable computers (from Gemperle, Kasabach, Stivoric, Bauer, and Martin, 1998).

stages of the design process. The CMU team plans to extend this research to include accessibility for different activities, weight distribution, thermal concerns, interaction issues, material preferences, and long-term effects to the body while using these wearables. In summary, the following hold:

- Static, anthropometric data exist; however, dynamic understanding and measurements of the human body do not. The CMU team has collected information that has aided designers in development of wearable systems. This Design for Wearability Study discussed above represents a start at putting this information together, organized, in one place, to be useful as a set of guidelines and a resource for designers that need to integrate issues of wearability into a design.
- The design guidelines discussed in this section of the chapter illustrate steps to take into consideration when designing something to exist on the human body. This set of guidelines presents a method of thinking about and understanding a wearable and its' wearability.
- Wearable technology should not compromise but enhance people. It is possible to create a wearable piece of digital technology that feels good. Design for wearability and the wearable forms provide both proof and process for this.

6.   CURRENT COMPUTATIONAL ACCESSORIES

In addition to the clothing that people wear every day, there are many types of fashion accessories that make people's outfits. Many of these accessories, such as watches, belts, and hats, are worn everyday as part of a person's outfit and are quite commonplace. Examples of these include miniature cellular phones worn as necklaces or pagers strapped to belts. The common trend in each of these types of fashion accessories is that not only are accessories being used today as fashion icons, but they are also being designed to serve a specific purpose. By embedding digital technology into these accessories, the metaphor for computational clothing is expanded into complete computational outfits. This not only expands the functionality of a person's computational apparel but also allows for an increased fashion sense and individuality that only accessories can create. When looking at computational accessory technology, it is helpful to partition the technology into three classifications since the devices in each of these categories are similar in design and purpose. The accessory categories for current technology include watches, personal communication devices, and personal aids.

6.1   Watches

Wristwatches are perhaps the most common fashion accessory for both men and women. They have been around for decades and were originally simple mechanical devices intended solely to indicate the time. However, as time passed and watches became more and more popular, they began to become fashion accessories as well. Currently Seiko has developed the 16-bit processor powered Ruputer Pro digital watch. This watch not only has all the features of common digital watches (time, data, stopwatch, and alarm), it also contains a full featured PDA and a game unit. The watch contains 2 megabytes of onboard flash memory, is Windows 95 linkable, and has all of the organizer functions of a Palm Pilot. In addition to this functionality, the Ruputer Pro will soon have infrared communication abilities with other Ruputers that would allow its users to play games. Another popular watch manufacturer, Swatch Telecom, is developing a digital watch that contains a cordless phone. Further in the future, Swatch is planning a version of the watch that not only tells time but will also function as a mobile cellular telephone. A similar venture by Timex and Motorola has yielded the Beepwear. This is a full-featured digital watch that also includes a full-featured pager. Users are able to receive complete text messages from both pages and voice mail. The watch alerts the user of new messages by either beeping or flashing. Finally, one of the most developed and sophisticated versions of the digital watch is the Casio Data Bank. This watch comes in a variety of styles and utilizes a touch screen interface. The touch screen not only displays the time and date but also has eight selectable icons that activate features such as a scheduler, which includes a calendar to remind the wearer of important dates, and a tele-memo, which stores 200 pages of names, phone numbers, addresses, appointments, and notes. The Casio Data Bank digital device also includes a business and executive mode that allow users to link companies to names and perform index searches of the data.

6.2   Personal Communications Devices

The second category of digital accessories consists of devices whose purpose is interpersonal communication. These devices include cellular phones, pagers, and mobile email devices, to name a few. While these devices are relatively new, compared to the digital watch, they are no less gaining widespread popularity as their costs decrease and their functionality increases. As these devices are becoming more lightweight and portable, more people are beginning to carry them as part of their everyday outfits. In some instances they are even beginning to wear these devices like fashion accessories. The most compelling example is that people are beginning to wear their cellular phones around their necks like necklaces. Designers of these technologies are aware of these trends and are starting to design their devices with this in mind. Their goals are to make these devices as common as everyday accessories.

Motorola, who has influenced the way people carry their cellular phones with their StarTAC line, has developed a new series of cellular phones that will be the smallest in the United States when they begin shipping. The new Motorola V series will weigh only 2.7 ounces and have up to 160 minutes of talk time. This device is only millimeters thicker than its actual battery and is able to be easily worn around the neck or clipped onto a belt. Also from Motorola is the Smart Pager SP1300. This device is the world's first electric organizer and pager to use a completely graphical user interface (GUI). The device consists of an all-touch-screen LCD control pad with 1 MB of onboard memory. It can also be hooked up to a PC to share and synchronize data. Another innovative personal communication device is the Accent developed by Phillips. This device attaches to a cellular phone and acts as an address book, email in-box, and fax machine. The device consists of a touch screen that allows all of these features to be accessed easily. The device also allows users to look up friends names and then automatically dial their numbers.

Finally, the most complete mobile personal communications system is the Kenwood RadCam. This device is the first walkie-talkie device to transmit not only sound but pictures. The RadCam is simply hooked up to a radio transmitter and the user can then send a live picture taken by the user to another person.

6.3 Personal Aids

The next category of devices can best be described as personal aids. They are also called personal digital assistants or PDAs. These are mobile devices that more and more people are starting to use and carry regularly. These devices are most often used as personal organizers, schedulers, notepads, and contact managers. There are a wide variety of these products on the market from a number of manufacturers. One example of such technology is the Cassiopeia E-10 from Casio. This small hand-held device performs all of the functions stated above plus has an optional modem for sending and receiving email and features voice recording. The E-10 provides all this functionality in a lightweight $3'' \times 5''$ device that fits in the palm of your hand. It features a $4''$ backlit screen and can be controlled with one hand. These types of PDAs are generally controlled by touch screens, pens, or buttons, and many of them have full-blown operating systems or are at least compatible with Widows 95 systems for data sharing.

Another personal assistant type system is the Garmin StreetPilot. This portable device uses global positioning satellite (GPS) technology to give the user real-time position and map data from wherever they are. Users can use data cards with the device to pull up detailed street maps and directions to destinations they enter. The device also tells users where the location of the nearest businesses, attraction, shopping, or food stores are. With this device, users should never be lost in a new city or not know how to get to shopping or food services.

Another emerging technology that people will be carrying in the near future are Subscriber Identity Module (SIM) smartcards. These are credit card size modules that can be connected to a cellular phone that allow it to become essentially a network computer. The smartcards contain user information and storage for electronic commerce transactions. With e-commerce becoming more and more popular as Internet technology develops, these smartcards will allow users to purchase items from anywhere at anytime. The key issues in this emerging technology include electronic payment techniques, data compression, and optimization and encryption of transactions. With over 150 million estimated cellular phone users worldwide by the year 2000, this technology is expected to become a $300 billion industry by the year 2002 according to the U.S. Commerce Department.

Along similar lines, Akyman Financial Services (AFS) of Australia has developed the AFS-800 wireless electronic funds transfer terminal. This hand-held device is an Electronic Funds Transfer at Point of Sale (EFTPos) system that allows retailers to guarantee fund transfers from customer bank accounts to merchants. Traditionally these types of terminals were only available in stores as credit card/debit card readers. However, this new technology contains user credit and debit card information, payment authorization functions, and person-to-person or smartcard-to-computer links. This will allow users to pay vendors and other users electronically and from anywhere at anytime.

Finally, Audible has designed a hand-held portable device called the Mobile Player that is able to play audio from the Internet and listen to it anywhere. The Mobile Player weighs less than 3.5 ounces and is able to store up to 2 hours of spoken audio. It also comes with a docking system that allows the user to transfer audio through a PC serial port. The device works by the user downloading audio programs from Audible's web site. They can then playback their selections how and when they want. The system includes a one-touch bookmarking system, the ability to fast-forward, reverse, pause, and skip from program to program easily, prompts that let you know exactly where you are in each program, and a rechargeable battery.

The military is currently working on the development of the Personal Information Carrier (PIC) or digital dog tag. The PIC is a small electronic storage device containing medical information about the wearer. While old military dog tags contained only five lines of information, the digital tags may contain volumes of multimedia information including medical history, X rays, and cardiograms. Using hand-held devices in the field, medics would be able to call this information up in real time for better treatment. A fully functional transmittable device is still years off, but this technology once developed in the military, could be adapted to civilian users and provide any information, medical or otherwise, in a portable, nonobstructive, and fashionable way. Another future device that could increase safety and well being of its users is the nose-on-a-chip developed by the Oak Ridge National Lab in Tennessee. This tiny digital silicon chip about the size of a dime, is capable of "smelling" natural gas leaks in stoves, heaters, and other appliances. It can also detect dangerous levels of carbon monoxide. This device can also be configured to notify the fire department when a leak is detected. This nose chip, which should be commercially available within two years, is inexpensive, requires low power, and is very sensitive. Along with gas detection capabilities, this device may someday also be configured to detect smoke and other harmful gases. By embedding this chip into worker uniforms, name tags, etc., this could be a lifesaving computational accessory.

In addition to the future safety technology, soon to be available as accessories are devices that are for entertainment and security. The LCI computer group is developing a Smartpen that electronically verifies a user's signature. The increase in credit card use and the rise in forgeries, has brought with it the need for commercial industries to constantly verify signatures. This Smartpen writes like a normal pen but uses sensors to detect the motion of the pen as the user signs his or her name to authenticate the signature. This computational accessory should be available in 2000 and would bring increased peace of mind to consumers and vendors alike.

Finally, in the entertainment domain, Panasonic is creating the first portable hand- held DVD player. This device weighs less than 3 pounds and has a screen about 6″ across. The color LCD has the same 16:9 aspect ratio of a cinema screen and supports a high resolution of 280,000 pixels and stereo sound. The player can play standard DVD movies and has a 2 hour battery life for mobile use.

7. COMPUTATIONAL ACCESSORIES IN THE FUTURE

In addition to the clothing that people wear every day, there are many types of fashion accessories that complete one's outfit. Many of these accessories, such as watches, belts, hats, and jewelry are quite commonplace. By embedding digital technology into these accessories, the metaphor for computational clothing is expanded into other realms of fashion and wearables. This not only expands the functionality of a person's computational apparel but also allows for an increased fashion sense and individuality that only accessories can create. When looking at future computational accessories it is helpful to separate them into classifications of functionality. The accessory classifications, for both current and future conceptual digital accessories, are personal communications, personal organization, personal entertainment, and personal body monitoring. Both current computational accessories and future concepts for computational accessories can be found in each of these classifications.

In the past fifty years designers have spent considerable time and energy speculating about what the future will hold, with particular interest in the future relationship between people and computers. Conceptualizing future computational accessories has become a booming industry that is driving the development of new mobile and wearable computers. Often the realm of designers and fine artists, conceptual computational accessories focus as much on functionality as they do on both fashion and human–computer interaction. Some exciting places to look for conceptual computational accessories include, for example, Philips Vision of the future ongoing project, the Industrial Designers Society of America, IDEA awards concept category, the LG electronics design competition, and various design schools and firms around the world.

In the context of computational accessories and wearables of the future, students at Carnegie Mellon University were recently asked to spend a semester developing concepts for future digital accessories—with style and attitude. Below is an outline of some of their projects.

- Snuggers (Figure 15.18), a concept created by Kat Cohen, is a wearable personal reminder designed expressly for preteens with that urge for independence.
- The Occhio concept (Figure 15.19), designed by Ignacio Filipino, is a fashionable virtual view monocle display.

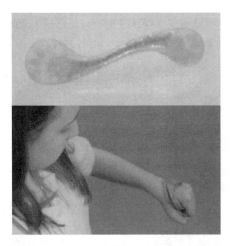

FIG. 15.18. "Snuggers" a wearable device, by Kat Cohen, CMU.

FIG. 15.19. "Occhio" wearable device, by Ignacio Filipino, CMU.

FIG. 15.20. "Kneph" wearable device, by Elizabeth Geuder, CMU.

FIG. 15.21. Wearable digital jewel pops, by Magnaniís, CMU.

- The Kneph neck jewel (Figure 15.20), created by Elizabeth Geuder, is a wearable personal recording device for creative endeavors.
- Nicole Magnaniís' wearable digital jewel pops open into a hand-held display (Figure 15.21).
- Jocelyn Pollackís' wrist worn personal assistant (Figure 15.22) can be easily removed and used as a phone.
- The Optra (Figure 15.23), created by Michael Benvenga, is a series of four wearable projectors that together allow the display of various sized information.

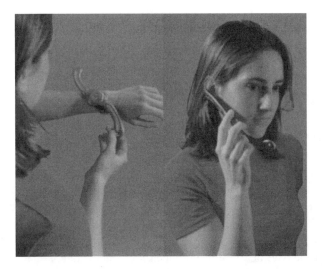

FIG. 15.22. Wrist worn personal asssitant, by Jocelyn Pollackís, CMU.

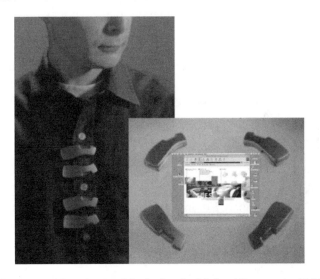

FIG. 15.23. "Optra" wearable device, by Michael Benvenga, CMU.

8. SUMMARY

As the number and complexity of wearable computing applications continues to grow, there will be increasing needs for systems that are faster, lighter, and have higher resolution displays. Better networking technology will also need to be developed to allow all users of wearable computers to have high bandwidth connections for real-time information gathering and collaboration. In addition to the technology advances that make users need to wear computers in everyday life, there is also the desire to have users want to wear their computers. In order to do this, wearable computing needs to be unobtrusive and socially acceptable. By making wearables smaller and lighter, or actually embedding them in clothing, users can conceal them easily and wear them comfortably.

To summarize, in this chapter we presented concepts related to the design and use of computational clothing and clothing accessories. As shown in this chapter, there are many application areas for this technology such as medicine, manufacturing, training, and recreation. Computational clothing will allow a much closer association of information with the user. By embedding sensors in the wearable to allow it to see what the user sees, hear what the user hears, sense the user's physical state, and analyze what the user is typing, an intelligent agent may be able to analyze what the user is doing and try to predict the resources he or she will need next or in the near future. Using this information, the agent may download files, reserve communications bandwidth, post reminders, or automatically send updates to colleagues to help facilitate the user's daily interactions. This intelligent wearable computer would be able to act as a personal assistant, who is always around, knows the user's personal preferences and tastes, and tries to streamline interactions with the rest of the world.

Acknowledgments   Dr. Woodrow Barfield would like to thank ONR (NOO0149710388) for an equipment grant that has supported research in the area of wearable computers. Dr. Steve Mann would like to thank Xybernaut Corp., Digital Equipment Corp., Waverider, ViA, Virtual-Vision, HP labs, Compaq, Kopin, Chuck Carter, Bob Kinney, and Thought Technologies Limited.

REFERENCES

Barfield, W., and Baird, K. (1998). Future Directions in Virtual Reality: Augmented Environments Through Wearable Computers, *VR'98 Seminar and Workshop on Virtual Reality*, Kuala Lumpur, April 24–15.

Bass, T. (1985). *The Eudaemonic pie*, Houghton Mifflin Company, Boston.

Bass, L., Kasabach, C., Martin, R., Siewiorek, D., Smailagic, D., and Stivoric, J. (1997). The Design of a Wearable Computer, *Proceedings of CHI' 97*, 139–146.

Craik, J. (1994). *The Face of Fashion*, Routledge Press.

Erwin, M. D., Kinchen, L. A., and Peters, K. A. (1979). *Clothing for Moderns*, Macmillan Publishing Co., Inc., New York.

Frankl, V. E. (1994). *Man's Search for Meaning*, Washington Square Press, New York.

Gemperle, F., Kasabach, C., Stivoric, J., Bauer, M., and Martin, R. (1998). Design for Wearability, *Second International Symposium on Wearable Computers*, Oct. 19–20, Pittsburgh, PA, 116–122.

Hall, E. T. (1982). *The Hidden Dimension*, Anchor Books.

Hatch, K. L. (1993). *Textile Science*, West Publishing Company, St. Paul, MN.

Kadolph, S. J. and Langford, A. L. (1998). *Textiles*, Prentice-Hall Inc., Englewood Cliffs, NJ.

Lind, E. J., Jayaraman, S., Rajamanickam, R., Eisler, R., and McKee, T. (1997). A Sensate Liner for Personnel Monitoring Applications, *First International Symposium on Wearable Computers*, Oct. 13–14, Cambridge, MA, 98–105.

Lyle, D. S. and Brinkley J. (1983). *Contemporory Clothing*, Bennett and McKnight Publishing Co., Chicago Illinois.

Mann, S. (1994). Mediated Reality, *TR 260, M.I.T. Media Lab Perceptual Computing Section*, Cambridge, MA.

Mann. S. (1996) 'Smart Clothing': Wearable Multimedia and 'Personal Imaging' to Restore the Balance between People and Their Intelligent Environments, *Proceedings, ACM Multimedia 96*, Boston, MA, Nov. 18–22, 163–174, http://wearcam.org/acm- mm96.htm.

Mann, S. (1997a). Wearable Computing: A First Step Toward Personal Imaging, *Computer*, Vol. 30.

Mann, S. (1997b). Eudaemonic Computing: Unobtrusive Embodiments of 'WearComp' for Everyday Use, *IEEE Proceedings of the First International Symposium on Wearable Computing*, Cambridge, MA, Oct. 13–14, http://wearcomp.org/eudaemonic.html.

Mann, S. (1997c). Smart Clothing: The Wearable Computer and WearCam, *Personal Technologies*, Vol. 1, No. 1, 21–27.

Mann, S. (1998). Humanistic Intelligence/Humanistic Computing: 'WearComp' as a New Framework for Intelligent Signal Processing, *Proceedings of the IEEE*, Vol. 86, No. 11, 2123–2151+cover, http://wearcam.org/procieee.htm.

Mcbriarity, J. P. and Henry, N. W. (Eds) (1992). *Performance of Protective Clothing*, ASTM STP 1133, ASTM, Philadelphia, PA.

McCormick, E. J. and Sanders, M. (1982). *Human Factors In Engineering and Design*. McGraw- Hill, Inc.

Paradiso, J. A. and Hu, E. (1997). Expressive Footware for Computer-Augmented Dance Performance, *First International Symposium on Wearable Computers*, Oct. 13–14, Cambridge, MA, 165–166.

Picard, R. and Healey, J. (1997), Affective Wearables, *First International Symposium on Wearable Computers*, Oct. 13–14, Cambridge, MA, 90–97.

Post, E. R. and Orth, M. (1997). Smart Fabric, or "Wearable Clothing," *First International Symposium on Wearable Computers*, Oct. 13–14, Cambridge, MA, 167–168.

Siegel, J. and Bauer, M. (1997). A Field Usability Evaluation of a Wearable System, *First International Symposium on Wearable Computers*, Oct. 13–14, Cambridge, MA. 18–22.

Spitzer, M. B., Rensing, R. R., McClelland, and P. Aquilino, (1997). Eye-Glass Systems for Wearable Computing, *First International Symposium on Wearable Computers*, Oct. 13–14, Cambridge, MA, 48–51.

Sutherland, I. (1968). "A Head-Mounted Three-Dimensional Display," *Proceedings, Fall Joint Computer Conference*, 757–764.

Tilley, A. R. (1993). *The Measure of Man and Woman*. Henry Dreyfuss and Associates, New York.

16

Situation Aware Computing
with Wearable Computers

Bernt Schiele, Thad Starner, Brad Rhodes,
Brian Clarkson, and Alex Pentland
Massachusetts Institute of Technology

1. MOTIVATION FOR CONTEXTUAL
AWARE COMPUTING

For most computer systems, even virtual reality systems, sensing techniques are a means of getting input directly from the user. However, wearable sensors and computers offer a unique opportunity to redirect sensing technology toward recovering more general user context. Wearable computers have the potential to "see" as the user sees, "hear" as the user hears, and experience the life of the user in a "first-person" sense. This increase in contextual and user information may lead to more intelligent and fluid interfaces that use the physical world as part of the interface.

Wearable computers are excellent platforms for contextually aware applications, but these applications are also necessary to use wearables to their fullest. Wearables are more than just highly portable computers; they perform useful work even while the wearer isn't directly interacting with the system. In such environments the user needs to concentrate on his or her environment, not on the computer interface, so the wearable needs to use information from the wearer's context to be the least distracting. For example, imagine an interface that is aware of the user's location: While being in the subway, the system might alert the user with a spoken summary of an e-mail. However, during a conversation the wearable computer may present the name of a potential caller unobtrusively in the user's head-up display, or simply forward the call to voicemail.

The importance of context in communication and interface cannot be overstated. Physical environment, time of day, mental state, and the model each conversant has of the other participants can be critical in conveying necessary information and mood. An anecdote from Nicholas Negroponte's book *Being Digital* [Negroponte, 1995] illustrates this point:

> Before dinner, we walked around Mr. Shikanai's famous outdoor art collection, which during the daytime doubles as the Hakone Open Air Museum. At dinner with Mr. and Mrs. Shikanai, we were joined by Mr. Shikanai's private male secretary who, quite significantly, spoke perfect English, as the Shikanais spoke none at all. The conversation was started by Wiesner, who expressed great interest in the work by Alexander Calder and told about both MIT's and his own personal experience with that great artist. The secretary listened to the story and then translated it from beginning to end, with Mr. Shikanai listening attentively. At the end, Mr. Shikanai reflected, paused, and then looked up at us and emitted a shogun-size "Ohhhh."

> The male secretary then translated: "Mr. Shikanai says that he too is very impressed with the work of Calder and Mr. Shikanai's most recent acquisitions were under the circumstances of . . . " Wait a minute. Where did all that come from?

> This continued for most of the meal. Wiesner would say something, it would be translated in full, and the reply would be more or less an "Ohhhh," which was then translated into a lengthy explanation. I said to myself that night, if I really want to build a personal computer, it has to be as good as Mr. Shikanai's secretary. It has to be able to expand and contract signals as a function of knowing me and my environment so intimately that I literally can be redundant on most occasions.

There are many subtleties to this story. For example, the "agent" (i.e., the secretary) *sensed* the physical location of the party and the particular object of interest, namely, the work by Calder. In addition, the agent could attend, parse, understand, and translate the English spoken by Wiesner, *augmenting* Mr. Shikanai's abilities. The agent also *predicted* what Mr. Shikanai's replies might be based on a *model* of his tastes and personal history. After Mr. Shikanai consented/specified the response "Ohhhh," the agent took an appropriate action, filling in details based on a model of Wiesner and Negroponte's interests and what they already knew. One can imagine that Mr. Shikanai's secretary uses his model of his employer to perform other functions as well. For example, he can remind Mr. Shikanai of information from past meetings or correspondences. The agent can prevent "information overload" by attending to complicated details and prioritizing information based on its relevancy. In addition, he has the knowledge and social grace to know when and how Mr. Shikanai should be interrupted for other real-time concerns such as a phone call or upcoming meeting. These kinds of interactions suggest the types of interfaces a contextually aware computer might assume.

While the computer interface described in *Being Digital* is more of a long-term goal than what can be addressed by current technology, many situationally aware applications are doable. This chapter summarizes several of the current wearable computing and augmented reality research projects at the MIT Media Laboratory that explore the dimensions of user and physical modeling. In particular, the Remembrance Agent and Augmented Reality Remembrance Agent describe applications made possible by contextual awareness. The DyPERS, Wearable Computer American Sign Language Recognizer, and DUCK! projects also have associated applications, but the emphasis for these projects is to push the sensor and context recognition technology to new limits. Finally, the Environmentally-A-Wearable project demonstrates new pattern recognition technology that can be used in the next generation of contextually aware applications. For more complete information on a particular project and related work, the reader is encouraged to refer to the original papers on these projects.

2. REMEMBRANCE AGENT

The Remembrance Agent (RA) is a program that continuously "watches over the shoulder" of the wearer of a wearable computer and displays one-line summaries of notes-files, old e-mail, papers, and other text information that might be relevant to the user's current context [Rhodes, 1997]. These summaries are listed in the bottom few lines of a head-up display, so the wearer can read the information with a quick glance. To retrieve the whole text described in a summary line, the wearer hits a quick chord on a chording keyboard.

The original RA was entirely text based. On the input side, the user would enter or read notes, papers, e-mail, or other text either on a wearable or a desktop computer. The RA would continuously watch a segment of the text being entered or read, and would find and suggest the "most relevant" documents from a set of pre-indexed text. Relevance was determined using text-retrieval techniques similar to those used in web search engines. While the wearable and desktop versions worked the same, the wearable version tended to allow more interactive, real-time usage. For example, someone taking notes on a conversation with the wearable RA is often able to connect the current conversation to previously taken notes, which might prompt more insightful questions.

The current version of the wearable RA still uses the text-based input but adds many other general fields by which a current context can be described. For example, a user's context might be described by a combination of the current time of day and day of the week (provided by the wearable's system clock), location (provided by an infrared beacon in the room), who is being spoken to (provided by an active badge), and the subject of the conversation (as indicated by the notes being taken). The suggestions provided by the RA are based by a combination of all these elements.

To ensure that the RA is useful in a wide variety of domains, the design makes as few assumptions as possible about the application domain. The information suggested can be any form of text or any information tagged with text, time, location, or person information. Similarly, few assumptions about the user's context are made. Often the RA could make more finely honed suggestions if a more specific domain were assumed. For example, if the RA were used by a Federal Express delivery person, many deductions could be made from routing and package information, and much more specific and potentially useful information could be suggested. However, such a deductive engine would be difficult to apply outside of the delivery domain. For the sake of making a general system, this research is attempting to push the envelope by producing as useful suggestions as possible while still making as few assumptions as possible about the application domain.

Overlay Versus Augmented Reality   The RA outputs suggestions on a head-up display (HUD), which in normal use provides some but not all features expected from an augmented reality interface. Most importantly, the HUD allows the RA to get the wearer's attention when presenting an important suggestion. This is an important distinction from a palm-top interface, where the display is only visible when the user thinks to look at it. The HUD also provides an overlay effect, where the wearer can both read the suggestion and view the real world at the same time. However, in normal use the RA does not register its annotations with specific objects or locations in the real world as one might expect from a full augmented reality system. In most cases such a "real-world fixed" display wouldn't even make sense, since suggestions often are conceptually relevant to the current situation without being relevant to a specific object or location.

2.1 Augmented Reality Remembrance Agent

One of the most distinctive advantages of wearable computing is the coupling of the virtual environment with the physical world. Thus, determining the presence and location of physical objects relative to the user is an important problem. Once an object is uniquely labeled, the user's wearable computer can note its presence or assign virtual properties to the object. Hypertext links, annotations, or Java-defined behaviors can be assigned to the object based on its physical location [Starner et al., 1997b; Nagao and Rekimoto, 1995]. This form of ubiquitous computing [Weiser, 1991] concentrates infrastructure mainly on the wearer as opposed to the environment, reducing costs and maintenance and avoiding some privacy issues. Mann [Mann, 1997] argues in favor of mobile, personal audio-visual augmentation in his wearable platform.

Objects can be identified in a number of different ways. With Radio Frequency Identification (RFID), a transmitter tag with a unique ID is attached to the object to be tracked [Hull et al., 1997]. This unique ID is sensed by special readers, which can have ranges from a few inches to several miles depending on the type and size of the tag. Unfortunately, this method requires a significant amount of physical infrastructure and maintenance for placing and reading the tags.

Computer vision provides several advantages over RFID. The most obvious is to obviate the need for expensive tags for the objects to be tracked. Another advantage of computer vision is that it can adapt to different scales and ranges. For example, the same hardware and/or software may recognize a thimble or a building depending on the distance of the camera to the object. Computer vision is also directed: If the computer identifies an object, the object is known to be in the field of view of the camera. By aligning the field of view of the eye with the field of view of the camera, the computer may observe the objects that are the focus of the user's attention.

FIG. 16.1. Multiple graphical overlays aligned through visual tag tracking. Such techniques as shown in the following three images can provide a dynamic, physically realized extension to the World Wide Web.

We have used computer vision identification to create a physically based hypertext demonstration platform [Starner et al., 1997b] as shown in Figure 16.1. The system was later extended by Jeff Levine as part of his Master's thesis [Levine, 1997]. Even though the system requires the processing power of an SGI, it maintains the feel of a wearable computer by sending video to and from the SGI and head-mount wirelessly. Visual "tags" uniquely identify each active object. These tags consist of two red squares bounding a pattern of green squares representing a binary number unique to that room. A similar identification system has been demonstrated by Nagao and Rekimoto [1995] for a tethered, hand-held system. These visual patterns are robust in the presence of similar background colors and can be distinguished from each other in the same visual field. Once an object is identified, text, graphics, or a texture mapped movie can be rendered on top of the user's visual field using a head-up display as shown in Figure 16.1. Since the visual tags have a known height and width, the visual tracking code can recover orientation and distance, providing 2.5D information to the graphics process. Thus, graphics objects can be rotated and zoomed to match their counterparts in the physical world. This system is used to give mini-tours of the laboratory space as shown in Figure 16.2. Active LED tags are shown in this sequence, though the passive tags work as well. Whenever the camera detects a tag, it renders a small red arrow on top of that object indicating a hyperlink. If the user is interested in that link and turns to see it, the object is labeled with text. Finally, if the user approaches the object, 3D graphics or a texture mapped movie are rendered on the object to demonstrate its function. Using this strategy, the user is not overwhelmed upon walking into a room but can explore interesting objects at leisure.

FIG. 16.2. When a tag is first located, a red arrow is used to indicate a hyperlink. If the user shows interest by staring at the object, the appropriate text labels are displayed. If the user approaches the object, 3D graphics or movie sequences are shown.

This physically based hypertext system has proven very stable and intuitive to use by visitors to the laboratory. However, can this system be generalized to work without explicit tagging of objects? To answer this question, the variant of the system is being used to visually identify buildings to create an augmented reality tourist agent for downtown Boston.

Using GPS and an inertial head tracking system, strong priors can be established on what buildings may be visible using a hand-built associative model of the city. This lessens the burden of the vision system from trying to distinguish between all potential objects the tourist may see over the day to the handful that might be currently visible. In addition, the inertial tracker can be used as a means of direct control for the user. For additional information, the tourist simply stares at the building of choice. The system recognizes this lack of head motion as a fixation point and attempts identification using computer vision, conditioned on location and head orientation. Currently, the multidimensional histogram techniques developed by Schiele [Schiele, 1997] are being used to experiment with visual identification. As vision software becomes stable, separate training and test "tours" of an area of Boston will be videotaped including fixation events as described above. The stream of synchronized GPS and head-tracking information will be recorded using a wearable computer. Three error measures can then be calculated. The first evaluates identification of fixation events. Next, assuming perfect recognition of fixation events, the error rate of the building recognition system will be calculated. Finally, the total combined accuracy of the system will be determined, using the same definition of accuracy as presented in the previous section. Depending on the results of these tests, experiments with additional sensors, such as a sonic range to target system, may be performed.

2.2 Dynamic Personal Enhanced Reality Agent

A recent extension of the above introduced augmented reality remembrance agent does not use a tag but a generic object recognizer in order to identify objects in the real world. The system, called "Dynamic Personal Enhanced Reality System" (DyPERS, [Schiele et al., 1999]), retrieves "media memories" based on associations with real objects the user encounters. These are evoked as audio and video clips relevant for the user and overlayed on top of real objects the user encounters. The system uses an audio–visual (A/V) association system with a wireless connection to a desktop computer. The user's visual and auditory scene is stored in real time by the system (upon request) and is then associated (by user input) with a snapshot of a visual object. The object acts as a key such that when the real-time vision system detects its presence in the scene again, DyPERS plays back the appropriate audio–visual sequence.

The system's building blocks are depicted in Figure 16.3. The audio–visual associative memory operates on a record-and-associate paradigm. Audio–visual clips are recorded by the push of a button and then associated

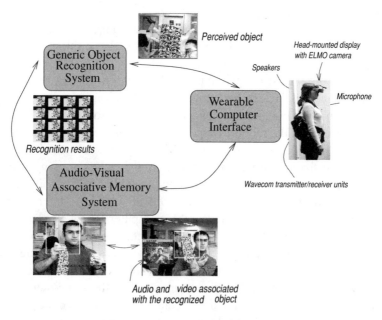

FIG. 16.3. DyPERS's architecture.

FIG. 16.4. Sample output through heads-up-display (HUD).

to an object of interest. Subsequently, the audio–visual associative memory module receives object labels along with confidence levels from the object recognition system. If the confidence is high enough, it retrieves from memory the audio–visual information associated with the object the user is currently looking at and overlays this information on the user's field of view.

Whenever the user is not recording or associating, the system is continuously running in a background mode trying to find objects in the field of view that have been associated to an A/V sequence. DyPERS thus acts as a parallel perceptual remembrance agent that is constantly trying to recognize and explain—by remembering associations—what the user is paying attention to. Figure 16.4 depicts an example of the overlay process. Here, in the top of the figure, an "expert" is demonstrating how to change the bag on a vacuum cleaner. The user records the process and then associates the explanation with the image of the vacuum's body. Thus, whenever the user looks at the vacuum (as in the bottom of the figure) he or she automatically sees an animation (overlaid on the left of his field of view) explaining how to change the dust bag. The recording, association, and retrieval processes are all performed online in a seamless manner.

An important part of the system is the generic object recognizer, based on a probabilistic recognition system. Objects are represented by multidimensional histograms of vector responses from local neighborhood operators. Simple matching of such histograms (using χ^2-statistics or intersection [Schiele, 1997]) can be used to determine the most probable object, independent of its position, scale, and image-plane rotation. Furthermore the approach is considerably robust to viewpoint changes. This technique has been extended to probabilistic object recognition [Schiele and Crowley,

FIG. 16.5. A DyPERS user listening to a guide during the gallery tour.

1996], in order to determine the probability of each object in an image only based on a small image region. Experiments showed that only a small portion of the image (between 15% and 30%) is needed to recognize 100 objects correctly in the presence of viewpoint changes and scale changes. The recognition system runs at approximately 10 Hz on a Silicon Graphics O2 machine using the OpenGL extension library for real-time image convolution.

Obviously, the discrimination of 100 objects is not enough to be of practical use in an unconstrained real-world scenario. However, by using information about the physical environment, including the location of the user, the time of day, and other available information, the number of possible objects can be significantly reduced. Furthermore, information about the user's current interests further reduces the number of interesting objects.

The current system has been used in a museum tour scenario: (See Figure 16.5.) A small gallery was created using twenty poster-sized images of various famous works ranging from the early sixteenth century to contemporary art. Three classes of users in different interaction modes were asked to walk through the gallery while a guide was reading a script that described the paintings individually. The guide presented biographical, stylistic, and other information for each of the paintings while the subjects either used DyPERS, took notes, or simply listened to the explanations. After the completion of the guide's presentation, the subjects were required to take a twenty-question multiple-choice test containing one query per painting presented. The users of the DyPERS system obtained slightly better results than the other test persons, indicating the possible usefulness of such a remembrance system.

Other applications of DyPERS using the record-and-associate paradigm are the following:

- Daily scheduling and to-do list can be stored and associated with the user's watch or other personal trigger object.
- A conversation can be recorded and associated with the individual's business card.
- A teacher records names of objects in a foreign language and associates them with the visual appearance of the object. A student could then use the system to learn the foreign language.
- A storyteller could read a picture book and associate each picture with its text passage. A child could then enjoy hearing the story by triggering the audio clips with different pages in the picture book.
- The system could be used for online instructions for an assembly task. An expert associates the image of the fully packaged item with animated instructions on how to open the box and lay out the components. Subsequently, when the vision system detects the components placed out as instructed, it triggers the subsequent assembly step.

Many of the listed scenarios are beyond the scope of this chapter. However, the list should convey to the reader the practical usefulness of a system such as DyPERS.

3. USER-OBSERVING WEARABLE CAMERAS

In the previous section, head-mounted camera systems face forward, trying to observe the same region as the user's eyes to identify objects in the user's environment. However, by changing the angle of the camera to point down, the user himself can be tracked. This novel viewpoint allows the user's hands, feet, torso, and even lips to be observed without the gloves or body suits associated with virtual reality gear. The hat-mount of Figure 16.7 provides a surprisingly stable mounting point for the camera with a built-in reference feature: the nose. Since the nose remains stable in the same area of the image and has a known color and size, it can serve as a calibration object for observing the rest of the body. Thus, the different lighting conditions of the mobile environment can be addressed.

In the following subsections, a user-observing wearable camera observes and detects the user's hand gestures in order to recognize American sign language. In Section 4 a user-observing camera in conjunction with a forward-pointing camera identifies the user's location as well as the user's task.

We are not aware of any other wearable computer systems under development that visually observe the user's body. However, room and desktop-based camera interaction systems are more common. Of recent interest is the desk-based sign language recognizer [Vogler and Metaxas, 1998], which uses three cameras to recover 3D movement of the arms similar what is described in the following subsection.

3.1   A Wearable Computer American Sign Language Recognizer

Starner proposed a system [Starner, 1995] for recognizing American Sign Language using colored gloves and a camera placed in the corner of the room. Mounting the camera onto the signer provides a unique view as shown in Figure 16.6. The eventual goal is to design a self-contained system into an ordinary-looking cap to (loosely) translate sign to English. The computer is installed in the back of the cap, the camera and speaker are hidden in the brim of the cap, and the headband is constructed of thin rechargeable batteries. Through the use of this wearable computer, a signer can converse with a nonsigner simply by donning the cap. While it is possible to create such a cap, the current system uses a tethered SGI for analysis.

The current system tracks the signer's hands by searching the camera image for blobs matching an a priori model of the subject's natural skin color. Second moment analysis [Horn, 1986] is performed on the blobs, which results in an eight-element feature vector for each hand: position, change in position, angle, eccentricity, mass, and magnitude of the first principal component of the blob. Tracking runs at 10 frames per second. Training and recognition occur using HMMs (hidden Markov models) [Young, 1993]. The system is evaluated using conventions established by

FIG. 16.6.   The baseball cap mounted camera and its perspective.

the speech recognition community. In this case, a database of 500 five-word sentences created from a forty-word lexicon [Humphries et al., 1990] is randomly divided into independent training and test sets. Accuracy is determined by the equation

$$Acc = \frac{N - S - D - I}{N},$$

where N is the total number of words, S is the number of words unrecognized or "substituted," D is the number of words deleted, and I is the number of words inserted. Using this measure, the system has been very successful with 98% accuracy with a grammar and 92% accuracy with no grammar. Details on this system and its evaluation can be found in Starner et al. [1998].

While this wearable system is being designed to be directly controlled by the user, the environment helps prototype wearable computing equipment and demonstrate a set of tools directed at recovering user context. Specifically, complex sets of time-varying signals (i.e., gestures) can be recognized from a self-observing body-mounted camera through the use of color blob analysis and HMMs. However, the user is constrained to looking straight ahead, and the system is tested and trained in the same space. What is necessary to generalize this system to identifying less constrained gestures in a mobile setting?

A current experiment, associated with the DUCK! environment below, uses the wearer's nose (as seen at the bottom of Figure 16.6) as a calibration object for adjusting the skin model during tracking. The nose provides a good model for the color of the user's skin and appears in a fixed place in the camera frame no matter how the user moves his or her head. Thus, as the user walks, the gesture tracker can continuously recalibrate to account for changes in lighting. Of course, such a tracker is subject to the caveat that it will not work in dark environments unless a light is provided in the cap. The evaluation of this new tracking system is simple: Use a video recorder to store the images from the cap camera during a normal day, run the tracker on the tapes, and count the number of "dropped" frames (not counting frames without sufficient illumination).

4. THE PATROL TASK

The "Patrol task" is an attempt to test techniques from the laboratory in less constrained environments. Patrol is a game played by MIT students every weekend in a campus building. The participants are divided into

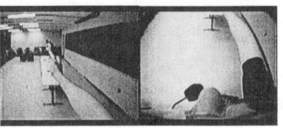

FIG. 16.7. (Left) The two-camera Patrol hat. (Right) The downward-
and forward-looking Patrol views.

teams denoted by colored headbands. Each participant starts with a rubber
suction dart gun and a small number of darts. The goal is to hunt the other
teams. If shot with a dart, the participant removes his headband, waits for
fighting to finish, and proceeds to the second floor before replacing his
headband and returning.

Originally, Patrol provided an entertaining way to test the robustness of
wearable computing techniques and apparatus for other projects, such as
hand tracking for the sign language recognizer described above. However,
it quickly became apparent that the gestures and actions in Patrol provided
a relatively well-defined language and goal structure in a very harsh "real-
life" sensing environment. As such, Patrol became a context-sensing project
within itself. The next subsections discuss current work on determining
player location and task using only on-body sensing apparatus.

Sensing for the Patrol task is performed by two hat-mounted wide-angle
cameras (Figure 16.7). The larger of the two cameras points downwards
to watch the hands and body. The smaller points forward to observe what
the user sees. Figure 16.7 shows sample images from the hat. While it is
possible to provide enough on-body computation to run feature detection in
real time, we currently record to videotape during the game for experimental
purposes.

4.1 Location

As mentioned earlier, user location often provides valuable clues to the
user's context. By gathering data over many days, the user's motions
throughout the day might be modeled. This model may then be used
to predict when the user will be in a certain location and for how long
[Orwant, 1993].

Today, most outdoor positioning is performed in relation to the Global Positioning System (GPS). Differential systems can obtain accuracies on the order of centimeters. Current indoor systems such as active badges [Want and Hopper, 1992; Lamming and Flynn, 1994] and beacon architectures [Long et al., 1996; Schilit 1995; Starner et al., 1997a] require increased infrastructure for higher accuracy, implying increased installation and maintenance. Here, we attempt to determine location based solely on the images provided by the Patrol hat cameras, which are fixed-cost on-body equipment.

The Patrol environment consists of fourteen rooms that are defined by their strategic importance to the players. The rooms' boundaries were not chosen to simplify the vision task but are based on the long-standing conventions of game play. The playing areas include hallways, stairwells, classrooms, and mirror image copies of these classrooms whose similarities and "institutional" decor make the recognition task difficult.

Hidden Markov models (HMMs) were chosen to represent the environment due to their potential language structure and excellent discrimination ability for varying time domain processes. For example, rooms may have distinct regions or lighting that can be modeled by the states in an HMM. In addition, the previous known location of the user helps to limit his or her current possible location. By observing the video stream over several minutes and knowing the physical layout of the building and the mean time spent in each area, many possible paths may be hypothesized and the most probable chosen based on the observed data. HMMs fully exploit these attributes. For a review of HMMs see Rabiner [1989].

As a first attempt, the means of the red, green, blue, and luminance pixel values of three image patches are used to construct a feature vector in real time. One patch is taken from the image of the forward-looking camera. This patch varies significantly due to the head motion of the player. The next patch represents the coloration of the floors and is derived from the downward-looking camera in the area just to the front of the player and out of range of average hand and foot motion. Finally, since the nose is always in the same place relative to the downward-looking camera, a patch is sampled from the nose, providing information about the lighting variations as the player moves through a room.

Approximately 45 minutes of annotated Patrol video were analyzed for this experiment (6 frames per second); 24.5 minutes of video, comprising 87 area transitions, are used for training the HMMs. As part of the training, a statistical (bigram) grammar is generated. This "grammar" is used in testing

TABLE 16.1
Patrol Area Recognition Accuracy

Method	Training Set	Test Set
2-state HMM	51.72%	21.82%
3-state HMM	68.97%	81.82%
4-state HMM	65.52%	76.36%
5-state HMM	79.31%	40.00%
Nearest Neighbor	−400%	−485.18%

to weight those rooms which are considered next based on the current hypothesized room. An independent 19.3 minutes of video, comprising 55 area transitions, are used for testing. Note that the computer must segment the video at the area transitions as well as label the areas properly.

Table 16.1 demonstrates the accuracies of the different methods tested. For informative purposes, accuracy rates are reported both for testing on the training data and the independent test set. Accuracy is calculated by $Acc = \frac{N-D-S-I}{N}$, where N is the total number of areas in the test set, D (deletions) is the number of area changes not detected, S (substitutions) is the number of areas falsely labeled, and I (insertions) is the number of area transitions falsely detected. Note that, since all errors are counted against the accuracy rate, it is possible to get large negative accuracies by having many insertions.

The simplest method for determining the current room is to determine the smallest Euclidean distance between a test feature vector with the means of the feature vectors comprising the different room examples in the training set. Given this nearest-neighbor method as a comparison, it is easy to see how the time duration and contextual properties of the HMMs improve recognition. Testing on the independent test set shows that the best model is a 3-state HMM, which achieves 82% accuracy. In some cases accuracy on the test data is better than the training data, probably due to changing video quality from falling battery voltage.

Another important attribute is how well the system determines when the player has entered a new area. Figure 16.8 compares the 3-state HMM and nearest-neighbor methods to the hand-labeled video. Different rooms are designated by two-letter identifiers. As can be seen, the 3-state HMM system tends to be within a few seconds of the correct transition boundaries while the nearest-neighbor system oscillates between many hypotheses.

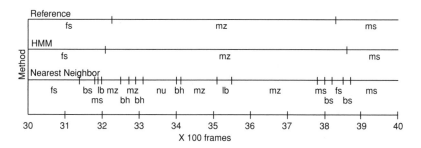

FIG. 16.8. Typical detection of Patrol area transitions.

As mentioned earlier, one of the strengths of the HMM system is that it can collect evidence over time to hypothesize the player's path through several areas. How much difference does this incorporation of context make on recognition? To determine this, the test set was segmented by hand, and each area was presented in isolation to the 3-state HMM system. At face value this should be a much easier task since the system does not have to segment the areas as well as recognize them. However, the system only achieved 49% accuracy on the test data and 78% accuracy on the training data. This result provides striking evidence of the importance of using context in this task and hints at the importance of context in other user activities.

4.2 User Tasks

By identifying the user's current task, the computer can assist actively in that task by displaying timely information or automatically reserving resources that may be needed [Feiner et al., 1993; Starner et al., 1997b]. However, a wearable computer might also take a more passive role, simply determining the importance of potential interruptions (phone, e-mail, paging, etc.) and presenting the interruption in the most socially graceful manner possible.

Here we describe an experiment to recognize the user tasks aiming, reloading, and "other" tasks. In order to recognize such user tasks we use a generic object recognition system recently proposed by Schiele and Crowley (see Schiele and Crowley [1996] for details). In the context of the Patrol data this system can be used for recognition of image patches that correspond to particular motions of a hand, the gun, a portion of an arm, or any part of the background. By feeding the calculated probabilities as feature vectors to a set of hidden Markov models, it is possible to recognize different user tasks such as aiming and reloading.

To use the recognition system we define a library of images grouped into images corresponding to the same action. Each image is split into 4 × 4 subimages used as an image patch database. In the experiment below we define three different image groups, one of each action so that the system calculates 3 groups ×16 = 48 probabilities at 10 Hz. These probabilities are then used as a feature vector for a set of HMMs trained to recognize different tasks of the user.

For two actions (aiming and reloading) we train a separate HMM containing 5 states on an annotated 2 min video segment containing 13 aiming actions and 6 reloading actions. Everything that is neither aiming nor reloading is modeled by a third class, the "other" class (10 sequences in total). The actions have been separated into a training set of 7 aiming actions, 4 reloading actions, and 3 other sequences for training of the HMMs. Interestingly, the actions are of very different length (between 2.25 s and 0.3 s). The remaining actions have been used as a test set. Table 16.2 shows the confusion matrix of the three action classes.

Aiming is relatively distinctive with respect to reloading and "other," since the arm is stretched out during aiming, which is probably the reason for the perfect recognition of the aiming sequences. However, reloading and "other" are difficult to distinguish, since the reloading action happens only in a very small region of the image (close to the body) and is sometimes barely visible.

These preliminary results are certainly encouraging, but they have been obtained for perfectly segmented data and a very small set of actions. However, an intrinsic property of HMMs is that they generalize to unsegmented data well. Furthermore the increase of the task vocabulary will enable the use of language and context models, which will help the recognition of single tasks.

TABLE 16.2
Confusion Matrix among Aiming, Reloading, and Other tasks

	Aiming	Reloading	"Other"
Aiming	6	0	0
Reloading	0	1	1
"Other"	0	1	6

4.3 Use of Patrol Context

While preliminary, the systems described above suggest interesting interfaces. By using head-up displays, the players could keep track of each other's locations. A strategist can deploy the team as appropriate for maintaining territory. If aim and reload gestures are recognized for a particular player, the computer can automatically alert nearby team members for aid.

Contextual information can be used more subtly as well. For example, if the computer recognizes that its wearer is in the middle of a skirmish, it should inhibit all interruptions and information. Similarly, a simple optical flow algorithm may be used to determine when the player is scouting a new area. Again, any interruption should be inhibited. On the other hand, when the user is "resurrecting" or waiting, the computer should provide as much information as possible to prepare the user for rejoining the game.

The model created by the HMM location system above can also be used for prediction. For example, the computer can weight the importance of incoming information depending on where it believes the player will move next. An encounter among other players several rooms away may be relevant if the player is moving rapidly in that direction. In addition, if the player is shot, the computer may predict the most likely next area for the enemy to visit and alert the player's team as appropriate. Such just-in-time information can be invaluable in such hectic situations.

5. ENVIRONMENTAL AWARENESS VIA AUDIO AND VIDEO

The Environmentally-A-Wearable (EW) uses auditory and visual cues to classify the user's environmental context. Like "the fly on the wall" (except now the fly is on your shoulder) it does not try to understand in detail every event that happens around the user. Instead, EW makes general evaluations of the auditory and visual ambiance and whether a particular environment is different or similar to an environment that the user was previously in. To use sight and sound is compelling because the user and the computer can potentially share perceptions of the environment. An immediate benefit is that the user naturally anticipates what the computer can and cannot observe.

An earlier version of the system [Clarkson and Pentland, 1998a], which analyzed audio alone, already was capable of differentiating speech from nonspeech. Such information can be used by a computer to decide if and how to present information to the user depending on whether the user is or is not involved in a conversation.

In order to make use of the audio–visual channel we construct detectors for specific events. Events can be simple, such as a bright light and loud sounds, or more complicated, such as speaker sounds and objects. Given a set of detectors higher order patterns can be observed. For example, a user's audio–visual environment can be broken into scenes (possibly overlapping) such as "talking to a person," "visiting the grocery store," "walking down a busy street," or "at the office" that are collections of specific events such as "footsteps," "car horns," "crosswalks," and "speech." We can recognize scenes by using detectors for low-level events that make up these scenes. This identifies a natural hierarchy in a person's audio–visual environment.

5.1   Sensors

An important question is: How exactly do we observe the auditory and visual environment of a person? There are many possibilities from close-talking microphones near the mouth to omnidirectional microphones fixed in clothing, and from cameras in eyeglasses that see what the user sees to wide-angle cameras that try to see everything at once. Choosing the appropriate setup is crucial for success. We concluded after some experimentation that in order to adequately sample the visual and aural environment of a mobile person, the sensors should be small and have a wide field of reception. In the EW, the environmental audio was collected with a lavalier microphone (the size of a pencil eraser) mounted on the shoulder and directed away from the user. The EW's video was collected with a miniature CCD camera (1/4″ diameter, 2″ long) attached to the user's back (pointing backwards). The camera was fitted with a 180° wide-angle lens giving an excellent view of the sky, ground, and horizon at all times (see Figure 16.9).

We chose features that are robust to errant noise like people passing by or small wavering in audio frequency. We want our features to respond only to obvious events such as walking into a building, crossing the street, and riding an elevator. Both the video and the audio features were calculated at a rate of 10 Hz, which is much faster than the rate at which the user's environment changes. This oversampling is advantageous for learning because it provides more data with which to make robust models.

Video   The visual field of the camera was divided into nine regions that correspond strongly to direction. From the pixel values (r, g, b) we

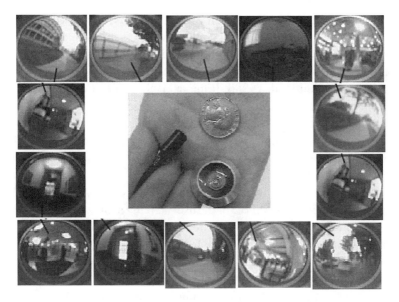

FIG. 16.9. The eyes and ears of the Environmentally-A-Wearable.

calculate the luminance $I = r + g + b$ as well as the chromatic channels $I_r = r/I$ and $I_g = g/I$. For each of the nine regions we calculate nine features including the three means and the six distinct covariances of I, I_r, and I_g. Hence, we are collapsing each region to a Gaussian in color space. This approximation lends robustness to small changes in the visual field, such as distant moving objects and small camera movements.

Audio Auditory features were extracted with 25 Mel-scaled filter banks. The triangle filters of the Mel-scaling give the same robustness to small variations in frequency (especially high frequencies), not to mention warping frequencies to a more perceptually meaningful scale.

5.2 From Events to Scenes

An individual's audio–visual environment is rich with repetition (going to work every morning, the weekly visit to the grocery store) that we can use to extract models for these situations. This means we do not need to make assumptions about which events can occur. We let the data speak for themselves by finding similar temporal patterns in the audio–visual data. This can be contrasted with the label–train–recognize approach taken by the speech recognition community.

The audio–visual data have their own set of units, which we call events, that are obtained by clustering the audio–video features in time. The EW clusters similar sequences of features into hidden Markov models (HMMs). These HMMs, which correspond to events, are later used to model scenes [Clarkson and Pentland, 1998b; Rabiner, 1989]. The HMM clustering algorithm can be directed to model the time series at varying time scales. Hence, by changing the time scale and repeating the clustering, we can build a hierarchy of events where each level of the hierarchy has a coarser time scale than the one below it.

For example, when we used a 3 s time scale for each event HMM, the emergent events are things such as closing doors, walking up stairs, and crosswalks. A specific example of the user arriving at his apartment building is shown in Figure 16.10. The figure shows the features (in the middle), segmentation (as dark vertical lines), and key frames for the sequence of events in question. The image in the middle represents the raw feature vectors (top 81 are video, bottom 25 are audio). Notice that you can see the sound of the user's steps in the audio features as vertical stripes (since audio features are just a form of spectrogram).

The EW takes these extracted events and learns their correlations in time. This allows the wearable to learn to recognize groups of events, which we call scenes. For example, suppose we wanted a model for a supermarket visit, or a walk down a busy street. The event clustering finds specific events such as supermarket music, cash register beeps, walking through

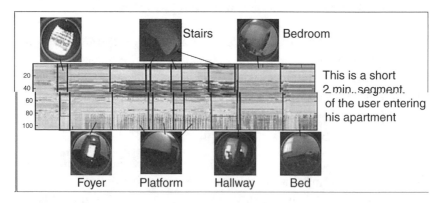

FIG. 16.10. Coming home: This example shows the user entering his apartment building, going up three stair cases, and arriving in his bedroom. The system's segmentation is depicted by the vertical lines along with key frames.

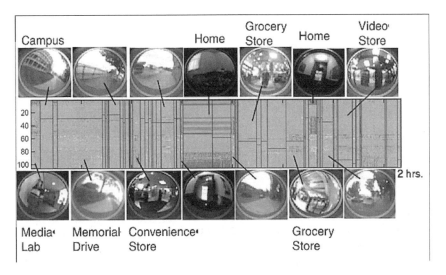

FIG. 16.11. The Scene Segmentation: Clustering the events in time gives a higher-level segmentation of the user's audio–visual history into scenes.

aisles, for the supermarket, and cars passing, crosswalks, and sidewalks for the busy street. By simply clustering raw audio–video features the system will not be able to capture the fact that events occur together to create scenes. So by clustering events themselves rather than low-level features, EW finds events that occur together and that therefore create a scene.

Figure 16.11 shows an example scene segmentation on roughly 2 hours of the user walking around the city and college campus. We evaluate performance by noting the correlation between our emergent models and a human-generated transcription. Each model plays the role of a hypothesis. A hypothesis is verified when its indexing correlates highly with a ground truth labeling. The table below shows some examples of events that matched closely with the event labeling:

Event Label	Office	Lobby	Bedroom	Cashier
Correlation Coeff.	0.9124	0.7914	0.8620	0.8325

The following table gives the correlations for some example scenes that matched with the scene labeling:

Scene Label	Student Dorms	Charles River	Necco Area	Sidewalk	Video Store
Correlation Coeff.	0.8024	0.6966	0.7495	0.7804	0.9802

From these results it is clear that unsupervised clustering of audio–video data is feasible and useful. Models that correlate highly with what humans consider to be meaningful events and scenes emerge from the raw data without prior knowledge engineering. Such information can be used to provide useful context for a variety of applications. Particularly interesting is the thought of learning the correlations between this environmental context and the context from explicit interaction with the wearable (for example the Remembrance Agent).

6. CONCLUSION

Wearable computers offer a new opportunity to sense a user's rich environment by becoming a platform for a wide range of cameras, microphones, and other sensors. At the same time, wearables need situationally aware applications more than traditional desktop computers, because wearables often need to operate in environments where the user is engaged in tasks other than interacting with the computer. This chapter has demonstrated several systems ranging from contextually aware applications to systems that give a wearable computer a rich, high-level understanding of the wearer's environment. These systems will in the future become the basis of new contextually aware applications.

REFERENCES

[Abowd et al., 1997] Abowd, G., Dey, A., Orr, R., and Brotherton, J. (1997). Context-awareness in wearable and ubiquitous computing. In *IEEE Intl. Symp. on Wearable Computers*. IEEE Computer Society.

[Azuma, 1997] Azuma, R. (1997). A survery of augmented reality. *Presence*, 6(4):355–386.

[Baum et al., 1996] Baum, W., Ettinger, G., White, S., Lozano-Pérez, T., Wells, W., and Kikinis, R. (1996). An automatic registration method for frameless stereotaxy, image guided surgery, and enhanced reality visualization. *IEEE Trans. Medical Imaging*, 15(2):129–140.

[Cho et al., 1997] Cho, Y., Park, J., and Neumann, U. (1997). Fast color fiducial detection and dynamic workspace extension in video see-through self-tracking augmented reality. In *Fifth Pacific Conference on Computer Graphics and Applications*.

[Clarkson and Pentland, 1998a] Clarkson, B. and Pentland, A. (1998a). Extracting context from environmental audio. In *Second Intl. Symposium on Wearable Computers*, pages 154–155.

[Clarkson and Pentland, 1998b] Clarkson, B. and Pentland, A. (1998b). Unsupervised clustering of ambulatory audio and video. Technical Report 471, MIT Media Lab, Perceptual Computing Group.

[Darrell et al., 1994] Darrell, T., Maes, P., Blumberg, B., and Pentland, A. (1994). A novel environment for situated vision and behavior. In *Proc. of CVPR-94 Workshop for Visual Behaviors*, pages 68–72, Seattle, Washington.

[Feiner et al., 1997] Feiner, S., MacIntyre, B., Hollerer, T., and Webster, T. (1997). A touring machine: Prototyping 3d mobile augmented reality systems for exploring the urban environment. In *IEEE Intl. Symp. on Wearable Computers*, Cambridge, MA.

[Feiner et al., 1993] Feiner, S., MacIntyre, B., and Seligmann, D. (1993). Knowledge-based augmented reality. *Communications of the ACM*, 36(7):52–62.

[Horn, 1986] Horn, B. (1986). *Robot Vision*. MIT Press, Cambridge, MA.

[Hull et al., 1997] Hull, R., Neaves, P., and Bedford-Roberts, J. (1997). Towards situaded computing. In *Proceedings of the First Intl. Symposium on Wearable Computers ISWC97*, Cambridge, MA.

[Humphries et al., 1990] Humphries, T., Padden, C., and O'Rourke, T. (1990). *A Basic Course in American sign Language*. T. J. Publ., Inc., Silver Spring, MD.

[Ishii and Ullmer, 1997] Ishii, H. and Ullmer, B. (1997). Tangible bits: Towards seamless interfaces between people, bits and atoms. In *Human Factors in Computing Systems: CHI'97 Conference Proceedings*, pages 234–241.

[Jebara et al., 1997] Jebara, T., Eyster, C., Weaver, J., Starner, T., and Pentland, A. (1997). Stochasticks: Augmenting the billiards experience with probabilistic vision and wearable computers. In *Preceedings of the First Intl. Symposium on Wearable Computers ISWC97*, Cambridge, MA.

[Kraut et al., 1996] Kraut, R., Miller, M., and Siegel, J. (1996). Collaboration in performance of physical tasks: Effects on outcomes and Communication. In *ACM Conference on Computer Supported Cooperative Work (CSCW)*, Boston, MA.

[Lamming and Flynn, 1994] Lamming, M. and Flynn, M. (1994). Forget-me-not: Intimate computing in support of human memory. In *FRIEND21: Intl. Symp. on Next Generation Human Interface*, pages 125–128, Meguro Gajoen, Japan.

[Levine, 1997] Levine, J. (1997). *Real-time target and pose recognition for 3-d graphical overlay*. Master's thesis, MIT, EECS.

[Long et al., 1996] Long, S., Kooper, R., Abowd, G., and Atkeson, C. (1996). Rapid prototyping of mobile context-aware applications: The cyberguide case study. In *MobiCom*. ACM Press.

[Mann, 1997] Mann, S. (1997). Wearable computing: A first step toward personal imaging. *IEEE Computer; http://wearcam.org/ieeecomputer.htm*, 30(2).

[Nagao and Rekimoto, 1995] Nagao, K. and Rekimoto, J. (1995). Ubiquitous talker: Spoken language interaction with real world objects. In *Proc. of Intl. Joint Conf. on Artifical Intelligence (IJCAI)*, pages 1284–1290, Montreal.

[Najjar et al., 1997] Najjar, L., Thompson, C., and Ockerman, J. (1997). A wearable computer for quality assurance inspectors in a food processing plant. In *IEEE Intl. Symp. on Wearable Computers*. IEEE Computer Society.

[Negroponte, 1995] Negroponte, N. (1995). *Being Digital*. Knopf.

[Ockerman et al., 1997] Ockerman, J., Najjar, L., and Thompson, C. (1997). Wearable computers for performance support. In *IEEE Intl. Symp. on Wearable Computers*. IEEE Computer Society.

[Orwant, 1993] Orwant, J. (1993). *Doppelganger goes to school: Machine learning for user modeling*. Master's thesis, MIT, Media Laboratory.

[Rabiner, 1989] Rabiner, L. (1989). A tutorial on hidden Markov models and selected applications in speech recongnition. *Proceedings of the IEEE*, 77(2):257–286.

[Rekimoto and Nagao, 1995] Rekimoto, J. and Nagao, K. (1995). The world through the computer: Computer augmented interaction with real world environments. *UIST'95*, pages 29–36.

[Rhodes, 1997] Rhodes, B. (1997). The wearable Remembrance Agent: A system for augmenting memory. *Personal Technologies*, 1(1):218–229.

[Rhodes and Starner, 1996] Rhodes, B. and Starner, T. (1996). Remembrance Agent: A continuously running automated imformation retrieval system. In *Proceedings of the First International Conference on the Practical Application of Intelligent Agents and Multi Agent Technology (PAAM'96)*, pages 487–495.

[Sawhney and Schmandt, 1998] Sawhney, N. and Schmandt, C. (1998). Speaking and listening on the run: Design for wearable audio computing. In *IEEE Intl. Symp. on Wearable Computers*.

[Schiele, 1997] Schiele, B. (1997). *Object recognition using multidimensional receptive field histograms*. PhD thesis, I.N.P. Grenoble. English translation.

[Schiele and Crowley, 2000] Schiele, B. and Crowley, J. (2000). Recognition without correspondence using multidimensional receptive field histograms. In *International Journal of Computer Vision* 36(1):31–50, Jan. 2000.

[Schiele et al., 1999] Schiele, B., Oliver, N., Jebara, T., and Pentland, A. (1999). An interactive computer vision system, DyPERS: dynamic and personal enhanced reality system. In *Intl. Conference on Computer Vision Systems*.

[Schilit, 1995] Schilit, W. (1995). *System architecture for context-aware mobile computing*. PhD thesis, Columbia University.

[Schmandt, 1994] Schmandt, C. (1994). *Voice Communication with Computers*. Van Nostrand Reinhold, New York.

[Sharma and Molineros, 1997] Sharma, R. and Molineros, J. (1997). Computer vision-based augmented reality for guiding manual assembly. *Presence*, 6(3).

[Smailagic and Martin, 1997] Smailagic, A. and Martin, R. (1997). Metronaut: A wearable computer with sensing and global communication campabilities. In *IEEE Intl. Symp. on Wearable Computers*. IEEE Computer Society Press.

[Smailagic and Siewiorek, 1994] Smailagic, A. and Siewiorek, D. (1994). The cmu mobile computers: A new generation of computer systems. In *COMPCON '94*. IEEE Computer Society Press.

[Starner 1995] Starner, T. (1995). *Visual recognition of American Sign Language using hidden Morkov models*. Master's thesis, MIT, Media Laboratory.

[Starner et al., 1997a] Starner, T., Kirsch, D., and Assefa, S. (1997a). The locust swarm: An environmentally-powered, networkless location and messaging system. Technical Report 431, MIT Media Lab, Perceptual Computing Group. Presented ISWC'97.

[Starner et al., 1997b] Starner, T., Mann, S., Rhodes, B., Levine, J., Healey, J., Kirsch, D., Picard, R., and Pentland, A. (1997b). Augmented reality through wearable computing. *presence*, 6(4): 386–398.

[Starner et al., 1998] Starner, T., Weaver, J., and Pentland, A. (1998). Real-time American Sign Language recognition using desk and wearable computer-based video. *IEEE Trans. Patt. Analy. and Mach. Intell*, 20(*):1371–****.

[Uenohara and Kanade, 1994] Uenohara, M. and Kanade, T. (1994). Vision-based object registration for real-time image overlay. Technical report, Carnegie Mellon University.

[Vogler and Metaxas, 1998] Vogler, C. and Metaxas, D. (1998). ASL recognition based on a coupling between HMMs and 3D motion analysis. In *ICCV*, Bombay.

[Want and Hopper, 1992] Want, R. and Hopper, A. (1992). Active badges and personal interactive computing objects. *IEEE Trans. on Consumer Electronics*, 38(1):10–20.

[Weiser, 1991] Weiser, M. (1991). The computer for the 21st century. *Scientific American*, 265(3):94–104.

[Wren et al., 1997] Wren, C., Azarbayejani, A., Darrell, T., and Pentland, A. (1997). Pfinder: Real-time tracking of the human body. *IEEE Trans. Patt. Analy. and Mach. Intell.*, 19(7):780–785.

[Young, 1993] Young, S. (1993). *HTK: Hidden Markov Model Toolkit V1.5*. Cambridge Univ. Eng. Dept. Speech Group and Entropic Research Lab. Inc., Washington DC.

17

Collaboration with Wearable Computers

Mark Billinghurst, Edward Miller, and Suzanne Weghorst

University of Washington

If, as it is said to be not unlikely in the near future, the principle of sight is applied to the telephone as well as that of sound, earth will be in truth a paradise, and distance will lose its enchantment by being abolished altogether.

—Arthur Strand, 1898

1. INTRODUCTION

A century has passed since magazine editor Arthur Strand imagined a world where people could communicate with anyone, anywhere, anytime [Strand 1898]. Today with the advent of portable computing and communications this is becoming possible. The nearly ubiquitous mobile phone allows people to have access to wearable collaborative audio spaces, extending their mouths and ears across continents. Videoconferencing and collaborative applications common on the desktop are appearing on the laptop, extending our sense of sight as well. Mobile computers and displays allow the use of visual and audio enhancements to further aid the communication process. As computing is applied to the task of communication the

539

Human–Computer Interface will give way to a Human–Human Interface mediated by computers.

As optimistic as this sounds, there are many problems that must be overcome before Strand's vision becomes a reality. Paramount among these is the task of providing universal access, enabling people to communicate wherever they are. One promising approach is through the newest generation of portable machines, wearable computers, coupled with improved wireless networking infrastructure. Worn on the body, wearable computers provide constant access to computing and communications resources. In general, a wearable computer may be defined as a computer that is subsumed into the personal space of the user, controlled by the wearer and has both operational and interactional constancy (i.e., is always on and always accessible) [Mann 97]. Wearables are typically composed of a belt or backpack computer, see-though or see-around head mounted display (HMD), wireless communications hardware such as a CDPD cellular modem, and a touchpad or chording keyboard input device. This configuration has been demonstrated in a number of real-world applications including aircraft maintenance [Esposito 97], navigational assistance [Feiner 97], and vehicle mechanics [Bass 97]. In such applications wearables have dramatically improved user performance, reducing task time by half in the case of vehicle inspection [Bass 97].

While wearable computers have been shown to be valuable for single–user applications, less research has been conducted on how they can enhance collaboration. This is despite the fact that many of the target application areas are those where the user could benefit from expert assistance, either local or remote. Several researchers have found that remote assistance significantly improves task performance in wearable applications [Siegal 95; Kraut 96]. However, these applications have involved connections between only one local and one remote user. Wearable computers can also be used to enhance communication among multiple remote people or between users at the same location.

The issue we are interested in addressing is how the computing power of a wearable computer can be used to support collaboration and communication. In particular we want to explore the following aspects:

- What visual and audio enhancements can be used to aid communication?
- How can a collaborative communications space be created among users?

- What is the effect of using wearable computers on communication between users?

These issues are becoming increasingly important as the telephone incorporates more computing power and as portable computers become more like telephones. A key question is whether or not it is necessary to use the visual and audio enhancements that wearable computers make possible: When do we need a wearable computer to mediate communication, and when is a conference phone call or shared whiteboard just as effective? In this chapter we discuss why wearable computers are an attractive platform for collaboration and we present several prototype collaborative interfaces. These prototypes show just some of the ways wearables could be used to support collaboration that goes beyond the conference phone call.

2. MOTIVATION: WHY COLLABORATION WITH WEARABLE COMPUTERS?

Certain attributes of wearable computers make them attractive as tools for collaboration. Fickas et al. [Fickas 97] identify the following key characteristics of wearable systems:

- *Hands-Free Operation:* Wearable computers can be used with one or no hands.
- *Mobility:* Wearable computers are not tethered, allowing the user to roam freely.
- *Augmented Reality:* See-through or see-around wearable displays allow the overlay of graphical information onto the real world.
- *Perception:* Wearable computers can be connected to sensors that measure aspects of the surrounding environment allowing the computer to respond in an intelligent and context-sensitive manner.

They suggest that augmented reality and the computer's ability to perceive aspects of its physical environment are the most novel aspects of wearable systems. These same attributes make wearable computers ideal platforms for computer supported collaborative work (CSCW) because they support two key aspects of collaborative interfaces: seamlessness and the ability to enhance reality.

2.1   Seamless Collaboration

A *seam* in an interface is a spatial, temporal, or functional constraint that forces the user to shift among a variety of spaces or modes of operation [Ishii 94]. For example, the seam between computer-based word processing and traditional pen and paper makes it difficult to produce digital copies of handwritten documents without a cumbersome translation step. Seams can be of two types:

- *Functional Seams:* Discontinuities between different functional workspaces, forcing the user to change modes of operation.
- *Cognitive Seams:* Discontinuities between existing and new work practices, forcing the user to learn new ways of working.

One of the most important functional seams is that between shared and interpersonal workspaces. The shared workspace is the common task area between collaborators, while the interpersonal space is the common communications space. In face-to-face conversation the shared workspace is often a subset of the interpersonal space, so there is a dynamic and easy change of focus between spaces using a variety of nonverbal cues. However, most CSCW systems have an arbitrary seam between the shared workspace and interpersonal space, for example, that between a shared whiteboard and a video window showing a collaborator (Figure 17.1). This prevents users who are looking at the shared whiteboard from maintaining eye contact

FIG. 17.1. The functional seam between a shared whiteboard and video window.

with their collaborators, an important nonverbal cue for conversation flow [Kleinke 86].

A common cognitive seam is that between computer-based and traditional tools. This seam causes the learning curve experienced by users who move from physical tools to their digital equivalents, such as the painter moving from canvas and oil to digital drawing tools. Grudin [Grudin 88] points out that CSCW tools are generally rejected when they force users to change the way they work; yet this is exactly what happens when computer-based collaborative interfaces make it difficult to use traditional tools in conjunction with the computer-based tools. There are many examples of seemingly effective collaborative interfaces that were rejected by users because of the additional learning associated with them and their lack of integration with current work methods. For example, Ehrlich describes an electronic calendar with a collaborative feature that checks other user's calendars before scheduling a meeting [Ehrlich 87]. For this to work effectively all of the people in the same group must use it, changing the way individuals work and forcing them to use an additional digital tool. The calendar was most useful for group managers and so was used by them as a collaboration tool, but it was ignored by most of the remaining group members.

The seam introduced by technologically mediated remote collaboration changes the nature of collaboration and produces communication behaviors that are different from face-to-face conversation. Sellen suggests that what makes the biggest difference is not the communication medium but whether the conversation is mediated or not [Sellen 95]. Comparing communication among audio-only, video-only, and face-to-face collaboration, Sellen found no difference in conversation structure between the audio- and video-only conditions. However, conversation structure in both these conditions differed significantly from face-to-face conversation. Even with no video delay, video-mediated conversation doesn't produce the same conversation style as face-to-face interaction [O'Malley 96]. This occurs because video cannot adequately convey the nonverbal signals so vital in face-to-face communication [Heath 91]. Thus, sharing the same physical space positively affects conversation in ways that is difficult to duplicate by remote means.

It is possible to design seamless interfaces that enhance collaboration. Ishii et al. developed the TeamWorkStation [Ishii 91] and ClearBoard [Ishii 92] interfaces, which use seamless design to remove the discontinuities in collaborative interfaces. TeamWorkStation removes the seam between the real world and shared workspace by combining video- and computer-based

FIG. 17.2. The ClearBoard seamless interface. (Image courtesy of H. Ishii, MIT Media Laboratory.)

tools. Video overlay on a collaborative whiteboard supports use of real-world and computer-based tools. ClearBoard addresses the seam between the individual and the shared workspace. By using work surfaces with large mirrors and applying video projection techniques, users can look directly at their workspace and see a projection of their collaborator behind it, as shown in Figure 17.2. Users can effectively and easily change focus, maintain eye contact, and use gaze awareness in collaboration; the result is an increased feeling of intimacy and copresence.

In order for an interface to minimize functional and cognitive seams it must have the following characteristics:

- It must support existing tools and work techniques.
- Users must be able to bring real-world objects into the interface.
- The shared workspace must be a subset of the interpersonal space.
- There must be audio and visual communication between participants.
- Collaborators must be able to maintain eye contact and gaze awareness.

Wearable computers exhibit these attributes and have the potential for seamless collaboration because of their mobility, augmented reality displays, and context-sensitive computing. The mobility of wearable computers means that they are able to support collaboration in the user's existing workplace. In contrast to many CSCW tools, the user does not have to go to a special workstation or teleconferencing room to collaborate. Wearable cameras and displays also enable users to bring remote collaborators into their workspace and send views of their workspace to distant collaborators. This has many potential applications for remote supervision or on-site training.

The use of see-through displays in wearable computing allows the overlay of computer graphics on the real world. This may be used by remote collaborators to annotate the user's view, or it may enhance face-to-face conversation by producing shared virtual models that can be manipulated by using traditional tools. Wearable augmented reality supports seamless collaboration with the real world, reducing the functional and cognitive discontinuities between participants. Context-sensitive computing means that the wearable computer can change functionality based on the surroundings or the user's behavior to seamlessly support the user's current tasks.

2.2 Enhancing Reality

Removing the seams in a collaborative interface is not enough. As Hollan and Stornetta point out [Hollan 92], CSCW interfaces may not be used if they provide the same experience as face-to-face communication; they must enable users to go "beyond being there" and enhance the collaborative experience.

Traditional CSCW research attempts to use computer and audio–visual equipment to provide a sense of remote presence. Measures of social presence [Short 76] and information richness [Draft 91] have been developed to characterize how closely telecommunication tools capture the essence of face-to-face communication. The hope is that collaborative interfaces will eventually be indistinguishable from actually being there.

Hollan and Stornetta suggest that this is the wrong approach. By considering face-to-face interaction as its own medium, it becomes apparent that this approach requires one medium to adapt to another, pitting the strengths of face-to-face collaboration against new CSCW interfaces. Mechanisms that are effective in face-to-face interactions may be awkward if they are replicated in an electronic medium, often making users reluctant to use the new medium. For example, the Cruiser video conferencing system was developed to replace face-to-face meetings and support remote awareness. However, the system was mostly used for brief conversations and to set up face-to-face meetings rather than replacing face-to-face collaboration [Fish 91]. In fact, it may be impossible for mediated collaborations to provide the same experience as face-to-face collaboration because of the nature of the medium [Heath 91]; however, it may be possible to compensate for these detrimental effects and to provide an even richer interaction environment.

Hollan and Stornetta argue that rather than using new media to imitate face-to-face collaboration, researchers should be considering what new attributes the media can offer that satisfy the needs of communication so well that people will use it regardless of physical proximity. A better way to develop interfaces for telecommunication is to focus on the *communication* aspect, not the *tele-* part. The main motivation should be developing tools that go beyond being there by identifying needs that are not met in face-to-face collaboration and evolving mechanisms that use new media to meet those needs.

Wearable computers are ideally suited for this approach. They allow normal face-to-face collaboration but enhance it with capabilities that satisfy previously unmet needs. Some of the limitations of normal face-to-face collaboration include the difficulty of archiving and retrieving conversations, accessing relevant external data, and producing supporting visual aids. Starner et al. [Starner 97] present single-user wearable applications that could be expanded to meet these needs, including:

- physically based hypertext in which graphics are overlaid over physical objects,
- a remembrance agent that continually searches the user's hard disk for information relevant to the current task and displays it in the user's field of view, and
- a face recognition tool that displays names and other information above people in the user's field of view.

Wearable computers are ideally suitable as a platform for collaborative interfaces because they enable the creation of collaborative interfaces that are seamless and enhance reality. Naturally, there are many different ways that these attributes can be used in real applications. We (and our collaborators) have developed several prototype interfaces to explore the effects of wearable technology on communication; in all cases we developed the application to explore how the characteristics of the interface affect the collaboration and which shared artifacts are needed to support communication. In the remainder of the chapter we present four interfaces, two for remote collaboration and two for collocated collaboration:

- Remote Collaboration:
  WearCom: An interface for multiparty conferencing that enables a user to see remote collaborators as virtual avatars surrounding them

in real space. Spatial visual and audio cues help overcome some of the limitations of current multiparty conferencing systems.

Block Party: An interface for collaboration between a user at a workstation and a user wearing a see-though head-mounted display. The workstation user can use virtual cues to aid the HMD user in performing a real-world task and can see video of the task environment.

- Collocated Collaboration:

Collaborative Web Space: An interface that allows people in the same location to view and interact with virtual Web pages floating about them in space. Users can collaboratively browse the Web while seeing the real world and use natural communication to talk about the pages they're viewing.

Shared Space: An interface that explores how augmented reality can enhance face-to-face collaboration and compares collaboration in this setting with that in an immersive virtual environment.

3. WEARABLE INTERFACES FOR REMOTE COLLABORATION

Remote collaboration is a particularly attractive application area for wearable computers. Many real-world tasks could be enhanced if people could easily access a remote expert and effectively convey information about their situation. Kraut et al. have found that users performing a bicycle repair task were able to complete repairs twice as fast when a remote expert provided assistance [Kraut 96]. In this case the user wore a head-mounted display and video camera, enabling him or her to share repair manual pages with the expert and give the expert a view of what the user was doing. The user didn't wear a computer but had wireless access to a computer displaying the repair manual pages. Similar results have been found by Siegal et al. [Siegal 95] and the British Telecom CamNet system [Garner 1997]. Figure 17.3 shows an emergency medic using CamNet to send images from a crash site back to the base hospital while talking to the waiting doctors about the patient's condition.

In these examples the wearable is just used as a two-way audio–visual display and input device. A key question is whether the visual and audio enhancements that the computing power of the wearable makes possible can improve collaboration. Do we need a wearable computer to mediate communication when a conference phone call may be just as effective?

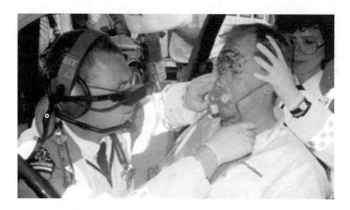

FIG. 17.3.   CamNet. (Image courtesy of British Telecom Laboratories Ltd.)

There is a large body of relevant research from the teleconferencing and CSCW fields that sheds light on this issue and suggest characteristics that wearable interfaces should have to enhance remote collaboration.

3.1   Related Work

Previous research on the roles of audio and visual cues in teleconferencing has produced mixed results. There have been many experiments conducted comparing face-to-face, audio-and-video, and audio-only communication conditions, as summarized by Sellen [Sellen 95]. While people generally do not prefer the audio-only condition, they are often able to perform tasks as effectively or almost as effectively as in the face-to-face or video conditions, suggesting that speech is the critical medium in teleconferencing [Whittaker 95].

Sellen reports that the main effect on collaborative performance is due to whether or not the collaboration is technologically mediated, not to the type of technology used. Naturally, this varies somewhat according to task. While Williams [Williams 77] reports that for cognitive problem-solving tasks face-to-face interaction is no better than speech-only communication, Chapanis [Chapanis 75] found that visual cues were important in tasks requiring negotiation.

There is, however, strong evidence that video transmits social cues and affective information, establishing "social presence" [Whittaker 97], although not as effectively as face-to-face interaction [Heath 91]. In general, the usefulness of video for transmitting nonverbal cues may be overestimated, and video may be better used to show the communication availability of others or views of shared workspaces [Whittaker 95]. Typically, when users do attempt nonverbal communication in a video conferencing environment their gestures are either distorted due to inadequate frame rates or must be wildly exaggerated to be recognized as the equivalent face-to-face gestures [Heath 91].

Based on these results, audio alone should be explored as a suitable medium for a shared communication space. An example of this, Thunderwire [Hindus 96], is a purely audio system that allows high-quality audio conferencing among multiple participants with the flip of a switch. In a three-month trial Hindus et al. found that audio can be sufficient for a usable communication space and that Thunderwire afforded a social space for its users. There were, however, several major problems with the approach, including:

- Users not being able to easily tell who else was within the space.
- Users not being able to use visual cues to determine one another's willingness to interact.

In addition, Thunderwire was rarely used by more than two or three users at once. With more users it becomes increasingly difficult to discriminate among speakers and there is a higher incidence of speaker overlap and interruptions. These problems are typical of audio-only spaces and suggest that while audio-only interfaces may be useful for small group interaction, they become less usable with more people present.

These shortcomings can be overcome through the use of visual and spatial cues. In face-to-face interaction, speech, gesture, body language, and other nonverbal cues combine to show attention and interest in collaborative conversations. However, the absence of spatial cues in most video conferencing systems means that users often find it difficult to know when people are paying attention to them, to hold side conversations, and to establish eye contact [Sellen 92]. This may explain the similarity in results between audio-only and video-and-audio teleconferencing conditions, and the differences they both exhibit from face-to-face results.

In collaborative "immersive" virtual environments spatial cues can enhance visual and audio cues in a natural way to aid communication [Benford 93]. The well-known "cocktail-party" effect shows that people can easily monitor several spatialized audio streams at once, selectively focusing on those of interest [Bregman 90]. Schmandt shows how a spatial sound system with nonspatial audio enhancements can allow a person to simultaneously listen to several sound sources [Schmandt 95]. Even a simple virtual avatar representation and spatial audio model of other users in the collaborative space enables users to discriminate among multiple speakers [Nakanishi 96]. Spatialized interactions are particularly valuable for governing interactions between groups of people, enabling crowds of people to inhabit the same virtual environment and interact in a way impossible in traditional video or audio conferencing [Benford 97].

These results suggest that, although it may be difficult to achieve, an ideal wearable interface for remote collaboration should have three elements:

- high-quality audio communication;
- visual representations of the collaborators; and
- an underlying spatial model for mediating interactions.

In the remainder of this section we describe two prototype interfaces we have developed that have these elements. The first is a wearable communications space designed to support communication among multiple wearable computer users and desktop computer users. The second is a task-based interface for collaboration between a wearable user and a remote expert on a desktop computer. The remote expert can manipulate virtual models in the wearable user's field of view to assist them in a real-world task.

3.2   WearCom: A Wearable Communication Space

Multiparty conferencing phone calls allow participants to share a collaborative audio space. However, as the number of participants increases it becomes increasingly difficult to distinguish among speakers, and communication may break down unless "asynchronous" speaker protocols are established. Providing the same spatial cues that people have in face-to-face conversations may reduce this problem. Taking advantage of the ability of current wearable computers to generate spatial audio and visual cues in real time, we have developed a prototype wearable communication space (WearCom) to evaluate the effects of spatialized audio and visual cues on communication.

One of the most important aspects of creating a collaborative communication interface is the visual and audio presentation of information. Most current wearable computers use see-through or see-around monoscopic displays in which command line or "desktop" interfaces are displayed. Combined with various spatial tracking options, however, visual and auditory information can be stabilized with respect to a variety of reference points—a concept first introduced for aircraft cockpit displays [Furness 88]. For ground-based body-worn systems, information can be presented in a combination of at least three ways:

Head-stabilized—information display is fixed relative to the user's field-of-view and doesn't change position as the user changes viewpoint orientation or position.

Body-stabilized—information display is fixed relative to the user's body position but adjusts within the field of view as the user changes viewpoint orientation. This requires the user's viewpoint orientation to be tracked relative to his or her body.

World-stabilized—information display is fixed to real-world locations and varies as the user changes viewpoint orientation and position. This requires the user's viewpoint position and orientation to be tracked relative to the environment.

Body- and world-stabilized displays are attractive for a number of reasons. As Reichlen [Reichlen 93] demonstrated, a body-stabilized information space can overcome the resolution limitations of head-mounted displays. In his work a user wears a head-mounted display while seated on a rotating chair. By tracking head orientation the user experiences a hemispherical information surround—in effect a "hundred million pixel display." World-stabilized information presentation enables annotation of the real world with context-dependent data, creating information enriched environments [Rekimoto 95] and increasing the intuitiveness of real-world tasks. For example, researchers at the University of North Carolina register virtual fetal ultrasound views on the womb to aid doctors in pregnancy planning [Bajura 92]. Despite these advantages, most wearables currently use only head-stabilized information display.

In our work we have chosen to begin with the simplest form of body-stabilized display, one which uses one degree of orientational freedom to give the user the impression that he or she is surrounded by a virtual cylinder of visual and auditory information. Figure 17.4 contrasts this with the traditional wearable display.

In the body-stabilized case we just track head motion about the vertical axis to change the user's view of the information space. Using only one

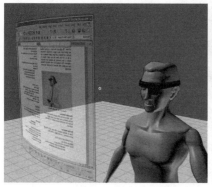

FIG. 17.4. A comparison of (left) head- and (right) body-stabilized
information displays.

degree of freedom has a number of advantages:

- Users cannot become easily disoriented.
- No additional input devices are needed to pan the display.
- It is natural to use, since most head and body motion is about the
  vertical axis.

Users can locate information more rapidly with this type of body-stabilized
information display than with a head-stabilized wearable information space
and can also more easily remember where information is within the display
space [Billinghurst 98].

With this display configuration it is possible to have remote collaborators
appear as virtual avatars or as live video streams distributed spatially about
the user (Figure 17.5). As they speak their audio streams can be spatialized
in real time so that they appear to emit from the corresponding avatar. Just
as in face-to-face collaboration, users can turn to face the collaborators
they want to talk to while still being aware of the other conversations taking place. Since the displays are see-through or see-around the user can
also see the real world at the same time, enabling the remote collaborators
to help them with real-world tasks. These remote users may also be using
wearable computers and head-mounted displays or could be interacting
via a desktop workstation. Naturally, the actual implementation of a system such as this depends on the amount of network bandwidth available.
With a high bandwidth wireless connection it may be possible to receive
and display several simultaneous audio and video streams. However, with

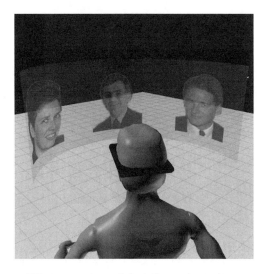

FIG. 17.5. A spatial conferencing space.

limited bandwidth it may be more appropriate to use static images and spatialize audio remotely.

3.2.1 Implementation

Our research is initially focused on collaboration between a single wearable computer user and several desktop PC users such as might be the case in a remote monitoring or technical assistance task. The aim is to develop software to support medium sized meetings (5–6 people) in a manner that is natural and intuitive to use. With this goal in mind we have implemented the prototype described above.

The wearable computer we use is a custom-built 586 PC with 20 Mb of RAM running Windows 95. A hand-held Logitech wireless radio trackball with three buttons is the primary input device and the display is a pair of Virtual i-O i-glasses!^{tm} converted into a monoscopic display (by the removal of the left eyepiece). The Virtual i-O head-mounted display can either be see-through or occluded, has a resolution of 262 by 230 pixels, and has a 30-degree field of view. The i-glasses! have stereo headphones and a sourceless inertial and gyroscopic three degree of freedom orientation tracker. A BreezeCom^{tm} wireless LAN is used to give 2 Mb/s Internet access up to 500 feet from a base station. The wearable also has a soundBlaster compatible sound board with head-mounted microphone.

FIG. 17.6. The wearable interface.

Figure 17.6 shows a user wearing the display and computer. The desktop PCs are standard Pentium class machines with internet connectivity and sound capability.

The conferencing space runs as a full-screen application that is initially blank until remote users connect. Our wearable computer has no graphics acceleration hardware, so the graphical interface was deliberately kept simple. When users join the conferencing space they are represented by blocks with 128×128 pixel texture-mapped pictures of themselves on them. The wearable computer does not have enough network bandwidth or computing power to support live video so these are static images that do not change throughout the collaboration. Although the resolution of the images is crude, it is sufficient to identify who the speakers are and, more importantly, where they are in relationship to the user. The wearable user's head is tracked so he or she can simply turn to face the speakers of interest. As the users face different participants the relative volume of each speaker's voice changes due to 3D sound spatialization. Users can also navigate through the virtual space; by rolling the trackball forwards or backwards their viewpoint is moved forwards or backwards along the direction they are looking. Since the virtual images are superimposed on the real world, when the user rolls the trackball it appears to them as though they are moving the virtual space around them, rather than navigating through the space. Figure 17.7 shows the interface from the wearable user's perspective. The interface was developed using Microsoft's Direct3D, DirectDraw, and DirectInput libraries from the DirectX suite.

Users at a desktop workstation interact with the conferencing space through an interface similar to that of the wearable user, although in this

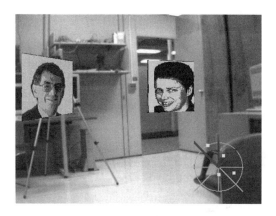

FIG. 17.7. The user's view.

case the application runs as a Windows application on the desktop. Users at the desktop machine wear head-mounted microphones to talk into the conferencing space, and they navigate through the space using the mouse. When the left mouse button is held down mouse movements rotate the head position; otherwise, they translate the user backwards and forwards in space. Mapping avatar orientation to mouse movements means that the desktop interface is not quite as intuitive as the wearable interface.

3.2.2 Software Architecture

The wearable and desktop interfaces use multicast sockets to communicate with each other. As shown in Figure 17.8, two multicast groups are used, one for user position and orientation, and one for audio communication. As users change their avatar position and orientation, values with unique avatar identify numbers are streamed to the position multicast group and rebroadcast to all the interested interfaces. This transformational data flows at a rate of 10 kb/s per user. Similarly when users speak, their speech is digitized, an avatar identity number is added, and the speech is then sent to the audio group to be rebroadcast. When the digitized speech arrives at the client computer, the audio identity number is used to find the speaker's position and spatialize the speech. In order for the audio to operate in full duplex mode it has to be captured at a rate of 8-bit 22 kHz, resulting in a data rate of 172 kb/s per user. All the connections to the multicast groups are bidirectional and users can connect and disconnect at will without affecting other users in the conferencing space.

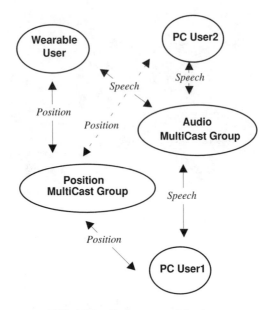

FIG. 17.8. Software architecture.

3.2.3 Initial User Experiences

We are in the process of conducting user trials to evaluate how the use of spatialized audio and visual representations in this interface affects communication among collaborators. Pilot observations have shown that users are able to easily discriminate among three speakers when their audio streams are spatialized, but not when nonspatialized audio is used. Users also prefer seeing a visual representation of their collaborators, as opposed to just hearing their speech. The visual spatial cues enhanced collaboration in the wearable computer even though static images were used; when wearable CPUs are able to support video texture mapping, collaboration may be enhanced further. Users also found that they could continue doing real-world tasks while talking to collaborators in the conferencing space because it was possible to move the conferencing space with the trackball so that collaborators weren't blocking critical portions of the user's field of view. These informal results show that the spatial audio and video cues generated by the wearable may indeed aid collaboration.

However, as more users connect to the conferencing space the need to spatialize multiple audio streams puts a severe load on the CPU, slowing down the graphics and head tracking. This makes it difficult for the wearable user to conference with more than two or three people simultaneously. This problem will be reduced as faster CPUs and hardware support for Direct3D graphics become available for wearable computers. Spatial culling of the audio streams could also be used to overcome this limitation.

3.3 Block Party: Remote Expert Assistance

Wearable computers can be even more useful for providing additional visual cues to enhance collaboration. This is particularly important for remote expert assistance where the expert may want to put virtual annotations on real objects or show the user how to perform a task using visual aids. For example, a remote physician could use virtual cues to help a medic operate a piece of equipment at the site of an accident, or to show where to perform a procedure. This capability arises from the use of see-through augmented reality displays in a wearable interface.

   To demonstrate an application of this we have developed a collaborative block-building application that allows a remote expert to assist a user in building a real model out of plastic bricks. The remote expert is seated at a desktop terminal, while the block builder uses a wearable see-though display, simulating a condition that could be common in the work place or factory environment of the future. The block builder wears a Virtual i-O head-mounted display, head position and orientation is tracked by a Polhemus Fastrak magnetic sensor [Polhemus 98], and he or she wears a small color video camera attached to the HMD to give the remote expert a view of what the wearer is doing (Figure 17.9). The Polhemus Fastrak is not a wireless sensor, but we used it in anticipation of future absolute position

FIG. 17.9. The block builder's interface.

and orientation tracking technology that is wearable. Companies such as InterSense [InterSense 98] are making positive steps in this direction.

The central issue we were seeking to address is finding what information must be exchanged between collaborators in order for them to easily understand one another and to work together effectively. We particularly want to discover if it is necessary to send full duplex high-bandwidth video and audio between collaborators, or if other lower-bandwidth information may work just as well. This is important for wearable interfaces where wireless bandwidth may be limited. For example, if graphical cues can aid collaboration as much as remote video, then the relevant graphics can be generated on-board the wearable computer and only position information needs to be sent between collaborators.

In this case we needed to provide interface elements that not only support communication but also enhance the real-world task through the use of virtual cues. To satisfy these requirements the expert's desktop interface has three components (Figure 17.10):

- a three-dimensional virtual model of the target object to be built,
- a simple shared 3D modeling package for building virtual models, and
- a video window showing the remote user's view and real-world task space.

The desktop user can interactively manipulate the 3D virtual model of the target object to get a clear understanding of how to construct the object,

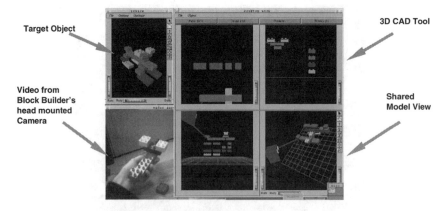

FIG. 17.10. The desktop user's interface, with remote video view and shared virtual environment.

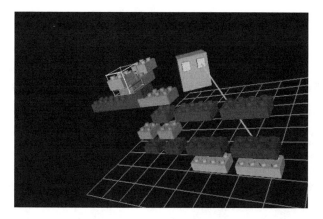

FIG. 17.11. The view through the augmented reality display.

and then use the 3D modeling tool to show the block builder how to put the real blocks together. The expert can also watch the video coming from the block builder's camera to monitor how he or she is performing the task. There is also an audio link between collaborators so they can speak freely about the task.

The block builder has a 3D view of the shared modeling environment shown on his or her (stereo) head-mounted display. Since this display is see-through the effect is of virtual blocks that float nearby the real workspace. Their head is tracked so they can freely move around and view the virtual blocks from any viewpoint. The block builder's view is shown in Figure 17.11. In both the expert's and block builder's interface their collaborator is represented by a simple virtual head avatar that moves according to how the respective real user moves. This avatar gives an indication of the coparticipant's viewpoint into the shared virtual space.

In designing this interface we wanted to explore how users worked together in a number of different conditions:

A) Audio but no video link or virtual cues: The remote expert can just talk to the block builder and there is no visual feedback from the real workspace.

B) Audio and video link but no virtual cues: The remote expert can see the block builder's workplace but can't provide virtual cues to assist their performance.

C) Audio and virtual cues but no video link: The remote expert can provide virtual cues to aid the block builder but can't receive any visual feedback.

D)   Audio and video link and virtual cues: The remote expert can see what the block builder is doing and can provide virtual cues to assist the builder's performance.

E)   Face to face: The two collaborators sit on opposite sides of a table, but only the expert can see the workstation monitor with the target model displayed.

These five conditions correspond to no visual communication, one-way visual communication from each collaborator, and full-duplex two-way visual communication between collaborators.

We were also interested in how avatar representation may affect performance. We explored using simple block avatars as shown in Figure 17.11 and avatars that had live video texture maps of the faces of the remote collaborators on them. We also provided virtual cues showing where the avatars were looking. When the desktop expert selects a block in the modeling interface a line is drawn from the expert's avatar to the selected block, showing that the expert's attention is focused on that block. Similarly, as the block builders look at different blocks, a line is drawn from their avatar to the block they are looking at.

3.3.1   User Experiences

We conducted two pilot trials with the block party interface. In the first, an experienced user acted as the remote expert while a novice wore the see-through head-mounted display and constructed the physical models. In this case we tested only the "video link and virtual cues" condition, with the desktop user being in a separate room from the block builder. Users reported that the addition of virtual cues helped them perform the block-building task. They liked being able to have their own viewpoint into the virtual space and being able to build with the real blocks, while at the same time being able to see what the remote expert was doing with the virtual blocks. Furthermore, the remote expert was easily able to see how well the block builder was following instructions by watching the view from the head-mounted camera.

In evaluating the use of texture-mapping live video on the collaborator's head avatar, however, we found no impact on performance. Although it may have been problematic that the (world-stabilized) avatars were not always in view, it appears that the nonverbal information provided by live video is not as important to this task as the gaze direction information provided simply by head avatar orientation. This finding reinforces the notion that avatar representation requirements are, to a large extent, task-specific.

In the second pilot study, six inexperienced users (three pairs), four men and two women aged 22 to 43 years old, attempted to build five different real models, counterbalanced across each of the conditions above. Equivalent models were used for each condition; all models contained the same number of bricks and were tested beforehand to make sure they took the same amount of time to construct. Subjects were screened for normal (natural or corrected) eyesight and hearing, and each subject was given a standardized test of spatial ability [ETS 76] prior to the experimental trials. Some of the subjects had prior experience with virtual environments, but none had used a see-through head-mounted display before.

One subject of each pair was chosen as the block builder while the other played the role of the desktop expert. Subjects kept these roles throughout the duration of the experiment. The desktop experts received about 15 minutes of training building practice with virtual objects until they felt comfortable with the interface. Then both the desktop and head-mounted display user practiced building objects in the audio-and-video condition until the HMD user felt comfortable with the augmented reality interface. For each condition, subjects were told to build each model as quickly as possible and were timed and measured for accuracy of the final plastic model. They were also given a postexperiment survey to record their subjective impressions of each condition.

In contrast to the previous study, the desktop virtual block-building interface proved to be problematic for some of these inexperienced subjects. In some cases the block builder had to wait while the expert put several virtual blocks into place; in others the remote participant simply abandoned the virtual interface in favor of purely verbal directions. This illustrates that an interface to provide remote performance assistance must be as fast and intuitive to use as possible; to be useful our interface should have allowed inexperienced users to assemble virtual blocks almost as quickly as it would take to assemble the Lego blocks in the real world.

Despite these problems and the small sample size, we found that the voice-only condition took the longest time and produced the most errors in the final objects. The virtual cues alone were useful, reducing the amount of errors made over the voice-only condition, but not improving the time taken because of the difficulty of rapidly assembling the virtual object. The use of remote video from the block-builder's head-worn camera, however, dramatically reduced task performance time and the number of errors made. The video enabled the expert to monitor the object as it was being assembled and provide immediate corrective feedback; even when they couldn't use virtual cues, experts were able to quickly tell the block builder how to assemble the object by watching the builder's actions.

These results are consistent with the users' subjective responses to the postexperiment survey questions. Users were asked to rate each condition on a scale of one to five for a number of different criteria. Figure 17.12 shows the average response for each condition to the question "How easily could you collaborate in this condition?," where 1 was "not easily" and 5 was "very easily." Users felt they could collaborate best in the mediated conditions with audio and video cues (B) and with audio, video, and virtual cues (D). This suggests that video feedback from the block builder to the remote expert was key. Similarly, in answer to the question "How easy was it to understand your partner in this condition?" (Figure 17.13) the

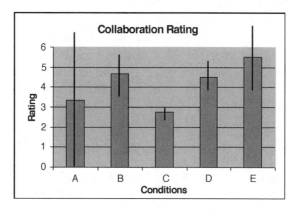

FIG. 17.12. User ratings of how easy it was to collaborate in each condition (1 = not easily, 7 = very easy).

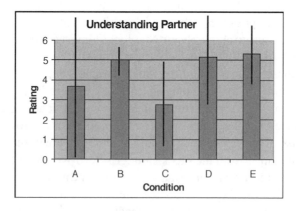

FIG. 17.13. User ratings of how easy it was to understand partner (1 = not easily, 7 = very easily).

mediated conditions that supported video feedback (B, D) were ranked the highest, almost as high as face-to-face collaboration (E).

In those conditions that did use virtual block cues, users found the virtual lines showing the collaborator's focus of attention to be very helpful, particularly for the augmented reality view. The expert could use this tool to indicate blocks they were about to move. For example, saying "pick up this block" while clicking on the virtual block produces a line to the block that makes it easy to spot.

3.4 Summary of Remote Collaboration Results

In this section we have described two ways in which wearable interfaces can aid remote collaboration: by providing a sense of social presence and by enhancing a real-world task. WearCom used spatial and audio cues to aid multiparty remote conferencing and created for the wearable user a sense of social presence similar to that of face-to-face collaboration. Preliminary results have found that spatialization does indeed allow people to communicate more naturally, suggesting a way in which wearables can be used to enhance conference phone calls.

Block Party went beyond this by using virtual cues to aid in a real-world task. The remote expert could use virtual blocks to show the local user how to put together the equivalent real blocks; the block builders could view these virtual blocks from any viewpoint to better understand how they were to fit the physical blocks together. Although this was effective for experienced users, novices had difficulty with the modeling interface, and video from the block-builder's viewpoint proved to be the most successful component of the communication interface. In the next section we show how wearables can be used to enhance face-to-face collaboration.

4. WEARABLE COMPUTERS FOR COLLOCATED COLLABORATION

The combination of augmented reality, mobility, and computer-enhanced perception provided by wearable computers also makes them useful for face-to-face collaboration. Augmented reality displays enable collocated users to view and interact with shared virtual information spaces while viewing the real world at the same time. This preserves the rich communications bandwidth that humans enjoy in face-to-face meetings, while adding virtual images normally impossible to see in the real world. Additionally, the mobility of wearables allows them to be used as tools to enhance opportunistic meetings. While most current collaborative interfaces require users to be seated at their desktop computer or in front of teleconferencing equipment, in reality the majority of collaborative meetings are spontaneous and unplanned, such as chance meetings in hallways or around the water fountain [Whittaker et al. 94].

In our research we have been focusing on the ability of wearable computers to support collaborative augmented reality, one of the most significant ways that wearables can enhance face-to-face collaboration. Fully immersive virtual environments can provide an extremely intuitive interface for collaborative interaction: The Greenspace [Mandeville et al. 96] and DIVE [Carlsson and Hagsand 93] projects, among others, have shown that remote users can collaborate in an immersive virtual environment as if they were in the same physical location. However, in these cases users are separated from the real world and their familiar tools, introducing huge functional and cognitive seams. Augmented reality has several advantages over fully immersive virtual environments, including:

- Participants can refer to notes, diagrams, books and other real objects while viewing virtual objects.
- Participants can use familiar real-world tools to manipulate the virtual objects, increasing the intuitiveness of the interface. For example, a real scalpel could be used to cut virtual skin in a surgical application.
- Participants can see each other's facial expressions, gestures, and body language, increasing the communication bandwidth.
- The entire environment doesn't need to be modeled, reducing the graphics rendering requirements.

Furthermore, there is evidence that augmented reality may be less provocative of "simulator sickness" than fully immersive virtual reality [Prothero 97], a required feature for any long-term collaborative work environment.

Schmalsteig et al. [Schmalsteig 96] identify five key characteristics of collaborative AR environments:

- *Virtuality:* Objects that don't exist in the real world can be viewed and examined.
- *Augmentation:* Real objects can be augmented by virtual annotations.

- *Cooperation:* Multiple users can see each other and cooperate in a natural way.
- *Independence:* Each user controls his or her own independent viewpoint.
- *Individuality:* Displayed data can be different for each viewer to support different roles.

In this section we describe two laboratory interface prototypes that show how collaborative augmented reality can enhance face-to-face communication. The first, a collaborative Web browser, enables users to load and place virtual Web pages around themselves in the real world and to jointly discuss and interact with them; users can see both the virtual Web pages and each other, so communication is natural and intuitive. The second example was designed to investigate how performance on a face-to-face collaboration task with a wearable AR interface compares with performance of collocated participants in a fully immersive virtual environment. Both of these wearable interfaces use see-through head-mounted displays with body tracking. We call these types of interfaces "Shared Space" interfaces because they allow multiple users in the same location to work in both the real and virtual world simultaneously, facilitating CSCW in a seamless manner.

4.1   Collaborative Web Space

We have developed a three-dimensional Web browser that enables multiple collocated users to collaboratively browse the World Wide Web. Users see each other and virtual Web pages floating in space around them. The effect is a body-centered information space that the user can easily and intuitively interact with. The Shared Space browser supports multiple users who can communicate about the Web pages shown, using natural voice and gesture. Figure 17.14 shows users interacting with the interface and Figure 17.15 shows the user's view. Potential applications include information delivery to mobile workers in medical, military, or manufacturing fields, as well as day-to-day information sharing by the average Web user.

In our laboratory, prototype users again wear see-through Virtual i-O stereoscopic head-mounted displays, and head orientation is tracked using a Polhemus Fastrak electromagnetic sensor. The interface is designed to be completely hands-free; pages are selected by looking at them and, once selected, can be attached to the user's viewpoint, zoomed in or out, iconified or expanded, or have additional links loaded with voice commands. Speaker-independent continuous speech recognition software is used to

FIG. 17.14. The Shared Space interface.

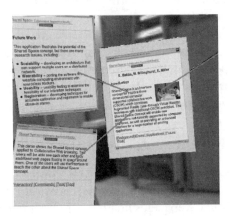

FIG. 17.15. The participant's view.

allow users to load and interact with Web pages using vocal commands. To support this, an HTML parser parses the Web pages to extract their HTML links and assign numbers to them. In this way users can load new links with numerical commands such as "Load link one." Each time a new page is loaded a new browser object is created and a symbolic graphical link to its parent page is displayed to facilitate the visualization of the Web pages. The voice recognition software recognizes 46 commands and control phrases with greater than 90% accuracy. A switched microphone is used so that participants can carry on normal conversation when not entering voice commands.

Two important aspects of the interface are gaze awareness and information privacy. Users need to know which page they are currently looking at as well as the pages their collaborators are looking at. This is especially difficult when there are multiple Web pages close to each other. The Virtual I-O head-mounted display has only a 30 degree field of view so it is tempting to overlap pages so that several can be seen at once. To address this problem, each Web page highlights when a user looks at it. Each page also has gaze icons attached to it for each user that highlights to show which users are looking at the page. In this way users can tell where their collaborators are looking. When each Web page is loaded it is initially visible only to the user that loaded it. Users can change page visibility from private to public with vocal commands; users can only see the public Web pages and their own private objects.

The collaborative Web interface uses a portable body-stabilized information space similar to that used in the WearCom interface. However, in this case all three degrees of head orientation are used, providing a virtual sphere of information. Even though the head-mounted display has only a limited field of view (30 degrees), the ability to track head orientation and place objects at fixed locations relative to the body effectively creates a 360-degree circumambient display. This overcomes the display size limitations of wearable displays. Since the displays are wearable, users can collaborate in any location, rather than needing to move to a particular computer, display, or physical environment. Finally, because interface objects are not attached to real-world locations, the registration requirements are not as stringent. One of the most challenging issues with augmented reality is image registration, particularly having images appear fixed with respect to the real world.

The augmented reality interface facilitates a high bandwidth of communication between users as well as natural 3D manipulation of the virtual images. The key characteristic of this interface is the ability to see the real world and collaborators at the same time as the virtual Web pages floating in space. This means that users can use natural speech and gesture to communicate with each other about the virtual information space. In informal trials, users found the interface intuitive and communication with the other participants seamless and natural. Collaboration could be left to normal social protocols rather than requiring mechanisms explicitly encoded in the interface. Unlike sharing a physical display, users with the wearable information space can restrict the ability of others to see information in their space. They were able to easily spatially organize Web pages in a manner that facilitated rapid recall, and the distinction between public and private information was found to be useful for collaborative information presentation.

4.2   Comparison between Augmented
and Immersive Collaboration

In the previous sections we have shown how wearable computers afford
seamless collaboration and can enhance the real world, both key elements
for effective CSCW interfaces. However, there are many questions that must
be answered in developing collaborative wearable systems. In this section
we describe a pilot study conducted to address one of these: Does the seamlessness between the real and virtual worlds really benefit performance?

4.2.1   The Effect of Seamlessness

If seamlessness does affect collaboration then there should be a difference seen in task performance on the same collaborative task performed in
an immersive virtual environment versus a shared space augmented reality
configuration. An immersive virtual world separates the user from the real
world entirely, creating a huge functional and cognitive seam. Users are not
able to use any of their traditional tools in an immersive environment and
must often learn new ways to interact with the environment. In contrast,
a collaborative augmented reality interface enhances rather than supplants
the real world. Differences in task performance between these two settings
could imply that the seamlessness inherent in wearable AR interfaces does
indeed affect collaboration.

To explore this we performed a simple pilot study using a Shared Space
interface [Billinghurst 96]. We developed a two-player game that involved
moving randomly distributed colored cubes or balls around a virtual space
and placing them in a target configuration. The two players each had a
different role. One was the "spotter" and could see all the virtual objects
(Figure 17.16). The role of this player was to search the space, find the
objects needed to complete a target configuration, and make them visible
using voice commands. The second player was the "picker." This player
had to find the objects that were made visible by the spotter, pick them up,
and drop them over the target configuration.

The role division between the players forced them to collaborate. Both
players were in the same room, wore stereoscopic Virtual i-O head-mounted
displays that could either be see-through or occluded, and had their head
and hand positions tracked by Polhemus electromagnetic trackers. When
the displays were used in see-through mode the virtual targets appeared
superimposed over the real world.

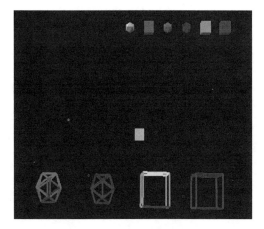

FIG. 17.16. The spotter's view, showing the target objects and attached icons.

Four experimental conditions were tested:

RW + RB: *Real World–Real Body:* In this case players could see the real world and each other; this corresponds to the ideal wearable augmented reality configuration.

RW: *Real World–No Body:* The HMD was see-through, but a sheet was dropped between the subjects; they could see the real world, but not each other.

VE: *Virtual World–No Body:* The glasses were in occluded mode, but the subjects did not have a representation of each other in the virtual room.

VE + VB: *Virtual World–Virtual Body:* The HMD was used in occluded mode, and the subjects were each represented by virtual avatars (Figure 17.17).

When a real or virtual body was present participants could use both voice and pointing gestures to show where objects are located. Without bodies they could use only voice communication. Real or virtual world reference objects could be used to aid target object location and communication. The virtual avatars used were simple block figures with a single arm, similar to those used in many collaborative virtual environments. Avatar position, orientation, and arm location matched that of the corresponding real person.

TABLE 17.1

Paired T-Test Results Comparing Performance in the Real World–Real Body
Condition to the Virtual Environment–Virtual Body Condition. The t-Value is
Significant at $p < 0.01$, $df = 9$, t critical $= 1.833$

| | *All Trials* | | |
Condition	*Mean*	*Var.*	*t-val*
RW + RB	83.4	325.0	−3.9*
VE + VB	102.2	816.9	

FIG. 17.17.   The immersive virtual environment with an avatar.

4.2.2 Results

Eighteen pairs of college students (twenty women, sixteen men) served
as subjects, each playing four games for each condition, for a total of sixteen
games per subject pair. Subjects were free to communicate within the constraints of each condition. Some subjects elected to use references to real
or virtual body cues to aid their performance, while others used alternative
strategies, such as specifying object location by clock or compass direction. For subjects who used body cues there was a significant difference
in performance across experimental conditions. These players completed
the game significantly faster in the real world–real body condition than in
the virtual world–virtual body condition (Table 17.1) and had quicker real
world–real body times than in each of the other conditions (Figure 17.18).

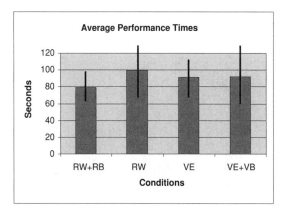

FIG. 17.18. Average task performance times.

Players were also given a postgame survey to determine their subjective evaluation of the conditions. For each condition, they were asked to rate how good they thought they were at playing the game (on a scale from 1 to 7, where 1 is "not good" and 7 is "very good"). Users felt they were best at playing the game in either the real world–real body case or the virtual world–virtual body case (Figure 17.19). There was a significant difference in self-ratings for this item across conditions, as indicated by a single-factor repeated measures ANOVA [$F(4,114) = 7.65$, $p < 0.0001$].

Users were also asked to rank the conditions according to how well they thought their pair had performed in each condition; the best condition was ranked 1 and the worst 5. The average rankings for each condition are shown in Figure 17.20. Using a Friedman two-way ANOVA, we again find a significant difference between rankings [$\chi^2 = 31.89$, $df = (4,21)$, $p < 0.0001$]. Users again thought they performed best in the real world–real body condition or virtual world–virtual body case, and all of the users who relied on body cues thought they performed best in the real world–real body case.

The significant performance difference among subjects that used body cues implies that the increased communications bandwidth facilitated by seeing the real world and a real collaborator may indeed aid task performance. The subject rankings imply that users may prefer collaboration in a setting where they can see their collaborators face to face, such as that provided by a wearable-computing platform.

In this section we have explored how wearables can enhance face-to-face collaboration. Although wearables could be used to improve such collaboration in a number of ways, we have focused on the ability of the wearable

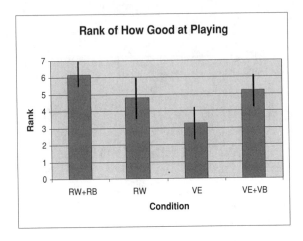

FIG. 17.19. Subject ratings of their task performance (1 = not good, 7 = very good).

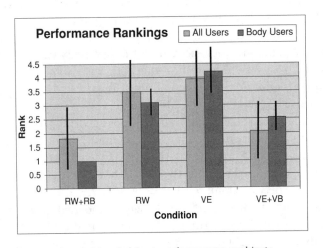

FIG. 17.20. Subject performance rankings.

to overlay shared virtual models onto the real world, creating collaborative augmented reality environments. The collaborative Web space shows the primary benefit of these environments: the ability of participants to use natural speech and gesture to interact with virtual models and each other, removing the seam between the real and virtual world. The collaborative game experiment showed that users prefer seeing real collaborators and suggests that task performance improves with the higher communications bandwidth afforded by shared augmented reality.

5. CONCLUSIONS

Wearable computers are ideal for Computer-Supported Collaborative Work because of the unique interface attributes they offer. The combination of augmented reality, mobility, and computer-enhanced perception enables wearables to overcome two major challenges of CSCW: seams and the need to enhance reality. In this chapter we have shown several ways in which wearable computers can be used to support collaboration. The common theme running through these interfaces is the use of spatial metaphors that are typically absent from traditional mobile communication interfaces. Figure 17.21 categorizes the interfaces we have presented according to the type of audio or visual spatialization used.

If these were to be placed on a continuum representing how closely they approached face-to-face collaboration, head-stabilized audio-only collaboration would be furthest from face-to-face collaboration while world-stabilized audio-and-video interfaces would be closest. The first two categories (audio, head-stabilized audio-and-video displays) have desktop equivalents in common audio or video conferencing applications. They require only a high bandwidth communications link, not the addition of any wearable computing power, and have been well studied in the CSCW literature. Many studies have shown the relative ineffectiveness of using video to enhance collaboration. Perhaps one reason for this is that non-spatialized video doesn't add many extra cues over nonspatialized audio, especially given typically less-than-real-time frame rates.

The types of interfaces that wearable computers can most contribute to are those in the latter two categories. In this case computing is necessary to provide real-time spatial audio and video enhancements. These types of collaborative wearable interfaces represent a paradigm shift in the way people interact with each other and information in a collaborative setting and have not been extensively studied. In the WearCom, Collaborative Web Space, and BlockParty interfaces we have demonstrated ways in which these spatial visual cues could be used to enhance communication.

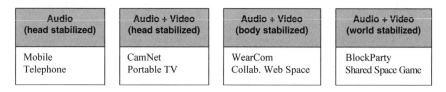

Audio (head stabilized)	Audio + Video (head stabilized)	Audio + Video (body stabilized)	Audio + Video (world stabilized)
Mobile Telephone	CamNet Portable TV	WearCom Collab. Web Space	BlockParty Shared Space Game

FIG. 17.21. Categories of mobile collaborative interfaces.

The prototype interfaces we have developed have been based on the following requirements:

- the need for an underlying spatial metaphor;
- a visual representation of remote collaborators;
- high-quality audio connections;
- an intuitive interface; and
- the need to minimize functional and cognitive seams.

In the case where the virtual collaboration interface was too cumbersome, the BlockParty interface, users found little additional collaborative value in the enhancements provided by the computer. However, in all the other examples the wearable interface enhanced face-to-face and remote collaboration beyond that achieved by unmediated communication. This preliminary work suggests that emerging wearable computer technologies may indeed have the potential to achieve Arthur Strand's dream of ubiquitous communication and collaboration.

6. FUTURE WORK

Considerably more work is required to establish the usefulness of wearables for CSCW. In our own work we have identified several pressing research and development needs:

- better objective measures of social discourse and collaboration;
- more research on the use of spatialized audio in collaborative environments;
- better understanding of VR vs. AR as a collaborative work tool;
- better understanding of real-world wearable collaborative applications and interface issues;
- development of low-cost hybrid wearable position and orientation tracking systems;
- adaptive transparency of HMD occluders to accommodate a wide variety of working conditions;
- better partitioning of what is needed locally vs. downloading globally;
- robust, lightweight wearable power supplies; and
- development of readily available high-bandwidth wireless communications networks.

Empirical studies also need to be conducted comparing collaboration in a wearable setting to other interfaces, establishing which of the attributes of wearables contribute most to facilitating collaboration and the types of collaborative applications wearables are most suited for. These studies should also provide requirement specifications for technology development, such as sourceless position and orientation trackers and improved displays. Finally, developers must use this knowledge to build wearable interfaces that overcome the limitations of current CSCW tools.

Acknowledgments   Portions of this work were completed while the first author was at British Telecom Laboratories. We would like to thank Jerry Bowskill, Jason Morphett, and our colleagues at British Telecom for many useful and productive conversations and Nick Dyer for producing the renderings used in some of the figures. Portions of this work were also supported by a Research and Technology Development grant from the Washington Technology Center.

REFERENCES

[Bajura 92] Bajura, M., Fuchs, H., Ohbuchi, R. "Merging Virtual Objects with the Real World: Seeing Ultrasound Imagery Within the Patient." In *Proceedings of SIGGRAPH '92*, Chicago, Illinois, July 26–31, 1992, New York:ACM, pp. 203–210.

[Bass 97] Bass, L., Kasabach, C., Martin, R., Siewiorek, D., Smailagic, A., Stivoric, J. The Design of a Wearable Computer. In *Proceedings of CHI '97*, Atlanta, Georgia. March 1997, New York: ACM, pp. 139–146.

[Benford 93] Benford, S. and Fahlen, L. A. Spatial Model of Interaction in Virtual Environments. In *Proceedings of the Third European Conference on Computer Supported Cooperative Work (ECSCW'93)*, Milano, Italy, September 1993.

[Benford 97] Benford, S., Greenhalgh, C., Lloyd, D. Crowded Collaborative Virtual Environments. In *Proceedings of CHI '97*, Atlanta, Georgia. March 1997, New York: ACM, pp. 59–66.

[Billinghurst 98] Billinghurst, M., Bowskill, J., Dyer, N., Morphett, J. An Evaluation of Wearable Information Spaces. In *Proceedings of VRAIS 98*, Altanta, Georgia, March 14–18, 1998, IEEE Press.

[Billinghurst 96] Billinghurst, M., Weghorst, S., Furness, T. Shared Space: Collaborative Augmented Reality. In *Proceedings of CVE '96 Workshop*, 19–20 September 1996, Nottingham, Great Britain.

[Bregman 90] Bregman, A. *Auditory Scene Analysis: The Perceptual Organization of Sound*. MIT Press, 1990.

[Carlsson 1993] Carlson, C. and Hagsand, O. DIVE—A Platform for Multi-User Virtual Environments. *Computers and Graphics*, Nov/Dec 1993, Vol. 17, No. 6, pp. 663–669.

[Chapanis 75] Chapanis, A. Interactive Human Communication. *Scientific American*, 1975, Vol. 232, pp. 36–42.

[Draft 91] Draft, R.L. and Lengel, R.H. Organizational Information Requirements, Media Richness, and Structural Design. *Management Science*, Vol. 32, 1991, pp. 554–571.

[Ehrlich 87] Ehrlich, S. Strategies for Encouraging Successful Adoption of Office Communication Systems. *ACM Trans. Off. Inf. Sys.*, Vol. 5, 1987, pp. 340–357.

[Esposito 97] Esposito, C. Wearable Computers: Field-Test Results and System Design Guidelines. In *Proceedings of Interact '97*, July 14th–18th, Sydney, Australia.

[ETS 76] Cube Comparisons Test, *Standard Ability Tests*, Educational Testing Service, Princeton, New Jersey, 1976.

[Feiner 97] Feiner, S., MacIntyre, B. Höllerer, T. A Touring Machine: Prototyping 3D Mobile Augmented Reality Systems for Exploring the Urban Environment. In *Proceedings of the International Symposium on Wearable Computers*, Cambridge, MA, October 13–14, 1997, Los Alamitos: IEEE Press, pp. 74–81.

[Fickas 97] S. Fickas, G. Kortuem, Z. Segall. Software Issues in Wearable Computing, Position Paper for the CHI '97 Workshop on Research Issues in Wearable Computers, March 23–24, 1997, Atlanta, GA. Available at <http://www.cs.uoregon.edu/research/wearables/Papers/>

[Fish 91] R.S. Fish, R.E. Kraut, R.W. Root, R. Rice. *Evaluating Video as a Technology for Informal Communication*. Bellcore Technical Memorandum, TM-ARH017505, 1991.

[Furness 88] T. Furness. 'Super Cockpit' Amplifies Pilot's Senses and Actions, *Government Computer News*. August 15, 1988, pp. 76–77.

[Garner 97] P. Garner, M. Collins, S.M. Webster, D.A. Rose. The Application of Telepresence in Medicine, *BT Technology Journal*, Vol 15, No 4, October 1997, pp.181–187.

[Grudin 88] J. Grudin. Why CSCW Applications Fail: Problems in the Design and Evaluation of Organizational Interfaces. In *Proceedings of CSCW '88*, Portland, Oregon, 1988, pp. 85–93.

[Heath 91] Heath, C., Luff, P. Disembodied Conduct: Communication Through Video in a Multimedia Environment. In *Proceedings of CHI '91 Human Factors in Computing Systems*, 1991, New York: ACM Press, pp. 99–103.

[Hindus 96] Hindus, D., Ackerman, M., Mainwaring, S., Starr, B. Thunderwire: A Field study of an Audio-Only Media Space. In *Proceedings of CSCW '96*, Nov. 16th–20th , Cambridge MA, 1996, New York: ACM Press.

[Hollan 92] J. Hollan, S. Stornetta. Beyond Being There. In *Proceedings of CHI '92*, May 3–7, 1992, pp.119–125.

[InterSense 98] InterSense Product Literature, InterSense Incorporated, 73 Second Avenue, Burlington, MA 01803, USA, http://www.isense.com/

[Ishii 91] H. Ishii and N. Miyake. Toward an Open WorkSpace: Computer and Video Fusion Approach of Teamworksation. *Communications of the ACM*, Dec 1991, Vol 34, No. 12, pp. 37–50.

[Ishii 92] H. Ishii, M. Kobayashi, J. Grudin. Integration of Inter-Personal Space and Shared Workspace: ClearBoard Design and Experiments. In *Proceedings of CSCW '92*, 1992, pp. 33–42.

[Ishii 94] H. Ishii, M. Kobayashi, K. Arita. Iterative Design of Seamless Collaboration Media. *Communications of the ACM*, Vol 37, No. 8, August 1994, pp. 83–97.

[Kleinke 86] C.L. Kleinke. Gaze and Eye Contact: A Research Review. *Psychological Bulletin 1986, 100*, pp.78–100.

[Kraut 96] Kraut, R., Miller, M., Siegal, J. Collaboration in Performance of Physical Tasks: Effects on Outcomes and Communication. In *Proceedings of CSCW '96*, Nov. 16th–20th, Cambridge MA, 1996, New York: ACM Press.

[Mandeville et al. 96] Mandeville, J., Davidson, J., Campbell, D., Dahl, A., Schwartz, P., Furness, T. A Shared Virtual Environment for Architectural Design Review. In *CVE '96 Workshop Proceedings*,19–20th September 1996, Nottingham.

[Mann 97] Mann, S. Smart Clothing: The "Wearable Computer" and WearCam. *Personal Technologies*, Vol. 1, No. 1, March 1997, Springer-Verlag.

[Nakanishi 96] Nakanishi, H., Yoshida, C., Nishimura, T., Ishida, T. FreeWalk: Supporting Casual Meetings in a Network. In *Proceedings of CSCW '96*, Nov. 16th–20th, Cambridge MA, 1996, New York: ACM Press, pp. 308–314.

[O'Malley 96] C. O'Malley, S. Langton, A. Anderson, G. Doherty-Sneddon, V. Bruce. Comparison of Face-to-Face and Video-Mediated Interaction. *Interacting with Computers*, Vol. 8 No. 2, 1996, pp. 177–192.

[Polhemus 98] Fastrak Product Literature, Polhemus Incorporated, PO Box 560 Colchester, VT 05446, USA, http://www.polhemus.com/.

[Prothero 97] Prothero, J.D., Draper, M.H., Furness, T.A., Parker, D.E., Wells, M.J. Do Visual Background Manipulations Reduce Simulator Sickness? In *Proceedings of the International Workshop on Motion Sickness*, May 26–28, 1997.

[Reichlen 93] Reichlen, B. SparcChair: One Hundred Million Pixel Display. In *Proceedings IEEE VRAIS '93*, Seattle WA, September 18–22, 1993, IEEE Press: Los Alamitos, pp. 300–307.

[Rekimoto 95] Rekimoto, J., Nagao, K. The World through the Computer: Computer Augmented Interaction with Real World Environments. In *Proceedings of User Interface Software and Technology '95 (UIST '95)*, November 1995, New York: ACM, pp. 29–36.

[Schmalsteig 1996] D. Schmalsteig, A. Fuhrmann, Z. Szalavari, M. Gervautz. Studierstube—An Environment for Collaboration in Augmented Reality. In *CVE '96 Workshop Proceedings*, 19–20th September 1996, Nottingham, Great Britain.

[Schmandt 95] Schmandt, C., Mullins, A. AudioStreamer: Exploiting Simultaneity for Listening. In *Proceedings of CHI 95 Conference Companion*, May 7–11, Denver Colorado, 1995, ACM: New York, pp. 218–219.

[Sellen 92] Sellen, A. Speech Patterns in Video-Mediated Conversations. In *Proceedings of CHI '92*, May 3–7, 1992, ACM: New York, pp. 49–59.

[Sellen 95] A. Sellen. Remote Conversations: The Effects of Mediating Talk with Technology. *Human Computer Interaction*, Vol. 10, No. 4, 1995, pp. 401–444.

[Short 76] J. Short, E. Williams, B. Christie. *The Social Psychology of Telecommunications*. London, Wiley, 1976.

[Siegal 95] Siegal, J., Kraut, R., John, B., Carley, K. An Empirical Study of Collaborative Wearable Computer Systems,. In *Proceedings of CHI 95 Conference Companion*, May 7–11, Denver Colorado, 1995, ACM: New York, pp. 312–313.

[Starner 97] T. Starner, S. Mann, B. Rhodes, J. Levine, J.Healey, D. Kirsch, R.W. Picard, A. Pentland. Augmented Reality through Wearable Computing. *Presence, Special Issue on Augmented Reality*, 1997.

[Strand 1898] Quoted in Mee, Arthur, *The Pleasure Telephone*. *The Strand Magazine*, pp. 339–369, 1898.

[Whittaker et al. 94] Whittaker, S., Frohlich, D., Daly-Jones, O. Informal Workplace Communication: What Is It Like and How Might We Support It? *Proceedings of the Conference on Computer Human Interaction (CHI '94)*, New York: ACM Press, 1994, pp. 131–137.

[Whittaker 1995] Whittaker, S. *Rethinking Video as a Technology for Interpersonal Communications: Theory and Design Implications*. Academic Press Limited, 1995.

[Whittaker 97] Whittaker, S., O'Connaill, B. The Role of Vision in Face-to-Face and Mediated Communication. In K. Finn, A. Sellen, S. Wilbur (Eds.), *Video-Mediated Communication*, Lawerance Erlbaum Associates, New Jersey, 1997, pp. 23–49.

[Williams 77] Williams, E. Experimental Comparisons of Face-to-Face and Mediated Communication. *Psychological Bulletin*, 1977, Vol. 16, pp. 963–976.

18

Tactual Displays for Sensory Substitution and Wearable Computers

Hong Z. Tan
Purdue University

Alex Pentland
Massachusetts Institute of Technology

1. INTRODUCTION

A major challenge in building practical wearable computer systems is the development of output devices to display or transmit information to the human user. Much effort has been devoted to visual displays that are lightweight and have high resolution. Such efforts are warranted since visual displays are still the dominant output devices used by most computing systems. Auditory displays are now becoming the norm of multimedia systems in addition to visual displays. Whereas vision is best suited for perceiving text and graphics, and audition for speech and music, the sense of touch is intimately involved in nonverbal communication. Whether it is a tap on the shoulder to get someone's attention or a firm handshake to convey trust, touch enables us to exchange information directly with people and the environment through physical contact. The skin is the largest organ of our body, yet only a small portion of it (i.e., the hands) is engaged in most human–computer interactions.

For a long time, the sense of touch has been regarded as the inferior sense as compared to vision or audition. However, the potential to receive information tactually is well illustrated by some natural (i.e., nondevice related) methods of tactual speech communication. Particularly noteworthy is the so-called Tadoma method that is employed by some individuals who are both deaf and blind. In Tadoma, one places a hand on the face and neck of a talker and monitors a variety of actions associated with speech production. Previous research has documented the remarkable abilities of experienced Tadoma users (e.g., Reed, Rabinowitz, Durlach, Braida, Conway-Fithian, & Schultz, 1985). Not only can these individuals converse with both familiar and unfamiliar talkers at high performance levels, but they pick up additional features such as the speaker's accent. The Tadoma method is a living proof that high information transmission is possible through the somatosensory system.

Earlier work on wearable tactual displays has concentrated on assisting the blind to see and the deaf to hear through their sense of touch (i.e., sensory substitution). For example, the Optacon (Linvill & Bliss, 1966) was initially developed in the 1960s for the daughter of one of its inventors as a reading aid for the blind. It converted images of printed materials to vibrational patterns on the index finger. The Optacon was probably one of the first commercially successful wearable tactual displays ever developed. With sufficient training, typical reading rates of 30–50 wpm can be achieved (Craig & Sherrick, 1982). Some exceptional subjects have demonstrated rates as high as 70–100 wpm (Craig, 1977). Recently, force-feedback systems have been developed for applications in teleoperation and virtual/augmented reality systems (Burdea, 1996). These systems can simulate forces that would have been generated by remote or virtual objects during manual manipulation, thereby enhancing an operator's sense of presence and task performance. Our current work is aimed at developing wearable tactual displays for general-purpose human–computer interactions. We are exploring new ways of utilizing the wearable tactual display technology developed for sensory substitution to convey information that is as intuitive as a sense of resistive force. Wearable computing also provides a unique environment for developing paradigms to distribute human–computer interfaces across the entire body and its various sensory channels.

The organization of this chapter is as follows. Section 2 defines terms that are used throughout this chapter. Section 3 presents a brief historical review of tactual displays that have been developed for sensory substitution. Section 4 discusses the requirements for wearable tactual displays. Section 5 describes a general-purpose tactual directional display that has been developed in our laboratory for wearable computing. Section 6 provides a summary.

2. DEFINITION OF TERMS

The term *haptics* refers to sensing and manipulation through the tactual sense. The human *tactual* sensory system is generally regarded as made up of two subsystems: the tactile and kinesthetic senses (see reviews by Loomis & Lederman, 1986; Clark & Horch, 1986). The *tactile* (or *cutaneous*) sense refers to the awareness of stimulation to the body surfaces mediated by sensors close to skin surfaces such as the mechanoreceptors. For example, when you touch a loudspeaker, your hands receive tactile stimulation. The *kinesthetic* sense (or *proprioception*) refers to the awareness of limb positions, movements, and muscle tensions mediated by sensors in the muscles, skin, and joints as well as a knowledge of motor commands sent to muscles (i.e., efference copy). For example, when you touch your nose with closed eyes, you rely on the kinesthetic sense to know where your fingertip is relative to your nose. When you hold an object in your hand, you use the kinesthetic sense to estimate the weight of the object. In both the nose touching and weight estimation cases, the tactile sense is also activated since the fingertip touches the nose, and the hand is in contact with the object. However, the information that is crucial to the successful execution of these tasks is derived primarily from the kinesthetic sense. In other words, the above distinction between tactile and kinesthetic senses is functional and task dependent.

The term *haptics* is often used in the literature. In this chapter, it refers to manipulation as well as perception through the tactual sense. An example of haptic interfaces is force-feedback joysticks for video games.

The term *display* refers to a human–machine interface that mainly transmits information from a machine to a human. The term *controller* refers to a human–machine interface that is mainly used by a human to control certain processes of a machine (Figure 18.1). The term *interface* is used in this chapter either to refer to a device that is both a display and a controller, or when such distinction is not important.

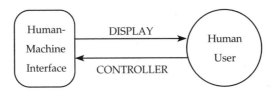

FIG. 18.1. Display vs. controller for human–machine interfaces.

3. TACTUAL DISPLAYS FOR SENSORY SUBSTITUTION

Most tactual communication systems for sensory substitution have been developed based upon two major principles: pictorial or frequency-to-place transformation. Devices for the blind tend to adopt the pictorial approach (i.e., direct translation of spatial–temporal visual information to the skin). Devices for the deaf are usually based on the cochlea model of speech (i.e., positional encoding of frequency information). Examples of pictorial tactual communication systems are the Optacon (OPtical-to-TActile-CONverter), the Optohapt (OPtical-TO-HAPTics), the TVSS (Tactile Vision Substitution System), and the Kinotact (KINesthetic, Optical and TACTile display).

The Optacon (Telesensory Corp, Mountain View, CA) is a direct translation reading aid for the blind that quantizes an area roughly the size of a letter into 144 black and white image points (24 rows and 6 columns) via photocells and displays these image points on a corresponding 24-by-6 array of vibrating pins (Linvill & Bliss, 1966). This portable system consisted of a small hand-held camera and a tactile display measuring 1.1 cm by 2.7 cm that fits under the fingertip of an index finger (Figure 18.2). Whenever a photocell detected a "black" spot, the corresponding pin vibrated. Reading speed with the Optacon varied from 10 to 100 wpm depending on the individual, the amount of training, and experience, with typical rates of 30–50 wpm (Craig & Sherrick, 1982).

The Optohapt consisted of a linear array of nine photocells that scanned vertically the output of an electric typewriter on paper and nine vibrators distributed along the body surface (Figure 18.3). The stimulation sites

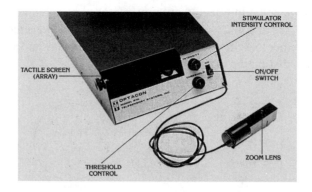

FIG. 18.2. The Optacon. Photograph courtesy of Telesensory Corp.

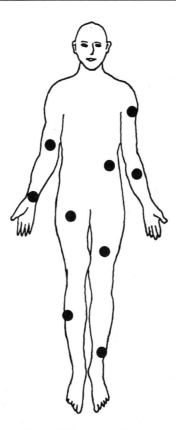

FIG. 18.3. The locations of the nine vibrators used in the Optohapt system. From Geldard (1966), reprinted by permission of Psychonomic Society, Inc.

were selected to avoid corresponding bodily points and were distributed
as widely as possible over the skin surface. It was found that "raw" letters of the alphabet were not the most efficacious symbols because they
lacked discriminability. It was suggested that the twenty-six most readily
discriminated signals (such as the period, the colon, a filled square, etc.)
be selected to encode letters of English (Geldard, 1966).

Whereas the Optacon and the Optohapt were aimed toward the transmission of text material, the TVSS was designed to transmit general visual
images. It consisted of a television camera controlled by the user, and a
20-by-20 matrix of solenoid vibrators (spaced 12 mm apart) mounted in the
back of a dental chair (Figure 18.4). Each solenoid would vibrate when the
corresponding region in the camera's viewfinder was illuminated. Initial
results showed that both sighted and blind subjects learned to recognize

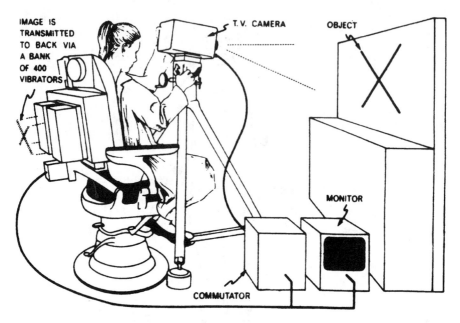

FIG. 18.4. The TVSS. From White et al. (1970), reprinted by permission of Psychonomic Society, Inc.

common objects (e.g., telephone, cup, etc.) and their arrangements in three-dimensional space. When given the control of the camera, the subjects also learned to externalize the objects presented tactually on their backs as being in front of them (Bach-y-Rita, 1972; White, Saunders, Scadden, Bach-Y-Rita, & Collins, 1970). Further investigation showed, however, that subjects had considerable difficulties in identifying internal details of a pattern, thus casting doubts on the skin's ability to identify complex visual patterns (White, 1973).

The Kinotact was very similar to TVSS in construction. It consisted of a 10-by-10 array of photocells, 100 switching circuits, and a corresponding 10-by-10 array of vibrators mounted on the back of a chair. Cutting off light to a particular photocell would switch on a corresponding vibrator. Block letters of the alphabet were used as stimuli in experiments during which subjects were required to identify letters based on the vibration patterns presented on their back. It was found that subjects could perform almost equally well whether or not the correspondence between columns of photocells and those of vibrators were in the same order (i.e., column 1 of photocells was connected to column 1 of vibrators, column 2 of photocells was connected to column 2 of vibrators, and so forth) or randomized (e.g., column 1 of photocells was connected to column 7 of vibrators, column 2 of photocells was connected to column 6 of vibrators, etc.), as long as there was a one-to-one correspondence between columns of photocells and vibrators. The amount of additional training for the subjects to learn the random mapping was minimal (Craig, 1973).

The results of studies with the Optohapt and Kinotact systems suggest that whereas a direct spatial–temporal mapping of visual images to vibrational patterns seems to be the most natural approach to take, encoding visual information in a way that results in highly discriminable vibrational patterns warrants higher performance levels at very little additional cost on training (Geldard, 1966; Craig, 1973).

In a typical tactual hearing aid using the frequency-to-place transformation model, the acoustic signal of speech is sent through an array of bandpass filters with increasing center frequencies. The outputs of these filters are rectified and used to modulate the amplitudes of a corresponding array of vibrators (Keidel, 1973). Examples of such systems range from the "Felix" system developed by Dr. Nobert Wiener in the early 1950s at the Research Laboratory of Electronics at MIT (Figure 18.5) to the modern day wearable version of Tactaid VII. The Tactaid VII system (Audiological Engineering Corp., Somerville, MA) consists of a small processing unit with an embedded microphone, which can be clipped to a belt or fit into a shirt pocket, and a harness with seven resonant vibrators (Figure 18.6). The harness can be worn on the forearm, the chest, the abdomen or around the neck. When used alone, it can convey useful information such as environmental sounds, but understanding speech is difficult. When used in conjunction with speechreading (i.e., lipreading), Tactaid VII provides a limited improvement to sentence reception accuracy with a typical increase of around 10% (the so-called ceiling effect) (Reed & Delhorne, 1995).

In general, performance levels with artificial tactual speech communication devices do not reach anywhere near that demonstrated by Tadoma users (Reed, Durlach, Delhorne, Rabinowitz, & Grant, 1989). Previous research has documented that these individuals can understand everyday speech at very high levels, allowing rich two-way conversation with both familiar and unfamiliar talkers (Figure 18.7). The information transfer rate for the Tadoma method has been estimated to be roughly 12 bits/sec, which is about half the rate of daily conversation conducted by hearing individuals (Reed, Durlach, & Delhorne, 1992). In contrast, the typical reading rates with the Optacon is about 30–50 wpm, or about 4–6.7 bits/sec in information transfer rate. The conversion from word rate to information transfer rate

FIG. 18.5.   Dr. Norbert Wiener with the "Felix" system. Photograph by Alfred Eisenstaedt, 1950.

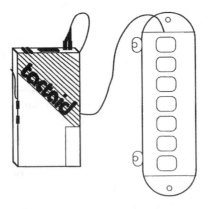

FIG. 18.6.   The Tactaid VII with its processor and the seven-vibrator array. Reprinted by permission from Weisenberger & Percy (1994). © 1994 Alexander Graham Bell Association for the Deaf, Inc.

FIG. 18.7. Two experienced Tadoma method users who are deaf and blind (Leonard Dowdy on the left and Raymond Boduch on the right) communicate with each other and Senior Research Scientist Nathaniel I. Durlach (center). Photograph by Hansi Durlach, 1980.

is based on two assumptions. First, according to Shannon (1951, Fig. 4), the uncertainty for strings of eight letters (including the 26 letters of the English alphabet and space) or more has an upper bound of 2 bits/sec. For simplicity, it is assumed that the test material is longer than eight letters. Second, it is assumed that the average word length is 4 letters. It follows that the information content in words is 2 bits/letter × 4 letter/word, or equivalently, 8 bits/word. The information rate for 30 wpm is, therefore, 8 bits/word × 30 words/minute, or equivalently, 4 bits/sec.

One problem with tactile aids is that they are composed of multiple stimulators that deliver "homogeneous" high-frequency vibrations to the tactual sensory system. In contrast, a talking face for Tadoma is perceptually rich, displaying various stimulation qualities (e.g., mouth opening, airflow, muscle tension around the cheeks, laryngeal vibration, etc.) that engage both the kinesthetic and tactile sensory systems. Recognition of the need

FIG. 18.8. The "reverse-typewriter" system. Photograph by James Bliss.

to develop devices that engage both the tactile and the kinesthetic senses is now prevalent. Examples of such displays are the "reverse-typewriter" system, the "OMAR" system, the MIT Morse code display, and the Tactuator, all of which stimulate the fingers.

The "reverse-typewriter" system was developed by Bliss (1961) prior to his invention of the Optacon. It was a pneumatic display that consisted of eight finger rests arranged in two groups on which the user could place the fingers of both hands in a manner similar to that of the "home" position of a typewriter (Figure 18.8). Each stimulator was capable of generating motions corresponding to the active movements of a typist's fingers in reaching the upper and lower rows on a keyboard. One experienced typist was trained to receive sequences of 30 symbols (the alphabet, comma, period, space, and upper case) and reached a maximum information transfer rate of 4.5 bits/sec (Bliss, 1961).

The "OMAR" system was a two-degree-of-freedom (up–down and front–back) finger stimulator that used motion, vibration, and stiffness cues to encode speech information (Figure 18.9). Initial experiments demonstrated that subjects were able to judge onset asynchronies of vibration and movement with this system (Eberhardt, Bernstein, Barac-Cikoja, Coulter, & Jordan, 1994).

FIG. 18.9.   The "OMAR" system. Photograph courtesy of Dr. Lynne
E. Bernstein. © 1998 House Ear Institute.

The MIT Morse code display was designed to move the fingertip of
one finger up and down in a way that was similar to the motions generated by ham radio operators using the straight keys for sending the code
(Figure 18.10). The ability of two experienced Morse code operators to
receive the code of everyday English sentences through this device was
estimated to be around 20 wpm, or 2.7 bits/sec (Tan, Durlach, Rabinowitz,
Reed, & Santos, 1997).

The Tactuator consisted of three independent, point-contact, one-degree-
of-freedom actuators interfaced individually with the fingerpads of the
thumb, the index finger, and the middle finger (Figure 18.11). Each movement channel is capable of delivering stimuli from absolute detection
threshold (i.e., the smallest displacement that can be detected by a human
observer) to approximately 50 dB SL (Sensation Level, defined relative to
detection threshold) throughout the frequency range from near DC to above
300 Hz, thereby encompassing the perceptual range from gross motion to
vibration (Tan & Rabinowitz, 1996). Information transfer rate with the
Tactuator was estimated to be about 12 bits/sec, which is roughly the same
as that achieved by Tadoma users in tactual speech communication (Tan,
Durlach, Reed, & Rabinowitz, 1998). This promising result was mainly

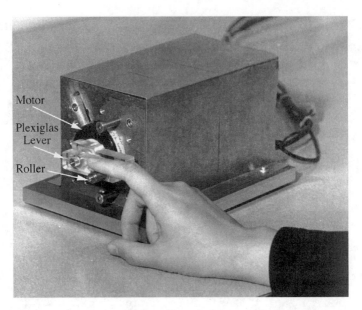

FIG. 18.10.   The one-degree-of-freedom (up and down) stimulator
used in the MIT Morse code study. From Tan et al. (1997), reprinted
by permission of Psychonomic Society, Inc.

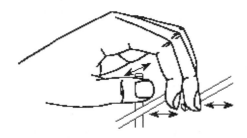

FIG. 18.11.   A diagram of the Tactuator. Reprinted by permission
of the ASME from Tan & Rabinowitz (1996).

attributed to the relative richness of the Tactuator as a tactual display (i.e.,
it used features such as finger location, motional and vibrational stimulation, etc. to convey tactual information).

Despite the recent promising results in the research laboratories, however, much work still remains before these experimental apparatus become
practical for the daily use by individuals with sensory impairments.

The above review, although brief and incomplete [e.g., we did not discuss electrocutaneous stimulation at all because of its tendency to induce

pain and discomfort (Rollman, 1973)], brings several conclusions that can be drawn from work in the area of sensory substitution. First of all, among the above-reviewed tactual communication systems developed for sensory substitution, the Optacon and the Tactaid VII (including its earlier versions) are the only commercially available, portable aids for the blind and the deaf, respectively. Laboratory apparatus are usually developed with the aim of precise stimulus control, not necessarily the device portability. Second, both the Optacon and the Tactaid VII require intensive and extensive training for their users. Only those individuals with severe sensory impairments are motivated enough to go through the training. Third, performance with tactile aids for the blind and/or the deaf do not match that achieved by people with normal sensory capabilities as far as pictorial and speech communication are concerned. Fourth, there remains much work in developing displays that deliver rich stimulation to the tactual sensory systems, and in devising coding schemes that best match the information content of stimulating signals with the capability of the somatosensory system.

4. CONSIDERATIONS FOR WEARABLE TACTUAL DISPLAYS

The development of tactual interfaces in general requires advancement in two areas. On the one hand, an understanding of the human somatosensory system enables us to associate physical stimulation parameters with well-defined percepts. On the other hand, advances in technologies make it possible for us to design apparatus that can deliver desired stimulation patterns. The development of wearable tactual displays presents additional challenges and opportunities.

The first consideration is the wearability of wearable tactual displays. Desktop-based displays such as the Tactuator and the PHANToM™ (Massie & Salisbury, 1994) are unlikely candidates for a wearable computer. Portable haptic displays, such as the exoskeleton system developed by EXOS Inc. for NASA astronauts, requires the user to carry the weight of the structure and absorb the excessive force at body sites strapped to the device. One human factor study recommends that such devices be worn for no more than an hour or two due to user fatigue (Tan, Srinivasan, Eberman, & Cheng, 1994). Given the state of current technology, vibrotactile displays are good candidates for wearable tactual displays for their light weight and low power consumption.

The second consideration is the body site to be stimulated by wearable tactual displays. The desire for high spatial resolution should be balanced with the accessibility and size of contact area. For example, the hands (especially the fingertips) are the best candidates for tactual displays in terms of sensory resolution. However, the hands are already engaged in many daily tasks, especially those involving human–computer interactions. The back has poorer spatial resolution, yet it is usually not engaged by any human–computer interfaces and can be easily accessed. Its relatively poor spatial resolution can be compensated for by its relatively large contact area.

The third consideration is the intended users of wearable tactual displays. We believe that with proper design, tactual displays should be useful for all users of wearable computers, whether they are sensory impaired or not. A good example of a well-designed universal adaptive structure is the street curb. Although originally conceived to provide access for individuals on wheelchairs, curbs are used by mothers with baby strollers, students on roller blades, and couriers with dollies. Tactual displays should be developed to enhance the functionality of wearable computers for all users.

The fourth consideration is the amount of training associated with the use of wearable tactual displays. A good wearable tactual display should minimize training by displaying information that is salient, intuitive, and easy to interpret through the sense of touch. For example, a buzzing on one's shoulder can attract immediate attention from the user. Increased pressure on one's back can signal something approaching from behind.

The fifth issue concerns the long-term wearability of tactual displays. For example, it is well known that all sensory systems adapt to the stimulation environment. For example, we cease to notice our clothes after we have put them on for a while although they are in constant contact with the skin surfaces. Static stimulation (such as a constant vibration) tends to "numb" the skin and "fades" after a short while. Dynamic stimuli (such as pulsing) are more likely to evoke the same perceptual intensity time after time.

Finally, a new paradigm is needed to seamlessly integrate tactual interfaces with visual and auditory displays. Given the existing well-developed visual and auditory interfaces, tactual interfaces would be most useful when used to supplement visual and auditory information, or to clarify it when vision or audition is overloaded. An example of such a system would be a tactile vest that redirects a driver's visual attention when a collision-warning alarm goes off.

In summary, a good wearable tactual display should be not only portable but wearable, should not interfere with the user's daily activities, should be useful for people with all degrees of sensory capabilities, should require minimum amount of training, should be resistant to sensory adaptation, and should be well integrated with existing visual and auditory interfaces.

5. A TACTILE DIRECTIONAL DISPLAY

Recently, a wearable tactile directional display has been developed at our research laboratory. It consists of an array of micromotors embedded in the back of a vest that delivers vibrational patterns to the back of the wearer. The display elicits salient and vivid movement sensations that are resistant to sensory adaptation. Informal tests with first-time users indicate that it requires no training and the interpretation of directional signals is highly consistent. Because the directional information is presented relative to the user's body coordinates, it eliminates the need to transform coordinates as is often the case with maps and building layouts. What's more, the perceived spatial resolution of the display can be manipulated by the signals delivered to the individual actuators and is not limited by the physical layout of actuator arrays.

All this is accomplished by taking advantage of a perceptual illusion called sensory saltation (also known affectionately as the "rabbit"), discovered by Dr. Frank Geldard and his colleagues at the Princeton Cutaneous Communication Laboratory in the early 1970s (Geldard, 1975). In a typical setup for eliciting the "rabbit," three mechanical stimulators are placed with equal distance on the forearm (Figure 18.12). Three brief pulses are delivered to the first stimulator closest to the wrist, followed by three more at the middle stimulator, followed by another three at the stimulator near

FIG. 18.12. A Norwegian cartoonist's illustration of sensory saltation. From Geldard (1975), reprinted by permission of Lawrence Erlbaum Associates, Inc. Publishers.

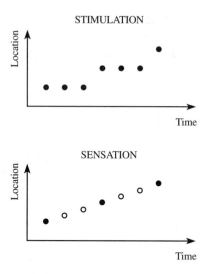

FIG. 18.13. An illustration of sensation vs. stimulation pattern for
sensory saltation. Open circles indicate perceived pulses at phantom locations between stimulators. Reprinted by permission from
Tan & Pentland (1997). © 1997 IEEE.

the elbow. The pulses are evenly spaced in time. Instead of feeling the
successive taps localized at the three stimulator sites, the observer is under
the impression that the pulses are distributed with more or less uniform
spacing from the wrist to the elbow (as illustrated in Figure 18.13). The
saltation effect can be elicited with mechanical, electrocutaneous, or thermal stimulation. The sensation is discontinuous and discrete as if a tiny
rabbit was hopping up the arm from the wrist to the elbow, leading to the
nickname "cutaneous rabbit." The same vivid hopping sensation can also
be elicited in vision and audition (Geldard, 1975).

An important feature of the cutaneous rabbit is its ability to simulate higher spatial resolution than the actual spacing of stimulators, yet
mimic the sensation produced by a veridical set of stimulators with the
same higher-density spacing (Cholewiak & Collins, 1995; Collins, 1996;
Cholewiak, Sherrick, & Collins, 1996). The perceived spacing of adjacent taps is inversely proportional to the number of pulses sent to each
stimulator. In theory, only two stimulators are needed in order to produce
the sensory saltation effect. Additional stimulators add redundancy and
robustness to the overall setup. Thus a sparse stimulator array can be effectively used to produce a dense perception. Another important feature of
the sensory saltation phenomenon is that the sensation remains vivid and

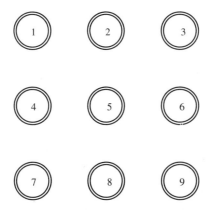

FIG. 18.14. Layout of the 3-by-3 two-dimensional "rabbit" display. Reprinted by permission from Tan & Pentland (1997). ©1997 IEEE.

does not fade away after repeated stimulation due to its discontinuity and discreteness.

Our wearable rabbit displays are implemented on the back of vests. An initial protocol consisted of 9 actuators arranged in a 3-by-3 array that measured 18 cm by 18 cm (Figure 18.14). The final version consisted of 16 actuators arranged in a 4-by-4 array covering an area of 18 cm by 18 cm. The main reasons for using a 4-by-4 array are to avoid direct stimulation to the spine area and to provide some additional redundancy. Each actuator is based on a micromotor with a biased load (MicroMo, Clearwater, Florida) that weighs 3.2 grams. The micromotor is mounted inside a square plastic tubing. The tubing is then attached to the garment with elastic bands. The state of the actuator (on or off) and its timing can be controlled by the parallel port of a personal computer through additional circuitry.

To test the effectiveness of our rabbit display, observers who are not aware of the sensory saltation phenomenon have been asked to wear the vests and report on the sensations associated with several stimulation patterns. One typical pattern consisted of three pulses sent to stimulator #8 (see Figure 18.14), followed by three pulses sent to #5, and followed by another three sent to #2. Most observers commented on a sensation of something "hopping" or "crawling" up the spine. When asked how many pulses were felt, most gave an answer between 6 and 8, thus indicating a perception of pulses in between stimulator locations. This is quite consistent with the classical definition of the sensory saltation phenomenon. In a two-dimensional pattern, three pulses of each were sent sequentially to stimulators in the order of #1, #2, #3, #6, #9, #8, #7, and #4. Instead of feeling a square, most observers reported that they felt a circular pattern. This was an interesting finding and its interpretation awaits further investigation of the sensory saltation phenomenon, especially in the case of a two-dimensional stimulator array.

In general, novice observers find it intuitive to perceive the directional information indicated by these patterns, and their interpretations of the signals are highly consistent. The signals can elicit vivid movement sensations (up, down, left, right, along a line of 45° or −45° incline). Because the directional cues are relative to the user's own body coordinate, no additional coordinate transformations are necessary. The fact that a circular pattern can be perceived suggests that other patterns might be "drawn" as well.

In an implementation of a navigational guidance system, the rabbit vest has been integrated with a driving simulation software system to provide directional cues to a driver. The overall system consists of an SGI, a PC, the vest, and input hardware that simulated the steering wheel, acceleration petal, and brake petal. The SGI is dedicated to the traffic simulation software SIRCA (Nissan Cambridge Basic Research, Cambridge, Massachusetts). The PC is used to interface with the hardware. The driver controls the steering wheel and acceleration and brake petals. The positions of these input devices, read by the game port of the PC and transmitted to the SGI through a serial link, are used to control a virtual car in SIRCA simulation. Depending on the location of the virtual car, surrounding traffic condition, road configuration, and the driver's intended destination, SIRCA generates one of the following directional commands: turn left, turn right, or go straight at the next intersection. This command, sent to the PC through a serial connection, triggers a preprogrammed signal pattern at the parallel port of the PC that controls the micromotors in the rabbit vest. The driver wearing the vest feels a left, right, or upward arrow on the back. Such a system demonstrates the feasibility and effectiveness of wearable tactile displays for navigation guidance.

6. SUMMARY

This chapter has presented a general review of tactual displays for sensory substitution and a description of a tactile directional display designed for wearable computing. The two-dimensional "rabbit" display has many features that make it an attractive candidate for displaying directional information in applications such as navigation guidance. One implementation has already demonstrated the usefulness of simple directional signals for providing directional cues to a driver. The possibilities of displaying more complicated patterns and using them to encode useful information have yet to be explored.

As wearable computing becomes more ubiquitous and distributed, a new generation of interfaces for wearable computers is emerging. In the future, smart clothing and furniture will become part of human–computer interfaces through contact sensing and display. By presenting our current efforts toward that goal, we hope to stimulate similar work on new and novel haptic displays that can work in concert with wearable visual and auditory displays.

Acknowledgments This work has been supported in part by British Telecom and in part by the Things That Think Consortium at the MIT Media Lab. Part of this work has been presented at the First International Symposium on Wearable Computers.

REFERENCES

Bach-y-Rita, P. (1972). *Brain Mechanisms in Sensory Substitution*. New York: Academic Press.

Bliss, J. C. (1961). *Communication via the kinesthetic and tactile senses*. Ph.D. Dissertation, Dept. of Electrical Engineering, M.I.T.

Burdea, G. C. (1996). *Force and Touch Feedback for Virtual Reality*. New York: John Wiley & Sons, Inc.

Cholewiak, R. W., & Collins, A. A. (1995). Exploring the conditions that generate a good vibrotactile line. Presented at the Psychonomic Society Meeting, Los Angeles, CA.

Cholewiak, R. W., Sherrick, C. E., & Collins, A. A. (1996). Studies of saltation. *Princeton Cutaneous Research Project (No. 62)*. Princeton University, Department of Psychology.

Clark, F. J., & Horch, K. W. (1986). Kinesthesia. In K. R. Boff, L. Kaufman, & J. P. Thomas (Eds.), *Handbook of Perception and Human Performance: Sensory Processes and Perception* (pp. 13/1–13/62). New York: Wiley.

Collins, A. A. (1996). Presentation at the Tactile Research Group Meeting of the Psychonomic Society Meetings.

Craig, J. C. (1973). Pictorial and abstract cutaneous displays. In F. A. Geldard (Ed.), *Cutaneous Communication Systems and Devices* (pp. 78–83). The Psychonomic Society, Inc.

Craig, J. C. (1977). Vibrotactile pattern perception: Extraordinary observers. *Science*, **196**, 450–452.

Craig, J. C., & Sherrick, C. E. (1982). Dynamic tactile displays. In W. Schiff & E. Foulke (Eds.), *Tactual Perception: A Sourcebook*. Cambridge University Press.

Eberhardt, S. P., Bernstein, L. E., Barac-Cikoja, D., Coulter, D. C., & Jordan, J. (1994). Inducing dynamic haptic perception by the hand: system description and some results. In *Proceedings of the American Society of Mechanical Engineers*, **DSC-1**, 345–351.

Geldard, F. A. (1966). Cutaneous coding of optical signals: The optohapt. *Perception & Psychophysics*, **1**, 377–381.

Geldard, F. A. (1975). *Sensory Saltation: Metastability in the Perceptual World*. Hillsdale, New Jersey: Lawrence Erlbaum Associates.

Keidel, W. D. (1973). The cochlear model in skin stimulation. In F. A. Geldard (Ed.), *Cutaneous Communication Systems and Devices* (pp. 27–32). The Psychonomic Society, Inc.

Linvill, J. G., & Bliss, J. C. (1966). A direct translation reading aid for the blind. *Proceedings of the Institute of Electrical and Electronics Engineers*, **54**, 40–51.

Loomis, J. M., & Lederman, S. J. (1986). Tactual perception. In K. R. Boff, L. Kaufman, & J. P. Thomas (Eds.), *Handbook of Perception and Human Performance: Cognitive Processes and Performance* (pp. 31/1–31/41). New York: Wiley.

Massie, T. H., & Salisbury, J. K. (1994). The PHANToM haptic interface: A device for probing virtual objects. In *Proceedings of the American Society of Mechanical Engineers*, **DSC-55**(1), 295–299.

Reed, C. M., & Delhorne, L. A. (1995). Current results of field study of adult users of tactile aids. *Seminars in Hearing*, **16**(4), 305–315.

Reed, C. M., Durlach, N. I., & Delhorne, L. A. (1992). The tactual reception of speech, fingerspelling, and sign language by the deaf-blind. *Digest of Technical Papers of the Society for Information Display International Symposium*, **XXIII**, 102–105.

Reed, C. M., Durlach, N. I., Delhorne, L. A., Rabinowitz, W. M., & Grant, K. W. (1989). Research on tactual communication of speech: Ideas, issues, and findings. *The Volta Review*, **91**, 65–78.

Reed, C. M., Rabinowitz, W. M., Durlach, N. I., Braida, L. D., Conway-Fithian, S., & Schultz, M. C. (1985). Research on the Tadoma method of speech communication. *Journal of the Acoustical Society of America*, **77**(1), 247–257.

Rollman, G. B. (1973). Electrocutaneous stimulation. In F. A. Geldard (Ed.), *Cutaneous Communication Systems and Devices* (pp. 38–51). The Psychonomic Society, Inc.

Shannon, C. E. (1951). Prediction and entropy of printed English. *Bell System Technical Journal*, **30**, 50–64.

Tan, H. Z., Durlach, N. I., Rabinowitz, W. M., Reed, C. M., & Santos, J. R. (1997). Reception of Morse code through motional, vibrotactile, and auditory stimulation. *Perception & Psychophysics*, **59**(7), 1004–1017.

Tan, H. Z., Durlach, N. I., Reed, C. M., & Rabinowitz, W. M. (1999). Information transmission with a multi-finger tactual display. *Perception & Psychophysics*, **61**(6), 993–1008.

Tan, H. Z., & Pentland, A. (1997). Tactual displays for wearable computing. *Digest of the First International Symposium on Wearable Computers*, 84–89. IEEE Computer Society.

Tan, H. Z., & Rabinowitz, W. M. (1996). A new multi-finger tactual display. In *Proceedings of the American Society of Mechanical Engineers*, **DSC-58**, 515–522.

Tan, H. Z., Srinivasan, M. A., Eberman, B., & Cheng, B. (1994). Human factors for the design of force-reflecting haptic interfaces. In *Proceedings of the American Society of Mechanical Engineers*, **DSC-55**(1), 353–359.

Weisenberger, J. M., & Percy, M. E. (1994). Use of the Tactaid II and Tactaid VII with children. *The Volta Review*, **96**(5), 41–57.

White, B. W. (1973). What other senses can tell us about cutaneous communication. In F. A. Geldard (Ed.), *Cutaneous Communication Systems and Devices* (pp. 15–19). The Psychonomic Society, Inc.

White, B. W., Saunders, F. A., Scadden, L., Bach-Y-Rita, P., & Collins, C. C. (1970). Seeing with the skin. *Perception & Psychophysics*, **7**(1), 23–27.

19

From "painting with lightvectors" to "painting with looks": Photographic/ videographic applications of WearComp–based augmented/mediated reality

Steve Mann

University of Toronto

1. PHOTOGRAPHIC ORIGINS OF WEARABLE COMPUTING AND AUGMENTED/MEDIATED REALITY IN THE 1970s AND 1980s

The original Personal Imaging application [Mann, 1997d] was an attempt to define a new genre of imaging and create a tool that could allow reality to be experienced with greater intensity and enjoyment than might otherwise be the case.

This effort also facilitated a new form of visual art called Lightspace Imaging (or Lightspace Rendering) in which the author chose a fixed point of view for the camera, and then, once the camera was secured on a tripod, the author walked around and used various sources of illumination to sequentially build up an image layer-upon-layer in a manner analogous to paint brushes upon canvas, and the cumulative effect embodied therein. Two early 1980s attempts, by the author, at creating expressive images using the personal imaging system developed by the author in the

(a)   (b)

FIG. 19.1. [Norman, 1992] Norman's criticism of the camera
arises from the fact that, in many ways, it diminishes our perception and enjoyment of reality. However, a goal of Personal
Imaging [Mann, 1997d], is to create something much more like
the sketch pad or artist's canvas than like the camera in its usual
context. The images produced as artifacts of Personal Imaging
are somewhere at the intersection of painting, computer graphics, and photography. (a) Notice how the broom appears to be
its own light source (e.g. self-illuminated), while the open door-
way appears to contain a light source emanating from within. The
rich tonal range and details of the door itself, although only visible
at a grazing viewing angle, are indicative of the affordances of
the Lightspace Rendering [Mann, 1992; Ryals, 1995] method.
(b) Hallways offer a unique perspective, which can also be illuminated expressively.

1970s and early 1980s are depicted in Figure 19.1. Throughout the 1980s,
a small number of other artists also used the author's apparatus to create various lightpaintings. However, due to the cumbersome nature of the
early WearComp hardware, etc., and the fact that much of the apparatus
was custom fit to the author, it was not widely used over an extended
period of time by others. However, the personal imaging system proved
to be a new and useful invention for a variety of photographic imaging
tasks.

To the extent that the artist's light sources were made far more powerful
than the natural ambient light levels, the artist had a tremendous degree
of control over the illumination in the scene. The resulting image was
therefore a depiction of what was actually present in the scene, together

with a potentially very visually rich illumination sculpture surrounding it. Typically the illumination sources that the artist carried were powered by batteries. (Gasoline powered light sources were found to be unsuitable in many environments such as indoor spaces where noise, exhaust, etc. were undesirable.) Therefore, owing to limitations on the output capabilities of these light sources, the art was practiced in spaces that could be darkened sufficiently, or, in the case of outdoor scenes, at times when the natural light levels were least.

In a typical application, the user positions the camera upon a hillside, or on the roof of a building, overlooking a portion of a city, usually having an assistant oversee the operation of this camera. The user may then roam about the city, walking down various streets, and use the light sources to illuminate various buildings one-at-a-time. Typically, in order that the wearable or portable light sources be of sufficient strength compared to the natural light in the scene (e.g., so that it is not necessary to shut off the electricity to the entire city to darken it sufficiently that the artist's light source be of greater relative brightness) some form of electronic flash is used as the light source. In some embodiments of the personal imaging invention, an FT-623 lamp (the most powerful lamp in the world, with output of 40 kJ) is used, housed in a lightweight 30 inch highly polished reflector, with a handle that allows it to be easily held in one hand and aimed (Figure 19.2).

1.1 Lightvector Amplification

The communications infrastructure is established such that the camera is only sensitive to light for a short time period (e.g., typically approximately 1/500 of a second), during the instant that the flash lamp produces light.

In using the personal imaging invention to selectively and sequentially illuminate portions of a scene or object, the user will typically point the source at some object in the scene in front of the camera, and issue a command through the wearable computer system. A simplified diagram of the architecture is depicted in Figure 19.3 The receiver at the camera is typically embodied in a communications protocol, which in newer embodiments of the invention runs over amateur packet radio, using a terminal node controller in KISS mode (TCP/IP). In the simple example illustrated here, the RECEIVER activates shutter solenoid S; what is depicted in this drawing is approximately typical of a 1940s press camera fitted with the standard 6 volt solenoid shutter release, while in actual practice there are no moving parts in the camera, and the shutter is implemented electronically.

FIG. 19.2. Wearable Computing and Augmented/Mediated Reality for Computer Supported Collaborative Photography: A portable electronic flashlamp is used to illuminate various buildings such as tall skyscrapers throughout a city. The viewfinder on the helmet displays material from a remotely mounted camera with computer generated text and graphics overlaid in the context of a collaborative telepresence environment. An assistant at the remote site wears a similar apparatus with a similar body-worn backpack-based processing system.

The camera is sometimes designed so that it provides a sync signal in advance of becoming sensitive to light, so that pulse compression may be used for the synchronization signal.

The wearable computer is generally distributed throughout a heavy black jacket, and the artist will typically wear black pants together with the jacket and hold the light source depicted in Figure 19.3(a) using a black glove,

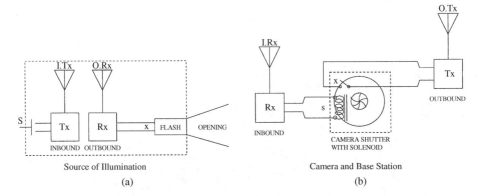

Source of Illumination
(a)

Camera and Base Station
(b)

FIG. 19.3. Artist's "paintbrush" for Computer Supported Collaborative Photography: The artist issues commands to a remote camera using a data entry device while monitoring the resulting pictures and overlaid text+graphics on a head-mounted display. Here a simplified diagram is used to illustrate signal routing. (a) When the artist issues a command by switch closures (S), a signal is sent through an INBOUND communications channel, depicted as transmitter I.Tx, to the central base station (b) and is received by the inbound receiver, denoted I.Rx. This initiates frame capture (depicted by solenoid S) with a computer system located at the base station. At the correct instant during frame capture, a signal (depicted by flash sync contacts X) is sent back by the camera's outbound transmit channel O.Tx, to the artist (a) and received by the artist's light source synchronization receiver, O.Rx. This activates FLASH through its synchronization contacts denoted X. Light then emerges through the OPENING and illuminates the scene at the exact instant during which the camera's sensor array is sensitive to light. A short time later, the image from the camera base station (b) is sent via the OUTBOUND channel to the artist (a) and is displayed on the artist's head-mounted display, overlaid with a calculated summation of previous differently illuminated images of the same scene and appropriate graphics for manipulation of the summation coefficients. Note the similarity of this communications architecture with that of Figure 9.6 in Chapter 9. This system is, in fact, typically implemented as an extension of the system depicted in Figure 9.5 of Chapter 9, allowing the artist to experience the effects of activating the flashlamp within a computer–mediated reality environment.

although this is not absolutely necessary. Accordingly, the housing of the lamp head will often be painted flat black.

In this manner a comparatively small lamp (small compared to the scale of a large city, e.g., a lamp and housing that can be held in one hand) may illuminate a large skyscraper or office tower in such a manner that the lamp appears, in the final image, to be the dominant light source, compared to fluorescent lights and the like that might have been left turned on upon the

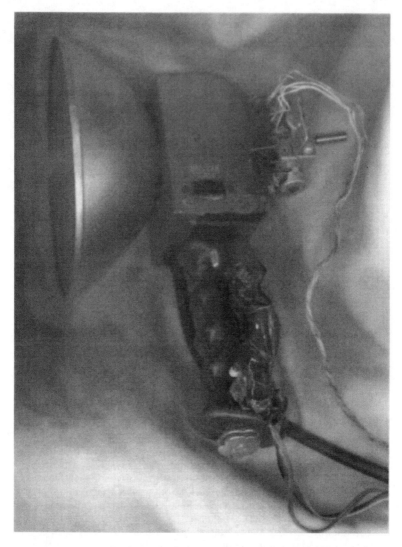

FIG. 19.4. **Typical "Keyboard" and "Mouse":** The portable electronic flashlamps used in conjunction with the invention are each equipped with a data entry device and cursor pointing device (useful in computer supported collaborative photography). Here a 1970s system is shown with five microswitches operable by the right hand while simultaneously holding and aiming the lamp. Thus it is possible that one can be walking and entering data at the same time, or even be climbing a ladder or rope, and stop to enter data. The pointing device (joystick) is operated with the left hand so that when "typing" and pointing, both hands are occupied. However, since the pointing device is not used frequently

various floors of the building, or to moonlight, or light from streetlamps, which cannot be easily turned off.

Typically, the artist's wearable computer system comprises a visual display capable of displaying the image from the camera (typically sent wirelessly over a data communications link from the computer that controls the camera). Typically, also, this display is updated with each new exposure. The wearable computer is generally controllable by the artist through a chording keyboard mounted into the handle of each light source, so that it is not necessary to carry a separate keyboard. In this manner, whichever light source the artist plugs into the body-worn system becomes the device for controlling the process. An example of an input device built into the handle of a smaller electronic flashlamp appears in Figure 19.4

Typically exposures are maintained as separate image files overlaid on the artist's screen (head-mounted display) together with the current view through the camera. Because the exposures are in separate image files the artist can selectively delete the most recent exposure, or any of the other exposures previously combined into the running "sum" on the head-mounted display ("sum" is used in quotes here because the actual entity, a summation in homomorhic vectorspace, will be described later). Additional graphic information is also overlaid to assist the artist in choice of weighting for manipulation of this "sum." This capability is quite useful, compared to the process of painting on canvas, where one must paint over mistakes rather than simply being able to turn off brushstrokes or adjust the intensity of brushstrokes after they are made. Furthermore, exposures to light can be adjusted either during the shooting or afterwards, and then recombined. The capability of doing this during the shooting is an important aspect of the personal imaging invention, because it allows the artist to capture additional exposures if necessary, and thus to remain at the site until a final picture is produced. The final picture as well as the underlying data set of separately adjustable exposures is typically sent wirelessly to other sites so that others (e.g., art directors or other collaborators) can manipulate the

---

the apparatus is usable with one hand most of the time. The lamp pictured in Figure 19.2 has a similar user interface ("keyboard" for right hand and "mouse" for left hand), except that the left-hand device is also built in proximity to a separate handle/grip to facilitate two-handed grasping of the lamp in conditions of high wind. (The 30 inch reflector acts like a sail, so it needs to be held with both hands in windy weather.) Otherwise the two lamps have the same user interface. Consistency of user interface was an important human-factors consideration.

various exposures and combine them in different ways and send comments back to the artist by email, as well as by overlaying graphics onto the artist's head-mounted display, which then becomes a collaborative space. In very recent embodiments (1990s) this has been facilitated through the World Wide Web. This additional communication facilitates the collection of additional exposures if it turns out that certain areas of the scene or object could be better served if they were more accurately described in the data set.

1.2   Lightstrokes and Lightvectors

Each of a collection of differently illuminated exposures of the same scene or object is called a lightstroke. In the context of Personal Imaging, a lightstroke is analogous to an artist's brushstroke, and it is the plurality of lightstrokes that are combined together that give the invention described here it's unique ability to capture the way that a scene or object responds to various forms of light. From each exposure, an estimate can be made of the actual quantity of light falling on the image sensor, by applying the inverse transfer function of the camera. Such an estimate is called a lightvector [Mann, 1992].

Furthermore, a particular lightstroke may be repeated (e.g., the same exposure may be repeated in almost exactly the same way, holding the light in the same position, each time a new lightstroke is acquired). These seemingly identical lightstrokes may collectively be used to obtain a better estimate of a lightvector, by averaging each of the lightvectors together to obtain a single lightvector of improved signal to noise ratio. This signal averaging technique may also be generalized to the extent that the lamp may be activated at various strengths, but otherwise held in the same position and pointed in the same direction at the scene. The result is to produce a lightvector that captures a broad dynamic range by using separate images that differ only in exposure level [Mann and Picard, 1994; Mann, 1993; Mann, html]. The underlying mathematical concepts of lightstrokes and lightvectors are described in Chapter 7 of this book.

1.3   Other Practical Issues of Painting with Lightvectors

Another aspect of the invention is that the photographer need not work in total darkness as is typically the case with ordinary lightpainting. With a typical electronic flash, and even with a mechanical shutter (as is used with photographic film), the shutter is open for only $1/500$ s or so for each "light-stroke." Thus the lightpainting can be done under normal lighting conditions (e.g., the room lights may often be left on). This aspect of the invention pertains to both traditional lightpainting (where the invention allows multiple flash-synched exposures to be made on the same piece of film), as well as to the use of separate recording media (e.g., separate film frames or electronic image captures) for each lightstroke. The invention makes use of innovative communications protocols and a user interface that maintain the illusion that the system is immune to ambient light, while requiring no new skills beyond that of traditional lightpainting. The communications protocols typically include a full-duplex radio communications link so that a button on the flash sends a signal to the camera to make the shutter open, and at the same time a radio wired to the flash sync contacts of the camera is already "listening" for when the shutter opens. The fact that the button is right on the flash gives the user the illusion that he or she is just pushing the lamp test button of a flash as in normal lightpainting, and the fact that there is really any communications link at all is hidden by this ergonomic user interface.

The invention also includes a variety of options for making the lightpainting task easier and more controlled. These include such innovations as a means for the photographer to determine if he or she can be "seen" by the camera (e.g., indicates extent of camera's coverage), various compositional aids, means of providing workspace illumination that has no effect on the picture, and some innovative light sources.

1.4 Computer Supported Collaborative Art

Finally, it may, at times, be desirable to have a real or virtual assistant at the camera, to direct or advise the artist. In this case, the artist's viewfinder, which presents an image from the perspective of the fixed camera, also affords the artist with a view of what the assistant sees. Similarly, it is advantageous at times that the assistant have a view from the perspective of the artist. To accomplish this, the artist has a second camera of a wearable form. Through this second camera, the artist allows the assistant to observe the scene from the artist's perspective. Thus the artist and assistant can collaborate by exchange of viewpoints, as if each had the eyes of the other. (Such a form of collaboration based on exchanged viewpoint is called seeing eye to eye [Mann, 1994a].) Moreover, through the use of shared cursors overlaid on this exchanged viewpoint, a useful form of computer supported collaborative photography has resulted.

The artist's camera is sometimes alternatively attached to and integrated with the light source (e.g., flash), in such a way that it provides a preview of the coverage of the flash. Thus when this camera output is sent to the artist's own wearable computer screen, a flash viewfinder results. The flash viewfinder allows the artist to aim the flash and allows the artist to see what is included within the cone of light that the flash will produce. Furthermore, when viewpoints are exchanged, the assistant at the main camera can see what the flash is pointed at prior to activation of the flash.

Typically there is a command that may be entered to switch between local mode (where the artist sees the flash viewfinder) and exchanged mode (where the artist sees out through the main camera and the assistant at the main camera sees out through the artist's eyes/flash viewfinder).

2.   PERSONAL IMAGING IN THE 1990s

2.1   Personal Imaging as a Tool for Photojournalists and Reporters

Throughout the 1990s the author experimented with personal imaging as a means of creating personal documentary, and in sharing this personal documentary video on the World Wide Web, in the form of *wearable wireless webcam* [Mann, 1994b].

On several occasions, new and interesting forms of collaboration emerged. On various occasions the author serendipitously encountered newsworthy events while wearing the apparatus. On some such occasions where the events were natural disasters or the like, there was no other coverage of these events (e.g., traditional journalists were unavailable to cover these events on the short notice involved, despite the great desire that there be coverage of these events).

An example of how the author functioned as a "roving reporter" is illustrated in Figure 19.5. This shows how Computer Supported Collaborative Photojournalism (CSCP) emerged from *wearable wireless webcam*. In actual practice, multiple images are "stitched together" to make a picture good enough for a full-page newspaper-size photograph despite the fact that each of the images has relatively low resolution. The manner in which pictures are combined will be illustrated later in this chapter.

In another example of CSCP the author encountered a flood in the basement of a building and notified the police of the flood. However, without really any conscious thought or effort, the fact that the author walked past the event and looked at it also resulted in its having been recorded and

FIG. 19.5. Serendipitously arising Computer Supported Collaborative Photojournalism (CSCP). Author encountered an event serendipitously through ordinary everyday activity. As it turned out later, the newspapers had very desperately wanted to get this event covered but could not reach any of their photojournalists in time to cover the event. The author, however, was able to offer hundreds of pictures of the event, wirelessly transmitted, while the event was still happening. Furthermore, a collaboration with a large number of remote viewers enabled a new form of CSCP.

transmitted wirelessly to remote sites. Thus the author was subsequently able to notify a newspaper of this transmission (wirelessly notifying the newspaper's editorial offices through email) and the very high resolution images of tremendously high dynamic range and tonal fidelity were retrieved by the newspaper's editorial office and published in the newspaper. The quality of the images was higher than is typical for the newspaper, suggesting that wearable wireless webcam can rival the photographic technical quality and resolution of professional photographers armed with the best cameras available on the market.

2.2 "Underwearables" for Covert Embodiments of Wearable Computing and Augmented/Mediated Reality

The early personal imaging systems of the 1970s and 1980s were characterized by a heavy and obtrusive nature. Due to much criticism that the author received in wearing these systems in day-to-day life, during the 1980s, a new generation of unobtrusive personal imaging system was developed during the 1990s, and a fully functional version was completed in 1995. Such an apparatus, known as an "underwearable" reality mediator, comprises an unobtrusive rig concealed in a pair of ordinary sunglasses (Figure 19.6(a)) connected to the underwearable computer (Figure 19.6(b)) Typical embodiments of the underwearable computer resemble an athletic undershirt (tank top) made of durable mesh fabric, upon which a lattice of webbing is

(a)

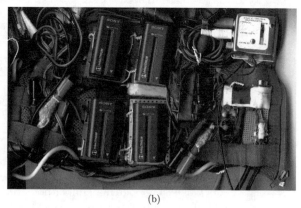
(b)

FIG. 19.6. The "underwearable" computer: (a) as worn by author without clothing to cover it, (b) close-up view showing webbing for routing of cabling.

sewn (Figure 19.6(b)). This facilitates quick reconfiguration in the layout of components and rerouting of cabling. Note that wire ties are not needed to fix cabling, as it is simply run through the webbing, which holds it in place. All power and signal connections are such that devices may be installed or removed without the use of any tools (such as a soldering iron) by simply removing the garment and spreading it out on a flat working surface.

Two examples of underwearables, as they normally appear when worn under clothing, are depicted in Figure 19.7, where the normal appearance

(a) (b)

FIG. 19.7. Covert embodiments of WearComp suitable for investigative documentary/photojournalism. Both incorporate fully functional UNIX-based computers concealed in the small of the back, with the rest of the peripherals, analog to digital converters, etc. also concealed under ordinary clothing. Both incorporate cameras concealed within the eyeglasses, used in the context of Personal Imaging. (a) Lightweight black and white version completed in 1995. This is also an ongoing project (e.g., implementation of full-color system in same size, weight, and degree of concealment is expected in 1998). (b) Full-color version completed in 1996 included special-purpose digital signal processing hardware based on an array of TMS 320 series processors connected to a UNIX-based host processor, concealed in the back of the underwearable. The cross-compiler for the TMS 320 series chips was run remotely on a Sun workstation, accessed wirelessly through radio and antennas concealed in the apparatus.

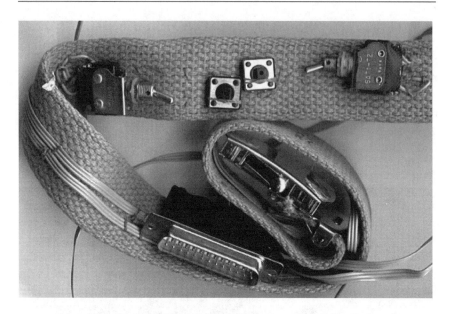

FIG. 19.8.   Covert belt-based input device operated by right hand, reaching behind back. Typically this device may be hidden underneath an untucked T-shirt or the like. The units that look like toggle switches are really spring-loaded extremely light-touch lever rockers. The DB25 connector used here is somewhat obsolete. A modern version would plug directly into the keyboard port of the body-worn computer rather than the serial port or parallel port.

is quite evident. Covert data-entry devices typically comprise switches located on the underwearable (undergarment) itself. These switches are easily actuated by pressing through clothing worn over the apparatus. Alternatively, a belt-mounted input device is used (Figure 19.8).

It should be noted that this class of system is more than just a wearable computer as might send and receive email, but, rather, it is a complete WearComp/WearCam personal imaging system, as defined in [Mann, 1997d]: It contains special-purpose image processing hardware [Mann, 1997b] and a complete video editing facility developed for the creation of the investigative metadocumentary ShootingBack (a documentary about making a documentary about video surveillance). The unobtrusive nature was necessary because it was found that the locations selected in which to shoot the documentary—establishments where video surveillance is used extensively (e.g., gambling casinos, department stores, banks, etc.)—also prohibit photography and video other than their own. Part of the purpose in constructing this apparatus was to challenge/investigate the nature of these one-sided establishments and shoot in establishments where photography/video was strictly prohibited.

Due to its ordinary appearance, the unobtrusive personal imaging system also suggests practical utility in everyday matters of personal life, such as personal documentary, personal safety, crime reduction, as well as investigative photojournalism.

3. DECONFIGURED EYE: ON BECOMING A CAMERA

It should be noted that the methodology of the new cinematographic and photographic genre characterized by personal imaging[Mann, 1997b] differs from current investigative journalism (e.g., miniature cameras hidden in the jewel of a tie clip, or in a baseball cap), in the sense that a long-term adaptation process, as described in Mann [1994a] (e.g., often taking place over a period of many years) makes the camera behave as a true extension of the mind and body and that the ability to augment, diminish, or otherwise alter the perception of reality is exploited fully, in the capturing of a much richer and more detailed perception of reality that captures the essence of wearer-involvement in his/her interaction with the world.

It should also be noted that the underlying principles of Mediated Reality (MR)[Mann, 1994a] differ from Augmented Reality (AR) where additional information is *added* onto the real world (e.g., through a display with a beamsplitter). Mediated Reality involves, in addition to the capability of augmenting reality, the capability of also diminishing or altering the perception of visual reality. Thus the personal imaging device must be fully immersive, at least over a certain *mediation zone* [Mann, 1994a]. A simple example of the utility of *diminished reality* is quite evident in the documentary video ShootingBack [Mann, 1997a, 1997c] when, for example, the author is asked to sign a bank withdrawal slip. Because of the deliberately diminished reality, it is necessary that the author bring his head very close to the written page (distance depending on the size of the lettering), in order to see it. A side effect of doing so is that video is produced in which the audience can also see the fine print, whereas shooting in a traditional investigative documentary style, this would not be so.

Once the Reality Mediator is worn for some time, and one becomes fully accustomed to experiencing the world through it, there is a certain synergy between human and machine that is not experienced with a traditional camera. More subtle differences between a recording made from the output of a reality mediator and that made from a conventional body-worn camera (such as might be hidden in the jewel of a tie clip) include the way that when one is talking to two people the closing of the loop forces one to turn one's head back and forth as one talks to one person and then to the other. This need arises from the limited peripheral vision the apparatus imposes on the wearer, which is yet another example of a deliberate diminishing of reality in order to heighten the experience of reality.

ShootingBack was shot through what amounted to "rot90" (90 degree rotating) eyeglasses, similar to George Stratton's upside-down glasses [Stratton, 1897] and Kohler's left–right reversing glasses[Kohler, 1964], but where the altering of visual reality was achieved through computational means rather than optical means. Furthermore, in addition to being another long-term psychophysical adaptation experiment, ShootingBack enabled new heights to be reached in concentration and seeing everything in a much more intensified way. Thus, just as copy editors often read their typed manuscripts in a mirror, so that they will spot all the typographical errors to which they had previously been *error blind*, in ShootingBack, the author learned how to see all over again. Similarly, just as the artist's sketch pad, because of its crudeness, forces us to concentrate and to really "see," Mediated Reality becomes an intensifying artifact in which the act of recording forces one to look and experience with more intensity and enjoyment than might otherwise be the case.

3.1   From "Fly on the Wall" Documentary to "Fly in the Eye" Personal Documentary

The Reality Mediator goes beyond merely functioning as a viewfinder to set the camera correctly because the human operator is completely in the loop (e.g., will trip and fall if it is not set correctly, since one will not be able to see properly), so that it transcends being a mere compositional tool, toward allowing the camera to "become" the eye of the wearer.

The nature of Personal Imaging is such that the combination of camera, processor, and display results in a visual adaptation process, causing it to function as a true extension of the mind and body: After an extended period of time, it does not seem to the wearer to be an external entity. Images captured on WearCam have a certain expressive quality that arises from having the wearer be "in the loop"—part of the complicated feedback process that synergizes both human and machine into a single unit.

3.2 Living in a 2-D World

In addition to adapting to transformations of the perceptual world, the author noticed a loss of the perception of 3-D depth. (A typical video camera lacks depth from stereo, depth from focus, etc.) In this sense, the author developed a "photographic mindset" in which an enhanced sense of awareness of light and shade, and of simple renaissance perspective, were attained. It was found [Mann, 1997b] that this effect persisted, even when the apparatus was removed, and that the effect would revisit in the form of 2-D "flashbacks," so that the world was seen in two ways, with a switching back and forth among the 2-D and 3-D interpretations. This discovery gave rise to the fingerpointing process (Figure 19.9) where it was found that pointing at objects was as though a 2-D plane projection. The notion of attaching a light to the finger arose out of various expressive lightpainting efforts, where the world was viewed as 2-D video, while a light source, attached to the finger, was moved around in 3-D space (Figure 19.10), while recording the process simultaneously on film (a beamsplitter being used to combine video and film cameras).

It was also found that performing tasks that required a great deal of hand-to-eye coordination caused a much more rapid learning of the mapping embodied in the reality mediator, than more passive tasks. Thus, for example, riding a bicycle with the apparatus on had a much lesser effect on

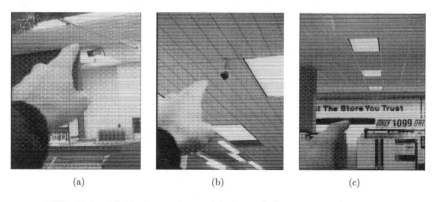

(a) (b) (c)

FIG. 19.9. Living in a 2-D world, through long-term adaptation. Fingerpointing from the perspective of *life through the screen*. Adaptation was in the "rot90" coordinate transformation described in the text. Notice how natural interactions with the world (such as pointing at objects) take place in 2-D projection on the image plane rather than in 3-D space.

(a)   (b)

FIG. 19.10.   Examples of tracing out a locus of points in 3-D space that are mapped onto a 2-D image. Here a small light source, attached to the author's finger, takes the form of a pointing device, which is used to outline objects in 3-D space, but falling upon their 2-D projection. (a) One of the early lightpaintings using this technique, 1986. (b) Image which won best color entry in the National Fuju Film competition, 1986. Here the method is perfected somewhat.

adaptation than reaching for objects such as while eating dinner, or playing sports, with the apparatus on.

Indeed, tasks like driving a car already incorporate a mediated reality of sorts—the process is mediated by the car itself (control of movement is by contact with a steering wheel, etc., rather than through direct contact with the world). Many of us are no doubt familiar with video games such as might be fitted with a steering wheel and computer screen, generating graphics images of a roadway. Thus it is not surprising that it is easier to do these tasks (such as driving) through a reality mediator than it is to do tasks like walking, eating, or playing sports through a reality mediator, since the former tasks already involve a mediated reality of sorts.

The finger operated cursor ("fingermouse") is an example of a task that is very direct and therefore requires an adaptation period prior to one becoming proficient.

In Mediated Reality the drawing takes place right on top of the video stream, so that registration is, for all practical purposes, exact to within the pixel resolution of the devices [Mann, 1994a], in contrast to the registration problem of Augmented Reality [Azuma, 1994]. This characteristic of Mediated Reality (perfect registration) has been suggested as a means of using the finger as a mouse to outline actual objects in the scene [Mann, 1997d]. This form of interaction with the real world, through the apparatus, is yet another example of the human–machine symbiosis that is at the core of Personal Imaging.

4. LOOKPAINTING: TOWARD DEVELOPING A NEW CAMERA USER INTERFACE

4.1 Introduction: What is Lookpainting?

One characteristic of the Reality Mediator is that, after wearing it for an extended period of time, one forgets (and so do other people) that one is wearing it, while at the same time, it provides an output signal of whatever passes from the real world through to the human visual system of the wearer [Mann, 1997b]. In addition to the personal documentary capabilities of this "video tap" into the visual system, there is the possibility of creating high-resolution environment maps by standing in one place and looking around. Assuming that one can look around faster than objects in the scene can change, images are typically found to be in approximately the same orbit of the projective group of coordinate transformations [Mann and Picard, 1995], modulo any changes in overall exposure level brought on by the automatic exposure mechanism of the camera [Mann, html].

4.2 Building Environment Maps by Looking Around

An environment map is a collection of images, seamlessly "stitched" together, into some unified representation of the quantity of light that has arrived from each angle in space, over the range of angles for which there exist measurement data. Examples of environment maps appear in Figure 19.11.

In constructing these environment maps, the computer performs basic calculations, which it is good at, while the human operator makes higher level decisions about artistic content, what is of greatest interest or importance, etc.

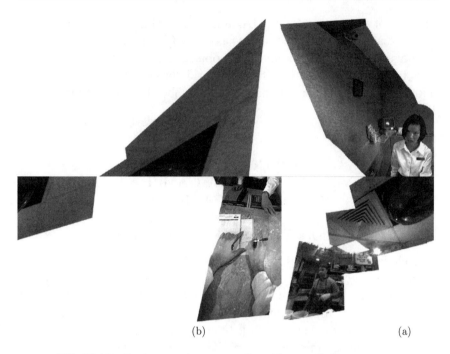

(b)   (a)

FIG. 19.11. Environment maps captured from a first-person perspective, through the process of looking around, are called "lookpaintings." Lookpaintings are characterized by irregularly shaped boundaries which capture the gaze pattern of the wearer. (a) Lookpainting made from 4 input images. Individual image boundaries are clearly visible. Note the unified perspective and the lack of distortion (e.g., lines on the ceiling tiles are almost perfectly straight, despite the extreme perspective). (b) Lookpainting made from 226 input images. This composite image illustrates the nature of first-person perspective. Note that both the author's hands are visible in the picture. Because the apparatus is wearable, it captures a new point of view, while at the same time capturing what is important in a scene. (Here, for example, it has included the video surveillance camera on the ceiling because the author has looked there.)

As described earlier, the human becomes at one with the machine (in this case the camera) through a long-term adaptation process, so that, as one experiences one's life through the apparatus (living in a computer-mediated world), the subject matter of interest is automatically captured by the human operator. Note that in this simple case, there is no Artificial Intelligence, but instead, there is a synergy between human and machine,

FIG. 19.12. This image depicts a group of people to whom the author is lecturing. Here the author is able to quickly sweep out the important details of this scene (namely all of the participants), while leaving out areas of the room where nobody is seated. Note how the image is close-cropped, leaving out the two empty chairs in the center, while at the same time extending out to include the feet of those sitting to the left and right of the empty chairs. This natural selection of subject matter happens often without conscious thought or effort, through the process of simply "looking around" at everyday scenes or objects, and is characteristic of the symbiotic relationship between human and machine that arises when the two become inextricably intertwined through a constancy of user interface extending over a time period of many years.

where the "intelligence" arises through having the human operator in the feedback loop of the overall image acquisition process.

Personal imaging, which is facilitated through wearable computing, image processing, machine vision, and computer-mediated reality, enables the user to effortlessly capture high-quality images of a new genre characterized not only by enhanced tonal range and spatial resolution but also by the ability to include and exclude areas of interest (Figure 19.12).

4.3 Automatic Generation of Photo Albums

It is not necessary to press any button on the "lookpainting camera" because it can automatically determine when images are in the same orbit of the projective group of coordinate transformations. This is done by analysis of the error terms in the coordinate transformation estimation algorithm [Mann and Picard, 1995], or by comparing, for each pair of images in the sequence, the error between the registered comparison frame $\hat{I}$ and the reference frame I, the error being given by $\sum_x(\hat{I}_i - p_{ij}I_j)^2$, with a threshold to determine whether or not they are in the same orbit. The theory is defined in Chapter 7 of this book. It is then assumed by the algorithm that, if more than a hundred or so images are captured within the same orbit, the subject matter is of sufficient interest to begin building a "look-painting." The process of building the lookpainting stops as soon as the incoming images are no longer in the same orbit, and if enough new images arrive to form a second orbit, a second lookpainting is generated, and so on. All of the lookpaintings are posted to a World Wide Web page. In this way, when, for example, one goes on vacation, one simply wears the glasses, and a photo album is generated automatically without the need for any conscious thought or effort.

5. COLLECTIVE CONNECTED HUMANISTIC INTELLIGENCE

Lookpainting provides more than just real-time "Wearable Wireless Webcam" style telepresence. Because the system is a fully functional networked computer, it also facilitates a new form of interaction. For example, while shopping at the grocery store, a spouse may not only visit vicariously (through the affordances and capabilities of Wearable Wireless Webcam) but may also interact with the wearer, such as by pointing at objects with a shared cursor. For example, a remote spouse may point to the milk to indicate a preference, or select fruits and vegetables over the wireless link. In this way, the capabilities of these shared environment maps are quite similar to the shared transparent whiteboard spaces of computer-supported cooperative living [Ishii nd Kobayashi, 1992; Ishii et al., 1993], except that the proposed methodology overlays the shared "work"space on the real world and the apparatus is wearable, wireless, and covert. Furthermore, typical uses of the apparatus are characterized by the domain of ordinary living (e.g., shopping, banking, or walking home late at night), rather than "working." Thus this new form of interaction is best described as *computer-supported cooperative living*. An example of a typical interaction using covert telepresence is illustrated in Figure 19.13.

One of the problems with a small covert personal imaging system incorporating a head-mounted display, which the author concealed into ordinary sunglasses, is distortion in the display system. Distortion in the context of

FIG. 19.13. Covert Computer-Supported Collaboration: The "lookpainting" algorithm uses a new result in algebraic projective geometry, combined with the Wyckoff principle (imaging done in the function space of the actual light falling on the image sensor) to build an extremely high-resolution/high-definition environment map from a collection of low-resolution/low-definition pictures of the same scene or objects. Here the author is wearing a covert personal imaging system (wearable computer with video cameras) in an establishment where photography is strictly prohibited. An environment map is generated merely by looking around the room. As a result of the wearer's natural gaze pattern, remote viewers on the World Wide Web may participate vicariously in the wearer's day-to-day life. Here a remote viewer, who was with the wearer during a previous bank transaction, recognizes the teller to the the wearer's right and sends a message back to the wearer indicating that more glances to the right are needed to make a better picture for later use as evidence. In this way the remote participant is as much the "photojournalist" as is the wearer.

a perspective display has been addressed in the literature [Barfield and Rosenberg, 1995; Rosenberg and Barfield, 1995] and is well understood. There are some differences, however, with a personal wearable unit, and these differences arise from the fact that it is worn over an extended period of time, often many years. It has been found that, after a long adaptation period, the author could cope with many of these distortions in ordinary day-to-day life, while operating within a projective coordinate space [Mann, 1994a], and that the distortion was eventually compensated for by the human visual system.

6. CONCLUSIONS: FROM PAINTING WITH LIGHTVECTORS TO PAINTING WITH LOOKS

Wearable tetherless computer mediated reality, with its orgins in a seemingly obscure visual aesthetic and photographic technique of the 1970s and 1980s, has evolved from computer supported collaborative photography, into computer supported collaborative videography, toward defining a new genre of personal documentary.

While the original "painting with lightvectors" application demonstrated the utility of wearable computing in an augmented/mediated reality environment, perhaps the real utility of this genre is in the production of high-quality image composites using a covert eyeglass-based reality mediator.

In summary, this *lookpainting* algorithm facilitates the use of a miniature covert personal imaging system to capture environment maps from which extremely high-resolution/high-definition pictures can be rendered. "Lookpainting" also provides a "look around" user interface, which is even more natural than the "point and click" user interface of modern cameras. Furthermore, lookpainting affords the user total control of the process, makes the process of capturing an image more engaging and fulfilling, and results in environment maps that can be shared remotely with others who have access to the World Wide Web or similar visual communications media. This application has also been successfully demonstrated for photojournalism, resulting in a useful tool for the reporter of the future. Moreover, lookpainting provides a new metaphor for *computer-supported cooperative living*, as well as a new era in which ordinary everyday experience can become newsworthy material—an era in which the job of the reporter may be spread out among ordinary people who will eventually, given enough people wearing enough units, encounter events far more newsworthy than those captured by a limited number of reporters covering only as much as can be so covered. Lastly, as mediated presence gives rise to mediated telepresence, a new communications medium arises with applications in personal safety and crime reduction as well as interactional capabilities that go far beyond portable internet video telephony.

Acknowledgments The author wishes to thank Simon Haykin, Charles Wyckoff, Chuck Carter, Kent Nickerson, Kris Popat, and Jonathan Rose for much in the way of useful feedback, constructive criticism, etc. as this work has evolved.

Additional thanks are owed to Kopin, Kodak, Digital Equipment Corporation, Compaq, VirtualVision, HP labs, Miyota, Chuck Carter, Bob Kinney, and Antonin Kimla.

REFERENCES

[Azuma, 1994] Azuma, R. (1994). Registration Errors in Augmented Reality: NSF/ARPA Science and Technology Center for Computer Graphics and Scientific Visualization. http://www.cs.unc.edu/~azuma/azuma_AR.html.

[Barfield and Rosenberg, 1995] Barfield, W. and Rosenberg, C. (1995). Judgments of azimuth and elevation as a function of monoscopic and binocular depth cues using a perspective display. *Human Factors*, 37:173–181.

[Ishii and Kobayashi, 1992] Ishii, H. and Kobayashi, M. (1992). Clearboard: A seamless media for shared drawing and conversation with eye-contact. In *Proceedings of Conference on Human Factors in Computing Systems (CHI '92)*, pages 525–532. ACM SIGCHI.

[Ishii et al., 1993] Ishii, H., Kobayashi, M., and Grudin, J. (1993). Integration of interpersonal space and shared workspace: Clearboard design and experiments. *ACM Transactions on Information Systems*, pages 349–375.

[Kohler, 1964] Kohler, I. (1964). *The Formation and Transformation of the Perceptual World*, volume 3(4) of *Psychological Issues*. International University Press, 227 West 13 Street, monograph 12.

[Mann, 1992] Mann, S. (1992). Lightspace. Unpublished report (Paper available from author). Submitted to SIGGRAPH 92. Also, see example images in http://wearcam.org/lightspace.

[Mann, 1993] Mann, S. (1993). Compositing multiple pictures of the same scene. In *Proceedings of the 46th Annual IS&T Conference*, Cambridge, Massachusetts. The Society of Imaging Science and Technology.

[Mann, 1994a] Mann, S. (1994a). 'Mediated Reality.' TR 260, M.I.T. Media Lab Perceptual Computing Section, Cambridge, Massachusetts, http://wearcam.org/mr.htm.

[Mann, 1994b] Mann, S. (1994b). Wearable Wireless Webcam. http://wearcam.org.

[Mann, html] Mann, S. (1994; http://hi.eecg.toronto.edu/icip96/index.html). "Pencigraphy" with AGC: Joint parameter estimation in both domain and range of functions in same orbit of the projective-Wyckoff group. Technical Report 384, MIT Media Lab, Cambridge, Massachusetts. Also appears in: *Proceedings of the IEEE International Conference on Image Processing (ICIP-96)*, Lausanne, Switzerland, September 16–19, 1996, pages 193–196.

[Mann, 1997a] Mann, S. (1997a). Humanistic intelligence. *Proceedings of Ars Electronica*. Invited plenary lecture, Sept. 10, http://wearcam.org/ars/ http//www.aec.at/fleshfactor.

[Mann, 1997b] Mann, S. (1997b). *Personal Imaging*. PhD thesis, Massachusetts Institute of Technology (MIT).

[Mann, 1997c] Mann, S. (1997c). Shootingback. *The Winners of the Prix Ars Electronica 1997*. http://www.rito.com/prix/winners.htm. See also http://wearcomp.org/shootingback.html.

[Mann, 1997d] Mann, S. (1997d). Wearable computing: A first step toward personal imaging. *IEEE Computer; http://wearcam.org/ieeecomputer.htm*, 30(2).

[Mann and Picard, 1994] Mann, S. and Picard, R. (1994). Being "undigital" with digital cameras: Extending dynamic range by combining differently exposed pictures. Technical Report 323, M.I.T.

Media Lab Perceptual Computing Section, Boston, Massachusetts. Also appears in *IS&T's 48th Annual Conference*, pages 422–428, May 1995.

[Mann and Picard, 1995] Mann, S. and Picard, R. W. (1995). Video orbits of the projective group; a simple approach to featureless estimation of parameters. TR 338, Massachusetts Institute of Technology, Cambridge, Massachusetts. Also appears in *IEEE Trans. Image Proc.*, Sept 1997, Vol. 6 No. 9.

[Norman, 1992] Norman, D. (1992). *Turn Signals Are the Facial Expressions of Automobiles*. Addison-Wesley.

[Rosenberg and Barfield, 1995] Rosenberg, C. and Barfield, W. (1995). Estimation of spatial distortion as a function of geometric parameters of perspective. *Transactions on Systems, Man, and Cybernetics*, 25(9):1323–1333.

[Ryals, 1995] Ryals, C. (1995). Lightspace: A new language of imaging. *PHOTO Electronic Imaging*, 38(2):14–16. http://www.peimag.com/ltspace.htm.

[Stratton, 1897] Stratton, G. M. (1897). Vision without inversion of the retinal image. *Psychological Review*.

20

Military Applications of Wearable Computers and Augmented Reality*

C. C. Tappert, A. S. Ruocco, K. A. Langdorf, F. J. Mabry, K. J. Heineman, T. A. Brick, D. M. Cross, and S. V. Pellissier

U.S. Military Academy, West Point

R. C. Kaste

U.S. Army Research Laboratory, Aberdeen Proving Ground

INTRODUCTION

Computers have evolved from huge mainframes to bulky workstations and desktops to slim notebooks and now to highly portable handheld or wearable computers. Wearable and handheld computers have great potential for military use, and there is currently great interest in what these highly portable devices can do. This chapter focuses on highly portable computing devices, those that are wearable (belt, wrist, and head-mounted) and handheld, and the term wearable computers (WCs) will often be used when

---

*The views expressed in this article are those of the authors and do not reflect the official policy or position of the United States Military Academy, Department of the Army, Department of Defense, or the United States Government.

referring to all these devices. The chapter briefly discusses input/output devices since the need for their miniaturization is one of the primary motivations for using WCs. It then summarizes the areas of military application of these devices: communications, position determination and map functions, report preparation and calculation, repair and maintenance, medical support, and the digitized battlefield. It also discusses the wearable computing equipment being developed for U.S. Army infantry units, a prototype of which was recently tested in a recent Advanced Warfighter Exercise.[*]

BACKGROUND: INPUT/OUTPUT DEVICES

Although miniaturization has allowed computers to become significantly smaller, the size of conventional input and output devices is now one of the main factors (other factors include thermal dissipation and power supply/battery size) limiting further miniaturization. Therefore, alternatives to the traditional keyboard and monitor input/output devices are being developed by the military and commercial companies and are competitively evolving in the civilian marketplace. These alternatives primarily concern pen computing, speech, and head-mounted displays.

Tablet digitizers (electronic tablets), available since the late 1950s, allow the capture of handwriting and drawing by accurately recording the x–y coordinate data of pen-tip movement. Pen computing arrived when transparent digitizers were combined with flat displays in the 1980s. This brought input and output into the same surface, providing immediate *electronic ink* feedback of the digitized writing and mimicking the familiar pen and paper paradigm to provide a "paperlike" interface. With a pen computer (or a pen-enabled computer), users can not only use the pen (writing stylus) as a mouse but also write or draw as they do with a pen on paper. There is no longer a need for a bulky keyboard since keyboard entry can be mimicked by touching sequences of buttons on a "soft" keyboard displayed on the screen or, alternatively, handdrawn characters can be automatically converted to ASCII code by handwriting recognition software. Available handwriting recognition products are highly accurate on careful handprinting and some products are available that recognize cursive script with accuracy dependent on the writing style and the

---

[*]The reader should be aware, however, that this field is changing so rapidly that there will be significant changes even within publishing cycles.

regularity and clarity of the writing (Tappert et al., 1990; Tappert & Ward, 1992).

Also in contrast to standard forms of input (i.e., keyboard and mouse), input based on speech recognition requires less operational work area and allows "hands-free" operation, eliminating the need for a keyboard. For speech input there are isolated-word and continuous-speech recognition systems. The simplest and most accurate systems operate only on words spoken in isolation (one at a time) and are text-dependent with a limited vocabulary. Since many military applications require only a small number of isolated commands or controls, these systems are relatively easy to create. Continuous speech systems are more complex and less accurate because they must handle the smearing of sounds across word boundaries (Padilla, 1997; Parsons, 1987).

Speech recognition systems currently available commercially include: Dragon Dictate Single, Kurzweil Voice Pad Pro, and IBM Voice Type—Simply Speaking. The first two handle vocabularies of 10,000 words, and the IBM system handles 22,000 words. These systems recognize connected speech, but they do require brief pauses between words. Recent work at Microsoft in audio recognition on a voice application program interface has demonstrated that continuous speech recognition is feasible on 200 Mhz + Pentium desktop systems (Microsoft, 1999.)

To be useful in battlefield applications, speech recognition hardware and software must provide acceptable performance under less than ideal conditions, such as speakers under stress, high levels of background noise, and all weather environments. In such situations, a large vocabulary system may, in fact, not be appropriate. More appropriate for the battlefield environment are small vocabulary, isolated-word (or short phrase) systems using noise canceling and helmet-mountable microphones, and a 100% failsafe system for critical commands (such as "FIRE!") would be imperative.

Speech output devices are also effective in WCs applications because they allow the user to listen, rather than divert visual attention to a monitor. Speech output can be created either synthetically or from recorded natural speech and such systems are readily available.

Another alternative to the traditional monitor is a Head-Mounted Display (HMD). The HMD mounts on the head like eyeglasses or goggles, provides the user with a large virtual display visible in one or both eyes, and is small and portable. Basically, it is a better trade-off of display size and quality versus equipment size than other display technologies, making it ideal for use with WCs. Many applications, including most of interest to the military, use see-through HMDs in which the display is either superimposed or, more usually, projected onto only a portion of the eyepiece so that the user's real-world vision is augmented by the ability to see the display. One such product currently available is *Virtual I-glasses! VPC* from Virtual i-O. HMDs used for virtual reality, on the other hand, attempt to totally immerse the user in the virtual environment, and they therefore completely block the user's normal vision.

In 1992, the Defense Advanced Research Projects Agency (DARPA) established the HMD Program with the goal of creating small, flat-panel, high-resolution displays that can be mounted in novel ways to present visual information to personnel on the battlefield. The emphasis of the HMD program is to provide information where ordinary displays are inappropriate or impractical: those requiring hands-free operations, where total immersion is important, and those involving unique viewing requirements, such as 3-D images, mapping graphical images onto real images, and night vision. The

FIG. 20.1. HMD for the U.S. Army Soldier Systems Command's Land Warrior.

FIG. 20.2.   U.S. Army Soldier Systems Command's Land Warrior.

11-ounce Combat Vehicle Crew (CVC) goggle, already developed under this program, incorporates a 1280 × 1024 active-matrix electroluminescent display into a standard U.S. Army sun, wind, and dust goggle. The first application of the CVC goggles will be for M1A2 tank commanders, while later plans include the Land Warrior (see Figures 20.1 and 20.2) and the Force XXI Soldier.

MILITARY APPLICATION AREAS

There are several aspects of WCs that relate to their use in the military. They must be rugged, and applicable military standards exist (e.g., MIL-STD-810E, Environmental Test Methods and Engineering Guidelines) to control tests and testing methods for rugged equipment (Hunter, 1996). They may have docking stations, cradles or holsters for recharging, uploading/downloading information, providing communication, or for securing the device and preventing damage. Battery life and other power related considerations are critical. Weight and cost are important. Finally, other features that arise from their mobility requirements include wireless communication and knowledge of current location.

With today's rapid technological advances, particularly in the area of small mobile computing devices, equipment designed and developed by the usual Department of Defense (DoD) method of "full military specification" is often obsolete before it is delivered. An alternative to the "old" method, and now a DoD initiative, is to purchase commercial off-the-shelf (COTS) equipment whenever possible and, only when necessary, modify it for military use (Herskovitz, 1994; McAuliffe, 1996). Sometimes COTS technology can be the starting point for military computers that are modified, for example, by making them more rugged (Fisher, 1996; Herskovitz, 1995). Because many current applications, such as those in the transportation industry, require rugged computing devices, additional ruggedization for military purposes may not always be necessary. Although the COTS initiative is causing changes in many organizations with, for example, NASA making extensive use of COTS subsystems (Rhea, 1996), some are saying that the initiative may be moving too fast and not giving proper consideration to unique military requirements (Wilson, 1996).

Commercially, although only about a half million wearable/handheld computers were sold in 1995 compared to 3.5 million notebook computers (Blodgett, 1996), their use and sale are growing rapidly and this growth is expected to continue. Pen computers, personal data assistants (PDAs), and other small computers are used extensively in industries such as transportation, healthcare, finance, construction, insurance, and police. Usually an organization begins to use these highly portable computers to gain an edge over their competitors.

As in the civilian sector, there are many military applications of PDAs/ WCs. Among these are special operations, maintenance, communications, language translation, position determination, map functions, report preparation and calculations, training, security, medical support, logistics support, distance learning, imagery gathering, law enforcement, and reconnaissance. Several of these aspects are discussed in more detail below. It should be noted that many of the advantages cited in individual sections apply throughout.

Land Warrior

As indicated previously, a WC can be worn like a belt, vest, or bandoleer, or shaped to fit into small spaces. Such a device can be used in battle dress or with other uniforms. With WCs there is no need to shuttle between computer workstation and operational site for information processing. These devices can also reduce the need for bulky paper documents.

The Land Warrior (see Figure 20.2) is an integrated fighting system for dismounted soldiers. Its purpose is to enhance the individual soldier's ability to survive in adverse conditions, acquire and engage targets, and become fully integrated in the digitized battlefield. It consists of four physical subsystems and a software subsystem (Aninger, 1996). The integrated helmet assembly subsystem consists of a helmet-mounted computer and sensor display by which the soldier can view digitized maps, intelligence information, and imagery from a thermal sight on his/her weapon. The equipment and protective clothing subsystem consist of a backpack frame that bends with the soldier's natural body movements and body armor that offers improved ballistic protection at a reduced weight. The weapon subsystem includes a thermal sight, video camera, and laser rangefinder/digital compass, which can be coupled with the Global Positioning System (GPS, described later) to provide extremely accurate positional and targeting information. The computer/radio subsystem, built into the backpack frame and integrated with GPS modules, consists of two radios (one for communicating with other soldiers within the squad, the other for communicating outside the squad) and a menu driven computer interface, controlled by a pointing device located on the chest strap. In sum, the Land Warrior program will enhance the soldier's battlefield capabilities through the development and integration of these subsystems, merging the soldier and technology into a cohesive, combat-effective system (Army, 1996).

Communications

One of the commander's main requirements is communications. That is, he must advise subordinates of changes in mission, support, and threat; also, he must inform superiors of changes in threat or timelines. The WC can simplify this procedure. Consider a company conducting a reconnaissance mission to note concentration, activity, and location of enemy defenders, washed out bridges, and/or battle damage. If the commander immediately reports the information by combat net radio, his emission alerts the enemy as to his presence; if he waits until returning to headquarters, his information will likely be stale. A remedy is to input the information into a WC, and at the first safe opportunity, use it to update the database on which such information resides. Communication can be facilitated by antenna arrays sewn into clothing associated with the WC.

The Army's Communications Electronics Command (CECOM) at Fort Monmouth predicts that by 1999 every soldier will be issued a small computer. They are currently experimenting with a rugged handheld computer called "The Grunt" that fits into the thigh pocket of the battle dress uniform (BDU). It communicates with satellites, runs complex software, and operates by fingertip on a touch-sensitive screen to eliminate the need for a keyboard (Matthews, 1994b). Such mobile computing devices in the hands of individual soldiers may be tomorrow's combat "force multipliers" (Glasser, 1997).

With the current trend toward multinational missions, communicating with automatic language translation becomes extremely important. Even when forces of various countries are separate, translation is vital in coordinating operations, both at headquarters, where such coordination is routinely performed, and also at the fighting level, where functions such as unit boundary development and field of fire coordination are addressed. There are several software options that can provide support in this area. For example, the Psion Series 3 PDA with Berlitz Interpreter permits translation of 28,000 words among five European languages (Psion, 1998).

Position Determination and Map Functions

In many operations it is necessary to know one's location accurately. For example, only when the location of a field artillery battery is properly surveyed and stored can gun orders be calculated. Moreover, one certainly would not want to call for artillery fire without accurately knowing one's own location! The Global Positioning System (GPS) was designed for and is operated by the U.S. Military. Twenty-four satellites orbit the earth in 12-hour orbits to provide the user with five to eight satellites visible from any point on earth. They send ephemeris data to GPS receivers over radio signals. GPS receivers require input from at least four satellites and convert the radio signals into x–y–z position and time. The primary function of GPS is navigation in three dimensions (latitude, longitude, and altitude). Velocity and acceleration are also easily computed.

Precise Positioning System (PPS) data can only be received by government-authorized users with cryptographic equipment and special receivers (20 meter accuracy). Standard Positioning System (SPS) signals, intentionally degraded by the DoD, provide decreased (100 meter) accuracy to civilian users that can be captured by most receivers without charge or restrictions. Differential GPS techniques can be used to correct bias errors at one location with measured bias errors at a known position, thereby increasing the accuracy of the data (differential SPS, for example, can obtain 1–10 meter accuracy).

As for the receiver, the Army is in the process of fielding the Precision Lightweight GPS Receiver (PLGR, AN/PSN-11). The PLGR is a small, handheld, GPS receiver featuring selective availability/anti-spoofing (SA/A-S) and anti-jam capability (see Figure 20.3). It provides precise

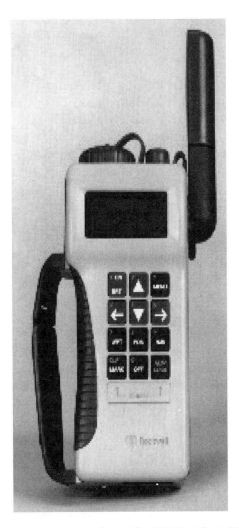

FIG. 20.3. Precision Lightweight GPS Receiver (PLGR).

positioning and timing solutions based upon signals received from the GPS satellite constellation. It is a five-channel receiver, capable of Precision Code (P Code) and Y Code (encrypted P Code) reception. Positioning solutions can be displayed in latitude, longitude, military grid reference system, Universal Transverse Mercator, British National Grid, and Irish Transverse Mercator Grid coordinates. It contains 49 map datums, can be programmed to support navigation, and has a built-in test feature.

Some commercial GPS products are stand-alone units used primarily for air, sea, and land navigation that receive the x–y–z location and time data and perhaps display the location on a chart or map. Some car rental agencies, for example, are beginning to add this equipment in their vehicles. Other GPS products are sold as add-on equipment for handheld and notebook computers.

Because the display of the PLGR is alpha-numeric only, the soldier must read the display and correlate it to his or her map. Alternatively, this can also be performed quickly and conveniently with FieldWorker software. Acting as a bridge between the GPS and a computer's operating system, it provides rapid flexible data collection, along with equipment linkage testing and GPS averaging (Apple, 1997). Along that same line, cadets at the United States Military Academy (USMA) recently participated in a project that involved taking data in real time from a PLGR and displaying the PLGR's location on a moving map displayed on a laptop computer screen. A map of West Point was digitized and a moving map software program that used Visual Basic was developed to display the user's location on the map. As the unit moved, the map was redrawn to indicate the movement, maintaining the cursor in a centralized position on the map.

Moving from one location to another is an integral part of operations such as attack and force consolidation. The manner and route of such movements should be determined in advance to avoid increased vulnerability to enemy fire and delays created by terrain, congestion, and excessively long routes. Route planning can be facilitated through WC capabilities that have been used in civilian applications. One software package, Microsoft's Automap Road Atlas, has 400,000 miles of North American highway data. On entering start and destination, Automap responds in seconds with the fastest, shortest, or preferred route; provides step-by-step instructions; and plots a customized map (Microsoft, 1998). Some research and development effort would be required to convert the commercial version, utilizing a network of distance and speed data, to military route planning. For example, a military version would have to store factors associated with a two-dimensional space, vehicle and surface types, weather conditions, etc.

Report Preparation and Calculations

A commander must routinely send reports throughout the chain of command with respect to various items such as target sightings, troop movement plans, ammunition availability, casualties, incoming fire, and maintenance requirements. A WC could help the commander by allowing him to compose such reports (or insert the date in formatted templates) in moments when his presence is not immediately required in battle management activities. The use of a WC also has the advantage of not constraining him to the location of his main battlefield computer/communication systems. A system such as the MessagePad, with compatible software, could easily be used for report preparation.

At some point, every commander finds some of his time expended by simple mathematical calculation. He must frequently perform tasks like determining north alignment for gun emplacements, estimating the time when his fuel will be expended, determining magnitudes of disparities between his operations orders and current status, estimating when he will be required to move his position, etc. These functions can be performed with a handheld calculator, but the WC would help the commander by facilitating the input and output.

Repair and Maintenance

See-through HMDs are currently being used for vehicle and other repair tasks. The Navy, for instance, is using a belt-mounted computer and HMD unit for helicopter maintenance, and the Army is putting 147 M1 tank maintenance manuals onto CD-ROMs for use in WCs. The user reads instructions and sees diagrams concerning the necessary repair in a display projected onto a small portion of his eyepiece and can work on the repair without shifting his focus of attention. Other uses of this equipment might allow a soldier coming up a hill to see in his eyepiece images transmitted from an unmanned aircraft of what lies on the other side of the hill (Matthews, 1994a). NASA has been using mobile computing since the first space shuttle flights. They currently use wireless units and pen computers to help with navigation and to help the crew monitor experiments and operations (MacNeill, 1996).

Medical Support

To an extent, the PDA has replaced reference books and patient index cards. Its main capability has been automation of ancillary tasks, giving the physician more time to practice medicine. The medically oriented PDA is now a specialized tool for managing large amounts of information. Educational Research Laboratories, Inc., has released a number of texts and handbooks in electronic format that can utilize the Newton's built-in search functions, thereby making it easy to cross-reference information. In the realm of patient management, PocketDoc by Physik, Inc., helps automate clinical experiences by allowing information to be handwritten directly into the database and handed off conveniently. There are other PDA options, including: Med-Notes, allowing quick point-of-care collection of patient information; Medicine Series, accessing the complete contents of references for medical residents; and PocketDoc Practitioner, a mobile record system designed to mimic the workflow of physicians (Mansell & Cogle, 1996).

On the battlefield, the first hour after a soldier is wounded is the most critical for patient care. Today's soldier can expect rapid transport from the battlefield to a hospital via aircraft; however, he or she must still wait for a medic to arrive, assess the situation, provide initial treatment, and have an aircraft dispatched. To expedite this process, one alternative is to bring the higher technology to the soldier.

In addition to answering the dismal question of whether a soldier is still alive, analysis of a soldier's vital signs, specifically heart rate and breath rate, serve as indicators of a soldier's immediate general health (e.g., current level of fatigue, stress, anxiety, etc.). If this analysis were done on a continuous basis and monitored remotely, it could assist superiors in assigning the best qualified personnel to a given mission. The Acoustic Sensor Division of the Army Research Labs (ARL) is developing a medical monitoring device that has a liquid filled acoustic sensor pad to detect these physiological sounds (Siuru, 1997).

The sensor pad has close to the same density of the human body and the surface material is close to that of human skin. Thus, when placed in contact with the human body, the pad provides good acoustic coupling between the person and the hydrophones embedded in the pad. The hydrophones produce a low amplitude analog signal that is amplified and passed through an analog-to-digital converter to create a digital representation of the physiological sounds. This signal is processed in the WC by digitally filtering to allow only the low-level heart and breath sounds to remain and performing Fourier analysis to separate heart and breathing signals for display if necessary.

Each soldier will be equipped with such a medical status monitor that constantly reads blood pressure, pulse, respiration, and blood oxygen level. A built-in expert system will determine when the soldier's condition warrants attention and automatically transmits an alarm, including GPS location and possibly injury severity.

Since there is a limit to the complex equipment a medic can carry, ARL and the Medical Advanced Technology Management Office are developing the Mobile Medical Mentoring Vehicle (M3V) (Balakirsky, 1996). Inside the M3V a physician's assistant and dispatcher are in charge of medical operations. The dispatcher alerts a medic in close proximity to the casualty over the Medic-cam (discussed below), notifying him of the location of the wounded, the conditions, and best route. In addition, actual coordinates are downloaded into the medic's WC. By integrating GPS with this route, the medic is given direction and distance indications to the wounded. The medic's video signal is compressed by his WC using a lossless scheme to avoid losing detail important for the physician. The medic's feed is controlled by a voice-activated interface and displayed on a monitor in the M3V. The assistant can retrieve an electronic inventory of what the medic is carrying and choose the most appropriate medication. The medic scans the barcode on the medication package, with the information automatically transmitted to the M3V, where it is incorporated into the treatment record.

Expertise from the M3V or even more distant medical centers will be observable through the use of the Medic-cam. The medical expert will remotely see and hear the patient and observe the output from the medical monitors. The Medic-cam comprises mainly a pencil video camera on glasses worn by the medic and electronics located on his load-bearing equipment. His voice is picked up by a flexible boom mike; audio reaches him through small earphones. A view screen positioned under the medic's eye allows him to see the image being transmitted. A WC, tactical communication links, and long-life batteries are currently under investigation for this system.

On a much larger scale, the Telemedicine Acquisition Initiative provides the Army Medical Department (AMEDD) with the latest communication capability to support field medicine from the forward edge of the battlefield to the continental United States. From reading diagnostic images to visually inspecting patient conditions with direct consultation links, AMEDD experts can participate in treating patients anywhere in the world.

Distributed Data Fusion

The military intelligence process concerns the collection, analysis, and interpretation of information to answer a request for specific intelligence concerning an area of interest. There are three corresponding information-processing tasks: data fusion, situation analysis, and countermeasure planning. Data fusion is the process of assembling a model of the domain of interest from disparate data sources and has three main characteristics: multi-sensor data fusion, uncertainty, and continuous process (Keene & Perre, 1990).

Data fusion represents the consolidation of data that are received from sensors. The data are consolidated, typically using some probabilistic model, to form a view of the area in question. One method of data fusion establishes a tree-structure hierarchy that identifies which levels perform which type of data reduction. Measures of effectiveness (MOEs) for data fusion concern the underlying probabilistic models being employed in the various fusion processes (Broman & Pack, 1994).

It is possible to use a "standardized template" based on the predicted data. These templates are based on previous studies of how forces typically deploy, such as predicting force structure entering your geographic area based on the sensor data (Mikulin & Elsaesser, 1994).

As described, data fusion is a one-way process; that is, data from sensors are taken in raw form and sent to the next level in the hierarchy. Each subsequent level continues to process these data. However, each level brings with it a degree of uncertainty. The top level, within the Army's All Source Information Center (ASIC), is the first time an assessment is made. This hierarchy is clearly delineated in Figure 20.4(a) (Keene & Perre, 1990).

Evaluations of the assessment are based on how well the assessment matches the reality. Yet, there is a substantial opportunity for the situation to change between the sensors acquiring data and the ASIC producing information. This is why the fusion process must be considered a continuous process. However, in order to stabilize the situation to allow a template to be utilized, a decision must be made to more or less freeze the input. At this point, the template may be employed as described by Mikulin and Elsaesser (1994). However, when model verification takes place, an inherent factor not mentioned in the referenced documents (Mikulin & Elsaesser, 1994) is establishing this "freeze" point. In other words, if a template matched the situation, did the template match because it accurately reflected the situation, or did the model match because the situation had not been allowed

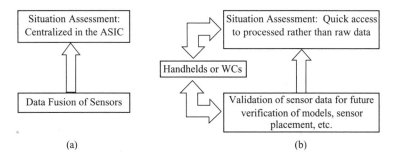

FIG. 20.4. Proposed modification to data fusion model: (a) the current hierarchy of one-way flow between hierarchy regions, and (b) utilizing handheld devices or WCs in a two-way linkage of hierarchy regions.

to develop to a point where a different template would have been more appropriate?

One reason there is an acceptance of the one-way nature of data fusion is the problem of integrating all forms of data. Human Intelligence (HUMINT), Communications Intelligence (COMINT), and Electronics Intelligence (ELINT) are fundamentally different types of sensors. Thus, the centralized assimilation in the ASIC seems the natural approach. Unfortunately, such a natural approach lends itself to having decisions made far removed from where data have their most reliable reading. Obviously, it is not possible to place a person at every sensor location. Yet the advent of WCs may allow for the insertion of decision makers at different points in the hierarchy, such as a two-way link within the hierarchy as depicted in Figure 20.4(b).

Figure 20.4(a) shows the traditional flow of data from the sensors to the ASIC. This is a one-way flow. After the ASIC makes an assessment, overall verification takes place outside the model. By using WCs, several possibilities identified in Figure 20.4(b) become evident. First, a decision can be made earlier in the process. For example, the information previously available only at the ASIC level can be transmitted to a handheld or WC operating in the "sensor" level of the hierarchy. Assimilated data from the ASIC can be based on data coming from the sensor at the fusion level of the handheld or WC. If the data are consistent, the ASIC can continue processing along its current models. However, if the data coming from the sensor show a significant difference, then the ASIC can immediately be told to recalculate for more suitable models based on current data. In this scheme, the handheld is bypassing some of the higher levels of the sensor region of the hierarchy. Thus, data from the lower portion of the hierarchy are weighed more heavily in this instance.

Such a weighing, though, is not represented by any of the measures identified in Broman and Pack (1994). Under the current MOEs, any data point is assumed to have the same value to the overall fusion process as any other data point. Obviously, then, the MOEs need to be adaptable to the potential of some data points being more "important" than others. Hence, the insertion of handhelds or WCs into the hierarchy influences the overall fusion process in terms of MOEs. How such MOEs would be modified, however, remains to be determined.

Related to weighing various sensor inputs is the coupling of confidence factors to possible template models. As described in Mikulin and Elsaesser (1994), a template is selected based on the results of computing several confidence factors. These confidence factors are algorithmically determined based on matches between the sensor-driven establishment of unit locations against a database-driven location of where a unit should be, based on the doctrinal template. The doctrinal template is established based on the correlation of the template to actual circumstances. This implies, then, that the doctrinal template is the end result of all the times that template has been selected. Thus, if a template is selected, and it does not match the reality of the situation, there are two negative effects. The first is that the wrong actions may be taken at the tactical level. For example, preparing for a unit of Type-A when it is actually a unit of Type-B. The second negative effect is that the template may be adversely adjusted due to an outlier data entry, that is, the wrong application.

Using WCs within the hierarchy can help eliminate such adverse effects. The WC can receive potential templates from the ASIC. Utilizing fuzzy-logic techniques, the WC can perhaps be used to backward engineer the data required for a template to have a high confidence factor. Again, the WC is used to bring the decision closer to the data-entry point in the fusion process.

Once the theoretical justification is made, an experimental environment would need to be established wherein testing resources (both hardware and software) could actually model the proposed hierarchy. This would be established based on using the WC as proscribed for a given level of the hierarchy. It would be best to assume the WC is not at the sensor level. If it were, one would have in effect substituted a person for the sensor, thus discounting the need for sensor fusion. By the same token, the WC would not be used within the ASIC, since other more powerful equipment is already at work on the assimilation of the data. Hence, the WC insertion into the hierarchy would have to be tested at some intermediate level (or levels).

As commonly shown in military exercises, the need to transmit data to the end-user is critical to swift decision making. This results in a two-way problem. The data being fused together results in a consolidation from thousands of points into a single, logical view of the battlefield. This view of the battlefield is then broken out and transmitted back down to units. The effective use of WCs, properly incorporated into the data fusion pyramid, can build the local picture by accessing data on its way up to the highest consolidation level. This will significantly reduce the time from data capture to information distribution.

Tactical Internet Applications

Just as the marketplace has generated new opportunities with respect to WCs for the military, it has also provided new ways to tie these computing elements together to develop new types of networked applications. How the military will actually link different information system components together and use the combined set of systems will be driven to some extent by commercial developments, particularly the Internet.

Presently information systems are constructed on the Internet utilizing a client/server architecture. This is a relatively recent shift in the general architecture of such systems. Beginning in the 1960s, central, time-shared computing systems used terminals (with varying levels of intelligence). This architecture continued into the era of the personal computer. Twenty-five years later, by approximately 1985, the general corporate computing architecture had begun the conversion to a PC/file server architecture. Then, within a short eight years (1985–93), that architecture was replaced by the present client/server architecture. This 25 to 8 ratio of change suggests that not only is more fundamental change coming, but that it is coming faster than ever before. In fact, after just four years, even the client/server architecture is now suggested as giving way to Network-centric application development (Netscape, 1996). The benefits and problems associated with the present architecture impact the way WCs will ultimately be interfaced and the context in which they will be used in corporate America and to some extent in the military.

A number of benefits have emerged from the client/server architecture that impact wearable computing applications and how they are interfaced and integrated into computing environments: *scalability* (there are no "power of computing" limitations that suddenly appear, as a solution is scaled up for use by more simultaneous users), *cross platform compatibility* (the same functionality is available on a wide variety of platforms), *interoperability* is established at a machine family level (i.e., multi platform support is designed in at a low level on a machine-specific basis rather than being added as conversion software or a gateway interface), *reusability*, *reliability* (with graceful degradation), and *extensible functionality* to new alternative uses.

This new architecture is not without its difficulties, among which are the unpredictability of user computing requirements (particularly when compared to prior architectures) and the need to support multiple protocols on each potential computing platform. To a significant degree, the problems that have been the good-news/bad-news of the commercial marketplace have carried over to the military computing context. Several of these topics are further expanded below.

Certainly, the most important development that will impact the migration of Internet technology to the tactical Internet is XML. In two years (1998–2000) the extensible markup language has become the *lingua franca* of Internet middleware computing (Cover, 2000). All current Army database server software vendors provide support for XML based import and export. The segregation of content data from presentation data is allowing PDA computing to incorporate integrated data sources and display that data in a manner that best uses the limited visual area common in WC's (Moeller, 1997).

Other emerging commercial areas of development with potential impact on military computing applications of WCs include: roaming; off-line use; on-demand access; push technology; digital TV regulatory announcements (digital TV by 2006); intelligent objects; content replication capabilities; dynamic recovery with real-time, expert system driven load reallocation; "Crossware" cross everything applications; "Netcasting" (server side) and recasting; client systems with multiple, cooperating, (possibly) distributed agents acting for the individual or the unit that the individual leads or represents (this is a different orientation of design and interpretation of intentionality than would traditionally be considered in a top-down design!); dynamic collaborative systems; and remote location configuration and site management.

Areas of potential interest that are suggested by military information systems development with respect to Internet and WCs include: dynamic evaluation of the proximate location of required minimal computational functionality (i.e., find the nearest, in time and/or space, system that can support the needed computation or interaction); maximum "thickness" of client software within a particular type of device; flexibility and reliability of Simple Network Management Protocol (SNMP) functions (as SNMP functionality expands, it may prove necessary to proxy some functions to avoid the occurrence of unnecessary or potentially dangerous communication traffic for a combat/military environment); and the need to bring new technologies that are applicable to a military computing environment quickly, and cleanly, in a maximally pervasive manner.

As discussed in the Data Fusion Section, communication delivery and enhancement concerns focused by the nature of military computing necessitate that communication support functions integrate data from a variety of sources such as traditional audio, visual (video in many forms), ground sensory systems, and satellite systems. Such data streams may be used to implement new communication supports: *virtual conversations* (for example, if a particular situation occurs, a prespecified command conversation is available for the soldier who is "on the spot!"); *distributed display of integrated information* from many sources (from real-time sensors on the ground to satellite downlinks, to audio [both real-time analog and digital], to video and artificial intelligence enhanced video, to virtual reality–like situational simulation based on available data and knowledge of enemy doctrine and tactical context); *temporally disjoint communication support* that would allow the review of recent similar communications from temporally indexed digital archives of analog communication); and Situation Report (SITREP) display of the combat zone with communication and status information from combat units being accessed in a "point and hear" or a "point and see" manner (with the level of "information freshness" also displayed); nontraditional forms of *command transmission* that might include a simulation of the action that is expected to occur and which is used by each level of the command chain to interpret their situation, needed resources, tactical alternatives, etc. This simulated course of battle command might be able to continue its "performance" by integrating real-time information about the actual battle and providing insights into how to respond to initial results even when communication with higher command elements is infeasible, difficult, or even dangerous.

When improvements to network-based applications force system rescaling, two strategies are possible (both may be employed simultaneously). First, assume use of "thin clients." That is, forcing a few servers to grow by a factor reflects the maximum dynamic user demand possible at any instant. Alternatively, assume use of "thick clients;" forcing all client machines to "grow a little," which may result in a "trickle down effect" to end-user systems, which is unacceptable. Interoperability should be achieved by vendors providing direct native support for a particular platform or cross platform support. Relative to the client/server architecture, support for nontraditional, non-PC devices should be done in such a way that new devices are fluidly integrated with those already in use in the military.

Wireless local area networks (LANs) are a new and exciting addition to the realm of networking, particularly with WCs. Wireless LANs offer cheaper installation costs than wired networks, as well as scalability and roaming. However, although they are easy to install and configure, their throughput may be too limiting to offer a suitable substitution to existing wired networks.

CONCLUSION

Much research is still needed, but new modes of human–computer interface will certainly be developed. For example, input may be via the wearer's camera tracking a "finger mouse." A worn computer that is constantly attentive to the wearer and the environment would assist the wearer with perceptual intelligence in daily activities. A sophisticated WC has potential as a "visual memory prosthetic" or perception enhancer, enabling the wearer to see things otherwise missed and to better remember items of note. Visual filters can be incorporated to study the otherwise unseeable; a freeze-frame capability, for example, would enable seeing propellers in motion. The user might transmit a sequence of images to show colleagues where he or she has been, in a kind of shared visual memory. HMDs could also computationally augment visual perception in real time. For example, virtual images could have attached labels for easier processing in a complex field environment. A community of wearers could have a networked "safety net" via the ability to see exactly what others see or by monitoring each other's physiology (Mann, 1997).

Voice recognition as an input alleviates the need for manual interfaces. Hands-free operation generally means less time is required for data entry and retrieval. Heads-up displays permit simultaneous operation of various options without keyboard or mouse. The screen can be variously visible or overlay the real-world scene for textual or graphical processing. Miniaturized into eyeglasses, the unobtrusive apparatus permits interaction both with the computer and other personnel. Many devices include connectors to support PC peripherals such as mice, keyboards, VGA video, serial ports, and audio I/O. Some permit plug-and-play enhancements.

Clothing is a natural way to carry a processing device, and by wearing the device it is always ready for use. Component miniaturization has enabled wearable systems that are almost invisible, so that the user can move and interact freely. A number of military functions could be performed more readily or efficiently with the use of WCs. These devices will continue to gain social acceptance due to improved miniaturization and the proliferation of similar devices, such as cellular phones and pagers.

Future network application system support "improvements" cannot dictate operating system updates that demand more powerful hardware. New functionality must not be achieved at the cost of continued compatibility conflicts with in-place technologies already deployed-and-trained in the active military forces. The more ubiquitous a system is to the military, the more the military will be wary of "improvements" or changes. Concerning the tactical internet, the importance of network-centric applications is being emphasized at the undergraduate level in many universities, including USMA. Here, for example, all plebes (freshmen students) develop their own home pages and many senior-level systems design projects involve the development of Internet or intranet-based applications and, in the near future, thin-client wearable/handheld computers will be used in many projects.

In March 1997, the Army conducted an Advanced Warfighter Exercise (AWE) at the National Training Center, Fort Irwin, CA (AUSA, 1997a; Time, 1997). The AWE put many of the Army's newest high-tech systems to the test. In particular, the light infantry soldiers were equipped with the dismounted soldier system unit (DSSU), an experimental WC and communications system, using a variety of commercial and government components. While the infantry soldiers reported that the DSSU enhanced their communications capabilities, they also found it heavy, bulky, and fragile (Griggs, 1997). The DSSU, however, is not intended as a fieldable system; instead, it is serving as a prototype for the fully integrated, smaller, and lighter Land Warrior system, which the Army plans to start fielding in the year 2000 (Aninger & Hubner, 1997). The initial outcome of the AWE test indicated a draw (Wilson, 1997), but the Army's leadership and doctrine writers will undoubtedly study this exercise well into the future.

General Hartzog, then Chief of the Army's Training and Doctrine Command, stated that the 21st Century Army must have information dominance on the battlefield (AUSA, 1997b). The AWE is at the heart of determining future requirements, and the Army plans to field a completely digitized division by 2000. That "Experimental Force" will test and measure the impact of information technology on future combat in terms of communications, computer hardware, and software (AUSA, 1997c).

The United States Military Academy has the mission of producing leaders capable of meeting the rapidly changing face of technology in today's Army. West Point student and faculty research can and should focus on the future needs of the Army. By using existing ties between the Academy, ARL, and other research centers, new ideas can be explored by initially developing them, and then turning the results over to the appropriate agencies for further development.

REFERENCES

Aninger, S., & Hubner, D. (1997), "Dismounted Soldier System and Land Warrior," *Backgrounder*, U.S. Army Soldier Systems Command, Public Affairs Office, Kansas Street, Natick, MA 01760, February 18.

Aninger, S. (1996), "Land Warrior Systems Description," *Backgrounder*, U.S. Army Soldier Systems Command, Public Affairs Office, Kansas Street, Natick, MA 01760, October 8.

Apple (1997), "Welcome to the Newton Site," On-Line: <http://newton.info.apple.com>.

Army (1996), "Land Warrior," *Army*, Vol. 46, No. 10, October, p. 252.

AUSA (1997a), "Land Warriors," *AUSA News*, Vol. 19, No. 6, April, p. 1.

AUSA (1997b), "Hartzog Previews Army XXI," *AUSA News*, Vol. 19, No. 6, April, p. 17.

AUSA (1997c), "Warfighting Experiment Uses Digitized Division," *AUSA News*, Vol. 19, No. 6, April, p. 16.

Balakirsky, S. (1996), "Medical Command and Control for Wounded on the Battlefield," ARL paper, Battlefield Systems International 96.

Blodgett, M. (1996), "Microsoft Joins PDA Market," *Computer World*, Vol. 30, No. 38, September 16, p. 2.

Broman, V., & Pack, J. (1994), "Measures of Effectiveness for the Distributed Data Fusion Problem," Technical Report 1648, Naval Command, Control and Ocean Surveillance Center, RDT&E Division, June. Available from Defense Technical Information Center.

Cover, Robin (2000). *The XML Cover Pages* Extensible Markup Language (XML). (at http://www.oasis-open.org/cover/xml.html).

Fisher, P. et al. (1996), "What You Need to Know about Ruggedized Computers," *Defense and Security Electronics*, March, pp. 7–8.

Glasser, L. A. (1997), "Advanced Displays: Windows into Information Warfare," On-Line: <http://eto.sysplan.com/ETO/ Articles/Article3.html>.

Griggs, J. (1997), "Light Infantry Tests DSSU Concepts," On-Line: <http://www.monroe.army.mil/pao/dssu.htm>.

Herskovitz, D. (1994), "Military Computing in the Year 2001 and Beyond," *J. Electronic Defense*, February, pp. 41.

Herskovitz, D. (1995), "A Sampling of Rugged Military Computers," *J. Electronic Defense*, July, pp. 60–64.

Hunter, D. (1996), "Rugged Pen Computers," *Pen Computing*, Vol. 3, No. 10, May/June, pp. 30–31, 49.

Keene, A. P., & Perre, M. (1990), "Data Fusion: A Preliminary Study," Report # FEL-90-B356, TNO Physics and Electronics Lab., The Hague, The Netherlands, December. Available from Defense Technical Information Center, 8725 John J. Kingman Road, Suite 0944, Fort Belvoir, VA 22060-6218.

MacNeill, D. (1996), "Pen Computers in Transportation," *Pen Computing*, Vol. 3, No. 10, May/June, pp. 18–24.

Mann, S. (1997), "Wearable Computing: A First Step Toward Personal Imaging," *Computer*, Vol. 30, No. 2, February, pp. 25–31.

Mansell, E., & Cogle, C. (1996), "What Can a Personal Digital Assistant Do for You," *Florida Family Physician*, Vol. 46, No. 1, January.

Matthews, W. (1994a), "A Grunt's 'Grunt'," *Air Force Times*, Vol. 55, No. 40, August 29, p. 32.

Matthews, W. (1994b), "Computer on a Hip," *Air Force Times*, Vol. 54, No. 48, July 4, p. 33.

McAuliffe, A. (1996), "Technology: Changing the Way the Army Does Business," *Military and Aerospace Electronics*, Vol. 7, No. 2, February, pp. 24–25.

Microsoft (1998), "Expedia," On-Line: <http://www.microsoft.com/expedia/cd.htm>.

Microsoft (1999), "Microsoft Speech Application Programming Interface Software Development Kit."

Mikulin, L., & Elsaesser, D. (1994), "Data Fusion and Correlation Techniques Testbed (DFACTT): Analysis Tools for Emitter Fix Clustering and Doctrinal Template Matching," Tech. Note 94-11, Defense Research Establishment Ottawa, December. Available from Defense Technical Information Center.

Moeller, Michael(1997), "Markup Language Takes HTML to Task." PC Week News 14/15 (April 14, 1997) 6.

Netscape (1996), Netscape Corporation, "The Netscape One Development Environment Vision and Product Roadmap," On-Line: <http://www.netscape.com/comprod/one/white_paper.html>.

Padilla, E. (1997), "Voice Recognition Technology Overview," Volume 3, January, On-Line: <http://pbol.com/eloquent/voice_ article.html>.

Parsons, T. (1987), *Voice and Speech Processing*, McGraw-Hill Inc.

Psion (1998), "Psion Computers—Berlitz Phrasebook & Berlitz Interpreter," On-Line: <http://www.psion.com/computers/psionscberlitz.html>.

Rhea, J. (1996), "Shuttle Replacement's Avionics Going COTS," *Military & Aerospace Elect*, Vol. 7, No. 8, August, p. 6.

Siuru, B. (1997), "Applying Acoustic Monitoring to Medical Diagnostics," *Sensors*, pp. 51–52, March.

Tappert, C. C., Suen, C. Y., & Wakahara, T. (1990), "The State-of-the-Art in On-line Handwriting Recognition," *IEEE Trans. on Pattern Analysis Machine Intel.*, Vol. PAMI-12, August, pp. 787–808.

Tappert, C. C., & Ward, J. R. (1992), "Pen Computing—Fad or Revolution?," *Information Display*, March, pp. 14–19.

Time (1997), "Wired for War," *Time*, Vol. 149, No. 13, March, pp. 72–73.

Wilson, G. (1997), "Cohen Likes What He Sees at NTC," *Army Times*, March 31, pp. 3 and 27.

Wilson, J. R. (1996), "Is DOD COTS Initiative Moving Too Fast?," *Military and Aerospace Electronics*, Vol. 7, No. 8, August, p. 6.

21

Medical Applications for Wearable Computing*

Richard M. Satava, MD FACS
Yale University School of Medicine

and

Defense Advanced Research Projects Agency (DARPA)

CDR Shaun B. Jones, MD
Uniformed Services University of Health Sciences

and

Defense Advanced Research Projects Agency (DARPA)

ABSTRACT

The delivery of health care has changed dramatically with the information revolution. New capabilities to acquire information about patients through noninvasive sensing and imaging modalities provides an increased level of knowledge about individual patients. Sophisticated computer programs for decision support provide the capability of local processing of the data, and portable computing and networking offer the opportunity to have this information available to the individual physicians and healthcare providers at the point of care in a timely fashion. Each of these areas—noninvasive sensing, local processing and point of service care—are enabled by

*The opinions or assertions contained herein are the private views of the authors and are not to be construed as official, or as reflecting the views of the Department of the Army, Department of the Navy, the Advanced Research Projects Agency, or the Department of Defense.

wearable computers and empower patients through acquisition and processing of the information and enhance physicians' ability to provide healthcare by having access to the information.

INTRODUCTION

Wearable computing can provide the solution to a number of needs over the entire spectrum of healthcare delivery, both for personal (or patient) and user (physician or healthcare provider) applications. For patient or personal requirements an arbitrary taxonomy for uses would include prevention, diagnosis, and treatment. For prevention, the wearing of smart clothing and accessories that contain noninvasive microsensors can continuously monitor and updated information of a person, the medication status of a patient, or the amount of physical stress of an athlete. (The term athlete will be used to encompass the entire spectrum of athletic activity, from simple individual health and fitness programs, to amateur "week- end warriors," to professional athletes). The following examples illustrate these three potential uses.

In the prevention arena, a person with a known family history of heart disease that is completely healthy could have continuous monitoring of vital signs and electrocardiogram (EKG) by a chest strap or tee shirt to look for early sings of heart problems before the person becomes symptomatic.

Today, a similar commercial product is being used to monitor respiratory rate on infants suspected of having a predisposition to sudden respiratory arrest while sleeping (Sudden Infant Death Syndrome—SIDS). A strap with sensors for respiratory motion is worn around the chest of the infant, and the motion sensors send the signals wirelessly to a bedside monitor. If the child stops breathing for a certain number of breaths, the monitor sends a signal to the receiving unit of the parent.

There is a different investigational system for a diabetic patient that can have noninvasive glucose and insulin levels closely followed and have the information sent to the wearable computer. After processing the information, if the blood sugar is too high a signal can be sent to an implantable pump to automatically infuse insulin to adjust the blood sugar level and prevent severe complications of diabetes.

In the future with technologies described below, an athlete, whether high school, Olympic, or professional sports level, can wear a full undergarment that would monitor both the vital signs and the stresses across key parts of the body (neck, shoulders, knees, ankles, etc). A further advantage is

FIG. 21.1. I-Port(early prototype)—a virtual reality training system for the military that incorporates vital signs monitors on the soldier to monitor health and performance. (Courtesy Dr. Stephen Jacobsen, Sarcos, Inc. Salt Lake City, UT.)

that the data can be remotely transmitted to the coach during practice (for training) or competition (to determine performance or fatigue). Today, the military is experimenting with a Personnel Status Monitor (PSM) system to monitor the performance of soldiers during training as well as to monitor health status on the battlefield (see below) for total situational awareness (Figure 21.1).

All of these systems have the common denominator of noninvasively acquiring data from sensors and conveying the information to a wearable computer for storage or immediate transmission. More importantly, as these needs are identified and met, the possibilities of numerous other areas will be discovered and expand the market ubiquitously; in so doing it will become cost effective to wear "smart clothing" and accessories controlled by a wearable computer.

The questions of security, privacy, and confidentiality always arise when there is the ubiquitous acquisition of data. The military has the requirement for stealth (low probability of detection, low probability of intercept, anti-jamming of signals), a much higher level of security than for personal use. The solution is that the information is handled by the methodology of "alert and query." The information is acquired by the microsensors, which have a wireless transmission distance of less than 1 meter. The signals are transmitted over this "wireless body local area network (LAN)" to the belt-worn computer. Decision support software analyzes the information, compares it to the person's baseline and most previous signals to determine if there are any significant changes. If and only if the information deviates substantially from the baseline or anticipated information will an "alert" be sent remotely to the closest medic from the wearable computer (through the secure battlefield information network, which employs various techniques such as spread spectrum, frequency-hopping, etc.). All other information will be entered into the data logger when there are changes that are within given predetermined parameters (documentation "by exception only"). For example, if a soldier has a heart rate of 70 beats per minute (bpm) and begins to run, the heart rate will be recorded only when it exceeds 70 bpm. Only if the rate exceeds 120 bpm and other parameters of possible injury (falling blood pressure, excessive respiratory rate, etc.) also are outside predetermined limits will the "alert" be invoked and sent to the medic. In civilian application, when parameters are exceeded, the information can be used to alert the individual, or under certain circumstances (nursing homes, assisted living, etc.) to the agreed upon healthcare provider.

The other component of monitoring, "query," provides the capability to search for the status of the individual wearing the system. In the military, the situation would be one in which a military commander would like to know the status of the troops during engagement; thus a query is sent to the specific soldier systems and the computer "reports back" to the commander. From the data it might be possible to determine the level of activity, performance, or fatigue of the soldier. Similar situations can be envisioned for elderly or remotely located persons, especially those who are handicapped by disease or debilitated from aging. Having the ability to quantify continuous performance will greatly aid in motivation, rehabilitation, and prevention. However, this new capability for continuous monitoring raises many nontechnical issues about security, privacy, and confidentiality that are not related to the technology, but rather are critical issues that can delay the use of the technology because of possible unethical or criminal misuse of the technology. Thus the issues of security, privacy, and confidentiality must be addressed in order for the technology to be used.

As indicated above, wearable computers are required for both the acquisition of information from patients, soldiers, or athletes and for the transmission to the person who needs the information such as physician, combat medic, or sports trainer. Previous generations of medical sensors are invasive, requiring insertion of a needle into the body or blood stream, such as electromyographical signals of muscle motion or central venous catheter placement into the heart to monitor heart function. Next generation biosensors for vital signs and physiologic and biochemical parameters will be noninvasive.

The prototype is the commercial product for oxygen saturation—the pulse oximeter. Using an infrared sensor worn on the finger, ear lobe, or other part of the body where there are blood vessels close to the skin, light is transmitted through the skin, with some being absorbed by the blood (specifically the saturated and unsaturated hemaglobin, which have very characteristic spectral signatures). The detector can thus "read out continuously" the varying concentrations of oxygen in the blood. In addition, the heart rate can be indirectly deduced from the variation in concentration over time. Hyperspectral analysis (in any number of parts of the electromatic spectrum) will provide the opportunity to "read" through the skin, by either transmission or reflection, many of the critical blood chemistry values or physiologic parameters that now require the drawing of blood or wearing of special monitors.

Another group of noninvasive sensors includes systems that employ impedance (such as thoracic impedance for cardio-pulmonary function) and acoustical sensor. Recent breakthroughs have enabled making miniature sensors reside in the micro electro-mechanical machine (MEMS) technologies. Just as micro-level circuitry permits computers to be hand held instead of room size (as the original ENIAC), the use of MEMS permits entire engines, valves, light-emitting diodes (LED), and computer circuitry to be "etched" onto a single microscopic size chip (see below). It is this revolution that will permit embedding biosensors virtually anywhere, because they are tiny, low power, and extremely inexpensive. This will result in ubiquitous monitoring without encumbering the patient, whether in a bedside stand, an instrument in the home, or in the clothing being worn.

While it is true that the information from smart clothing can be transmitted (under the concept of mobile communications and ubiquitous computing) to a center for processing and use, the power of wearable computing lies in making this same information available at the "point of service." In medicine this would mean having the information available to the doctor or nurse at the bedside, in the emergency room, intensive care unit, operating room, or patient room so the physician can make the decision at that moment rather than having to call to the laboratory for test results, go to the radiology suite to see the x-ray images, or go from each patient's room back to the nursing station for information. Just as important is having the information available in the nonhospital situation, such as outpatient clinics, nursing homes, or within the home of a patient with a chronic health problem. The benefit is that patients no longer have to make repeated visits back to the hospital or clinic just to have their blood pressure checked or their blood sugar taken.

Nurses in charge at nursing homes make "rounds" about once an eight-hour shift to check patients, assess their health, and record their vital signs. Many of these patients are invalid but are quite active within the nursing home. By having wearable sensor systems, a continuous monitoring of their vital signs wherever they are in the nursing home would provide the charge nurse a continuous monitoring of the patients with automatic alerts to tell the nurse a patient is beginning to have a problem long before there would be serious consequences. Since there are not enough nurses to give nursing home patients the level of attention that is ideal, wearable computing could greatly enhance the ability for monitoring and prevention. Just as the physician must care for patients in the clinic, hospital or outpatient office so too is the nurse moving throughout the nursing home providing care. Having a wearable computer with a head-mounted display or augmented reality glasses (such as those by MicroOptic Corporation—see below), the information about patients would be available to the nurse wherever he or she may be in the home.

Another extreme necessity for medical information is on the battlefield, where the medic needs to know as soon as a soldier is wounded and where he is located. The PSM system, which is described in detail below, emphasizes the value of bringing the information to a person in a very mobile and flexible environment. One could postulate that the information would be just as valuable to a factory supervisor to help maintain the health of the workforce, especially in dangerous or hazardous working conditions. Finally, during athletic training or competition, a trainer or coach would have the ability to monitor the athletes on the track, court, or playing filed in real time, literally at the athlete's side, in order to improve training or optimize performance. Thus the power of wearable computing is that it provides orders of magnitude more value when the acquisition, processing, and decision-making capabilities are available at the point of need to allow timely and accurate assessment and accurate results.

In addition to the power of the "point of service" concept, there is also another inherent capability of such systems. As sensors become microscopic, embedded, and cheap, sensor information now becomes ubiquitous. The new paradigm for decision making will be multiparameter analysis, when the change-over-time of multiple parameters are instantly compared to the individual's own baseline values and integrated against an individualized predictive model (concurrently updated based upon the person's activities and performance). Multiple-points-over-time permits trending of data in real time, and integrating the multiple points supports decision making that is not possible with single points of reference. For example, until recently, it was not possible to accurately predict when a person who is on a respirator in an intensive care unit could be safely removed from the breathing apparatus. Measurements of vital capacity, oxygen saturation, carbon dioxide content, etc. alone did not permit accurate determination of the patient's ability to breathe on their own once removed from the machine. However, at the Latter Day Saints Hospital in Salt Lake City, the surgical intensive care specialists developed a decision support software program based upon multiple-parameters-over-time principles listed above; the result was over 96% accuracy in predicting successful disconnection from the ventilator machine (1). Thus, the devices are only one component for successful decision; the incorporation of numerous noninvasive micro-sensors and sophisticated "local processing" of the information amplifies the power of wearable computing.

CURRENT SYSTEMS
FOR MEDICAL APPLICATION

In order to implement the concepts above, numerous advanced technologies need to be developed as well as integrated into a transparent system that allows the user to perform day-to-day activities without being aware of the system that is supporting these activities. Most people have no idea of how their television works; they just press a button and the program comes on to the screen. So too must wearable computing, with all the computing, networking, and telecommunication support, become absolutely invisible to the wearer. A number of new systems are emerging that exemplify the application of wearables. These are all in early prototype; however, they illustrate a few of the needs within the medical field. Many systems have been pioneered for military applications and the systems reflect the unique military need; however, it must be emphasized that there is no difference between a wounded soldier on the battlefield or MASH tent hospital and a farmer injured in a remote village, an elderly patient in a nursing home, or perhaps even an astronaut in a space station. The systems developed will support health, fitness, countermeasures, and medical support in any and all of these environments.

In the need for acquisition of medical information, the military has been developing the personnel status monitor (PSM) system to monitor the health of soldiers on the battlefield and to predict when a soldier is either injured or fatigued (2). The system, which is being developed by Sarcos, Inc., of Salt Lake City, Utah consists of three components: 1. global positioning satellite (GPS) location, 2. computer and communications, and 3. vital signs sensors. The GPS gives the location of the soldier on the battlefield at all times; should an injury occur the closest medic could instantly locate the soldier's position. The computer and communications systems can process the data (location and vital signs) and transmit the information. The vital signs monitors can continuously monitor the health status of the soldier. The current vital signs to be monitored are pulse rate, 2-lead EKG, respiration, temperature, oxygen saturation, shivering (for hypothermia), and motion. The physical configuration of the system is that of suspenders (web gear) that contain the GPS location, computer, and transmitters (Figure 21.2). Multiple vital signs sensors are located within

PSM Demo System Description

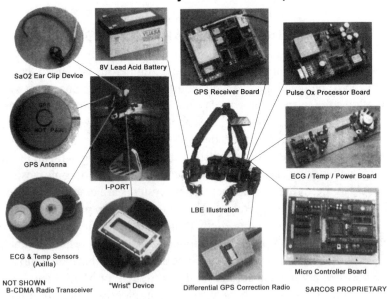

FIG. 21.2. The PSM as a system, illustrating the numerous components that must be integrated to achieve a robust, wearable system. (Courtesy Dr. Stephen Jacobsen, Sarcos, Inc. Salt Lake City, UT.)

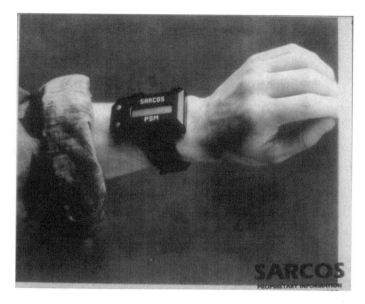

FIG. 21.3. The PSM "wristwatch" display for the individual soldier. (Courtesy Dr. Stephen Jacobsen, Sarcos, Inc. Salt Lake City, UT.)

a chest band. There is a "wristwatch" display device that shows the soldier his or her vital signs and location (Figure 21.3). The medic carries a hand-held device that shows soldiers' location on a map grid, direction from medic to the soldier, and text readout of vital signs. Information is relayed back to a full computer workstation in the command and control area in rear echelons, allowing the information to be distributed throughout the battlefield.

The system is in continuous monitor mode; however, information is only stored as changes from baseline. In addition information is transmitted in only an alert or query mode. This emphasizes the need for powerful computing at the "local level" so only a minimal of information is needed to be brought to the attention of the user or transmitted remotely. The essential parameters for vital signs are monitored, and only if there is a significant change is the information transmitted. The medic or commanders also have the option to query the location and vital signs of soldiers when necessary. Sophisticated algorithms are being developed to analyze the vital signs to determine if injury has occurred, the extent of injury, and whether the soldier is alive or dead. These pioneering efforts point to a way to relieve the "information overload" that would result from accumulating such a

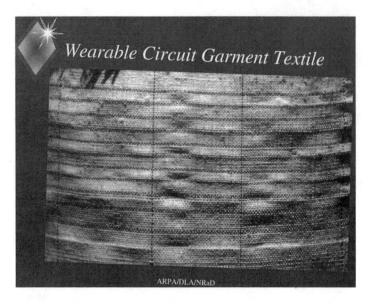

FIG. 21.4. The "smart tee shirt" in the first generation alpha prototype. (Courtesy of Dr. Eric Lind, Naval Research and Development
Lab (SPAWAR), San Diego, CA.)

massive amount of information about an individual. By local processing,
only relevant changes will be recorded and transmitted.

The sensate liner or "smart tee shirt" (3) being developed by the U.S.
military is a skintight expandable sleeveless tee shirt (the next generation
will be a full tee shirt) similar to current clothing worn for health and
fitness exercise but of a construction technique that also contains fiberoptic
and piezoelectric fibers woven into the fabric (Figure 21.4). These will
detect penetration (such as wounding to a soldier) or stretch for monitoring
respiratory rate. Totally independent micro-miniaturized vital signs sensors
(developed by Oak Ridge National Labs, Tennessee) will be embroidered
into the material. These systems contain the full system: (4) microsensor
for the vital signs (heart rate, EKG, oxygen saturation, respiratory rate,
etc.), A-D converters, signal processors, CPU, RF transmitter, antenna,
and battery supply, and they can transmit the information approximately 3
meters (Figure 21.5). The concept is to provide a "body local area network"
or body-LAN, which can wirelessly communicate data to the computer that
the soldier will be wearing. This will leave the soldier unencumbered and
will feed data into the PSM system.

In the area of display systems for the information, the use of augmented
reality systems may play a role for medical practice. In the 21st Century

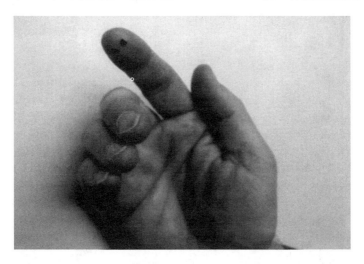

FIG. 21.5.   Microsensor for vital signs small enough to be embroidered into a garment. (Courtesy of Dr. Tom Ferrell, Oak Ridge National Labs, TN.)

Land Warrior, there is a face shield for eye protection from laser blinding
(Figure 21.6); this same protection can be used for an augmented reality display of real-time critical information to the soldier. There is a new commercial product by Micro Optical Corporation that provides a low-resolution
image directly on the lens of a pair of glasses (Figure 21.7). This extremely
lightweight and unencumbering system could easily be the method of providing instant access (in combination with the wearable computer) to the
healthcare provider at any location during the course of their varied duties.

In July 1996, the PSM system was tested in the jungle terrain of Florida
during Ranger Training by the U.S. Army. Five soldiers wore the system
continuously for five days with continuous display of their vital signs and
position, their exercise included a neck deep river crossing. Following the
five days there were no signs of skin irritation or chafing. The system has
proven to be robust in the most severe of conditions and the military is
continuing to develop the system for full deployment. The commercial
applications of such a system will be many, including monitoring in the
hospital in critical situations such as the emergency room, Intensive Care
Units, and Operating Rooms, in nursing homes, and even for home health
care to permit early discharge of patients from the hospital.

Special environments that are remote or hazardous will also require
monitoring. One particular application will be to monitor the health of the
astronauts and other space voyagers as we begin to build the International

FIG. 21.6.   The 21st Century Land Warrior system worn by the experimental fighting forces for testing and evaluation, exhibiting the
eye protection that will also be used as a head-mounted display.
(Courtesy of Carol Campbell, 21st Century Land Warrior program.)

Space Station and make preparations for Mission to Mars. Currently the
astronauts take special time to apply monitors connected to wires and do
single measurements of vitals signs. Such a wearable system with multiple
microsensors would be able to continuously monitor their health without
interfering with their multitude of duties. Other equally hazardous environments include mining, diving operations, extreme or professional sports,
fire fighting, search and rescue, and expeditionary missions. In May 1998,
the Yale–NASA Everest Extreme Expedition joined the 1998 American

FIG. 21.7. Miniature head-mounted display system as a pair of glasses. (Courtesy of Mark Spitzer, Micro Optical Corporation.)

Everest Expedition in order to provide medical support on Mt. Everest to the climbers (the closest medical support is a three-day trek away) by establishing a telemedicine clinic at Everest Base Camp at 17,700 ft. Daily morning "medical rounds" through video tele-consultation was performed, checking the status of the climbers as well as learning of the latest scientific information gained about high-altitude physiology. In addition, a personal status monitoring system developed by Dr. Michael Hawley of MIT Media Lab, called the BioPack, was utilized to monitor the climbers while climbing from base camp at 17,700 ft altitude to Camp 1. Similar to the PSM, the heart rate, respiration, temperature, and oxygen saturation were monitored and transmitted back during the climb to Camp 1. Whenever individuals are placed at risk, noninvasive monitoring and wearable computing will provide an additional margin of safety.

CONCLUSION

The healthcare system is enormous, encompassing millions of workers (over 660,000 physicians in the United States alone) and an annual budget exceeding one trillion U.S. dollars. Wearable computing can provide a giant step forward in improving the overall quality of healthcare by providing continuously monitoring critical vital signs on selected individuals or by providing access to healthcare information by the physician or healthcare provider at the "point of service" in an realtime manner that allows for decision making immediately. Other selected environments, such as those that are remote or hazardous, can benefit from the identical capabilities. The concept of wearable computing forms the infrastructure that will revolutionize healthcare by integrating the full spectrum of healthcare through information technology. With these qualities, healthcare can be improved by orders of magnitude during the future generations.

REFERENCES

1. Rudowski R, East TD, Gardner RM. Current status of mechanical ventilation decision support system: a review. *Int J Cli Monit Comput*, 1996, Aug; 13(3):157–166.
2. Satava RM. Virtual reality and telepresence for military medicine. *Ann Acad Med Singapore*, 1997, Jan; 26(1):118–20.
3. Lind EJ, Jayaraman S, Park S, Rajamanickam R, Eisler R, Burghart G, McKee T. A sensate liner for biomedical monitoring applications. In Westwood JD, Hoffman HM, Strendney D, Weghorst SJ (Eds). *Medicine Meets Virtual Reality: Art, Science, Technology: Health Care Revolution*. Amsterdam: IOS Press, pp. 258–264, 1998.
4. Ferrell TL. Miniaturized biosensor/transmitter systems. *http://web nt.sainc.com/arpa/abmt/oakridge.html*

22

Constructing Wearable Computers for Maintenance Applications

Len Bass, Dan Siewiorek, Malcolm Bauer,
Randy Casciola, Chris Kasabach, Richard
Martin, Jane Siegel, Asim Smailagic,
and John Stivoric
Carnegie Mellon University

1. INTRODUCTION

The maintenance of large vehicles (airplanes, trains, and tractors) provides difficult problems for computing devices due both to environmental and human factors. The environment has extremes of temperature and light, dirt and grease are common, and tools such as computers must be very robust. The technicians who perform the maintenance must have the mobility to move around, over, under, and inside the vehicle and must have their hands free much of the time. Maintenance is an activity that is performed both solo and with collaboration and the individuals who perform it tend to have little computer sophistication.

Since 1993, the Wearable Computer Laboratory at Carnegie Mellon University has been constructing and testing a variety of different hardware and software systems in a variety of different maintenance contexts. Five different disciplines have been involved in these designs: user interface designers, industrial designers, and software, electrical, and mechanical engineers. These disciplines cover a set of skills from sensitivity to the interaction of human and device (the user interface and industrial designers) to knowledge of the technology necessary to construct small, sophisticated computer-based systems (the software, electrical, and mechanical engineers). Some of the systems constructed are body worn and some are hand held. All have limited capability for input and output and are designed for the maintenance environment.

This chapter will describe some systems that have been developed during the wearable work and how and why the systems were designed as they were. The manner in which the various disciplines interact and the constraints they place on each other will also be explored. We begin by describing the problems associated with utilizing wearable computers during the maintenance of large vehicles. We next describe the organization context in which the systems described were conducted and the problem domain. Next we describe the particular systems developed. Finally, we describe the design problem from the point of view of the hardware designers, the user interface designers, and the software designers.

2. WEARABLE COMPUTERS FOR MAINTENANCE TECHNICIANS

Wearable computers deal in information rather than programs, becoming tools in the user's environment much like a pencil or a reference book. The wearable computer provides automatic, portable access to information. Furthermore, the information can be automatically accumulated by the system as the user interacts with and modifies the environment, thereby eliminating the costly and error-prone process of information acquisition. Much like personal computers allow accountants and bookkeepers to merge their information space with their workspace (i.e., a sheet of paper) wearable computers allow mobile processing and the superposition of information on the user's work space.

Wearable computers make it possible to get the right information to the right person in the right place at the right time. In this chapter, the people we discuss are maintenance technicians. For them, the right place is often environmentally challenging and is usually changing in various locations around their work place, and the right time is when their primary task is maintenance, not interacting with a computer. These characteristics make the problem of providing computer support especially difficult.

Maintenance applications are characterized by a large volume of information that varies slowly over time. For example, even simple commercial or military aircraft will have over 100,000 manual pages. One standard checklist has 625 items. One normal procedure includes over twenty steps. These procedures are currently included in paper-based manuals that the technicians physically carry to the aircraft. Due to operational changes and upgrades, half of these pages are made obsolete every six months. Rather than distribute CD-ROMs for each maintenance person and run the risk of a maintenance procedure being performed on obsolete information, maintenance facilities usually maintain a centralized database to which maintenance personnel make inquiries for the relevant manual sections on demand. A typical request consists of approximately ten pages of text and schematic drawings. Changes to the centralized information base can occur on a weekly basis.

There are times, however, when an individual requires assistance from experienced personnel. Historically this assistance has been provided by an apprenticeship program wherein a novice observes and works with an experienced worker. Today, with down-sizing and productivity improvement goals, teams of people are geographically distributed and yet are expected to pool their knowledge to solve immediate problems. A simple example of this is the "Help Desk" wherein an experienced person is contacted for audio and visual assistance in solving a problem. The Help Desk can service many people in the field simultaneously.

The Challenge of Wearable Computer Design

The objective of wearable computer design is to merge the user's information space with his or her work space. The wearable computer should offer seamless integration of information processing tools with the existing work environment. To accomplish this, the wearable system must offer functionality in a natural and unobtrusive manner, allowing the user to dedicate all of his or her attention to the task at hand with no distraction provided by the system itself. Conventional methods of interaction, including the keyboard, mouse, joystick, and monitor, all require some fixed physical relationship between user and device, which can considerably reduce the efficiency of the wearable system.

Among the most challenging questions facing mobile system designers is that of human interface design. As computing devices move from the desktop to more mobile environments, many conventions of human interfacing must be reconsidered for their effectiveness. How does the mobile system user supply input while performing tasks that preclude the use of a keyboard? What layout of visual information most effectively describes system state or task-related data?

To maximize the effectiveness of wearable systems in mobile computing environments, interface design must be carefully matched with user tasks. By constructing mental models of user actions, interface elements may be chosen and tuned to meet the software and hardware requirements of specific procedures.

3.   CMU ORGANIZATIONAL CONTEXT

The CMU wearable computers are conceived, designed, and fabricated in an interdisciplinary course with over two dozen students, more than two thirds of whom are undergraduates. The teaching staff for the course is composed of an electronics engineer, software engineer, industrial designer, and a human–computer interaction designer. In this course, students learn to work in interdisciplinary teams to deliver products to clients on time. The brief cycle time of these products is ideally suited to the academic semester. Figure 22.1 illustrates the iterative nature of user-centered design to elicit feedback during the course. Student designers initially visit the user site for

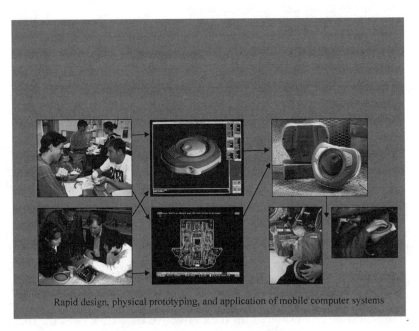

Rapid design, physical prototyping, and application of mobile computer systems

FIG. 22.1.   Iterative prototyping cycle.

a walk-through of the intended application. A second visit after a month of design elicits responses to story boards of the use of the artifact and the information content on the computer screen. After the second month a software mock-up of the system running on a previous generation wearable computer is evaluated in the end-user's application. During the third month, a prototype of the system receives a further user critique. The final system is delivered after the fourth month for field trial evaluation.

The design is guided by an Interdisciplinary Concurrent Design Methodology (ICDM) that has evolved through a dozen generations of wearable computers [Smailagic, Siewiorek, Anderson, Kasabach, Martin and Stivoric, 1995]. The goal of the design methodology is to allow as much concurrency as possible in the design process. Concurrency is sought in both time and resources. Time is divided into phases. Activities within a phase proceed in parallel but are synchronized at phase boundaries. Resources consist of personnel, hardware platforms, and communications. Personnel resources are dynamically allocated to groups that focus on specific problems. Hardware development platforms include workstations for initial design, personal computers for development, and the final target system. Communications allow design groups and individuals to communicate between the synchronization points.

System engineering is performed by the class as a whole and then the various disciplines perform detailed design and implementation. During the whole process, the four disciplines interact along well-defined design boundaries as shown in Figure 22.2. The hardware design must merge with the mechanical/industrial engineering design so that the hardware fits within the case and so that sensors can provide input and output for the hardware. The hardware must merge with the software design so that adequate resources are available for the necessary functions of the software and so that software drivers are available for the hardware. The software design must merge with the user interface design so that input/output can be performed and so that the users have available the functions necessary to perform their task. Finally, the user interface design must merge with the mechanical/industrial design to enable the interactions between the system and the user with the particular electro-mechanical user interface devices utilized and to support ergonomic requirements of the user.

For the remainder of this chapter, we will first describe the maintenance problem and its various activities; we then give a short explanation of four of the systems we have developed and which maintenance activities they support. Finally, we will describe the systems from the point of view of each of the disciplines and discuss how each discipline was constrained by the

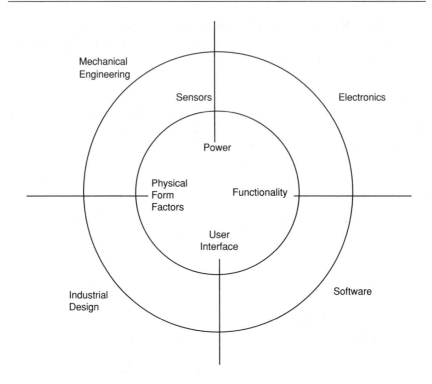

FIG. 22.2. The interaction of the various disciplines involved in constructing wearable computers.

disciplines with which it directly interacts. We conclude by summarizing the ways in which wearable computer design differs from normal desktop design.

4. MAINTENANCE AS AN ACTIVITY

Maintenance has four types of activities: inspection, preventive maintenance, troubleshooting, and repair. Surrounding these activities is a collection of short- and long-term information flows. Preventive Maintenance (PM) is a routine activity, but the technician may not have internalized the procedures due, perhaps, to the infrequency of the task. The technician consults on-line manual pages to complete the task. In troubleshooting, the source of an unexpected problem is determined by consulting reference material and human experts. Short-term information flows are used to govern job assignments on a particular day or collaboration in the event of a problem. Long-term information flows are used to govern such things as corporate history with respect to a class of problems, design changes, and trouble reports. We first discuss the types of activities in more detail, then the short- and long-term information flows, and finally, our systems with respect to the portion of the maintenance process they were intended to support.

Inspection Task

A maintenance technician usually performs an inspection on a vehicle at scheduled periods. During this activity, the vehicle is examined and various defects such as corrosion on the skin or defective transmission that will need repair or further analysis are inspected and recorded. Two of our example studies, VuMan3 and Navigator 2, were identification tasks. In the VuMan3 case, over 600 items were examined and, for each item, an entry was made in a checklist that indicates the current status of these items. We called this type of inspection a one-dimensional inspection since the results were entries into a prespecified checklist. In the Navigator 2 case, an inspection was made of an aircraft skin and various defects were noted. We called this type of inspection two dimensional since the notation included the location of the defect on the skin as well as the type of defect.

Fault Isolation Task

Fault isolation occurs when the vehicle is demonstrating incorrect behavior but the cause of this behavior is not known. The technician will examine specific procedures that give a set of tests to make in particular cases or, if the procedures do not suffice, will look at representations of particular subsystems of the vehicle such as a schematic drawing of the electrical system.

Repair Task

The replacement of a component takes place once the component has been identified as faulty. The identification can occur in either the inspection task or the fault isolation task. The technician follows a set of procedures (of the form "remove cover, loosen bolt A in diagram 1 one half turn . . . ") that are both textual and graphic. The graphics are intended to illustrate the proper performance of a particular procedure and tend to be stylized representations of a visible component (as opposed to a schematic).

Information Recording Task

The technician is expected at various stages during the activities of a normal day to report status and incident information as well as to make suggestions for engineering or procedural changes. Thus, for example, a report at the end of a shift would contain activities performed during the shift, problems encountered during the shift, and activities that are being carried forward to the incoming shift. This information is usually in the form of structured text. That is, there is a form with entries for information and the information put into each entry tends to be unstructured.

5. SYSTEMS CONSTRUCTED

Wearable computers are used during the maintenance process both to retrieve information and procedures for the technician and to provide structured information about the vehicle on which maintenance is being performed. They are not used to provide free-form input. They are also used as a communications device, especially when collaboration is required. That is, wearable computers are intended to support maintenance technicians while they are interacting with a vehicle: performing an inspection, troubleshooting a problem, or replacing a part. During these operations, the technician must be mobile and must have hands mostly free. These requirements lead to a wearable computer that is not suitable for the entry of free text. Indeed, much of the challenge of designing wearable computers is to construct a usable system without the necessity for the entry of free text.

The retrieval of information is performed during any type of maintenance operation. If an inspection is being performed, then the information that is retrieved consists of the checklist (whether one dimensional or two) and information about the vehicle being inspected. If the technician is troubleshooting a problem, or replacing a part then the information retrieved consists of troubleshooting flow charts for diagnosis or procedures for replacement. It can also consist of various engineering change notices or problem reports filed by either the engineering department or by other sites where maintenance occurs.

A final aspect of the process of using wearable computers in maintenance applications is their use in collaborative settings. Some tasks are inherently collaborative, such as observing the effect of flipping a switch in the cockpit when the switch controls a light at the rear of an aircraft. Others become collaborative because assistance is needed in the performance of a single-person task such as troubleshooting or replacing a complex part. In support of collaboration, the wearable becomes primarily a communication device but, again, the limitation on input devices causes restrictions compared to collaboration on a desktop.

Because of the requirements for mobility and mostly hands-free operation, the wearable computer is not suitable for all tasks. For example, shift reports are routinely filed that document particular problems and open issues during a shift. This is a free text document where neither mobility nor hands-free operation is required during its creation. Thus, it can easily be generated from a desktop; therefore this is not a task for a wearable. The line between the tasks that are performed with a wearable and those performed at a desktop is thus an important one to establish. The wearable must communicate with desktop machines both for reporting and retrieval purposes but the tasks to be accomplished with each type of machine must be explicitly specified.

With this background, we will now report how our various projects were integrated into the normal process of maintenance.

VuMan 3 and Navigator 2

Both VuMan 3 and Navigator 2 were designed to perform inspections. They both occupied the same position in the maintenance process—recording the identification of imperfect parts (in the case of Navigator 2, the parts were skin panels). In the use of the VuMan 3, the inspection step was one of the first steps in checking out the vehicle. In the use of the Navigator 2, the inspection step came after the vehicle was prepared for inspection and stripped of paint. In both cases, job orders were used to instruct the technician which vehicle to inspect and the data recorded from the inspection had to be fed back into the job order system so that parts for repairs could be made available and the repairs scheduled. Also, in both cases, the technician was required to provide personal identification to the computer so that responsibility for the inspection task could be tracked. Figures 22.3 and 22.4 show the VuMan 3 and Navigator 2 systems in use. Bass, Kasabach, Martin, Siewiorek, Smailagic and Stivoric [1997] describe the design of the VuMan 3 in more detail. The technicians were required to move around the vehicle and record defects they discovered. This required (in the VuMan 3 case) viewing the top, bottom, sides, and interior of a vehicle. In the Navigator 2 case, only the exterior of the vehicle was inspected but this inspection required the technician to work on the top of a "cherry picker" lift and make very detailed visual and tactile inspections.

FIG. 22.3.   VuMan 3 system in use.

In neither of these systems was there any retrieval of procedural or troubleshooting flowcharts or any provision for collaboration. They were designed strictly to support the inspection process and to interface with the other facets of the repair process in a limited fashion.

Adtranz

The Adtranz project used a hand-held computer to support the replacement process. The procedures necessary to do the replacement were retrieved onto the computer and were available to the technicians as they went through the process. The vehicles being repaired were always in a fixed location on a track and the supports for the track provided sufficient

FIG. 22.4. Navigator 2 system in use.

area to set down a computer and still have it visible. Thus, the Adtranz system was a mobile computer that was not worn but was placed in the work place. The other process issues were the same however in that information was retrieved and the technicians had the ability to navigate through it to follow particular procedures. Figure 22.5 shows the Adtranz technicians under the vehicle.

C-130

The C-130 project was designed to use collaboration to facilitate training. Inexperienced users were being trained to perform a cockpit inspection and the trainers were located in the more spacious cargo hold of a

FIG. 22.5. Adtranz technicians performing maintenance.

C-130. Ultimately, there is no reason why the trainers need to be located in the aircraft and the ultimate goal of this experiment was to provide more training with the same number of trainers.

As an example of the C-130 project, a student loaded the inspection procedures and performed the inspection task. The instructor looked over the shoulder (through a small video camera) and offered advice when problems occurred. The advice was demonstrated by indicating areas on the video image that was being shared through a whiteboard. Because of the limited input devices on the wearable computer, the instructor managed the sharing session and used the whiteboard. The student's use of the whiteboard was limited to observation. Figure 22.6 shows the C-130 system in use.

The limitations of input devices, such as the inability of the student to point to a particular area of the screen caused by the use of a dial as the main input device, affected the division of tasks between the wearable and the desktop systems. The desktop systems managed the normal job order process and were used by the instructors in C-130 project to observe the student's behavior.

Table 22.1 outlines the functionality of the Maintenance and Collaboration systems.

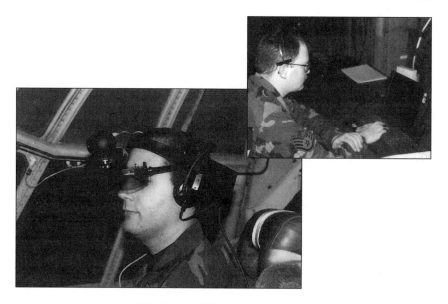

FIG. 22.6. C-130 system in use.

6. HARDWARE DESIGN

The wearable computers can be classified into two generic classes of systems: 1. custom designed (VuMan3) and 2. designed by composition (Navigator 2, AdTranz, and C-130), using mainly off-the-shelf components. In addition to custom-designed electronics and mechanical enclosure/ interface, VuMan3 also adopted an embedded, custom-designed approach to the software system. In Navigator 2, a modular "mix-and-match" hardware architecture allowed multiple configurations, increasing the general purpose nature of the system.

VuMan3

The VuMan3 computer (Figure 22.7) is a $5'' \times 6.25'' \times 2''$ unit weighing less then two pounds including a rotary dial input device integrated with an environmental sealed housing, a Private Eye (manufactured by Reflection Tech) display with a customized headband, and a smart docking station, which monitors the use of the NiCd rechargeable batteries and also acts as a communication link to a host computer system, to which the inspection data are uploaded.

TABLE 22.1
Example CMU Experimental Wearable Computer Systems

Application	Sensing (input)	Database	Computing	Communications	HCI	Actuation (output)
MAINTENANCE						
Vehicle inspection (VuMan 3)	Dial	One-dimensional checklist	20 MHz 386	Serial line	Fill in form	Display
Aircraft inspection (Navigator 2)	Audio microphone	Two-dimensional schematic	100 MHz 486	Serial line	Speech controlled icons	Display
COLLABORATION						
Train maintenance (Adtranz)	Audio microphone	• Trouble shooting flow charts • Engineering drawing • Repair procedures • Parts list	133 MHz 486	Spread spectrum radio	Shared white board/ Two-way phone	Two-part control/ data display
Help Desk (C-130)	• Audio microphone • Video camera	• Repair procedures • Trouble shooting flow charts	100 MHz 486	Spread spectrum radio	Instructor controlled white board	Display pointer

676

FIG. 22.7. VuMan 3.

The VuMan 3 wearable computer electronics consists of two custom-designed printed circuit boards: a motherboard containing a 386EX processor, which also includes power management capability, 1.25 MB of SRAM, 128 KB EPROM, real-time clock, silicon serial number chip; the second board is a detachable PCMCIA Controller board, allowing application program memory cards and peripheral hardware cards to be connected to the wearable computer through two PCMCIA card slots. The power control block consists of a PIC16LC71 microcontroller, which manages three separate power supplies, and two sets of batteries. The main batteries are eight rechargeable NiCd batteries, while power is provided to the power control, clock, and static RAM by two lithium batteries when the NiCd batteries are discharged. The input control consists of another PIC microcontroller, which reads the status of the dial and the select button and transmits that information to the processor. The peripheral block is managed by an Altera FLEXlogic FPGA, which provides an interface between the processor bus and the following three components: the Private

Eye display, a DS1302 timekeeping chip, and a DS2401 silicon serial number.

Explicit design considerations included:

- Increased modularity in the design so that design fragments could be reused. This included input controller, power control, and PCMCIA slot controller.
- A set of primary and secondary design decisions were defined, and an iterative approach adopted.
- A new manufacturing technology, surface mounting of components, was introduced, which provided for a more compact design.
- The introduction of programmable microcontrollers for the input block (allows reconfiguration of the electronics with dial, mouse, force resistive device) and power management (selectively turns off unused chips) provided a higher degree of flexibility than previous designs.

The VuMan 3 dial input device was designed to be easy to learn and use as well as easily modifiable and expandable. In addition, it was low cost and power efficient. The system consists of a rotary switch and three buttons and a microcontroller, which translates the dial turns and button presses into data in a form the microprocessor (an Intel 386 EX) can accept.

The dial itself is a 16-position binary coded rotary switch, which outputs a four-digit Gray code representing the switch's position. A PIC16LC71 microcontroller accepts input from the rotary switch and pushbuttons, and uses this information to transmit user input to the microprocessor through a serial port.

The housing fabrication included molding and machining. The housing design followed an evolutionary path, from initial drawing and mock-ups, through refined stereolithography (SLA) model, and 3D CAD Proengineer model. Upon receiving the lot of parts from the manufacturer, the necessary postproduction processes were applied. Once postproduction was complete, each part was detailed, including spraying EMI shielding, adding color and/or texture to a part, anodizing or coating aluminum, and silk-screening graphics onto the parts.

Navigator 2

Navigator 2 (Figure 22.8) included a novel dual architecture (486 application processor and speech recognition digital signal processor), spread spectrum radio, and VGA head-mounted display. The Navigator 2 semi-custom

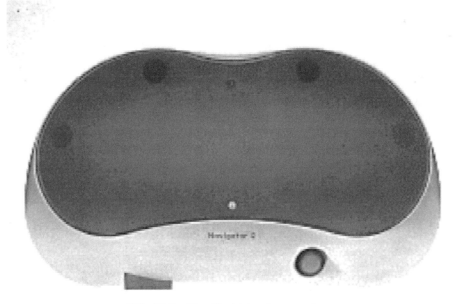

FIG. 22.8. The Navigator 2 computer.

electronic design included two major electronic boards: a custom designed system board and a 486-based processor board. The custom-designed system board captured all glue logic functions and also provided support for two PCMCIA card slots (accommodating a speech recognition card and hard disk). The major design considerations taken into account when evaluating the processor/motherboard subsystem were:

- The modular hardware architecture must support customization of input/output devices.
- The subsystem must satisfy specific constraints on the following wearable computer attributes: size, weight, form of input/output, and battery consumption.

The Navigator 2 was built to run a voice-controlled aircraft inspection application. The speech recognition system was based on a commercially available system from TERI, with a secondary manually controlled cursor, offering complete control over the application in a hands-free manner and allowing the operator to perform an inspection with minimal interference from the wearable system. Entire aircraft manuals, or portions there of, could be brought on-site as needed, using wireless communication. The results of inspection were downloaded to a maintenance logistic computer.

Adtranz and C-130

The Adtranz application was supported by a commercially available pen-based computer enhanced with a spread spectrum radio, voice transmission, image capture, and support for a VGA head-mounted display. The computer unit featured a 50 MHz 486DX2 processor; 12 MB RAM memory; 170 MB hard disk; two PCMCIA Type II slots; one serial, one parallel, and one infrared port; and grayscale 640 × 480 display. The PCMCIA slots were occupied by an AT&T WaveLAN card and Wave Jammer sound card, used for voice communication.

The C-130 application also used a commercially available pen-based computer but it was adapted so that a dial provided all of the necessary input capability except for the video camera and associated video capture card.

7. USER INTERFACE DESIGN

Several principles underlie the user interface design of the "normal" wearable system. These are:

- Simplicity of function. One of the final portions of the user interface design process is to jettison any functions that are not absolutely required. The user interface is designed, as much as possible, to be feature free.
- No textual input. The user interfaces are designed so that they can be operated without the use of a keyboard or a keyboard surrogate. Those functions that require alpha-numeric input do not drive the interface design but are integrated, as well as possible, into an interface that would suffice if that function were not included. Also, functions that seemingly require alpha-numeric input can often be performed through selection.
- Controlled navigation. In order to keep the interface simple, several basic strategies are used to navigate through the interface:

  Hierarchical navigation paths. The base navigation strategy throughout the user interface is for the user to ascend or descend a hierarchy. At each interior node, the children nodes are available for choice, as is the ability to retrace the path through the hierarchy. The spread at each interior node is limited to seven or eight choices. At the leaf nodes, the user is furnished with a procedure, can select an entry in a checklist, or indicate the location of an imperfection. At any point in a descent through the hierarchy, a help function is available that enables browsing through information relative to the current task or position in the hierarchy. Both Figures 22.10 and 22.11 demonstrate control of navigation paths. In neither case is it possible for the user to move to an arbitrary portion of the interface. They movement is restricted to up or down a hierarchy or through a link to a different mode of interaction.

Clear identification of control and content. Control actions (except for those that can be keyed from content) are clearly differentiated from the content of the data. This differentiation is accomplished by different fonts, different screen areas, and different button shapes. Those control actions that are connected with data content, such as hypertext links, are embedded in the data but are differentiated using various cues.

The availability of current location in the hierarchy on the display. In order to avoid disorientation within the hierarchy, the current position is always available to the user.

We will now describe the user interfaces for each project and see how these design rules were actually used.

VuMan 3

VuMan 3 was designed for streamlining Limited Technical Inspections (LTI) of amphibious tractors for the U.S. Marines at Camp Pendleton, California. The LTI is a 600-element, 50-page checklist that usually takes four to six hours to complete. The inspection includes an item for each part of the vehicle (e.g., front left track, rear axle, windshield wipers).

The inspector selects one of four possible options about the status of the item: Serviceable, Unserviceable, Missing, or On Equipment Repair Order (ERO). Further explanatory comments about the item are selected (e.g., the part is unserviceable due to four missing bolts). The LTI check list consists of a number of sections, with about one hundred items in each section. The user sequences through each item by using a dial to select "next item," or "next field."

Upon completion of the inspection, the results of the LTI are typed into a logistics computer from which work orders are issued and parts are ordered. Performing actual full LTI inspections with and without VuMan 3 were evaluated. The results from six inspections indicated a 40% average saving in actual inspection time and virtual elimination of the data entry

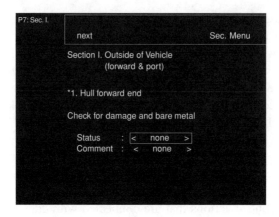

FIG. 22.9.   VuMan3 user interface.

time into the logistics computer. There was a 70% savings in time from the beginning of the inspection until the data were in the logistic computer.

The VuMan3 has a low-resolution display (CGA) and, consequently, a purely textual interface. Figure 22.9 shows a sample screen from the user interface. The user navigates through a geographically organized hierarchy: top, bottom, front, rear; then left, right, and more detail. Eventually, at the node leafs, individual components are identified. There are over 600 of these components. Each component is indicated to be "serviceable" or "unserviceable." If it is serviceable then no further information is given. If it is unserviceable then one of a small list of reasons is the next screen.

The user can return up the hierarchy by choosing the category name in the upper right corner, or sequence to the next selection in an ordering of the components. Once a component is marked as serviceable or unserviceable, the next selection in the sequence is automatically displayed for the user. Furthermore, each component has a probability associated with it of being serviceable and the cursor is positioned over the most likely response for that component.

The relationship between the user interface design principles and the VuMan3 user interface is:

- Simplicity of function. The only functions available to the user were to fill out a checklist for one of two vehicles, to transfer checklist data to another computer, to enter identification information both for the vehicle and for the inspector, and to see a screen that describes the VuMan3 project.
- No textual input. The identification information required entering numbers. A special dialogue was developed to enable the entering

of numeric information using the dial as an input device. This was cumbersome for the users but only needed to be performed once per inspection.

- Controlled navigation. The interface was arranged as a hierarchy. The top level consisted of a menu that gave a choice of function. Once the inspection function was chosen, then the component being inspected was navigated to via its location on the vehicle. At each stage, the user could go up one level of the hierarchy.

Navigator 2

Navigator 2 was developed for recording information during detailed inspection of the outer surfaces of KC-135 aerial refueling tankers at McClellan Air Force Base, Sacramento, California. The sheet metal inspection required 30 to 36 hours. Upon completion the inspector entered each defect into a forms-based database from which work orders were generated. Six inspections by three inspectors were evaluated both before and after the introduction of Navigator 2. A 50% average reduction was observed in the time to record inspection information (for an overall reduction of 18% in inspection time) and almost two orders of magnitude reduction in time to enter inspection information into the logistics computer (from over three hours to two minutes).

The Navigator 2 uses a display with a VGA resolution and, consequently, a graphical interface is possible. Figure 22.10 shows an interface from the use of Navigator 2 in sheet metal inspection at McClellan Air Force Base. The inspector records each imperfection in the skin at the corresponding location on the display. The type of the imperfection is also recorded.

The user navigated to the display corresponding to the portion of the skin currently being inspected. This navigation was partially textual based on buttons (choose aircraft type to be inspected) and partially graphical based on side perspectives of the aircraft (choose area of aircraft currently being inspected). The navigation could be performed either through a joystick input device or through the use of speech input. The speech input was exactly the text that would be selected. The positioning of the imperfection was done solely through the joystick since speech is not well suited for the pointing necessary to indicate the position of the imperfection. Once the imperfection had been positioned, a menu is placed on the display so that the type of imperfection (corrosion, scratch, etc.) could be specified. The user could navigate to the main selection screen by selecting the "Main menu" option on all of the screens. One level up in the hierarchy could also be achieved through a single selection.

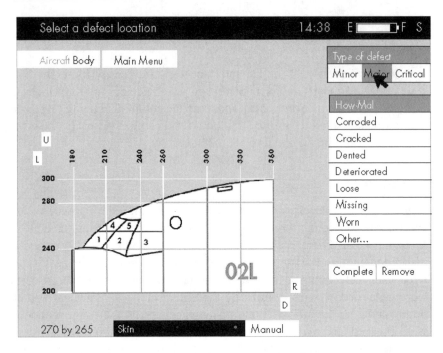

FIG. 22.10.   Navigator 2 user interface.

The relationship between the user interface design principles and the
Navigator 2 user interface is:

- Simplicity of function. The only functions available to the user were
  to enter skin imperfections for one of four aircraft, to transfer data
  to another computer, to enter identification information both for the
  vehicle and for the inspector, and to see a screen that describes the
  Navigator 2 project.
- No textual input. The identification information required entering
  numbers. A special dialogue was developed to enable the entering of
  numeric information using the joystick as an input device. This was
  cumbersome for the users but only needed to be performed once per
  inspection.
- Controlled navigation. The interface was arranged as a hierarchy. The
  top level consisted of a menu that gave a choice of function. Once
  the inspection function and then the vehicle were chosen, the area of
  the skin inspected was navigated to via selecting an area of the aircraft
  to expand. Once an imperfection was indicated, the user had to select

one of the allowable types of imperfections. At each stage, the user could go up one level of the hierarchy or return to the main menu.

Adtranz

The Adtranz system is the only system of the ones we are reporting that allows textual input. The basic input mechanism is a pen and this allowed the technician to do free text queries. Within the interface, the classes of functions that are available to the technician (procedures, schematic diagrams, searching an information base for similar problems) were available on the control or left side of the screen. The right side of the screen was reserved for the content.

The content area, occupying the left two thirds of the display, contained documentation and user collaboration. The control area allowed the user to select documents, set bookmarks, enter alarms, etc. The bottom of the display contained a menu bar that allowed access to the major usage modes at any time. The major usage modes included login/out, reference, bookmark, troubleshoot, annotate, and collaboration.

The user could select a vehicle area and a list of vehicle systems in that area was displayed. The user could select a system (such as electrical, pneumatic, propulsion, etc.) that, in turn, displayed a list of devices associated with that system. Selection of a device generated more specific information such as preventive maintenance schedule, preventive maintenance procedures, troubleshooting information, and device description. At the top level of the reference mode the user could also view preventive maintenance schedules and procedures. The user could place a bookmark on any of the reference pages. The bookmark appeared in the control area for quick selection and "page flipping."

The troubleshooting mode was designed to maximize the computer's searching capability, thus minimizing the user's time to find suggestions on probable causes of malfunctions. The user could select from a list of alarms. A list of possible causes then appeared in the content area. When the user selected a possible cause, more details were provided including manual reference pages and schematic drawings. A schematic drawing could be browsed by zooming in/out, panning, tracing circuit paths (e.g., selecting a component highlights other components to which it is attached), and displaying text associated with the schematic. Annotations could be personal, public to a site, or public to all sites.

In the collaboration mode, the user could seek help from other personnel. Users could collaborate by a whiteboard wherein all members of a session can view the same content area including a picture captured by a camera at one site. The whiteboard allowed annotation. Information, annotation, and speech was transmitted over a wireless network. It was also possible to initiate a telephone call to personnel who do not have wireless access.

The relationship between the user interface design principles and the Adtranz user interface is:

- Simplicity of function. This interface was actually more complicated than most of the interfaces we are discussing. The user had a collection of functionality available and needs to understand the different types of functionality available. On the other hand, since reporting on results and the filling out of the various reports was done on a desktop computer, there was no requirement for identification or extensive free text. The query facility was implemented via a simple text box.
- No textual input. As already discussed, a free text query facility took advantage of the ability of the user to enter text via a pen interface. No other textual input was required.
- Controlled navigation. The top level functional choices were kept at the left and, at any time, the user could exit the current operation and invoke one of these functions. Navigation within a function was accomplished via links within the content.

C130

A multimedia system with head-mounted display and wireless communications provided access to electronic maintenance manuals and remote access to a human help desk expert on the C-130 flight line at the 911th Air National Guard, Pittsburgh, PA.

The C130 system used the dial as an input device and supported collaboration. Figure 22.11 shows an example of the user interface for this system. This interface supports a checklist application in a training context. The assumption was that the person doing the inspection is being trained and that a remote expert was "looking over the shoulder" with a video camera and assisting where necessary with the inspection.

The interface was organized as a hierarchy with a sequential list of inspection steps embedded in the hierarchy. Each screen gave the ability to go to the next or previous step of the inspection. Failures could be indicated within the instructions.

Managing the collaboration was the task of the remote expert who was assumed to be in front of a workstation. A collaboration consisted of the novice verbally stating something about a problem and the expert indicating

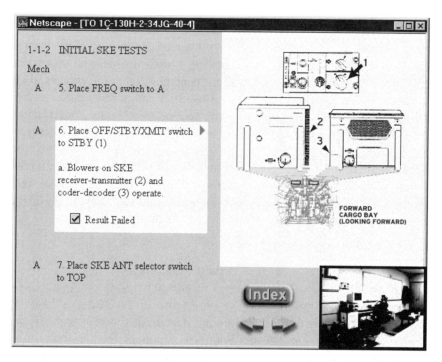

FIG. 22.11. C-130 user interface.

with the cursor and describing verbally the actions to be taken. Since the expert was at a desktop, moving a cursor to any position on the field of view was not a problem. The novice was wearing the computer and using a dial as input and so arbitrary cursor movement was not possible.

The relationship between the user interface design principles and the C130 user interface is:

- Simplicity of function. The wearable user had two main functions: navigating through the checklist and initiating a collaboration. The navigation through the checklist followed the principles that we have already discussed. Initiating the collaboration was a single function and each novice is preassigned an expert. Thus, there was no need for elaborate session management at the wearable computer. The expert managed several novices and controled the sessions.
- No textual input. The user interface was controllable with the dial and required no textual input.
- Controlled navigation. All navigational links were either through the natural sequence of the checklist or through a simple menu.

The Influence of the Other Disciplines

In all of these systems, all disciplines influenced the user interface through the system level choices of input device, output device, and task to be performed. The interfaces were also influenced through the choices made during software design. The VuMan3 used an embedded system and so all user interface interactions had to be supported by code written by the development team. This led to a simple collection of interactions. Navigator 2 was based on a simple window system and some of the interaction between the user interface designers and the software designers dealt with issues of fonts and button shapes. Both Adtranz and the C130 system were constructed using World Wide Web browsers and took advantage of the hypertext linking facility supported by that software.

8.   SOFTWARE DESIGN

The basic software solution is given in Figure 22.12. The user interacts with one of a variety of input and output devices attached to the software input/output component. The input/output component interacts with a middleware component that keeps the system specific information and this, in turn, interacts with a collection of local and remote databases. The arrows represent both data and control connections. They are always data connections and, in the case where the components are combined into a single process, they also represent control connections.

The database component was decomposed as in Figure 22.13. Although this is a simple and fairly generic structure, several design decisions are already apparent. Some of these decisions and concerns arising from them were:

- Separation of the input/output from the remainder of the software deferred a decision as to which actual input and output devices are going to be used. It also provided a container in which either standard

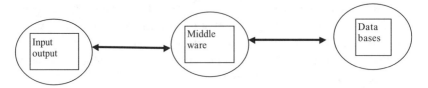

FIG. 22.12.   Basic software solution.

FIG. 22.13. Decomposition of database component.

user interface mechanisms could be used (such as in the Navigator 2, C130, or Adtranz systems) or the user interface mechanisms could be optimized for performance (such as in the VuMan3).

- The primary purpose of the software was to move data between the user and the databases. The middleware provides the mechanisms to accomplish this movement.
- Maintaining consistency between the information in the database and the information with which the user is interacting was a concern for the system. Active database mechanisms such as triggers could be used to inform the user of changes in the data. Transaction oriented strategies such as commit could be used to ensure consistency of the data in the database. The middleware was responsible for the coordination of these mechanisms.
- Variation in the type and number of databases was encapsulated in the database component. This encapsulation allowed legacy databases as well as internally generated ones to be incorporated into the software.

Decisions made during the system architecture stage determined whether the software was contained totally within the wearable computer or whether network connections were assumed. The user interface component clearly must be contained within the wearable computer. The other two components could be either within the wearable computer or distributed across a network. We will see examples of both cases. We now will examine our four systems and see how they actually were implemented and why.

VuMan3

The system characteristics of the VuMan3 were:

- stand-alone computer system
- one-dimensional inspection where the inspector input information via checking items on a checklist

- a 386 processor
- program and data entry via a flash memory

The VuMan3 was a system with limited memory, limited processor power, and limited scope. The database component was preloaded with the checklist to be filled out; the responses to the questions were saved in the database until they were communicated, via a serial line, to a desktop system. Since the VuMan3 was self-contained, the data in the database never got out of synchronization with the data observed by the user.

The internal format of the questions was created via an off-line tool that we will describe later in this section. The responses were kept in an internal format until one of the middleware functions sent them through the serial line to the job control system of the U.S. Marines. The middleware component for the VuMan3 consists of a collection of procedures for moving data and for performing the menu selections. Some of the specific functions these procedures accomplished were:

- Interpret input selection and either send data to database to record selection or retrieve next menu contents from the database.
- Move data from the database to the serial port.
- Save user or vehicle identification in the database.
- Set format to be either "left eye" or "right eye."

The user interface component performed the following functions:

- Send the data to the display. The display had a direct memory connection to improve performance and so the sending of the data to the display was a nontrivial portion of the code
- Prepare the data for display. This involved generating a bit map with the screen image on it and preparing the characters based on a special font used in the VuMan3 software. It also involved knowing whether the display was currently over the left eye or the right eye and presenting the image in the correct orientation.

Preparing the questions for use by the VuMan3 software was the responsibility of a tool that was used prior to the use of the VuMan3 by the maintenance personnel. This tool took as input questions in the checklist for the technician and structured them in a hierarchy that simplified the navigation and prepared them in the internal format. This format was then loaded on flash memory for use during the maintenance procedure.

Navigator 2

The system characteristics of the Navigator 2 were:

- stand-alone system
- 486 processor
- used in two-dimensional inspections
- program and data entry through loading them into a Type 2 PCMCIA rotating disk

The data in the database consisted, essentially, of the same two portions as in the VuMan3. The difference is that a two-dimensional inspection does not have a checklist but instead has a visual representation of the aircraft. The data stored in the database consisted of a collection of bit maps that represented segments of the aircraft's exterior.

The middleware for the Navigator 2 had a more complicated task than for the VuMan3. It must correlate an input selection with the correct location on the exterior of the aircraft. Each exterior segment consisted of a data object that contained not only the bit map but also location information for the segment. The segments were organized in a hierarchy so that the user could navigate easily to a leaf segment. The hierarchy pointers were also kept in the segment object.

Another function of the middleware for the Navigator 2 was to map between the coordinate space of the aircraft and that of the display. This enables a selection on the display to be positioned correctly on the aircraft for the reporting function. The middleware also controlled the displaying of the menus that enumerated types and the grouping of imperfections. A group is a collection of imperfections in the same general area of the exterior. All of this information was collected into an imperfection object that was stored in the database.

The user interface for Navigator 2 was constructed utilizing a commercial window package that was compatible with MSDOS. The decision to use MSDOS was made because (at the time) the only PCMCIA speech recognition system (TERI) had only MSDOS drivers. The window package provided only line and curve drawing facilities and a user interface widget package was constructed that provided the ability to specify a curve in terms of a collection of polygons and determine whether the current mouse position was within the curve. The same polygon recognition was used for the menu items and for determining the segment of the exterior to expand during the initial positioning. Thus, one of the elements of the segment description was the polygons used to describe that segment. The speech recognition software is also included in the user interface component. This software provided discrete word recognition for a small (less than 100 word) vocabulary.

C130 and Adtranz

From a software perspective, the C130 and Adtranz systems were essentially identical. The system characteristics were:

- distributed system using wireless, spread spectrum LAN
- 486 processor
- collaborative systems for inspection, troubleshooting, and repair
- programs and data entered over the network

With these two systems, the hardware provided sufficient resource so that a full functioned operating system (Windows 95) could be used on the wearable portion. A standard desktop system or laptop was used for the distributed portion of the system. The database and middleware components of the software were resident on the desktop or laptop. The software design was based on the WWW and the available servers and browsers. Special "plug-ins" were written in Java to provide collaborative drawing services. The database component consisted of a collection of databases, both newly created for the particular project and preexisting. The C130 project had databases that contained checklist information; the Adtranz project had databases that contained replacement procedures, electronic diagrams, and various textual databases.

In both cases, the middleware included a WWW server that provided interfaces to the databases and generation of HTML with the information from the databases included. The middleware also included the communication aspects of the collaboration software. The user interface component consisted of a standard WWW browser. Microsoft's Internet Explorer was used for the C130 project because it was easy to integrate the dial as an input device. Internet Explorer allows Tab and Shift-Tab as mechanisms for navigating through the links on a page. A clockwise rotation of the dial was mapped into the Tab and counterclockwise into the Shift-Tab and, consequently, the browser could be used without modification. Collaboration on the C130 was controlled by the person at the desktop since the dial is not a direct access device.

The Adtranz software used Netscape's Navigator. Collaboration was accomplished through custom written plug-ins. Since the hardware used for Adtranz had a pointing device, either of the collaborators could control the cursor in the collaboration. The specially written plug-in had a limited function, object-oriented, drawing package.

Effect of the Other Disciplines

As we saw from Figure 22.2, the software and the electronics must negotiate to determine the functionality available. The VuMan3 was an embedded system and, hence, the functionality of the system was quite limited. This was caused by the power requirements of a more powerful processor. The user interface of VuMan3 was a textual checklist. This was caused by the resolution of the display. The constraints on the user interface, in the VuMan3 case, were caused by the power requirements of the electronics.

Navigator 2, on the other hand, was not as constrained by the electronics. There was sufficient processor power, memory, and disk space to allow some choice in the operating system. The choice of operating system was constrained by the electronics available to support the use of speech. In this case, it was the user interface modality that influenced the electronics and which, in turn, constrained the software.

Adtranz and the C130 project were not constrained by the electronics. The software design was chosen primarily to enable the utilization of the collection of commercially available components, both for the display and for the communication. In this case, it was the software that constrained the electronics to be sufficiently powerful to support a full functioned operating system that, in turn, supported the WWW software.

9. SUMMARY

Wearable computers introduce new design problems in several areas:

- The means with which a user interacts with the computer system is fundamentally different from a desktop in a wearable context. We have seen several different types of interaction within the systems we have described.
- The environment of use of a wearable computer is also fundamentally different from a desktop. Temperature variations, foreign substances

in the work place, and the possibility that the users may be wearing gloves contribute to these differences.

- The software and the user interface can be both simplified and made more powerful because of the limitations on function that are possible if the wearable computer is viewed as supporting a limited portion of the maintenance process rather than the whole process.

- The industrial design and mechanical engineering features of the wearable system are fundamentally different because of the wearing of the computer. Heat dissipation becomes more difficult when a person is wearing a computer.

On the other hand, despite these design problems, it is clear that in the maintenance context, wearable computers provide a solution to the problems of dealing with environmental problems and task demands that cannot be provided by other types of computing platforms.

Acknowledgments   We would like to gratefully acknowledge the support received from the Defense Advanced Research Projects Agency under the supervision of Dick Urban as well as support from the Daimler-Benz Corporation.

REFERENCES

[Bass et al, 1997]. Bass, L., Kasabach, C., Martin, R., Siewiorek, D., Smailagic, A., and Stivoric, J., "The Design of a Wearable Computer," Proceedings of CHI97, Addison-Wesley, pp. 139–146.

[Martin, 1994]. Martin, T., "Evaluation and Reduction of Power Consumption in the Navigator Wearable Computer," Engineering Design Research Center Technical Report, Carnegie Mellon University, July 1994.

[Siewiorek, 1994]. Siewiorek, D. P., A Smailagic, J. C. Y. Lee, and A. R. A. Tabatabai, "Interdisciplinary Concurrent Design Methodology as Applied to the Navigator Wearable Computer System," Journal of Computer and Software Engineering, Vol. 2, No. 3, pp. 259–292, 1994.

[Smailagic and Siewiorek, 1996]. Smailagic, A., and D. P. Siewiorek, "Modalities of Interaction with CMU Wearable Computers," IEEE Personal Communications, Vol. 3, No. 1, pp. 14–25, February 1996.

[Finger, 1996]. Finger, S., M. Terk, F. Prinz, D. P. Siewiorek, A. Smailagic, J. Stivoric, and E. Subrahmanian, "Rapid Design and Manufacture of Wearable Computers," Communications of the ACM, Vol. 39, No. 2, February 1996.

[Smailagic, 1995]. Smailagic, A., D. P. Siewiorek, D. Anderson, C. Kasabach, T. Martin, and J. Stivoric, "Benchmarking an Interdisciplinary Concurrent Design Methodology for Electronic/Mechanical Systems," 32nd Design Automation Conference, pp. 514–519, 1995.

23

Applications of Wearable Computers and Augmented Reality to Manufacturing

Woodrow Barfield, Kevin Baird, John Shewchuk, and George Ioannou
Virginia Tech

1. INTRODUCTION

This chapter discusses applications of wearable computers and augmented reality to manufacturing. Specifically, the potential impact on manufacturing of an information processing technology that is mobile, always on, and always available to shop floor workers represents the theme of this chapter. The reader should note that this chapter does not present material on immersive virtual environments where observers are completely surrounded by computer simulations of manufacturing facilities or processes. Instead, we are interested in real-world facilities and processes, and the augmentation of these facilities with computer-generated information. Finally, while auditory systems and haptic feedback can form part of a wearable computer system, in this chapter the main focus is on wearable computers using visual displays.

In the manufacturing domain, information technologies are used to design and test parts, evaluate and control the flow of material between machining centers, and supervise system performance. However, even though there have been major advances in the design of software and computing

FIG. 23.1.   Example of maintenance worker in confined space using a wearable computer. (Picture courtesy of Carnegie Mellon University, http://www.cs.cmu.edu/afs/cs/project/vuman/www/home.html.)

resources used for CAD/CAM purposes, to access these resources the operator must travel to where the computing resources are located within the manufacturing environment. Having to access computing resources at fixed locations and often at fixed times, rather than being able to access them at any time and at any place can be: (1) a detriment to performance and (2) contrary to how operators actually need to access information on the shop floor. For example, in manufacturing environments jobs are often done in confined spaces where it may not be possible to use a desktop computer to access needed information (Figure 23.1). Furthermore, the task may require the use of both hands such that it may not be possible to use a keyboard or other standard input device to query a database. In addition, in manufacturing environments, operators often do work that requires them to move around the workspace or facility, requiring that time critical information is delivered at the right place and at the right time. The standard ways of accessing information for this case, either paper diagrams or instructions, or via a desktop computer, are useful in some situations, but they fail to allow the operator access to information at any time and at any place. For these reasons, many researchers are advocating wearable computers as a way to allow operators continual access to information in manufacturing environments.

As indicated above, the "wearable" aspect of the technology we discuss will allow users to access information at any time or any place within the manufacturing environment. Furthermore, such systems will allow objects within the world to be "augmented" with text, graphics, or sensor information. For example, using a see-through display, an operator may view material requirements for a particular part superimposed on a machining center or view a wire diagram superimposed in the environment where the information is relevant. The ability to augment objects within the world with computer-generated information is made possible with video-based or see-through head-mounted displays (HMDs). It is interesting to note that each type of visual display technology has costs and benefits associated with their use. For example, video represents a lower resolution image than possible when viewing the world directly with the eyes; however, video can be manipulated in ways that real-world images cannot.

2. WEARABLE COMPUTER SYSTEMS

Wearable computers are fully functional, self-powered, self-contained computers that are worn on (or carried by) the user's body (Barfield and Baird, 1998) (see Figure 23.2). With the development of wearable computers has come the evolution of wearable displays, input devices, peripherals, and

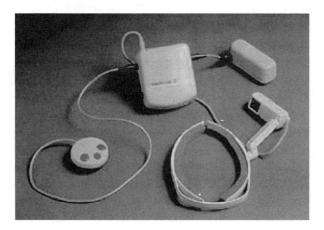

FIG. 23.2. VuMan system. (Image courtesy of Carnegie Mellon University.)

networking tools. The main components of a wearable computer are the computer unit and housing, the power supply, the output devices, and the input devices. The computer unit houses the CPU, motherboard, hard drive, memory, and all other components necessary to process and output data. The processing power and storage capability of a wearable computer is generally comparable to that of current laptop technology. The power supplies used by the wearable computer are usually lithium-ion based battery packs that are worn on the body and connected to the computer. Their charges can generally last for 2–4 hours before they need to be recharged. However, because the battery life before recharging is rather short, research into using alternative power supplies is underway, including using energy from the human body as a power supply, a technique that some day may become a viable method (Starner, 1996).

2.1   Augmented Reality

One of the main benefits associated with using a wearable computer is the ability to access information as the operator moves around the manufacturing environment. When a head-mounted display is used to project information in the environment, this capability is referred to as "augmented or merged reality" (Figure 23.3) (Feiner, MacIntyre, and Seligmann, 1993; Bajura, Fuchs, and Ohbuchi, 1992; Barfield, Rosenberg, and Lotens, 1995). Essentially this capability allows real objects to be augmented with information. Augmented reality environments can be created using desktop computers, or in the case of wearable computers, with a computer worn by the operator (allowing mobility within the manufacturing environment). Wearable computers thus extend the applicability of augmented reality to

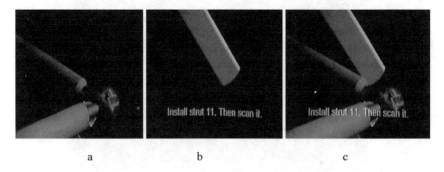

a   b   c

FIG. 23.3.   Real piece (a), virtual image (b), and augmented image (c). (Picture courtesy of Computer Graphics and User Interfaces Lab at Columbia University.)

situations where accessing information using a desktop computer is not feasible.

In order to generate augmented reality environments with sufficient fidelity and mobility for manufacturing applications, several issues need to be addressed. In terms of the presentation of visual information, displays need to be of high enough resolution and contrast to accurately and realistically present virtual objects as well as readable text under a variety of visual conditions. For example, in a manufacturing environment, workers perform tasks in very bright conditions. In this condition, the ability of the HMD to present virtual images of sufficient brightness compared to the ambient illumination is an important issue impacting operator performance. Current HMD technology falls short on this requirement—most commercial HMDs allow VGA resolution and medium fields of view in monochrome or color but with a limited range of brightness. However, as display technology continues to advance, better displays with the capability to present a broader range of brightness levels will become available.

The two major types of HMDs currently used with wearable computers for augmented reality are opaque HMDs and see-through HMDs (Azuma, 1997). Furthermore, see-through and opaque HMDs can present images to either one eye or to both eyes. If one eye is stimulated, the display is termed a "monocular" display; if the same image is shown to both eyes the display is termed "biocular"; and if slightly offset images are shown to each eye the display is termed "binocular." With monocular HMDs, the virtual and real-world images are often seen as one seamless environment, that is, the user fuses both images together as one (Starner, Mann, Rhodes, Levine, Healey, Kirsch, Picard, and Pentland, 1997). Further, video based see-through displays use CCD cameras mounted near the user's eyes to project video of the environment on the wearable display. The computer then merges (using luminance or chroma keying) the video image with the computer-generated image to create a video-based augmented-reality environment (Edwards, Rolland, and Keller, 1992). Both HMD types are commonly used for a variety of tasks and have different advantages and disadvantages associated with them (Holloway, 1997).

2.2 Input and Output Devices

In the manufacturing environment, the operator is often using one or both hands to perform a task; therefore, input devices that are used with wearable computers need to be selected with this requirement in mind. To meet this requirement, speech recognition is one input technique that is normally a component of commercially available wearable computer systems. Other input devices used with wearable computers include hand-held keyboards, wrist keyboards, or track pads. Each of these techniques allows the user to input data, while allowing at least one hand to be free for object manipulation. In addition, more advanced systems in the prototype stage include gesture based input, EEG, or EMG input; each technique allows the operator the ability to access information and interact with a computer, while also allowing the operator the ability to manipulate an object.

One of the latest fields of research in the area of output devices is tactual or force feedback displays (Tan and Pentland, 1997). These haptic devices allow the user to receive force feedback output from a variety of sources. This capability allows the user to actually feel virtual objects and manipulate them by touch. Furthermore, this is an emerging technology that will be instrumental in enhancing the realism of the augmented environment for manufacturing applications. Tactual displays have previously been used for scientific visualization in virtual environments by chemists and engineers to improve perception and understanding of force fields and of world models populated with impenetrable objects (Brooks et al., 1990). In addition, force feedback acts as a powerful addition to augmented reality simulations for problems that involve understanding of 3D structure, shape, or fit, such as in assembly tasks.

As indicated above, the input devices that have evolved for use with wearable computers are very diverse. Growth in the popularity of wearable computers has sparked an increasing amount of research in the design and evaluation of input devices (Thomas et al., 1997). For data entry or text input, body-mounted keyboards, speech recognition software, or hand-held keyboards are often used. Devices like IBM's Intellipoint, track balls, data gloves, and the "Twiddler" (a hand-held corded keyboard that allows for one-handed data input) are used to take the place of a mouse to move a cursor and select options or manipulate visual data. The common factors considered in the design of these input devices are that they all must be unobtrusive and they should allow at least one free hand for the manipulation of objects.

In addition to tactual displays, the use of audio displays that allow sound to be spatialized are being developed for wearable applications. Designers will soon be able to pair spatialized sound to virtual representations of parts or machining processes when appropriate to make the virtual experience even more realistic to the user. It is the successful combination of these various types of input/output devices that make augmented reality more realistic to the user, which allows better performance compared to conventional methods of interacting with data.

3. MANUFACTURING PROCESSES
AND WEARABLE COMPUTERS

The manufacturing facility of the future will be vastly different from that which we are used to seeing today. Along with new advances in manufacturing technology, there will be a corresponding need to allow people to interface with data and information associated with manufacturing processes and computer-aided design models. Some of these data may include information about the three-dimensional (3D) structure of parts, assembly instructions, or facility layout, or information about the flow of material through a manufacturing facility. In each of these cases, the visualization of information or processes and the need to manipulate data provide the common theme. The use of augmented reality and wearable computer technology for the visualization of information at a particular time and place within the manufacturing environment, as well as the direct manipulation of data associated with manufacturing processes, has tremendous potential for the manufacturing domain, from the initial design stages of a product or structure, to manufacture and postdelivery support.

Using augmented reality, the ability to visualize three-dimensional data or textual information in the environment, provides the user an intuitive means to interact with information, explore structures, parts, or data, in a way that has not been previously available. The ability to visualize and interact with an environment that has been augmented with virtual images could have a great impact on the way current manufacturing processes are accomplished and could provide solutions to the manufacturing problems of the future.

In this chapter, we present a few of the main areas where wearable computers can impact manufacturing. These areas, which manifest themselves in a variety of manufacturing activities (to be described in Section 3.1), are as follows:

- *Training*: Wearable computers will allow operators to be trained on-site as new techniques and information concerning processes and parts are developed and downloaded.
- *Knowledge enhancement*: Manufacturing is a data-rich environment in which operators must access information about many system variables at various locations within a facility. Wearable computers will allow information about machines, tools, parts, materials, processes, procedures, and system status to be accessed anywhere, at any time.

- *Simulation*: Wearable computers will allow operators to simulate processes and to visualize the results superimposed on the actual facility. Such a capability will allow engineering analysis results to be accessed when and where they are needed in the manufacturing facility.
- *Hands-free interaction*: Using wearable computers, workers will be able to project information in the form of text or graphics in their work environment while concurrently interacting with real-world equipment, tools, and parts.

Wearable computers and augmented reality can play a role in different types of manufacturing systems. In job shops, for example, processing and material handling activities are usually performed manually. The provision of the required information for such activities (e.g., information specifying where to move a part to) can be accomplished via augmented reality and wearable computers. Another example is flexible manufacturing systems (FMSs), collections of numerically controlled machines operating under hierarchical computer control and connected by an automated material handling system (Figure 23.4). As shown by the figure, there are numerous

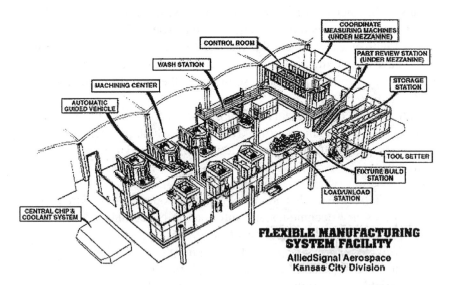

FIG. 23.4. Drawing of a flexible manufacturing system showing various stations and processes (http://www.os.kcp.com/home/map/fmsimage.html).

locations within the FMS where operators need to access information and thus the capabilities of a wearable computer. For example, in an FMS, training for operators would be extremely beneficial for tasks such as maintenance and repair of equipment. If such training were performed on-site, the operator would be able to respond to problems in an efficient manner and receive the most recent maintenance and repair updates by downloading information from a database. Knowledge enhancement would be especially beneficial to operators in an FMS. Directly projecting information at the appropriate site about quality control, or the availability of tools, material, and parts, would surely enhance operator performance. Also, the ability to perform a simulation showing, for example, the flow of material through the FMS would be especially useful in allowing the operator to form an accurate mental model of system functioning. Finally, with the appropriate input device, the operator could not only access information about the state of the FMS but also have both hands free to manipulate objects at the same time.

3.1 The Process of Manufacturing

Manufacturing is the process of converting input items (raw materials, purchased semifinished or finished goods) into finished products of greater value via a sequence of operations. Operations in which the item's characteristics are changed are called manufacturing operations: Such changes result from application of one or more manufacturing processes. *Fabrication* processes are those which change an item directly (e.g., machining), while *assembly* processes are those which change an item by combining it with other items. In order to ensure the items conform to specified quality requirements at each stage of processing, *inspection and testing* operations are performed. Finally, *material handling* operations are needed to move the items to the various pieces of equipment where manufacturing and inspection operations are performed.

For manufacturing to be performed, a product must first be designed. Together, we can consider product design, fabrication, assembly, inspection and testing, and material handling as the principle *activities* necessary for manufacturing to occur. A great many planning activities (process planning, production planning, scheduling, etc.) are of course also necessary in order to be able to determine *how* to use a manufacturing facility (collection and logical arrangement of equipment, tools, machines, and humans) for producing specified products in specified quantities. These activities, however, are in general performed in environments where traditional computational devices and data-presentation techniques can be easily utilized, and hence they offer less potential for application of augmented reality via wearable computers. Consequently, we will focus on the aforementioned five manufacturing activities in this chapter.

With the advent of new technologies, manufacturing activities are undergoing rapid change with respect to the way they are performed. The use of augmented reality and wearable computers is one of the emerging technologies that will change the way these activities are executed, by allowing the operator to access information when and where it is needed. Table 23.1 lists typical applications of wearable computers to manufacturing activities.

As previously specified, the first step in the manufacturing process is product design. Computer-aided design (CAD) models have traditionally

TABLE 23.1
Applications of Wearable Computers to Manufacturing Activities

Manufacturing Activity	Task	Application for Wearable Computers
Product Design	Using CAD tools for prototyping parts and assemblies.	View CAD models in 3D on the actual workspace in the environment.
Fabrication	Machining parts manually or supervising machining operations.	Provide instructions or diagrams to workers for machining parts, or supervising the machining processes.
Assembly	Joining parts together in the correct order and orientation.	Provide instructions and/or diagrams to workers for correctly matching and assembling the various parts.
Inspection and Testing	Visually inspecting items for defects.	Help inspectors locate where visual inspection is to be performed and/or where measurements are to be taken.
	Taking measurements to establish properties (dimensional, geometry, hardness, etc.).	Provide evaluation and disposition instructions.
Material Handling	Locating equipment in the facility to obtain a suitable material flow pattern.	Allow designers to place virtual machines on the shop floor and run simulations in real time, aiding in facility layout and redesign.
	Moving parts, tools, fixtures, etc. from one location to another.	Provide location and handling information (e.g., fragile item) to workers.

been very useful in the design of all types of manufactured products. The ability to use a computer to generate 3D representations of parts and then assemble them together in a computer environment has saved companies sizable amounts of time and money. Instead of incurring the time and cost of fabricating a part and then fitting the part to another real part, designers can quickly and inexpensively prototype parts and fit them together without ever actually producing the items. This saves time and scrap in the redesign stage, and it minimizes losses if the parts are not successful and/or do not go into full-scale production.

Although the use of CAD tools in manufacturing has become widely used and very successful, there is still the lack of true realism using current techniques. While computers can render CAD images in 3D, they are still projected on a 2D monitor and are only able to be manipulated in two dimensions on the screen with traditional input devices. Being able to project a 3D CAD image of a mating part on an actual product would greatly increase the designer's understanding of how the part would fit, look, and act during assembly. The information the designers would receive and the realism they would experience would be far greater than anything available with current CAD tools. To realize the potential of wearable augmented reality environments for the study of CAD product models, appropriate data interfaces between CAD packages and augmented reality environments are necessary. The development of these interfaces will allow the complete and accurate representation of product data in three-dimensional space and will serve as the means for integrating the associated development and analysis tools.

One critical step for achieving the high productivity, quality, and flexibility demanded of modern manufacturing enterprises is to ensure that the flow of information is integrated across the various manufacturing activities. This requires the development and adoption of standards for representing various manufacturing objects, such as parts. These standards will allow neutral interfaces to support system development efforts. Standards have been developed in the CAD domain, particularly for mechanical parts. Examples include the Initial Graphical Exchange Specification (IGES) and the Standard for The Exchange of Product (STEP) model data. The focus of the development in this field should be the development of an integrated object-oriented product model (application protocol) for the viewing of three-dimensional parts (i.e., viewing the virtual representation of a part) or final assemblies in augmented environments. This model could be based on widely accepted product data representation standards.

Once the product design activity is completed, fabrication operations are performed to change the dimensions, geometry, physical, chemical, and/or mechanical properties of items by application of force or energy. Categories of fabrication processes include machining, bulk deformation (rolling, forging, extrusion, drawing), casting, sheet metal forming, nontraditional machining (water-jet cutting, electric-discharge machining, etc.), heat treatment, surface coating, and cleaning (Kalpakjian, 1995). To perform fabrication operations, workers may be required to either produce parts by hand (e.g., sheet metal bending, grinding, drilling processes) or supervise and monitor fabrication machinery. In the latter case, workers are often required to add materials to the machinery at appropriate times and/or monitor the process to ensure it remains in control. In both cases, augmented reality and wearable computing could aid workers performing these tasks.

For workers performing manual fabrication operations, augmented reality could be used to provide textual or auditory instructions for machining the part, or a wire-frame overlay could be superimposed on the part letting the worker know how far to bend a sheet of metal, where to drill a hole, etc. Augmenting the actual workpiece allows the task to be completed more quickly and accurately than if the worker had to continuously refer to a paper diagram or mark off drilling points prior to performing the task. For fabrication tasks requiring an operator to supervise machinery or add quantities of material to the process at intervals, augmented reality can be used as an intuitive interface between the operator and the machining process. With the appropriate information connection between the machine and operator's wearable computer, the machine can prompt the operator when to add certain materials, what stage the process is in, and if any errors or problems are occurring and how to fix them. Currently, in order to perform these tasks, operators must memorize or estimate when and how much material should be added, continuously monitor displays for errors, and know how to fix errors once they occur.

Once parts are fabricated, they must be assembled into final products via assembly operations. Categories of assembly processes include bonding, mechanical fastening, and welding (Boothroyd and Alting, 1992). One of the crucial prerequisites for the successful production of complex products in discrete parts manufacturing facilities is the correct and efficient mating of components for final assembly. The relative placement and interaction between such components and subassemblies is determined both by the functionality of the design, cost, and spatial constraints. On the other hand, the drive for compact designs and increased functions of modern products restricts the feasible solution space, causing inconsistencies and errors (Groover, 1996).

Augmented reality technology offers the means for identifying such inefficiencies early in the design cycle by using virtual representations of complex product assemblies that can be manipulated. Allowing the designer to fit and view a number of virtual parts on an actual workpiece would give them a better understanding of how the final assembly would look and fit before the real part is fastened, perhaps permanently. This would allow timely alterations and optimization of designs and would proactively support effective production on the shop floor. Assembly visualization via augmented reality could be used in several applications. In automobile manufacturing, for example, the same chassis is employed by several product models, which incorporate very different additional components.

In addition to allowing assembly tasks to be previewed, augmented reality could also aid in assembly itself. Complex, multistep assembly and maintenance operations generally require the operators to alternate their attention between the assembly instructions and part being assembled. This constant change of attention consumes valuable time, especially when the instructions are not conveniently placed relative to the operator. An example of this is a mechanic working on an engine in tight quarters where there is no room for written instructions. The constant change of attention is also a frequent cause of repetitive motion and strain injuries. These injuries are a source of lost productivity when the operator slows down or requires sick leave and is replaced with less-efficient temporary personnel. Augmented reality technologies can provide annotated instructions in real time, superimposed on the operator's field of view. Wearable computers and augmented reality could be implemented to aid workers in assembly by displaying instructions and diagrams to the workers. Users could look at the piece and see a diagram, text, or hear instructions telling them what to do next and how. As with fabrication operations, these augmented instructions, displayed in the worker's environment near or on the workspace, would enhance performance and allow workers to be more flexible in the types of tasks they could do. The operator could then concentrate on the task at hand without having to change positions to receive the next set of instructions. The work being done at Boeing (Section 4.1) is a good example of the technology being used in this area. Workers have been able to successfully use wearable computers and augmented reality to aid in aircraft part assembly as well as wire board construction.

Often after fabrication and assembly, *inspection* and *testing* operations are required to ensure that items are manufactured to the desired specifications and will function as intended (Groover, 1996). Though such operations do not add value to the item, they are still considered part of the manufacturing process. Industrial inspection (using human inspectors) is still widely used to assure proper levels of quality in manufacturing (Kleiner and Drury, 1993). Augmented-reality-based inspection is an area that has great potential in helping inspectors do their job. When performing flaw detection (a type of inspection task), workers have to check items for various types of defects at various locations. Traditionally, they have used inspection diagrams and written instructions to determine what types of defects to look for and where. This typically includes looking up the first type of defect and figuring out where on the piece it would be, followed by doing the inspection, and then recording if the defect is there or not on a log sheet. Augmented reality systems can improve the accuracy of the human inspector and the quality of the production system by superimposing in-spection aids on the product as the inspector moves around the item. Using a wearable computer, the inspector could start the task, have the computer display the first defect to look for, and then by holding the piece over a wire frame or by tracking the piece, the augmented reality system could highlight where the defect occurred. Once the first inspection is done, the inspector could use voice input to indicate whether or not the defect was there, and then instruct the augmented reality system to move to the next inspection task. This would eliminate the need for the inspector to move back and forth between tasks and would keep a computer-based log of inspections. The use of wearable computers and augmented reality for this type of job is already beginning to be seen on the shop floor. One example is the use of this technology for quality assurance in a food processing plant (Section 4.3).

The final activity required for manufacturing is material handling. Ma-terial handling refers to the physical transportation of raw materials, work-in-process, and finished goods between the manufacturing resources, as well as to the transfer of raw materials from storage to production areas, or of finished goods from the production shop to warehouses (Allegri, 1984). Consider a conveyor-based material handling system. In such systems op-erators need to access information at multiple stations. The acquisition of modern material handling equipment requires heavy investment and cap-ital commitment from manufacturing enterprises. In most cases, the new system replaces manual and outdated equipment. The visualization of the effect of implementing the changes in the system design to accommodate modern automated material handling designs is critical for the successful transformation of the shop floor operations.

4. RECENT APPLICATIONS OF WEARABLE COMPUTERS IN MANUFACTURING

The utility of wearable computers and augmented reality are already beginning to be seen by the manufacturing industry. Companies are starting to implement these powerful tools on their shop floors, and more companies are beginning to look into the technology. There are several current cases in industry where augmented reality and wearable computers are being used to increase productivity and help workers perform manufacturing tasks. Some of these examples are as follows.

4.1 Aircraft Assembly

Engineers at Boeing have implemented wearable computers and augmented reality to aid workers in the assembly of airplanes (Caudell and Mizell, 1992). Boeing's augmented reality project was designed to display instructions and diagrams supporting assembly directly in the environment where the assembly task is being performed. Specifically, Boeing employs wearable to render wire frame diagrams or text instructions at arm's length in the user's work environment (see Chapter 14 by Mizell). Since the computer is not required to render complex images, wearable computers can present images of sufficient quality for the task. One of the main challenges associated with using augmented reality and a wearable for this application is registering the user's position relative to the workpiece so that the diagram stays put when the user moves his head. To solve this problem, Boeing engineers are working on a real-time videometric tracker.

4.2 Manual Inspection

Inspection is a task that can benefit substantially from the use of wearable computers and augmented reality technology. Inspection is performed on incoming raw materials, when parts are moved from one production department to another, at the completion of processing, and before shipping the final assembled product to the customer. In each of these cases, the operator can benefit from a mobile inspection platform capable of providing information at the right place and at the right time.

As an example of the use of wearable computers for inspection, researchers at Carnegie Mellon University (CMU) have developed the Navigator 2 wearable computer system (http://www.cs.cmu.edu/afs/cs/project/

vuman/www/home.html) (see Smailagic and Siewiorek, 1996 for related work). In this case the inspector's task is to examine the "skin" of an aircraft for cracks and corrosion. Other types of maintenance and repair tasks in manufacturing include inspecting for tolerances, appearance, performance, and reliability. The task investigated by the CMU researchers requires extensive physical activity and travel (in the context of the airplane fuselage) by the inspector, thus explaining its relevance for wearable computer technology. The primary input device used by the inspector is a joystick that allows input in the X and Y directions. In addition, as noted by the CMU team, the joystick is used in conjunction with speech to mark discrepancies. Field evaluations for aircraft inspection have indicated a large savings in inspection time and reduced inspection data entry time during the inspection task.

4.3   Quality Assurance

Recently, engineers at the Georgia Tech Research Institute were asked to improve the performance of quality assurance inspectors in a food processing plant (Najjar et al. 1997). For this task, they decided to use a wearable, voice-operated computer with a wireless network. The quality assurance workers were required to walk around a very large plant and take food samples at various processing points. To perform the quality assurance task, the workers selected samples, evaluated them, and then used a pen and a clipboard to record their results. At the end of their shift, the inspectors were required to submit reports, which were then typed into a networked computer to be read later by a member of the management team. To improve the performance of this operation, the inspection team used a wearable computer with voice recognition input and a monocular HMD connected to a wireless network. With this system, the inspector is able to go to a station, select a sample, and input inspection data about the sample into a computer database. In addition, the inspector can add verbal comments into the networked wearable computer using speech input. The monocular display also shows the inspector what's been entered and what measurements have been taken. Once the inspection task is finished, the data are transmitted to a central computer over the wireless network.

4.4   Construction

The Computer Graphics and User Interfaces Lab at Columbia University is working on a project using augmented reality for construction tasks (Feiner, MacIntyre, and Seligman, 1993). They use an optical see-through HMD system to display instructions and assembly information to the user performing a "space frame" construction task. The space frames are composed of cylindrical shapes and spherical nodes of similar sizes and shapes. Due to the similarities in size and shape between parts, it is possible for the operator to assemble the components incorrectly, which could lead to structural breakdown. The purpose of the augmented reality system is twofold: (1) to guide workers through the assembly of the space frame and (2) to assist the workers in inspecting the space frame to make sure the proper parts are assembled in the proper locations. The system works by first directing the worker to a collection of parts and informing the worker which parts to select by displaying text instructions and playing a sound file with verbal instructions. The worker then scans the part with a bar code reader to ensure that the correct part was selected. Following this, the system directs the worker to install the part by showing a 3D virtual image of where it must go on the structure, along with audio instructions on how to install it. Finally, the worker scans the component with the tracked bar code to ensure that it was installed correctly and in the right place.

5. CONCLUSIONS

Many payoffs are expected to occur as a result of applying wearable computer and augmented reality technologies to manufacturing. The use of wearable computers and augmented reality technologies, coupled with the ability to use dynamic body/head positioning over large sites within a manufacturing facility, will allow computer-generated images to be superimposed over real-world images at that site. Such a capability will fundamentally change the way we think about delivering information in these environments. For example, plans could be provided and specifications overlaid on top of items, eliminating the need for hard copies and the ensuing lack of understanding by the user. Engineers will be able to employ wearable computers and augmented reality to improve individual worker performance and quality, as opposed to using computers primarily to process information between management and workers. An example of this would be a virtual simulation done on-site to provide manufacturing system designers the ability to model and analyze complex manufacturing facilities and their operation. This in turn would allow them to utilize a level of detail for design that has not previously been possible. An order-of-magnitude increase in robustness for shop floor control strategies is another example of the possible implications of the use of wearable computer technology in the manufacturing environment. Finally, as intelligence is implemented into manufacturing environments in the form of powerful microprocessors and intelligent software agents, operators using wearable computers will be able to access the appropriate intelligent help and assistance to improve not only their performance but overall system performance as well.

Acknowledgement   Woodrow Barfield gratefully acknowledges ONR (N000149710388) for providing the funds to purchase wearable computer, haptic, and virtual environment equipment.

REFERENCES

Allegri, T. M. Sr., *Material Handling: Principles and Practices*, Van Nostrand Reinold, New York, 1984.

Azuma, R. T. (1997). A Survey of Augmented Reality. *Presence: Teleoperators and Virtual Environments, Vol. 6*, pp. 355–385.

Bajura, M., Fuchs, H., and Ohbuchi, R. (1992). Merging Virtual Objects with the Real World: Seeing Ultrasound Imagery within the Patient. *Computer Graphics, Vol. 26, No. 2*, pp. 203–210.

Barfield, W. and Baird, K., Future Directions in Virtual Reality: Augmented Environments Through Wearable Computers, *VR'98 Seminar and Workshop on Virtual Reality*, Kuala Lumpur, April 24–15, 1998.

Barfield, W., Rosenberg, C., and Lotens, W., Augmented-Reality Displays, in Barfield, W., and Furness, T. (editors), *Virtual Environments and Advanced Interface Design*, Oxford University Press, 1995.

Boothroyd, G. and Alting, L. (1992). Design for Assembly and Disassembly. Keynote Paper, *Annals of CIRP, 41, 2*, 625–636.

Brooks, F. P., Jr., M. Ouh-Young, J. J. Batter, and P. J. Kilpatrick, Project GROPE: Haptic Displays for Scientific Visualization, *Computer Graphics: Proc. SIGGRAPH '90*, August 1990, 177–185, Dallas, TX.

Caudell, T. and Mizell, D., Augmented Reality: An Application of Heads-Up Display Technology to Manual Manufacturing Processes, in *Proceedings of the Hawaii Int. Conf. on Sys. Sci.*, Hawaii, January, 1992.

Edwards, E. K., Rolland, J. P., and Keller, K. P. (1992). Video See-Through Design for Merging of Real and Virtual Environments, *Proceedings IEEE Virtual Reality Annual International Symposium*, Seattle WA, Sept. 18–22, pp. 223–233.

Feiner, S., MacIntyre, B., and Seligmann, D. (1993). Knowledge Based Augmented Reality, *Communications of the ACM, Vol. 36, No. 7*, pp. 53–62.

Groover, M. P., *Fundamentals of Modern Manufacturing: Materials, Processes, and Systems*, Prentice-Hall, Englewood Cliffs, NJ, 1996.

Holloway, R. L. (1997). Registation Error Analysis for Augmented Reality, *Presence: Teleoperators and Virtual Environments. Vol. 6*, pp. 413–432.

Kalpakjian, S., *Manufacturing Engineering and Technology*, 3rd ed., Addison-Wesley, 1995.

Kleiner, B. M. and Drury, C. G. (1993). Design and Evaluation of an Inspection Training Program. *Applied Ergonomics, Vol. 24, No. 2*, pp. 75–82.

Najjar, L. J., Thompson, J. C., and Ockerman, J. J. (1997). A Wearable Computer for Quality Assurance Inspectors in a Food Processing Plant, *Proceedings of the First International Conference on Wearable Computers*, Cambridge, MA, pp. 163–164.

Smailagic, A. and Siewiorek, D. P. (1996). Modalities of Interaction with CMU Wearable Computers, *IEEE Personal Communications, Vol. 3, No. 1*, pp. 14–25.

Starner, T. (1996). Human Powered Wearable Computing, *IBM Systems Journal, Vol. 35 (3)*.

Starner, T., Mann, S., Rhodes, B., Levine, J., Healey, J., Kirsch, D., Picard, R., and Pentland, A. (1997). Augmented Reality Through Wearable Computing. *Presence 6(4)*, pp. 384–398.

Tan, H. Z., and Pentland, A., Tactual Displays for Wearable Computing, *IEEE First International Symposium on Wearable Computers*, Oct. 13–14, Cambridge, MA, pp. 84–89, 1997.

Thomas, B., Tyerman, S., and Grimmer, K., Evaluation of Three Input Mechanisms for Wearable Computers, *IEEE First International Symposium on Wearable Computers*, Oct. 13–14, Cambridge, MA, pp. 2–9, 1997.

ADDITIONAL REFERENCES

Baran, Nick. Get Smart-Wear a PC. *BYTE Magazine*. March 1996, http://www.byte.com/art/9603/sec4/art10.html.

The DeVry Student Chapter I.E.E.E. Official Wearable Computer Homepage. June 12, 1998, http://www.devrycols.edu/ieee/wearable.html.

Intelligent Machines Branch, GTRI. Electronic Performance Support System. Multimedia Information in Mobile Environments Laboratory, May 19, 1998, http://mime1.gtri.gatech.edu/mime/epss/default.html.

Jastrzembski, Mike. CMU Wearable Computers, Carnegie Mellon University. July 19, 1997, http://www.cs.cmu.edu/afs/cs/project/vuman/www/general.html.

Mann, Steve. "Smart" Clothing. MIT Media Lab, Feb 14, 1996, http://www.wearcam.org/smart_clothing/smart_clothing.html.

Oeler, Kurt. Now, computers are wearable. CNET News.com. October 2, 1997, http://www.news.com/News/Item/0%2C4%2C14850%2C00.html.

Sibley, Kathleen. Computers face a ready-to-wear future. *Computer Dealer News (CDN)*. December 15, 1997, http://www.plesman.com/archive/cdn/97YCD16.html.

Shemeligian, Bob. Dress for success with wearable computers. *Las Vegas Sun*. November 19, 1997, http://www.lasvegassun.com/sunbin/stories/text/1997/nov/19/506514146.html.

Wearable Audio Computing, http://www.pcs.ellemtel.net/claes/WAC.html.

24

Computer Networks for Wearable Computing

Rick LaRowe and Chip Elliott
GTE Internetworking–BBN Technologies

INTRODUCTION

Networking is as important to wearable computing as it is to traditional computing. In fact, the desire to minimize power consumption and the size and weight of wearable computers only amplifies the need for access to external information sources. Some wearable applications require access to large databases, perhaps storing design schematics for an aircraft being repaired. Other users, such as package delivery workers, are involved in data collection activities, where the rapid transfer of collected information back to headquarters enhances business objectives. Similar application requirements might be found in the growing home health care market, where remote medical monitoring is expected to become more common. And of course, wearable computer users "on the go" demand access to their everyday electronic communications resources, such as email and access to the WWW.

Networks for wearable computing can be divided into two broad classes:

1. *Off-Body*—these are networks that connect the wearable computer to other systems not worn or carried by the wearable computer user.

715

2. *On-* or *Near-Body*—these are networks that connect the computers, peripherals, sensors, and other devices that comprise a single user's wearable computer system.

Connecting a wearable computer to an *off-body* network demands a solution different from that used for connecting desktop machines. The wearable user cannot be tethered with an Ethernet or telephone/modem cable and still maintain all the benefits expected of a wearable computing system. Wireless solutions are needed to support the freedom of movement in the environment that motivates wearable computing. Other factors (minimum bandwidth needs, operational or mobility range, tolerable infrastructure and usage costs, and the usual wearable constraints of power, size, and weight) tend to vary with the specific application.

On-body or *Near-Body* networks are also important to wearable computing. These networks tie together the computers, peripherals, sensors, and other devices that are carried/worn on or relatively near the body, and they contribute significantly to the ergonomics of the wearable system.

An on-body network can be thought of as a new level in the traditional WAN-MAN-LAN network hierarchy: the *Personal Area Network*, or PAN. The concept of a PAN is more general than wearable computing but is especially important to wearable systems because PANs enable the use of devices (CPU and peripherals) that are as small as possible and conveniently placed, worn, or carried on the body.

Figure 24.1 depicts a military use of wearable computing with both on- and off-body wireless networking. The Wireless LAN (an off-body network) is used for communication amongst the squad, to support the exchange of intelligence and status information. (An ad hoc wireless network is formed among squad members.) The Wireless PANs are used as an on-body (or near-body) network to support communication among the devices (computer, display, inertial tracking unit, etc.) carried by each individual soldier. The cloud surrounding the solider in the forefront of the figure represents the PAN (on-body) network coupling the devices that comprise that individual soldier's wearable system. The cloud in the upper-right portion of the system represents the off-body wireless LAN that networks the individual soldiers together so that they may communicate.

In this chapter, we review the technology requirements and alternatives for both on- and off-body communications for wearable computing systems.

FIG. 24.1. Wearable computing example with both on- and off-body networking.

OFF-BODY COMMUNICATION OPTIONS

Advances in mobile personal communications and wireless LAN technology are the primary technology sources for off-body networking in a wearable computing system. The mobile personal communications industry brings forward analog and digital cellular telephony, the paging infrastructure, and commercial wide area messaging services. Wireless LAN technology can also be exploited for off-body communications in many environments, depending upon the needs of the wearable application.

The current state-of-the-art does not present a single ideal off-body communications solution for wearable computing. Options differ in respects that demand careful consideration to system level requirements. Data rates, availability, operational range, infrastructure requirements, cost/pricing plans, and other issues must be considered when selecting from the technology options available today.

The future will offer improved off-body communications for the wearable computing user. In particular, we anticipate growth in the use of *tiered* wireless networks combining integrated voice, data, and other multimedia services. Tiered networks weave different types of radio technology—each suitable for short, medium, or long-haul communication—into a single, seamless communications medium. These networks, discussed later in this chapter, will prove ideal for satisfying the needs of many augmented reality applications.

Technology Selection Considerations

Starting with an applications-level needs-based understanding of the off-body communications requirement is the best way to evaluate and contrast the alternative technologies.

Mobility/Coverage Requirement

By their very nature, wearable computing applications tend to have some mobility requirement. However, the degree of mobility can vary significantly between application domains. A warehouse worker need only maintain off-body network connectivity within the warehouse. A delivery person must maintain connectivity within a metropolitan area. And a traveling field service technician might require nationwide connectivity.

The connectivity issue may extend beyond the needs of a single user, taking into account corporate-level standardization and technology integration. For example, while a delivery company serving only Washington, DC might chose the limited coverage Metricom network for its off-body communications, a larger company employing delivery workers across the country would probably desire an alternative solution with broader coverage (e.g., Ardis or CDPD).

Infrastructure Investment Considerations

Some off-body networking solutions require an investment in fixed infrastructure to support communications. Use of IEEE 802.11 wireless LAN technology on a factory floor, for example, would require the installation of the base stations within the factory. A large company might deploy a private nationwide radio network for off-body communications using one of the licensed FCC bands. Federal Express, for example, owns a large nationwide private radio network.

Data Rate/Bandwidth Requirements

Different applications of wearable computing have different requirements with respect to bandwidth. The aircraft service technician attempting to interactively access design schematics or even short animations and video clips from a remote database will require a speedy connection (but this communication need is primarily unidirectional, to the mobile user). A delivery worker periodically transmitting a few-hundred-byte transaction record can suffice with much lower data rates. Other users may have intermediate data requirements, perhaps wanting to browse the Internet in order to access information stored at a corporate web site. Still other applications will demand high data rates from the mobile user to the remote site; an example might involve the transmission of imagery taken from a body-worn camera.

As a rough guide, we have listed (in Table 24.1) the *nominal bandwidth* for each of the major off-body technologies. It is extremely important to realize that these bandwidths are (very optimistic) upper bounds on the actual end-user throughput. In practice, the actual throughput will be much less than the advertised value. For instance, a Push-to-Talk radio with a nominal bandwidth of 9600 bps (bits per second) may deliver something closer to 600 bps in practice. The discrepancy between the nominal and actual values are caused by a wide range of issues including the failure to account for FEC (forward error correction) and control overhead in the nominal figure, the need for retransmissions, channel access problems, and the interactions between IP (Internet Protocol) and the link-layer protocols. These mechanisms are quite complex and technology-specific; curious readers may find Tisal (1997) interesting for its detailed description of the mechanisms employed in one popular cell-phone technology (GSM).

As a final aside, the nominal bandwidths for many technologies are quickly increasing as technology advances and as infrastructure providers upgrade their equipment.

Radio Ranges

Different types of radios can operate over widely differing ranges. Typical ranges run from about 1 meter for body-area radios, to roughly 100 meters for wireless LANs, to tens of kilometers for cellular systems, trunked radio systems, and paging systems, up to hundreds or thousands of kilometers for LEO or GEO satellite systems.

TABLE 24.1
Attributes of Available Off-Body Networking Options[a]

Techno-logy	Variant	Two-Way?	Max. Nominal Bandwidth (at time of writing)	Shared or Dedicated Channel	IP Capable?	How Secure?	Est. Battery Life	Roaming Area
Wireless LAN	802.11 Radio	Yes	2 Mbps	Shared	Yes	Fair	Hours	In building
	Proprietary Radio / IR	Yes	10 Mbps	Shared	Yes	Good	Hours	In room
Cordless Telephony	Fixed base station	Yes	Voice Only - No data	Dedicated	No	None to Poor	Hours	In building
	PHS, DECT	Yes	32 kbps	Dedicated	Yes	Very Good	Hours	Metro+
Cellular Telephony and New SMR	AMPS	Yes	14.4 kbps	Dedicated	Yes	None	Hours	Metro+
	CDPD	Yes	19.2 kbps	Shared	Yes	Very Good	Hours	Metro+
	GSM	Yes	9.6 kbps	Dedicated	Yes	Very Good	Hours	Metro to Continent
	GSM Short Messaging	No	? 160 char. messages	Shared	No	Very Good	Hours	Metro+
	CDMA	Yes	14.4 kbps	Dedicated	Yes	Good	Hours	Metro+
	Motorola iDEN	Yes	9.6 kbps	Dedicated	Yes	Fair	Hours	Country
Paging	Tone or Numeric	No	1200 bps	Shared	No	None	Months	Country
	Alphanumeric	No	2400 bps	Shared	No	None	Months	Country
	Two-Way	Yes	6400 bps	Shared	No	None	Hours	Metro+
Trunked Radio Networks	Typical Push-to-Talk	Yes	9600 bps	Shared	Seldom	None	Hours	Metro
	RAM	Yes	8000 bps	Shared	Optional	Poor	Hours	Metro
	Ardis	Yes	19.2 kbps	Shared	No	Poor	Hours	Metro
Wireless MAN	Metricom	Yes	28 kbps	Appears Dedicated	Yes	Very Good	Hours	Metro
LEO Satellite	Iridium	Yes	2400 bps	Dedicated	Yes	Not Public	Hours	World
	Teledesic	Yes	64 Mbps	Dedicated	Yes	Very Good	Not Feasible	World
GEO Satellite	DirectPC	No	12 Mbps	Shared	Yes	Not Public	Not Feasible	Continent
	VSAT	Yes	2 Mbs	Shared	Yes	Varies	Not Feasible	Country or World

[a] Shaded options are not really viable for wearable computing applications.

Shared/Dedicated Channel

Radio channels are inherently shared at the physical level, in that multiple transmitters within range of each other will interfere with each other's transmissions. At the link level, however, a clear distinction emerges between systems that allow multiple transmitters to share a channel (i.e., a frequency, spreading code, etc.) on a per-frame basis and those that allocate the entire channel to a given transmitter for the duration of a session. The first type resembles an Ethernet; the second resembles a telephone circuit.

Each approach has strengths and weaknesses and must be evaluated in light of the desired application. The shared versus dedicated channel distinction must, however, be kept clearly in mind. For instance, CDPD cellular data service at 19.2 kbps (kilobits per second) nominal throughput appears faster than circuit-switched GSM cellular data service at 9.6 kbps. However the CDPD bandwidth must be shared among all the users in a given sector. With ten active users in the sector, the bandwidth available to any one of them would be no more than 1920 bps, and very likely less due to contention. The GSM service is not shared, and hence it is not affected by the number of users in the sector; however, if there are too many users, no new users will be allowed to enter the network.

IP Architecture

Figure 24.2 illustrates the three basic options for integrating an on-body network into the global Internet (or a private IP intranet). A number of other approaches are also possible, but most are essentially just variations

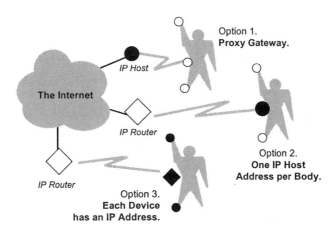

FIG. 24.2. Integration of PAN and global Internet options.

or combinations of these options. Each option has good and bad points, which are briefly discussed in turn.

1.   *Option 1:* No device on the body has an IP address. All communications across the body, and across the off-body link as well, are carried in a non-IP format. These are translated to IP at a Proxy Gateway computer, which then acts as the application-level go-between that glues the body network into the Internet. This approach has the benefits of being simple, requiring little in the way of IP address administration, avoiding the expense of sending IP headers across scarce radio resources, and avoiding the power and computational resources needed to run an IP stack on the body. It has the drawback of requiring a specialized proxy gateway, which tends to act as a bottleneck to the introduction of new features. All in all, however, this option is the most commonly adopted; examples include most radio messaging applications and the Unwired Planet browser model.

2.   *Option 2:* Each body is assigned a single IP host address. Here we have shown this address as belonging to the body "hub" that contains an off-body radio. IP datagrams can flow directly between the Internet and the body hub over the off-body network; but communications across the body network are not IP. This "middle way" has the advantage of eliminating the proxy gateway (or rather of merging its functionality into the body hub) in favor of providing a full-fledged Internet connection to the body, but still not requiring every device to be able to run an IP stack. It retains the basic drawbacks of a proxy gateway, however.

3.   *Option 3:* Every wearable device is assigned its own IP host address, and IP datagrams can flow directly between any body device and the Internet. In general this solution requires the body network to have a distinct IP network address, and the body "hub" to have at least minimal IP router functionality. This has the advantage of fully integrating the body network, and its devices, into the global Internet. The prime disadvantages, though, are that every device must have enough power and computational resources to be able to run an IP stack, that IP headers must be conveyed across perhaps thin radio links, and that it requires a fairly large number of IP addresses.

As noted earlier, variations and combinations of these options are also feasible. In particular, it is fairly easy to imagine systems combining options two and three, where the more capable devices in the PAN run IP and the simpler devices (e.g., personal peripherals or sensors) run a lightweight non-IP protocol. However, in general, the system developer does not get a free choice among these options. Many current off-body network technologies either cannot efficiently carry IP datagrams (e.g., paging networks) or implicitly assume that the device accessing this network is a host computer, rather than a router (e.g., CDPD).

How Secure?

"Security" is a word with many meanings. To a service provider, fraud prevention is perhaps the most important aspect of security; hence modern telephone systems include relatively elaborate authentication mechanisms to cut down on cloning and theft of services. In the context of off-body networking for wearable computing, however, privacy is likely to be the primary concern.

Unfortunately many of available technologies provide little or no assurance of privacy at the link level. The worst offenders are the paging technologies, push-to-talk, and the old (current) AMPS cellular telephone technology. Here one must assume that anyone nearby can intercept all communications and (if they desire) inject spurious traffic. The other available technologies are adequate for most purposes, though the exact details vary as to what features they provide. If not, privacy can be supplied via higher-level protocols such as IPSEC (Kent, 1997).

Finally, commercial radios—although perhaps meeting commercial needs for security, depending on the application—do not meet the unique battlefield needs for security such as Low Probability of Intercept (LPI) or Low Probability of Detection (LPD), nor do they have any good defense against jamming. Hence most tactical applications of off-body communication will need to employ special military radios.

Battery Life

Table 24.1 includes an order-of-magnitude estimate of battery life for each technology. These estimates can serve only as very rough guides, however, for several reasons:

- Traffic patterns drive power usage. Every well-engineered radio system supplies a "sleep" mode that draws very little power when not actually transmitting or receiving. Hence the battery life depends strongly on the transmit–receive duty cycle. Our rough estimates assume continuous activity evenly split between transmit and receive, except for one-way pagers, which assume a light, receive-only duty cycle.

- For many technologies, actual power requirements depend strongly on the path loss between the wearable unit and its base station.
- Standard technologies are not designed for wearable networking and hence are less battery-efficient than could be achieved in custom solutions. As a concrete example, the current "best" battery for an IBM ThinkPad computer provides 45 watt-hours. Assuming a radio modem that requires only 250 mW (for near communications), one should be able to achieve a radio subsystem that can run nearly 180 hours, or 22.5 work days, between recharges. Off-the-shelf solutions can typically provide only a few hours at most.

Roaming Area

It is extremely difficult to quantify the transmission range of a given off-body communication system. Any given technology's range will vary widely depending on a number of factors such as the antenna position (e.g., whether the wearer is kneeling or lying down), wall materials, nearby interference, and so forth. In addition, many wireless technologies allow a mobile unit to roam from one base station to another while maintaining full communications. This can extend the range throughout an entire building for wireless LANs, across an entire continental range for GSM telephony, or indeed across most or all of the world with LEO satellite systems.

Costs in Dollars, Power, Size, and Weight

Every off-body solution has costs associated with it, not only in terms of the monetary costs of the access device and the service (if any), but also the power consumption and the physical size and weight of the device. The effect of these issues on their respective system budgets can be a critical factor in the selection of an off-body networking solution. It is difficult to give any general guidance on this such issues, since they are typically dominated by the details of a given product's requirements.

Current Off-Body Communications Solutions

Currently available off-body communications solutions for wearable computing fall into five classes: Wireless LAN based technology, cellular telephone based technology, pager based technology, specialty commercial wireless providers, and custom networks based on the use of licensed portions of the spectrum.

It is worth noting that none of the technologies discussed in this section is specifically designed to serve the needs of wearable computing. Rather, these technologies are being adapted to fill the need. This technology adaptation model is likely to continue until wearable computing becomes more mainstream.

The reader interested in pursuing the technologies discussed in this section in greater detail might find a recent text by Dayem (1997) a worthwhile starting point. The state of Wireless LAN technologies was recently addressed in a survey article by LaMaire, Krishna, Bhagwat, and Panian (1996), which also provides a good literature review starting point. A similar, though slightly older, survey article discusses the state of cellular and cordless telephony technology (Padget, Gunther, and Hattori, 1995).

Wireless LAN Technologies (Radio and Infrared)

Wireless LAN (WLAN) products can be used to support off-body communications, but these solutions generally require that some infrastructure be installed to support the communications at the other side of the data link. There are numerous WLAN products available; most employ low-power spread-spectrum RF operating in the unlicensed ISM (Industrial Scientific Medical) bands.

The recently approved IEEE 802.11 standard defines both direct sequence and frequency hopping spread-spectrum WLAN solutions in the 2.4 GHz ISM band. Three different physical media are specified: infrared (IR), frequency-hopping radio, and direct-spread radio. Data rates of 1 and 2 Mbs (megabits per second) are supported, with a 10 Mbs variant in the works, making products that support this standard very attractive for wearable applications that require high data rates. The range of these products is typically limited to a few hundred to three thousand feet at the most, depending upon the physical environment. Please see the relevant IEEE standard, listed in the references section, for details.

Cordless Telephony Technologies—DECT and PHS

Although conventional cordless telephones would appear to provide a good base technology for certain types of off-body networking, it seems that no manufacturer of American cordless phones provides any sort of data capability. However, the overseas developments are more promising.

Both the Japanese Personal Handy-Phone System (PHS) and the European Digitally Enhanced Cordless Telecommunications (DECT) systems offer a useful blend of traditional cordless telephony with cellular telephony. When in range of the owner's base station, the phone acts as a cordless phone. When out of range, it uses public base stations and hence resembles a cellular phone. Public base stations are deployed much more densely than is typical for a conventional cellular system, which means that the phone uses, on average, less power. At the time of writing, the PHS and DECT systems offered a nominal 32 kbps circuit-switched data service. Plans are afoot for both a higher-speed circuit-switched service and for a shared service similar to CDPD.

Cellular Phone Based Technologies

The cellular telephone infrastructure is the source of several technology solutions for off-body communication. At the most mundane level, a simple cellular modem can be used over the analog AMPS network, providing a 14.4 or even 28.8 kbps circuit-switched data link across most of the United States. CDPD is available in many areas; these systems also exploit the AMPS network but use idle channels to transmit individual packets, providing a packet-oriented service rather than a connection-oriented service.

Digital cellular is coming to the United States with the introduction of the PCS frequency bands. Two of the wideband blocks have been auctioned, and these are expected to be deployed for use in digital telephony. Competing standards in this area leave open the question of whether these systems will be TDMA (Time Division Multiple Access) or CDMA (Code Division Multiple Access) based. Perhaps the more significant question concerns that of data services, which have not yet appeared but are expected by many. The form that these data services will take is a significant question.

See Mazda (1996), Tisal (1997), and Redl, Weber, and Oliphant (1995) for useful introductions to various types of cellular telephony.

Pager Technologies

A number of different paging technologies exist, together with a confusing melange of national standards, proprietary protocols, and ongoing system upgrades. From this chaos has arisen a range of systems, both public and private, with scopes extending all the way from private, single-building systems to full continental, satellite-based public systems. At present, paging functionality is blurring into a variety of related technologies, most prominently the modern cellular telephony technologies, the FLEX standards, and the private RAM and Ardis networks.

Paging is appealing mainly for its very low battery consumption and the ease of acquiring frequencies for private use. The designers of paging equipment have focused their talents on developing very small, very low power systems. Unfortunately, most of this effort has gone into receive-only systems; the two-way paging systems, which are most useful for data networking, do not provide enough bandwidth or enough battery life to be serious technology contenders for networking. However, they do have the advantage that it is easy and cheap to acquire paging frequencies for local use, and this may prove the decisive factor for some applications.

Please see Mazda (1996) for a relatively thorough, although somewhat dated and Euro-centric, tutorial on popular pager technologies.

Trunked Radio Network Technologies

Trunked Radio Networks are another area in which technology is rapidly mutating from the old-fashioned "push to talk" taxicab radios into modern systems that more closely resemble cellular telephony or two-way paging. We have lumped three widely deployed technologies into this category: "push to talk," RAM, and Ardis. All appear to offer fair possibilities for off-body networking.

Taxicab dispatch radios are perhaps the most familiar form of trunked radios. A simple data modem attached to such radios provides a rudimentary networking functionality. The very high bit error rate on these channels typically reduces their nominal 9600 bps throughput to something closer to 600 bps. In the past, the data services provided over push-to-talk radios have not been IP compliant, nor have they been connected to a larger (e.g., nationwide) network. There is, however, no reason why they could not be.

The RAM and Ardis networks are public access networks that provide national messaging services via proprietary radio modems that can be inserted in a variety of hand-held devices, including PCs. Historically these have not been IP compliant; instead they provided email gateways into the Internet email infrastructure. Recently, however, RAM has introduced an Internet access service. Thus far, the radio speeds have not been high enough to support a rich set of services, especially since all radios in an area must share the channel. Even so, these technologies appear suitable for off-body communication.

Trunked radio network technologies are proprietary, and not documented in the open literature. Please contact the relevant manufacturers if you wish detailed information about their systems.

Wireless Metropolitan Area Network (WMAN) Technologies

A WMAN is a data network that covers an entire metropolitan area. It can be mentally pictured as a network that provides Wireless-LAN-like services, namely Internet access, with the coverage of a cellular voice network. There is currently only one pure WMAN technology, namely the Metricom network, although CDPD comes close.

Metricom provides an Internet-access service at end-user speeds that typically run at 28.8 kbps or greater. Although implemented as a multihop radio network, the service appears to the user very much like a circuit-switched dial-up line. This service is eminently suitable for off-body networking, both in terms of technology and in terms of pricing. However Metricom's coverage is currently limited to a handful of metropolitan areas.

Satellite Networks

We live in a golden age of satellite schemes. At the time of writing, 34 data-capable satellite networks are being proposed: 20 narrowband data systems, some of which (*) also support paging (APMT*, ASC, E-Sat, ECCO*, Ellipso*, FAISAT, GE Starsys*, Gemnet*, Globalstar*, ICO, Inmarsat*, Iridium*, Leo-One*, Movisat, MSAT, Odyssey*, Optus*, Orbcomm*, Thuraya*, VITAsat); 12 broadband data systems (Astrolink, Celestri, Cyberstar, Expressway, GE*Star, KaStar, M-Star, M2A, Millennium, SkyBridge, Spaceway, Teledesic); and 2 voice systems that also support paging (AceS, Satphone).

Unfortunately, and for a variety of reasons, none of these seems a good match for wearable networking.

LEO Satellite Technologies Low Earth Orbiting (LEO) satellite constellations appear poised to provide the next big breakthrough in wide area voice and data networking. For our purposes, LEO systems fall into two basic categories. The voice-oriented systems, such as Iridium, are designed to provide a service akin to cellular telephony; they provide a low data rate circuit-switched service to mobile handsets. The data-oriented systems, such as Teledesic, will provide a high-bandwidth circuit-switched service to fixed earth stations.

Sadly, neither type seems very useful for off-body communication. The high-bandwidth service draws too much power and needs too large an antenna. (Of course, a hybrid system that uses some form of wireless link from the body to a fixed earth station is possible; but it seems an implausible answer to most needs.) The voice-oriented systems will probably provide too little bandwidth at too high a cost for most needs.

GEO Satellite Technologies   There is a similarly large field of candidates for geo synchronous (GEO) satellite systems, ranging from the well-established VSAT systems to various novel propositions. GEO systems fall naturally into two broad categories: those in which the earth stations can receive only, and those that allow two-way communication. Unfortunately neither category seems well suited for off-body networking, due to power and antenna requirements.

LEO and GEO satellite proposals are changing too quickly to be well documented. The interested reader might consult magazine articles such as Montgomery (1997) and Evans (1998) for a general overview. Detailed information is generally proprietary and not disclosed.

The Future of Wireless Networking

Wireless communication technology is evolving very quickly at present, and it is difficult to make any useful predictions for even the near-term future. We have chosen three major technology thrusts for a brief review. These will at least give a general flavor of current trends.

LMDS and MMDS   These two terms each involve a US-regulated frequency range, other similar ranges, and an expected set of access techniques and uses for those bands—and as such have rather blurry meanings! The acronyms stand for *Local* and *Metro Multipoint Distribution System*, respectively. Each uses frequencies above 10 GHz to distribute some type of data in a metro area. Offered services vary from Wireless Local Loop to Wireless ATM to digital "wireless cable" TV. In general, these frequencies require fairly large, nonmobile antennas and line-of-sight connectivity from transmitter to receiver. Hence they do not appear especially attractive for off-body wearable communications.

Multihop "Ad Hoc" Radio Networks   This has been an area of ongoing research from the early 1970s, which now appears to be (finally!) bearing fruit. The basic idea is independent of underlying radio technology or frequencies: Each radio in a network is "smart" (i.e., contains a router that automatically discovers its neighbors, determines routes, and forwards packets appropriately). A collection of these nodes thereby self-organizes into a multihop radio network, in which packets are automatically relayed hop-by-hop, from radio to radio, on their way from the source to the destination. In military usage, the entire network may be freestanding; in commercial scenarios, it is connected at multiple locations into a faster, wired infrastructure to form a portion of the global Internet. Metricom is the leading commercial provider of such services. This technology appears to have great promise for off-body communication as it removes the need for a fixed communications infrastructure near the body. It also provides a number of other intriguing benefits (e.g., a series of short hops requires less transmitter power than a single long-range transmission and hence can help conserve battery life).

Unregulated Bands   In the past few years, the US FCC has opened up several RF bands for unlicensed use. The 900 MHz and 2.4 GHz bands are now widely used for wireless LANs. More recently the FCC has allocated 300 MHz of spectrum at 5.15–5.35 GHz and 5.725–5.825 GHz to the Unlicensed National Information Infrastructure Band (U-NII). This band also appears to be extremely useful for off-body communication, though it is not yet clear what forms of radio devices will be developed for this band.

PERSONAL AREA NETWORKS

A *Personal Area Network*, or PAN, is the next step downward in the WAN-MAN-LAN network hierarchy. A PAN is used to interconnect the computers and devices (peripherals, sensors, actuators) used by an individual within their immediate proximity (0–10 feet). These devices may be carried or worn on the person or in some cases placed nearby. The PAN provides connectivity "for the last ten feet."

The need for PAN technology is very real. A basic wearable computing system comprises a main central processing module, a separate keypad or pointing device, a display device, a voice input device, and a speaker for audio output. In addition to this basic set of devices, application specific devices commonly added to the wearable system include bar-code scanners, wearable printers, geo-location devices, off-body network connections, health monitoring sensors, etc. This plethora of devices must be somehow interconnected.

To effectively support wearable computing, a number of requirements must be satisfied by a Personal Area Networking technology:

- Wireless without line-of-sight restrictions: Wired solutions are simply too encumbering given the desire to wear the system, and any line-of-sight restriction would greatly restrict the placement of devices on the body relative to one another.
- Extremely low power consumption: Every device must be self-powered, so the PAN must consume minimal amounts of power.
- Small and lightweight: Wearable computing dictates that devices be small and lightweight, so the PAN component of each device must be a minor contributor to its size.
- Network-based solution supporting multiple devices: Wearable systems often comprise multiple devices, some of which demand bidirectional communications capability.
- Cross-network interference tolerance: Every wearable user will have his or her own PAN, and these must continue to operate when in proximity to one another.
- Broad generality, with easy integration into simple and complex devices: The range of devices that may be included in a wearable system is great (e.g., from Pentium-based PCs to simple sensors), yet all must communicate over the same PAN infrastructure.
- Low cost: Some of the devices that are to be included in the wearable computing system PAN are very inexpensive (e.g., sensors, pointing devices, etc.), so the PAN interconnect must be low cost in order to avoid becoming the principal cost component.

Perhaps surprisingly, there currently exists no PAN solution that satisfies all these requirements. There are, however, three PAN technology alternatives that can be deployed, with some limited degree of success, in a wearable system today: 1) wire, 2) infrared based on the IrDA standards, and 3) RF-based "point solutions."

In this chapter, we also discuss two new technologies designed explicitly to satisfy the needs of Personal Area Networks. These are MIT/IBM's Near-Field PAN technology and GTE Internetworking's BodyLAN™ technology.

Currently Deployable PAN Solutions

Wired PAN solutions have been around for many years, though they have not generally been thought of as networks. For example, the maze of cables and wires emanating from the back of the typical desktop workstation or PC to connect the keyboard, mouse, microphone, speakers, monitor, modem, and local printer comprises a PAN. This solution has also been applied in wearable systems; witness the serial cable from the hand-held computer to the hip-worn printer the next time you return a rental car to Hertz! The MIT wearable computing group provides an excellent example of the use of wired PAN solutions to interconnect the various input and output devices that comprise their systems. Steve Mann (1996, 1997) presents a nice history, summarizing his work at MIT, with excellent photos of wired PANs in use.

The USB (Universal Serial Bus) represents an example of the latest "standard" wired PAN technology for the desktop. The USB Specification has gained widespread acceptance from companies such as Compaq, Digital, IBM, Intel, Microsoft, and others, as the universal interconnect for PC peripherals. Through a tiered star topology (see Figure 24.3), the USB standard is able to support an aggregate data rate of 12 Mb/s shared by up to 127 devices.

USB supports isochronous communication links, which can be especially useful in wearable systems such as sensors (e.g., health monitoring or environmental) that must be sampled at very regular intervals. USB also supports asynchronous communications modes, which are well suited to input devices such as keyboards, barcode scanners, and microphones,

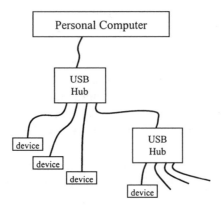

FIG. 24.3.   USB tiered star topology of hubs and devices.

where the communications requirements come at unforeseen times driven by the human interactions with the system. Coupled with the support for both isochronous and asynchronous communications modes, USB offers features such as hot plug-and-play support and built-in power distribution for lower-powered devices.

Despite the long list of useful capabilities, USB and other wired PAN solutions fail to meet the very first requirement identified for wearable computing: wireless! Wires running about the body are cumbersome and limit the effectiveness of the wearable system. It's also interesting to note that the power consumption of wired solutions may in some cases be as great or even greater than that of a well-designed wireless solution.

Infrared technology, especially in the form of the IrDA (Infrared Data Association) family of specifications, has gained fairly wide adoption in the portable PC industry. IrDA uses free-space optics in the infrared band. (The peak wavelength for the IrDA physical layer is between 0.85 and 0.9 nanometers.) An output driver with an LED (Light Emitting Diode) is used for the transmitter, and an IR detector is used for reception. Integrated IR transducers are commercially available for several sources.

The IrDA physical specification defines several point-to-point links that operate at speeds ranging from 2.4 kb/s to 4 Mb/s. A data link specification based on HDLC (High-Level Data Link Control), an ISO standard (1991), and a link management protocol add to the standard protocol stack.

PAN solutions based on the IrDA standards are less than satisfactory for most wearable computing applications. (This applies to IR in general, including diffuse IR.) Perhaps the most significant difficulty is the physical line-of-sight requirement needed to support communication between two devices. (The IrDA physical link specification actually demands operation of the link when the devices are aimed within $+/-30$ degrees of each other for most of the supported data rates.) This clearly prevents placement of wearable devices in many convenient and comfortable locations on the body, such as in a pocket or embedded within an article of clothing. Infrared also performs poorly in bright sunlight and/or dusty conditions, which can be the norm for many wearable-computing applications.

The IrDA protocol stack, as defined in the specifications, includes addressing support and all the essentials needed to support multinode networking. Unfortunately, the fundamental nature of the physical media, which demands that communicating devices lie within a direct line of sight of each other, has limited IrDA use in this multinode fashion. In fact, IrDA Lite implementations, designed to support many of the sorts of embedded applications that might be applicable to wearable computing, typically eliminate the multinode networking capability.

Virtually any of the off-body communications technologies discussed earlier can be exploited to implement a PAN. Just as one could use an Ethernet LAN running TCP/IP to connect standard desktop peripherals (keyboard, mouse, speakers, etc.) to a PC motherboard, an IEEE 802.11 Wireless LAN could be used to do the same for a wearable computing system. Symbol, for example, uses wireless LAN (WLAN) technology as a means to connect some of its barcode scanner products (e.g., the LS3070) to a nearby PC or Point-of-Sale (POS) terminal.

The problem with the use of wireless LAN technology to implement a PAN is that the design trade-off points are mismatched. Wireless LAN technology is designed to support high data rates over as great a range as feasible without consuming too much power and space. The primary constraints are FCC Part 15.247 regulations governing spread spectrum operations in the 2.4 GHz ISM (Industrial-Scientific-Medical) frequency band. IEEE 802.11 WLAN technology provides data rates of up to 2 Mbps (and even higher speed variants are now being developed), with a range of 1,000 feet or more.

PAN applications demand minimal power consumption and size, as well as low cost. Data rates and range can be traded to better optimize these goals. Further, a PAN solution must satisfy the requirement of tolerating cross-network interference, which is generally not an overwhelming issue for Wireless LANs. (Cross-network interference solutions requiring user-driven channel selection complicate the usage model and thus are not as desirable as more transparent solutions.)

Another form of "point solution" that has been applied to PAN problems is the direct incorporation of a low-power RF link by a product vendor to support communication between a pair of devices. Low-power RF chips, available from companies such as RF Microdevices, RF Monolithics, and others, can be directly incorporated into a pair of products to support a communications link. The problem with this strategy is that the generated solutions do not satisfy the broad set of PAN requirements outlined above, but some narrow set of requirements derived by the specific product specifications. Multinode networking is not available, reuse of the solution is missing, there is no path for certain interoperability with products of other vendors, cost (especially for the design cycle) is high, and often, power consumption and size are greater than desired. Relying on point solutions to the PAN problem is akin to every vendor developing their own peripheral interface technology for every new product; it's simply a bad idea.

IBM Personal Area Networking: Near-Field Intrabody Communication

At the IBM Alameda Research Center, Thomas Zimmerman (1996) is continuing PAN research he began as a graduate student working with Professor Neil Gershenfeld at the MIT Media Laboratory. The IBM approach is unique in that it exploits near-field communication technology as an alternative to physical wire, infrared light, and RF.

The essence of the near-field approach is the use of the human body tissue as a sort of "wet wire" over which data can be communicated at very low power levels. Electrostatic coupling is used to tie the transceiver device to the human body, which acts as the transmission medium. This is depicted in Figure 24.4, where the electrodes at the transmitter and receiver essentially form a pair of capacitors (one electrode in each logical capacitor is the body surface itself) "wired" together through human body tissue.

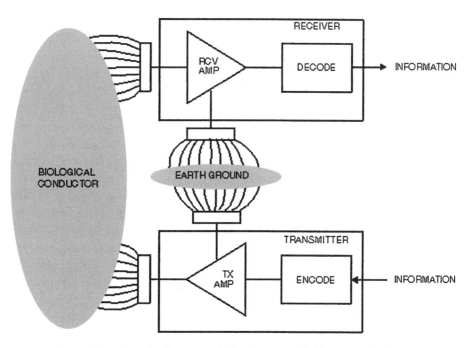

FIG. 24.4. Block diagram of the IBM near-field PAN solution. Copyright 1996 International Business Machines Corporation. Reprinted with permission from IBM Systems Journal, Vol. 36, No. 3/4.

The basic principal of communication through this channel is the detection of impedance mismatches. One of the transmitter electrodes is aimed toward the body; it has lower impedance than the one aimed outwards toward the "earth ground" environment. This arrangement allows the transmitter to impose an oscillating potential on the body, relative to the earth ground. The same impedance asymmetry can be found at the receiver electrodes and is the key to the receiver's ability to detect transmitted signals.

As might be expected, if the impedance differential grows too small, as might happen if the body capacitance to the environment shrinks, the ability to communicate will degrade. Zimmerman (1996) reports, for example, that in one experiment the difference between standing barefoot or not translated into 12 dB for wrist-mounted devices. The same sort of analysis leads to the conclusion that the best place for near-field PAN devices is the feet, since they provide an ideal location for putting large electrodes in close proximity to the body and the earth ground, respectively.

The IBM near-field PAN solution has a number of characteristics that make it especially attractive to some applications. Susceptibility to eavesdropping, for example, is very low with near-field communication, since near-field signal strength decreases with distance cubed (whereas an RF signal propagates with signals that decrease with distance squared). In order to "listen in," an eavesdropper would need to be physically very close to the PAN user. This same characteristic of the near-field approach simplifies the problem of dealing with interference from other PAN users, or other external interference sources.

The near-field PAN solution can be designed to operate at very low frequencies (.1 to 1 MHz), thus contributing to power savings. As an example, the IBM prototype (discussed shortly) operates at 330 kilohertz at 30 volts with a 10-picofarad electrode capacitance, consuming 1.5 milliwatts discharging the electrode capacitance. Zimmerman claims that most of this energy is recycled through use of a resonant inductance-capacitance (LC) tank circuit.

The combination of low power levels and cubic signal drop-off rate of near-field communications also means that the IBM PAN solution is able to avoid dealing with regulatory issues. The prototype system has a field strength of 350 picovolts per meter at 300 meters, which is 86 dB below that allowed by the FCC.

A near-field PAN prototype has been built by Zimmerman (1996) at IBM Almaden Research, as a proof-of-concept demonstration vehicle. The design is based on a low-power Microchip PIC 16C71 microcontroller, with its built-in 8-bit A/D converter. A few off-the-shelf components, including a current amplifier able to detect the tiny received displacement currents (e.g., 50 picoamperes, 330 kHz), brings the cost of the components to less than $10 in quantity. The entire solution, which is currently about the size of a thick credit card, could easily be integrated into a single CMOS device.

The IBM prototype supports two modulation schemes: on–off keying (also known as Amplitude Shift Keying, or ASK) and direct sequence spread spectrum. These two strategies differ in terms of their susceptibility to noise; one would expect to get greater signal to noise ratios with the spread-sequence spread-spectrum approach, due to the process gain of the spreading. The prototype supports a 2400 bps data rate, though numerical analysis shows that the concept theoretically supports up to 417 kbps maximum channel capacity.

To demonstrate their near-field PAN concept, IBM developed a "business card handshake" demonstration. In this demonstration, PAN devices placed on the floor simulate those that might be someday embedded within a shoe insert. A man and a woman, each with such a device, can then shake hands, closing a circuit that allows picoamp signals to pass from the transmitting device to the receiving device. The signals encode a business card; these data are thus transferred from the woman to the man simply by shaking hands! In this demonstration, the "wet wire" extends from the transmitter at the foot of the woman, through her body to her hand, to the man's hand and through his body, to the receiver at his foot. A serial link from the foot receiver to a laptop is used to transfer and then display the woman's business card.

The MIT and IBM research, as reported publicly, has not yet addressed the issue of multidevice networking for Personal Area Networks. In order to provide a complete PAN solution, this aspect of the problem would of course need to be addressed.

The human body can be thought of as one large broadcast medium (much like a bus or the RF space), since any signals transmitted through the body can be received (to some extent) at any other location on the body. Thus, some multiple access protocol must be implemented. CSMA (Carrier Sense Multiple Access) with collision detection (as in Ethernet) or collision avoidance (as in IEEE 802.11) might be employed. Alternatively, a TDMA (Time Division Multiple Access) scheme such as that employed in BodyLAN™ (discussed later) might be used, or perhaps a CDMA (Code Division Multiple Access) scheme such as that used in some digital cellular phone networks. See Sklar (1988) for a good introduction to multiple access protocols. In any of these schemes, multiple devices will need to receive messages on the network at the same time; in some cases there may also be multiple transmissions. The published literature on the IBM PAN does not address how impedance balances are maintained between transmitter and receiver in such an environment.

More recently (Post et al. 1997), the results of continuing PAN research at MIT have been reported. In addition to demonstration of an FSK-based modulation scheme, providing an improved data rate of 9600 bps, the latest work demonstrates wireless power distribution through exploitation of the same technique. They report that under reasonable conditions, they can take 200 mW of 1 MHz AC at the hand and deliver, wirelessly, approximately 20 mW of rectified, filtered DC to the foot. They are currently working on a way to duplex both data and power over the same medium (i.e., the body).

There are of course disadvantages to the near-field approach to implementing a PAN solution. Most significant among them is the need for the physical conduit (i.e., the human body) between the two communicating devices. There are applications where simple proximity is the preferred model; for example, when I approach my desk I'd like the PDA in my shirt pocket to automatically synchronize with my desktop PC, without the need to establish physical contact. The ability to locate sensors in the physical environment that automatically "attach" to the PAN when nearby is also attractive, especially when actual physical contact is not a requirement.

GTE Internetworking Personal Area Networking: BodyLAN™

The BBN Technologies division of GTE Internetworking has been developing an RF-based Personal Area Network technology called BodyLAN™. BodyLAN™ has the following characteristics specifically targeted to satisfy PAN requirements: low power consumption (approximately 5.4 mA at the base network data rate of 32 kbps), small size (two small IC chips), wireless without line of sight limitations, networks of multiple devices (up to 128), short range (2–10 meters), interference tolerant, cross-product generality, and easy integration into OEM products and solutions.

At the media access (MAC) level, BodyLAN™ is a star topology, comprised of a single Hub device surrounded by up to 127 PEA (Personal

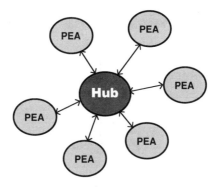

FIG. 24.5.   BodyLAN MAC-level star topology.

Electronic Accessory) devices. This is depicted graphically in Figure 24.5. The Hub is responsible for orchestrating the BodyLAN[TM] network; all MAC- level communications is between the Hub and a PEA. The Hub manages the timing of the BodyLAN network, allocates available bandwidth among the PEAs participating in the PAN, and supports the attachment and detachment of PEAs from the PAN.

It is expected that the Hub device be one that the user normally carries in a physically central location. Typically, the Hub will reside in a wearable computer or perhaps in a simple pager-like device. (Of course a pager is just a simple wearable computer with an off-body communications link.) The Hub could, however, be just about anything.

PEAs may vary dramatically in terms of their complexity. A very simple PEA might be a movement sensor built of just an accelerometer, an 8-bit microcontroller, and a BodyLAN[TM] interface. A barcode scanner and its microcontroller exemplify an intermediate complexity PEA. And at the more complex end are PEAs that represent PDAs, cellular telephones, or even desktop PCs and workstations that join the PAN when the wearable system comes into their proximity.

The critical BodyLAN[TM] core is embodied in its digital control logic component. The digital control logic architecture includes patented protocol innovations specifically designed to minimize power consumption, cost, and size. The result is extremely low power consumption that is a direct function of the device data rate.

At the foundation of BodyLAN[TM] is a TDMA (Time Division Multiple Access) protocol. As depicted in Figure 24.6, TDMA is an access protocol in which networked devices share the physical media (RF spectrum) by "taking turns" in time. Due to the star topology of BodyLAN[TM], the Hub

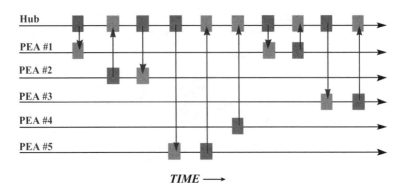

FIG. 24.6.   Abstract depiction of TDMA timing structure.

is involved in all communications, but PEAs only communicate when it is their turn.

The TDMA implementation for the BodyLAN™ PAN is unique in that the very short distance between devices (generally two meters or less) means that propagation delays are minimal. This in turn enables a network with very fine grained network synchronization. Then, with close synchronization, BodyLAN™ PEAs are able to control RF transceiver power at the individual 8-bit burst level. In particular, a BodyLAN™ device transmits an 8-bit burst exactly when there is a device (Hub or PEA) expecting to receive that burst, and a BodyLAN™ device enables its receiver exactly when it expects to receive a burst. Much of the digital circuitry can also be powered off when idle. Realizing that powering a receiver consumes as much current as a short-range transmitter, the ability to shut down the receiver except when actually receiving data is critical to reducing the average current consumption of the BodyLAN™ network device.

A second important innovation employed by the BodyLAN™ architecture is network support with a dynamic attachment mechanism. Dynamic attachment eliminates the need for low-level network management; devices join a BodyLAN™ network simply by coming into range, asking for permission to join, and then establishing communication with application software. This wireless variant of the "Plug and Play" concept is referred to as "Come Close and Play."

A simplified depiction of the BodyLAN™ TDMA timing structure is shown in Figure 24.7. At the highest level of abstraction, time is divided into five kinds of network communication events: synchronization *beacons*, bandwidth allocation *tokens, status response* messages, *physical data unit* transfer blocks, and *acknowledgments*.

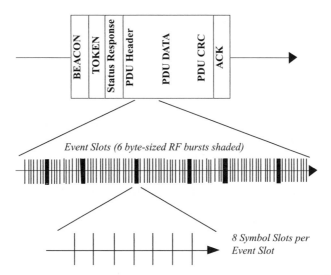

FIG. 24.7. Simplified depiction of the high-level BodyLAN^TM TDMA structure.

Beacons are special signals transmitted by the Hub periodically to establish the timing of the network. PEAs are able to distinguish beacons from noise and other transmissions and establish fine-grained network synchronization by continually adjusting their local reference clocks to match the timing of the Hub clock established by the beacons.

Tokens are special messages transmitted by the Hub at fixed locations in the TDMA plan. Unless they've arranged to do otherwise, PEAs awaken to listen to tokens, as these indicate which PEA is involved in the next data transmission. The token also indicates the direction (Hub to PEA versus PEA to Hub) of the following communication. The token additionally includes an indication of from which PEA the Hub wishes to receive status.

Status Response messages are sent by PEAs upon request from a Hub (see above) and indicate whether the PEA has data to transmit to the Hub. The response is used to help manage the allocation of incoming data bandwidth, as well as to support an assortment of other network management tasks.

Physical Data Unit (PDU) transfer blocks are used to communicate application and system messages to and from a PEA. Each PDU includes a small header, which includes a sequence bit and a length field, as well as a trailing CRC (cyclic redundancy check). The header and CRC support reliable ordered delivery of the PDU data blocks, which vary in length up to 62 bytes.

The final element of the high-level TDMA structure is the Acknowledgment message, which is a reverse-direction transmission from the PDU transfer. The acknowledgment is used as part of the single-bit sliding window protocol implemented by the Digital Control Logic.

Each of the five high-level communication events comprises a structured series of event slots, a subset of which are used to transmit an RF energy burst. The depiction in Figure 24.7 shows some portion of the PDU transfer event decomposed into a series of event slots where six RF bursts happen to be transmitted. The spacing of these bursts is based upon Optical Orthogonal Code sets, which helps deal with the issue of cross-network interference. Each RF burst comprises eight symbol slots, each of which is used to communicate a bit of digital information. Bit spreading is done at the burst level, providing a simple form of interleaved forward error correction for improved bit error rate performance.

Through use of a combination of techniques, multiple BodyLAN™ networks may efficiently operate within range of one another. One of these techniques is based on the OOC-based burst spreading. This technique can alternatively be thought of as a form of Code Division Multiple Access (CDMA) or a form of time-based spread-spectrum. See Sklar (1988) for an overview of CDMA and Dixon (1994) for a review of spread-spectrum communications. In addition to the OOC-based burst spreading, BodyLAN™ employs frequency agility as a means to separate potential cross-network interference sources.

The interface between the host microcontroller or microprocessor and the BodyLAN™ digital control logic is based on a small set of *commands* and *responses*. The host issues commands to initialize the network, send messages, make status inquiries, and perform other assorted management functions. Response communications from the digital logic are used to feed incoming messages to the host and to provide the host with status information as requested.

The host processor, which is anything from a simple microcontroller to a high-performance microprocessor, runs the BodyLAN™ device driver and protocol software. A depiction of the implementation/protocol stack, including the hardware layers, is shown in Figure 24.8.

BodyLAN™ software is divided into a small platform-dependent component and a few machine-independent pieces. These include basic support for communication over BodyLAN™, support for dynamic network management, and network interfaces to higher-level protocols or applications.

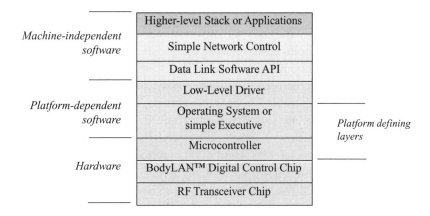

FIG. 24.8. Implementation/protocol stack (including hardware).

Much of the traditional MAC-layer and data link layer protocols are implemented by the BodyLAN™ digital control logic chip hardware, so a simple platform-independent software API provides access to these mechanisms to the higher-level stack components.

The BodyLAN™ software core is designed for use on simple microcontrollers, and so it does not include support for higher-level standards such as TCP/IP. These can easily be implemented, as can other protocols or device driver abstractions, on top of the basic BodyLAN™ software core.

Forthcoming Products, Consortia, and Standards

The need for PAN technologies has begun to gain visibility and attention, as potential users search for solutions to their Personal Area Networking requirements. While we cannot yet select from among an assortment of suitable options, we can expect new products and technologies to emerge from the research labs and industry over the next several years. Further, since Personal Area Networking is a relatively new technology area, advances are likely to be rapid in the short term, with new innovations offering improved performance (relative to PAN needs) introduced frequently.

Unannounced products are rumored to be forthcoming from several technology companies. We can expect public announcements in the near term, perhaps by the time this book is published.

Two important industry consortia have formed that relate to Personal Area Networking. The First is the HomeRF Working Group (http://www.homerf.org/), a group including Intel, Microsoft, Compaq,

Phillips, and others. The stated mission of HomeRF is "to enable the existence of a broad range of interoperable consumer devices, by establishing an open industry specification for unlicensed RF digital communications for PCs and consumer devices anywhere, in and around the home." Though not specifically targeted for PAN applications, HomeRF technology may prove applicable and usable for wearable computing applications.

The second relevant industry consortium is known as Bluetooth (http://www.bluetooth.com). Bluetooth is led by Ericsson, IBM, Intel, Nokia, and Toshiba. A laundry list of other companies have also joined the Bluetooth Special Interest Group. Bluetooth will enable "seamless voice and data transmission via wireless, short-range radio, . . . will allow users to connect a wide range of devices easily and quickly, without the need for cables, expanding communications capabilities for mobile computers, mobile phones and other mobile devices, both in and out of the office." Though aimed primarily at the interconnection of devices carried by a typical business professional, Bluetooth appears to have applicability to the entire spectrum of PAN applications in wearable computing.

In addition to the industry consortia, in March, 1998, the IEEE P802.11 Working Group (*http://grouper.ieee.org/groups/802/11/index.html*) announced the formation of a study group to identify a project for standardization of a LAN for wireless communications for wearable computing devices. This study will examine the requirements for Wireless Personal Area Networking (WPAN) of PCs, peripherals, and consumer electronic devices to communicate and interoperate with one another. The Study Group has been soliciting industry input on market requirements and technical solutions for a WPAN with 0 to 10 meter range, data rates of less than 1 Mbps, low power consumption, a size less than 0.5 cubic inches, and low cost relative to target device.

REFERENCES

Dayem, R. A. (1997). *Mobile Data and Wireless LAN Technologies*. Upper Saddle River, NJ: Prentice-Hall.

Dixon, R. C. (1994). *Spread Spectrum Systems with Commercial Applications*, Third Edition. New York: John Wiley & Sons, Inc.

Evans, J. (1998). New Satellites for Personal Communications. *Scientific American*, April 1998, 71–96.

IEEE, 1997. *IEEE Std 802.11-1997 Wireless Local Area Networks, 802.11-1997* Information technology—Telecommunications and information exchange between systems—Local and metropolitan area networks—Specific requirements—Part 11: Wireless LAN Medium Access Control (MAC) and Physical Layer (PHY) Specification.

ISO, 1991. ISO 4335 *High Level Data Link Control (HDLC) Procedures—Elements of Procedures*. 1991-09-15.

Kent, S. (1997). *Security Architecture for the Internet Protocol*, Internet Draft draft-ietf-ipsec-arch-sec-02.txt.

LaMaire, R. O., Krishna, A., Bhagwat, P., & Panian, J. (1996). Wireless LANs and Mobile Networking: Standards and Future Directions. *IEEE Communications Magazine, 34:8*, 86–94.

Mann, S. (1996). Smart Clothing: The Shift to Wearable Computing. *Communications of the ACM, 39:8*, 23–24.

Mann, S. (1997). Wearable Computing: A First Step Toward Personal Imaging. *IEEE Computer, 30:2*, 25–32.

Mazda, F. (1996). *Mobile Communications*. Oxford: Focal Press.

Montgomery, J. (1997). The Orbiting Internet: Fiber in the Sky. *Byte*, November 1997.

Padgett, J. E., Günther, C. G., & Hattori, T. (1995). Overview of Wireless Personal Communications. *IEEE Communications Magazine, 33:1*, 28–41.

Post, R., Reynolds, M., Gray, M., Paradiso, J., & Gershenfeld, N. (1997). *Proceedings of the First International Symposium on Wearable Computers, IEEE*.

Redl, S., Weber, M., & Oliphant, M. (1995). *An Introduction to GSM*. Norwood, MA: Artech House.

Sklar, B. (1988). *Digital Communications Fundamentals and Applications*. Englewood Cliffs, New Jersey: Prentice-Hall.

Sreetharan, M., & Kumar, R. (1996). *Cellular Digital Packet Data*. Norwood, MA: Artech House.

Tisal, J. (1997). *GSM Cellular Radio Telephony*. Chichester: Wiley.

Zimmerman, T. G. (1996). Personal Area Networks: Near-Field Intrabody Communication. *IBM Systems Journal, 35:3/4*.

25

Computing Under the Skin

Dwight Holland[1], Dawn J. Roberson[2], and Woodrow Barfield[3]

[1]*University of Virginia School of Medicine/Virginia Tech*
[2]*Rehabilitation Institute of Chicago*
[3]*Virginia Tech*

1. INTRODUCTION

The preceding chapters in this book have discussed in great depth the design and use of wearable computers and augmented reality displays. The chapters on wearable computers have emphasized computational resources that are carried by the user, external to the surface of the skin, and not directly integrated with the wearer's biological systems. However, based on developments in microelectronics, sensor technology, and medicine, it is actually possible to apply computing resources under the surface of the skin and in some cases to integrate digital technology with the user's physiological systems. Such capabilities will allow computing technology to monitor and control various physiological processes, or to act as a sensory and motor prosthesis. In fact, due to advances in technology that have occurred in medicine over the past thirty years, there have already been major developments that relate to placing computational resources under the skin (e.g., cardiac pacemakers/defibrillators). This chapter reviews past developments that fall in the general area of "computing under the skin," while also discussing the recent advances in digital technology that are allowing smaller and more powerful circuitry to be placed under the skin.

It is interesting to note that human beings have evolved with numerous biological "computers" that are either imbedded in the skin itself or located under the surface of our skin. Just consider the human brain, and the auditory, touch, olfactory, and visual sensors as computers we were born with (Barfield, Hendrix, Bjorneseth, Kaczmarak, and Lotens, 1995). Nature has provided us with an incredible range of wearable biological processors that perform many essential activities such as monitoring our internal physiological state or assisting us in monitoring the state of the world. From the perspective of human physiology, the idea of placing computing resources under the surface of the skin cannot be considered a new idea as nature has already accomplished this feat. However, what this chapter focuses on, what represents the motivation for this chapter, is the idea that we can use man-made computing and sensor technology to enhance and in some cases replace the biological "computers" that nature has provided us. Such capabilities will lead to many interesting developments in the next century as neural implants, nanotechnology, materials science, therapeutic genetic alterations, and prostheses are likely to slowly blur the distinction between a "pure" human and a human–machine combination commonly referred to as a "cyborg."

In all likelihood, if Moore's Law continues to prevail (i.e., that computing power doubles approximately every 1.5 years), in the next one hundred years or so there will be advances in human–computer interfacing that even a reasonably forwarding-looking book chapter such as this cannot accurately foresee. Furthermore, even if the gains in computing power predicted by Moore's Law fizzles to a slower pace around 2020—as some predict will be the case for technical reasons—the advances simply due to gains in other areas such as sensors and materials science, combined with enhanced computing power and other high technology innovations, will change the way we think about ourselves in terms of what it means to be "human" in the next century. Some authors have already began writing about the extrapolation of such computing power and human development for the future from an evolutionary perspective in ways that are surprising and thought provoking (Kurzweil, 1999; Paul and Cox, 1996; Moravec, 1999). They predict that human beings will quickly veer toward a "cyberevolution" type of track.

While the medical aspects of computing under the skin will motivate much of the research in this area, there will also be numerous nonmedical applications. For example, Professor Warwick at the University of Reading recently underwent a surgical procedure in which a transponder was implanted in his forearm (Warwick, 1999). The transponder consisted of a glass capsule, which contained an electromagnetic coil and a number of silicon chips. The transponder was approximately 23 millimeters long and 3 millimeters in diameter. The circuitry was able to transmit a specific 64-bit signal. A receiver located outside the body was able to detect the unique identifying signal and generate commands via a computer to operate a range of devices—such as doors, lights, heaters, or other computers. Which devices are operated and which are not depends on the requirements for the individual transmitting the signal. The idea of implanting microchips under the skin for pets has been around for years. The microchip that is implanted under the pet's skin bears a unique identification number. The chip is encased in a "biocompatible" protein layer and is injected into the skin between the animal's shoulder blades. The number can be read through the pet's skin in a fraction of a second by a scanner that uses a radio signal. Some 18,000 animal shelters, veterinarians, and animal control offices in the United States have the scanners. The chip's protein casing anchors it in place underneath the animal's skin, ensuring that it does not move or pass out of the animal's body. The protein anchor also keeps the area under the skin where the chip is located from becoming infected. Since the system does not have moving parts to wear down, or a power source to replace, it can theoretically last for decades. So far, some 80,000 chip-bearing pets have been reunited with their owners.

Recent advances within the molecular chemistry and biological communities married with computer chip technology and the science of reading DNA sequences are leading us to the possibility that a person's genetic code might be placed on a rather small wearable chip or in a subdermally implanted chip. Such "gene chips" with the correct information on them could be combined with recent advances in the new field of "pharmacogenetics" to better predict how an individual will respond to certain drug therapies. Researchers at Georgetown University are pioneering this expanding pharmacogenetic knowledge base by showing that different individuals respond to a drug in a variety of ways. Some common drugs used by many of the population where this has been found to be the case include codeine (a pain killer) and prozac (an antidepressant), among others. Knowing which genetic profiles respond most optimally to a specific drug therapy, and which do not, could save lives and money by improving prescribing efficiencies within the healthcare system. Such chips might aid future doctors in the diagnosis or more optimal treatment of certain diseases. These chips known as "gene-reading arrays" could possibly drive the revolution currently occurring in genetics research far past the current scope of the Human

Genome Project. (The goal of the Human Genome Project is to map the genetic code for all of the approximately 100,000 human genes by 2005.)

To take a liberal interpretation of what the term "computing under the skin" may imply, there are two main approaches that we delineate. First, due to the rapid advances in microelectronics, digital devices may either be located under the skin or directly interfaced with the body. Breakthrough technology relating to computing under the skin includes insulin pumps for diabetics, cardiac monitor/defibrillator devices, cochlear implants to aid the hearing impaired, and deep brain stimulators to mitigate some of the effects of Parkinson's disease, among many others. Many of these devices are described in a general sense as "prostheses" and are discussed throughout this chapter. Secondly, computational processes occurring outside of the body (e.g., MRI, ultrasound) may be used to apply processes or treatments on or inside the body. This is an area of explosive growth in the application of medical technology, and the surface of current and future developments in this area has barely been scratched. Considering the case where computer technology outside the body directly influences processes or organs under the skin, one example is the manipulation of a large data set generated from Magnetic Resonance Imaging (MRI) studies. MRI can be crudely described as a technique whereby different image "slices" of the targeted tissue are taken, and the various slices are combined computationally to produce a "whole picture" that has the capability for two- and three-dimensional manipulation of the data set. Such manipulation of MRI data can suggest therapies to be applied to organs or processes under the skin's surface. Every day that goes by results in more medical specialties utilizing these data set manipulations and applications of medical informatics for better medical decision making.

An important question arises with respect to what the wearers of some of the potential technologies discussed in this chapter really do need in order to live better lives. Often researchers think they know what other people such as those with Spinal Cord Injury (SCI) or other neurologically compromised patients need. On occasion, researchers need to ask the patients what *they* want and need from the scientific and engineering communities in terms of technology to enhance their lives. At the 1994 Neural Prosthesis IV Conference, Peter Axelson, having lived with SCI for nineteen years iterated what he thinks the research priorities should be. He and other SCI persons like him have suggested that their priorities for a higher quality of life are good skin condition, bladder function, reproductive function, bowel function and standing function (Axelson, 1996). He later goes on to say that ambulating may produce benefits that contribute to other higher priorities; however, it by itself is a lower priority.

In the introduction we briefly introduced past and likely future developments that relate to placing computing resources under the skin. As Warwick points out, the potential of such technology is enormous. Warwick indicated that it is quite possible from a technological standpoint in the near future for an implant to replace an access, credit, or bankcard. Such an implant could carry large amounts of data on an individual regarding an insurance number, blood type, and various medical conditions. These data could be updated and added to whenever necessary. There would be a host of legal and confidentiality issues, along with the necessary legal protections to be added to the mix, but the technology to accomplish such implants is rapidly evolving. One clear benefit of this technology is that a person would not have to be conscious for emergency medical technicians or rescuers to begin the most appropriate treatment.

2. SENSORY PROTHESES

One of the areas where enhanced computing resources have been applied to humans is in the area of prosthetics. A prosthetic device, as defined by the Bantam Medical dictionary is "any artificial device that is attached to the body as an aid." This very simple definition encompasses a wide range of devices from highly complex electronic equipment to strictly cosmetic prosthetics. This might include every type of device from rudimentary visual devices, cochlear implants, and artificial hands to penile implants. The overview of prostheses in this chapter will emphasize some of the technological aspects of prosthetics, delving into the history, materials, and limiting factors of prostheses where applicable. These products have four major users: those with some type of nervous system communication deficit, individuals with endocrinological disorders (with diabetes as the prototype), amputees, and those individuals that are in need of some kind of surgical/radiologic intervention.

2.1 Hearing

The human ear is a wonderfully intricate mechanism that uses a variety of mechanical and frequency-based methods to transfer sound from outside the body to an electrochemical signal that is fed to the brain via the vestibulocochlear (VIII) cranial nerve. When a component of this system breaks down, hearing loss or distortion results. Clearly, knowledge of the nerve pathways and hearing mechanisms of action must be reasonably well understood for the appropriate interventions to be developed and utilized. If the damaged/destroyed bony or tissue component of the hearing mechanism occurs before the cochlear (auditory) nerve, many times a cochlear implant may assist in restoring functional hearing.

The cochlea is the spiral-shaped organ that analyzes sound and transduces it into an electrical signal. Simply put, different frequencies of sound cause different areas of the basilar membrane of the cochlea to vibrate. These vibrations cause specialized neural sensory hair cells attached to the tectorial membrane of the cochlea to distort in a complex manner. This mechanical movement causes the hair cells' neurons to fire, thereby sending electrical signals down the neurons' axons. The axons then join to form the cochlear (auditory) nerve as it exits the Organ of Corti near the internal acoustic meatus (Loizou, 1998). The brain then eventually receives this combined signal in the brainstem and passes it on to the temporal lobe area of the brain known as the Auditory Cortex. The brain interprets this biologically generated signal as frequency- and amplitude-based sound. Frequency components correspond to the sensation of "pitch," while the amplitude of the signal is perceived as "loudness."

If these cochlear hair cells get damaged or destroyed, they do not grow back properly and permanent hearing damage or specific frequency-dependent hearing losses may occur. This has been commonly referred to as "nerve deafness." For hearing impairment due to hair cell damage, in theory the key to improving hearing would be to somewhat augment the frequencies where there has been damage with a high technology "hearing aid," while avoiding amplification of those frequencies at which the person has normal, or close-to-normal, hearing. Often the hearing aid cannot overcome the type or magnitude of damage, resulting in only partial improvement in hearing function (the concept here is that amplifying noise only leads to more perceived noise).

The situation for individuals that have some type of congenital or acquired mechanical deafness is usually different. The challenge with regard to computing under the skin for deaf individuals is to properly artificially stimulate the cochlear with an implant in a manner similar to that provided by a normal-functioning cochlea. The basic design of a cochlear implant includes a microphone to collect sound, a signal processing device to break out the signal into appropriate inputs, and an implanted electrode that actually electrically simulates the hair cell neurons. The simplest and earliest type of cochlear implant was the single-channel implant. The single-channel implant provided stimulation at a single site ("channel") in the cochlea. These designs were simple and low cost, and they provided barely adequate auditory input for some individuals. In one brand (House/3M), the signal was bandpass filtered (340–2700 Hz) and then used to modulate a 16 kHz carrier signal. That filtered signal was then transmitted to the implant. Another brand (Vienna/3M) preamplified the sound, and then compressed it (adjustable by the individual), whereupon the signal was then fed through a frequency-equalization filter (100–4000 kHz). This signal was then amplitude modulated for transmission to the implanted device, which demodulated it and fed the result ultimately to the cochlear nerve. Some Vienna/3M patients were able to recognize speech. The more modern implants use similar technology and signal processing approaches but have more signal processing sophistication and channels for cochlear stimulation input.

The modern multichannel cochlear implants stimulate multiple sites in the cochlea using an array of electrodes. This method attempts to more closely replicate and simulate the frequency-based stimulation that the hair cells normally provide to the cochlear nerve. The arrays are arranged so that high frequency sounds stimulate some of the same neurons that "high frequency" hair cells should. Because of the wide variety of technical approaches to such stimulation, several signal processing choices have evolved on which to base such stimulation. The signal processing strategies that have evolved are roughly divided into two basic categories: feature extraction and waveform analyses. In feature extraction, a characteristic spectral feature will generate a unique input pattern to the neurons. For waveform approaches, the signal is processed into its frequency components and then analyzed using a variety of signal processing steps. The output from these steps is then provided to the neurons as input.

In the present day, there are cochlear implants (24 electrodes) that can significantly improve a person's ability to recognize speech. Despite the improvements in this technology, only about 15% of the patients with cochlear implants can understand words without the additional cues generated from lip reading. Interestingly, people with cochlear implants quite often cannot understand the cochlear-implant-generated words if the words are not in the individual's native language. This is one reason that prior language skills are deemed important to be able to decipher the processed signals entering the brain from these implants. Similarly, there are some healthcare providers that encourage the use of cochlear implants for extremely young children, hoping that their developing language center(s) will be able to learn the "electronic language" that they are hearing.

A decision to be made with cochlear implant design in general is the choice of the transmission datalink. That is, should there be a direct transcutaneous connection of the entire system with the resultant possibilities of infection, or should the link be completely implanted, with power and signal transmitted across the skin in a percutaneous fashion? Either choice has several positive and negative attributes, but if the choice is to decrease infection risks, a percutaneous approach is desirable. However, if a percutaneous component needs repair, surgery is necessary. Choices such as these reflect a more general question with regard to implantable devices and computing under the skin: How is access to the computational device or its power source best handled when there is some level of risk associated with direct access to that device for the person wearing it? These are very important usability design considerations that are often handled at the end of the design process as an afterthought, instead of being thoughtfully considered at the front end of the device's systems engineering design process to enhance device performance and lower the risk to the individual over the implant's useable lifetime.

2.2 Vision

A potentially important application of computing under the skin would be the development of a retinal prosthesis for those individuals that have lost the capability to see from retinal disease. The retina is the biologic transducer of light energy that is converted into signals that the human brain perceives as vision. The nervous tissue contained in the retina is a direct "outcropping" of the human brain from an embryologic and nervous system "wiring" standpoint. The anatomy of the human retina is a bit unusual, and perhaps somewhat counterintuitive. The actual transducers of the light signals, the rods and cones, are located at the back of the retina and receive nutrition from the choroidal area of the eye. The layer that appears forward (anterior) of the rods and cones in the retina contains the horizontal, amacrine and bipolar cells. Next, the ganglion cell layer that follows in the most forward (or epiretinal) location known as the "inner retina." The ganglion cells in that layer quickly join to form the optic nerve, which then traverses to the back of the brain in the area known as the occipital lobe. Minor splays of the optic nerve pass off the main nerve bundle to various other stations in the brain, such as those mediating the circadian rhythms and the sense of balance and motion. At each retinal level there is much processing of visual information such that the visual system functions in a highly integrated complex fashion not only within its own modality of function, but also as it interacts with other parts of the nervous system.

There are two important blinding diseases of the eye, Retinitis Pigmentosa and Macular Degeneration, that are especially germane to this chapter on applications of computing under the skin. Retinitis Pigmentosa is an inherited degenerative disease of the eye that blinds approximately 40,000 individuals each year in the United States, while Macular Degeneration causes blindness in about 100,000 people each year. The macula of the eye is where the vast majority of the most sensitive photoreceptors to fine detail and color reside. These photoreceptors are known as the "cones" and they reside along with the night vision capable "rods" of the eye as part of a structure known as the retina. Retinitis Pigmentosa and Macular Degeneration damage the cone photoreceptors and surrounding tissue leading to visual disability. There is relatively little that physicians can do to stop these disease processes from resulting in blindness. The retina is a paper-thin, very fragile projection of brain tissue onto the back of the eye. Once damaged, it is difficult if not impossible to repair.

In the early 1990s, researchers began to think seriously about searching for ways to stimulate the cells in the retina that were not affected by these debilitating conditions. They reasoned that after light passed through the eye's optics and epiretinal tissue that perhaps it was not necessary for the light to get all of the way to the damaged photoreceptor layer in the back of the eye ("outer retina") in order to create a stimulus that might be transduced by the other undamaged cells of the retina that were still intact. Normally, by the time the analog light signal has made it to the ganglion cells, it has been sampled and converted to nerve impulses by the previous complex cell interactions in the deeper retina. The ganglion cells feed processed signals to the optic nerve whose approximately one million fibers carry the highly modulated signal to the other areas of the brain.

Two pioneering researchers, Drs. Gene de Juan and Mark Humayun (see Dr. Humayun's 1994 Ph.D. dissertation) demonstrated that the more anterior retinal ganglion cells could be electrically stimulated to create an impression of light in the cortex without requiring deeper penetration to the photoreceptor layer at the back of the retina. The researchers had previously noted that 30 to 80 percent of the nonphotoreceptor retinal neurons were intact in Retinitis Pigmentosa eyes and that they might be utilized for electrical signal transmission that arose from light sources. The devil was in the details, however, as getting any light signal converted properly to a retinal electrical signal was certainly not going to be a trivial task. This appeared to be a task for a retinal prostheses with a large amount of built-in computer image processing circuitry.

An early glimpse of what might eventually be done with electrical signals simulating the anterior regions of the retina was accomplished by Drs. Humayun and Juan as they stimulated an alert blind patient's ganglion cells with a probe inserted into the ocular cavity and placed onto the retinal surface. Their results showed that a blind individual person could be made to "see" a point of light at a variety of retinal locations. Further studies showed that simultaneous electrical stimulation of retinal areas only one millimeter apart resulted in two discernable light sources. This was encouraging since it suggested that more sophisticated multiposition electrical stimulation of the retina might be a viable approach to aiding the visual function of people with damaged photorecptor layers. Later work (Humayun et al., 1996) revealed that a 5 × 5 electrode array placed on the retina could be used to create the percept of a line.

This line of retinal research has continued with a productive collaboration between a team of physicians and engineers between two universities. Drs. Juan and Humayun (now at Johns Hopkins) have lead the biological and clinical studies, while Drs. Wental Liu and Elliot McGucken focused on the engineering photoreceptor and electronics design work that must be perfected to continue the retinal prostheses enhancements being tested. As Lui et al. (1997) describe it, the Dual Unit Intraocular Visual Prostheses consists of two basic components. The first part is the Photosensing, Processing, and Stimulating (PPS) chip mounted directly behind the cornea, which is capable of processing external visual stimulation and generating the required current pulses. These pulses are then transmitted to the second part of the device consisting of a 5 × 5 kapton/polyimenide/gold or platinum electrode array implanted directly on the damaged retinal surface and transmits a carefully measured pulse of electrical energy to the retina (see Figure 25.1). The results thus far indicate that the patient can "see" a very primitive image with this prosthetic retinal computing device "under the skin." However, for a 25-electrode array (5 × 5 square) some confusion exists with the interpretation of simple images such that the results must be considered crude, but promising. For example, in a 1996 test, the patient tested could not distinguish the letter "U" when it was presented, but perceived an "H" instead.

Despite these technical problems, some researchers are beginning to speculate on what configuration the near-term visual prosthetic is likely to evolve toward. In an interview with Chase (1999), John Wyatt and Joseph Rizzo suggested that near-horizon technologies will enable the analog visual world to be digitized through a small camera interfaced with an easily obtained Charge Coupled Devices (CCD), or the newer Active Pixel Sensor

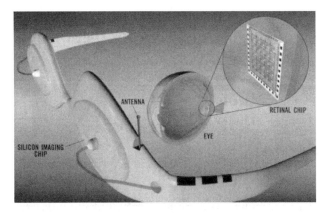

Retinal Prosthesis Project

Johns Hopkins University
North Carolina State University

FIG. 25.1. Summary of Artificial Retina System. Light energy is captured by a video camera, which transmits the video image to a retinal implant. The retinal implant next converts this signal to an electrical impulse, which then stimulates the ganglion cells of the retina. (From W. Liu, North Carolina State University.)

Technology (APS). A very fast but lightweight wearable computer will process the digital image in real time, which will then be transmitted, perhaps by a laser, to the implant on the surface of the retina. This implant will likely be a silicon chip packed full of transistors in an epiretinal configuration. The side facing the camera toward the front of the eye will contain photovoltaic cells, while the other side attached directly to the anterior (inner) layer of the retina will be laced with many electrodes that will stimulate the ganglionic cells to fire off an artificial image via the optic nerve to the brain.

The previously noted U/H letter misidentification result for surface stimulation of the retina was not entirely surprising since the higher-level processing that usually occurs in the undamaged retina has been bypassed with direct stimulation of the inner retinal ganglionic cells. Actually, the graded potential electrical activity response of the photoreceptor and bipolar cells that are accessible to direct stimulation in the subretinal space (behind the retina) is better understood with good one-to-one stimulus-to-signal spatial correspondence input as compared to the more highly processed and frequency-modulated circuitry of the ganglion cells and nerve fiber layer that exists in the inner retina. Given a clearer understanding of the basic properties and responses of the subretinal photoreceptor cells leads to the other retinal stimulation approach where the outer retina, or remnants thereof, are selected for stimulation with subretinal implants.

One of the leading proponents of the subretinal approach is Dr. Alan Chow, who has been working diligently on a "subretinal implant" that will not require any external camera, transistors, or power sources for lasers/radio signals. Chow's idea is to convert light energy into electrical signals in the subretinal space just as a solar cell converts light energy into electricity. The proof-of-concept "Artificial Silicon Retina" subretinal implant is a thin 25 micrometers thick wafer packed full of 70,000 microphotodiodes with a diameter of only 1–3 millimeters. Tests have been carried out on subretinal electrical stimulation of the intact rabbit retina, which resulted in measurable evoked visual cortical potentials (Chow and Chow, 1997). Such experiments on the intact rabbit retinas revealed that the implant placed in the subretinal space retained a stable position and provided signals to the other retinal structures when light energy was incident. However, the signals were not always as expected and the retinal tissues directly overlying the implant area were noted to have a degree of fibrosis and thinning associated with the implant (Peyman et al., 1998). This result calls into question a host of issues and research needs regarding the biocompatibility of subretinal implants.

All of the highlighted technologies for enhancing the blind person's ability to have rudimentary visual function are in very preliminary, yet interesting and critical stages of development. As computing technology, neuroscience, and microelectronics all improve, so will the possibility that blind individuals will be able to "see" basic shapes in the not-to-distant future. Much research is needed to further continue the promising work that these early pioneers have begun.

2.3 Touch

In the case of prostheses, the sense of touch is very important for feedback to a nervous system for better limb control and to avoid dangerous stimuli from the environment. To replace tactile sensation ("touch"), or to provide a semblance of tactile feedback, is a daunting task. There are some force transducers that are available that are sensitive enough, but they are not really "tactile" sensors. In many cases, the classical force transducers are too large, bulky, or too fragile for the high range of forces generated for the most common target area—the hand. In this case the tactile sensors that provide the most promise use digital circuits (Beebe et al., 1998). The benefits of these sensors are that they are small, have reasonably linear responses, and are of fairly rugged design. These sensors do not determine direction of force, which proves to be significant for sensation, but do provide good values for amount of force generated. At present for prosthetic limbs, feedback to the individual on the amount of force measured is provided graphically through an oscilloscope or similar instrument, although the ideal situation would be to provide feedback through proprioception (Hansen et al., 1997). Better feedback in theory could mean enhanced control of whatever body part is being manipulated. These sensors are currently worn externally, and it is not likely that they will be implanted into intact limbs, but rather used on prosthetic limbs.

Some scientists such as Dr. Donald Lyons, a Hampton University physics professor, are taking the first steps toward the goal of providing some modicum of sensation to prosthetic limbs through artificial proprioception (Figure 25.2). With a NASA Glenn Research Center grant, Dr. Lyons is attempting to evaluate the scientific and engineering problems associated with building an artificial peripheral nervous system that transmits signals to the rest of the body by way of a fiber-optic network of light impulses (Lyons, 1993; Lyons and Ndlela; 1996). The goal is for a person with an artificial limb to be able to tell where a limb is touched, and how much pressure has been put on that limb. Optical fibers are already widely used for complex data communications transmission, so it is reasonable that optical fibers might also be used to transmit pressure information to a person's peripheral nervous system with an array of optical fibers embedded in an artificial limb.

As currently envisioned the optical fibers in the limb would only be the diameter of a human hair and would be interfaced with a controlling computer that would send out the pulses of light that would travel the entire length of the limb. The fibers might contain thousands of sensors that are each designed to evaluate how much light is traveling in the optical fiber. The more pressure that the sensor has applied, the less light signal it would transmit back to the controlling computer. The signal would be highly processed and integrated with information coming from other regions of the limb to form a coherent whole. The microcomputer could then take those myriad signals and translate them into an electrical signal. That signal properly modulated and interfaced with the peripheral human nervous system in the shoulder or upper leg regions could then be interpreted as to where, and with how much pressure, the artificial limb was touched. Indeed, this is a very complex and challenging "computing under the skin" problem to be further investigated.

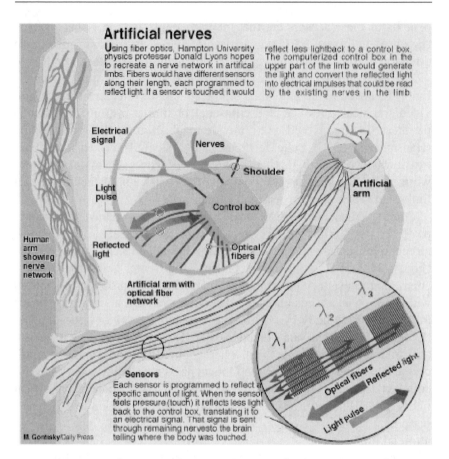

Artificial nerves

Using fiber optics, Hampton University physics professor Donald Lyons hopes to recreate a nerve network in artificial limbs. Fibers would have different sensors along their length, each programmed to reflect light. If a sensor is touched, it would reflect less light back to a control box. The computerized control box in the upper part of the limb would generate the light and convert the reflected light into electrical impulses that could be read by the existing nerves in the limb.

Electrical signal

Nerves

Shoulder

Light pulse

Artificial arm

Control box

Human arm showing nerve network

Reflected light

Optical fibers

Artificial arm with optical fiber network

Sensors

Each sensor is programmed to reflect a specific amount of light. When the sensor feels pressure (touch) it reflects less light back to the control box, translating it to an electrical signal. That signal is sent through remaining nerves to the brain telling where the body was touched.

M. Gomisky/Daily Press

λ_3

λ_2

λ_1

Optical fibers

Reflected light

Light pulse

FIG. 25.2. Illustration showing an arm artificial nerve network. (Picture courtesy of Dr. Donald Lyons.)

Loss of touch and pressure sensation may occur as a result of spinal cord injury, brachial plexus trauma, diabetes, or Hansen's disease. Not only does this sensory deficit interfere with fine control of grasp and pinch forces even when muscle function is retained, but lack of protective sensation increases risk of injury to the hand. Further, when disease or spinal-cord trauma robs patients of their sense of touch, they not only lose grasp control but also risk hand injury. Medical engineers at Stanford University and the Department of Veterans Affairs are hoping that will change (Jaffe, 1995; Sabelman et al., 1999). They are currently exploring an avenue of research that, by using implantable touch sensors, some feeling to nerve-deafened fingers could be restored (Figure 25.3). To pursue this idea, they built a functional 20X scale model of an inductively coupled

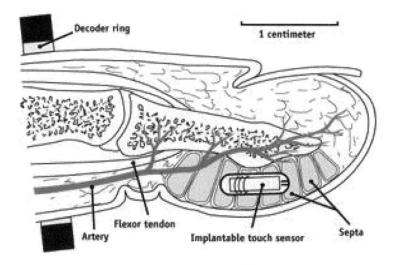

FIG. 25.3. Concepts for externally mounting sensor arrays on the
hand with implantable sensor. (From Sabelman, Kovacs, Hentz,
Rosen, Kurjan, Merritt, Scott, Gervin, and Teachout, 1999, VA Rehabilitation R&D Center, Palo Alto, CA, http:/guide.stanford.edu/
Publications/dev5.html.)

fingertip force transducer (miniaturization to the 1 mm × 4 mm size required for implantation was too expensive given their resources). Interestingly, the state-of-the-art of implantable sensors has advanced to the
point that transducers can be encased in glass tubes small enough to inject
through a large-bore hypodermic needle (Sabelman et al., 1999). The technology that they propose to adapt for deep fingertip pressure transducers
includes the following: a pressure-sensing diaphragm, amplifier, signal and
excitation voltage-conditioning circuits, and output transmitter that will be
constrained to fit within a 3 mm tube of less than 10 mm length. Sensing
pressure through miniature transducers, they would radio signals to nearby
rings, which would then contract accordingly to give the touch feedback to
the patient. The rings would also power the implants via inductive power
transmission.

3. LIMB PROTHESES

The current "average" artificial hand for amputees has not changed very
much in fifty years. They are still controlled by moving shoulder muscles
attached to cables. Recent advances have reduced the cost and weight of

these devices. Technical improvements have also increased the efficiency of the artificial hand, all the while requiring less shoulder force to generate the same hand force. The majority of these prosthetic hand devices still fall short of what is possible to develop with the present state-of-the-art in microcomputer technology. Dr. Williams Craelius in the biomedical engineering department at Rutgers University has recently been working toward a tendon activated prosthetic system since earlier research teams have found it a challenge to isolate the exact muscle groups involved with individual finger motion. With more research and development, a dexterous prostheses may enable amputees to accomplish a variety of tasks that have been previously impossible with the standard primitive clamplike function of many current hand prostheses.

More modern prosthetics has been created with basic shoulder and elbow motions. One prototype bionic arm/shoulder combination has been under evaluation at the Princess Margaret Rose Orthopedic Hospital in Edinburgh, Scotland. The arm called EMAS (Edinburgh Modular Arm System) uses electronics for control of the battery-powered motors with epicyclical gearboxes. The smallest motors and gearboxes power the fingers and thumb functions, elbow, and shoulder. With this system, very basic arm, hand, and shoulder functionality is possible.

3.1 Functional Neural Stimulation/ Functional Electrical Stimulation

Functional Neural Stimulation (FNS), also known as Functional Electrical Stimulation (FES), are the terms used to describe current research that involves electrically stimulating a muscle, or collection of muscles, to cause movement of the limb. This is a very promising line of research for computing under the skin in that it simulates what the CNS does for intentional movement, that is, it sends a "contract" command signal down select motor neurons ("motoneuron") causing contraction of the muscle cells controlled by those motoneurons. Voluntary movement utilizes a well-supported concept called the "size principle," which states that neurons are recruited in order of increasing diameter. Smaller diameter neurons tend to have a smaller number of muscle cells attached thereby generating less force, while larger diameter neurons tend to control a larger number of muscle cells. This increasing diameter sequence of recruitment allows for fine control, among other characteristics. Each motor unit (combination of a single motoneuron and the muscle cells it controls) can be fired independently of all others, and the signal to the motoneuron causes all the motor unit's muscle cells to contract (the "all or nothing" firing concept). The central nervous system (CNS) can control muscle force generation via two methods, by recruiting more motor units, or by increasing the firing frequency of the motor units already active. In this manner, the CNS can generate the appropriate force over an interval of time.

External stimulation of motoneurons causes the largest motoneurons to fire at the same time. This is the opposite order of recruitment as compared to natural motoneuron recruitment. External stimulation will cause an almost continuous contraction of the motor neurons firing, leading to fatigue in less than two minutes, which is not a desirable state for moment-to-moment control and functionality. Studies designed to explore selectively activating specific motor neurons is currently an area of active research (for example Hoffer et al., 1997; Kagaya et al., 1997; Parrini et al., 1997; Yoshida et al., 1997). Using feedback loops with computer control to activate the desired muscle groups is a goal, and many research efforts are underway with that as the long-term objective. As might be surmised from comments in this section, many pieces of the neuromuscular feedback control loop are not particularly well understood and this must change before computer-controlled FES/FNS will have much utility.

3.2 Stimulation Methodologies

There are two major methodologies to stimulate muscle, percutaneous (through the skin) and with the use of implanted electrodes. The benefits for the percutaneous method are that surgery is not required and it is easy to access hardware for changes. Drawbacks include increased current requirements that often damage the underlying tissue, discomfort if sensation remains, and the lack of consistency in electrode placement. Implanted electrode stimulation requires surgery and all of the inherent risks associated with that, and additional surgery if problems develop or to perform routine maintenance of the hardware implanted. Implantable stimulation tends to generate a very repeatable signal over the long term with minimal tissue damage.

Another method being studied in a variety of settings to cause muscle contraction is through the use of very strong well-placed magnetic fields. Functional Magnetic Stimulation (FMS) tests have been applied to stimulate phrenic nerves to provide stimulus for breathing in spinally compromised patients (Similowski et al., 1989). Other researchers have yielded positive results for thoracic spinal nerve stimulation (Chokroverty et al.,

1995), motor cortex for respiration function (Lissens, 1994), and to assist in restoring cough in patients with tetraplegia (Lin et al., 1998). The restoration of cough and breathing functions without the assistance from other devices/methods might well lower the overall mortality of these populations from a variety of lung infections that plague this community of patients.

Replacing the upper extremity with normal function for the amputee or replacing the function of the arm for individuals with a damaged neural system is not possible with today's technologies. Transplantation of a compatible upper extremity on humans has only very recently been attempted with the outcome of such surgeries in the future still in question. Tissue rejection and immune attacks on the transplanted material make the use of powerful immune suppressants mandatory, despite the best compatibility in terms of transplant selection. The side effects from these drugs are very serious, so that it is not likely that transplants alone will solve the problem of restoring upper extremity function. Interestingly, robotic hands/arms have been created that can accomplish very specific highly skilled tasks, but broadening the scope of performance of these devices and integrating them with an intact nervous system has not yet been accomplished. Research continues in these areas.

We have noted that what is currently available to the amputee can be designed to be cosmetically acceptable and lightweight but clearly does not come close to simulating the impressive dexterity of the human hand. Most amputees in the United States choose to utilize hooks for the hand, primarily due to their functional abilities. Amputees elsewhere tend to choose prosthetics that are lifelike in appearance but provide almost no function. The functionality available depends upon whether the amputation occurred above or below the elbow. For the below-elbow amputee, there are three main classes of prosthetics to choose from, a hook, a hand or a nonstandard device. Hooks are body powered and they frequently open and close by movement of a shoulder muscle. Hands may be either body powered, as in the hook, or use electric signals for control. Hands controlled by myoelectric signals are available, and recent designs include sensors to detect when a gripped object is slipping. The sensors feed that information to a wearable computer chip inside the artificial hand, which adjusts the grip. The unaffected hand normally accomplishes rotation of the prosthetic arm.

Above-elbow amputees have another choice in that they may choose body-powered or electric elbows to attach to their prosthetic hand. As with the hand, the body-powered elbows extend or contract based on the position of a particular upper arm muscle. A myoelectric signal from a surviving muscle or a switch may be used to control electric elbows. Any of the choices for hand may be attached to the prosthetic elbow. The combination of body-powered elbow and computer-controlled hand may be customized to specific needs and/or requirements of the individual. A tactile sensor in an artificial hand holding a cup of coffee can detect slippage and increase the pressure on the cup, preventing spillage and a potential burn. It may also be used to insert a disk into a computer, using the speed of the computer to aid the individual in more independent living.

In the future, it is entirely plausible that signals from the upper arm nerve bundles, or muscles (or both), will be routed through a wearable computer that will in turn activate a rather sophisticated neural network-based arm/hand response that will be able to perform many functional tasks that are important for daily living. While playing Mozart on the piano is unlikely, very beneficial daily living functionality is within the realm of the possible in the next ten to fifteen years at the present rate of technology development.

Individuals with intact limbs, but a damaged CNS, have a different set of issues to consider. The extent of the CNS compromise or disease plays a significant role in options to consider. As the innervation of the shoulder and arm muscles varies at different levels of the spinal cord, damage to the lower spinal levels provides more options, as more of the nervous system is intact and more natural motor control remains. Due to the variability of the damage and specific actions desired by the user, each implementation is custom designed. For those with C5 or C6 level of injury, elbow flexion and shoulder control remains intact generally, but elbow extension could be minimal or lost. Crago et al. (1998) have developed and tested an elbow extension prostheses to allow for arm extension in specific positions for these levels of spinal injury. One of the individuals tested had already used an implanted hand-grasp neuroprostheses for over ten years with this system. The triceps (elbow extension) was stimulated via implanted electrodes in either of two situations—a switch is turned on or the arm is lifted above a particular threshold level. The controller turning on the stimulator is worn externally and in this case the signal is transmitted via a RF (Radio Frequency) link.

The FDA recently approved its first implanted system that included the FNS technique, known as the Neurocontrol Freehand System. This system includes implanted stimulating electrodes on muscles in the lower arm and hand, sensory electrodes near the shoulder, a shoulder-mounted joystick (controller), and stimulator, all implanted. Power and control signals are transmitted via RF signals across the skin to the stimulator, located in the chest region. The only external portions of this system are the transmitting coil and the processing unit. For some individuals a "tendon transfer" is necessary for hand grasp. In that case, a functioning muscle (or portion of one) is attached to a muscle that is nonfunctioning to augment contraction. The system allows for two functions, palmar grasp, where the tips of the thumb and fingers touch (or close to it), and lateral grasp, where all fingers curl in towards the palm, which can be used to hold silverware or a floppy disk. Many of the options available for the upper extremity can be used on the lower extremity. The leg has significantly less degrees of freedom for mobility than the arm but has a significantly greater strength. These characteristics must be taken into account for the replacement or augmentation of function.

The below-knee amputee has many options for a functional prosthetic, although few (if any) have limb motion control. There are three basic designs: the solid ankle cushion heel, the single-axis design, and the multi-axis design. Each has its strengths and weaknesses and there is not a "best choice" for every amputee. Some prosthetics require less energy to use, others are designed for the more active individual, while others are very inexpensive or long-lasting. A few manufacturers claim that their prosthetics store energy and then return expended energy in different phases of the walking cycle. Reliability, function, weight, and costs are all factors in deciding which artificial foot to use. None use implanted chips of any kind at this time.

For those with a spinal cord compromise in lower areas of the spinal cord, an implantable system to assist in transfers (from bed to chair and such) has been under development since the 1960s. These individuals often have complete use of their upper limbs, but they are usually severely compromised in the lower back or leg regions. The technology is very similar to that of the upper extremity, the main difference here being the number of muscles involved and the amount of current required to active the appropriate muscles. Leg extension (standing) involves very large muscles and must be carefully and simultaneously activated on both legs. This technology allows for limited standing and walking with assistance, but a return to normal gait activity remains for future development. In the early 1980s, there was a lot of publicity concerning FNS and its apparently promising future. Progress has been made, but FNS still does not live up to the media hype that was generated.

A rudimentary nonimplanted system for walking has been FDA approved since the mid-1980s (Graupe, 1999). A 16-channel implanted FES walking system is a representative example of what has been tested (Sharma et al., 1998). Two surgeries were required for implanting the electrodes, one for the muscles on the front and one for the muscles on the back of the legs. Nerve cuff electrodes were used to stimulate the appropriate muscles, allowing for adequate contraction. A preprogrammed menu allowed the user to choose from a selection of stimulation patterns for standing, walking, sitting, and exercise functions.

The main technical difficulty with these systems at present is that the artificial stimulation of the muscle groups in the legs does not have the fine neuromuscular integration that the gait and balance systems provide in the intact spinal cord, where various signals are fed back through to the neuromuscular control centers. The result is a jerky, unbalanced, robotic style of gait that is far from satisfactory. Dramatic improvements are needed to integrate balance information into the leg muscle gait programs before a true unaided walking capability is achieved. The equilibrium/muscle activation feedback loop in such artificial systems would require a very sophisticated wearable computer that could not only fire the large muscle groups properly but adjust the pattern of firing very quickly (most likely at the tens of milliseconds level) and precisely for the smaller groups as well. This would allow for the slight moment-to-moment balance corrections to gait so that the person using the system does not topple over. It would be very important for the "smart" wearable computer system to account for seemingly unimportant factors in the environment such as mild wind gusts. These are not a problem for an intact neuromuscular control system but would represent a serious programming challenge indeed for the person using an artificial gait generating wearable computer program. This raises the issue with regard to the requirement for some type of accelerometer measurements that can be integrated into the gait program that would serve some of the functions of the neurovestibular system where balance control and recovery is concerned.

4. OTHER PROSTHESES

4.1 Pacemakers

Improvements for wearable computational cardiac pacemaker technology has proceeded at a respectable rate. In some situations, the implanted pacemaker provides stimulation for a majority of the heartbeats of the wearer. For other heart conditions, the implanted instrument is often more of a defibrillator than a pacemaker. These instruments, termed Implantable

Ventricular Cardioverter Fibrillators (known as "ICDs"), detect abnormal heartbeats and provide stimulation to the wearer when abnormal tachycardias are detected. As such, detection and discrimination between normal and life-threatening cardiac situations is required of the device. While these devices perform admirably a great majority of the time, occasional inappropriate shock therapy still remains a problem (Grimm et al., 1992). These inappropriately delivered shocks may compromise normal sinus rhythm, as well as reduce battery life and create some patient anxiety.

Current computations that the wearable ICDs can perform are based on comparisons of heart waveform morphology by evaluating the anticipated heart electrical signal shapes with the actual ones recorded by the ICD. These pacemaker devices may utilize a host of computational techniques to evaluate the signals including: frequency domain analyses (spectral comparisons between measured and anticipated spectrum), amplitude evaluations, or zero-crossings. The most sophisticated devices use a combination of the above, such as comparing frequency shapes on filtered data. Such higher tech signal analysis comparisons prove to be more robust, but increase the power drain, which results in the need to replace or recharge the implanted batteries. Such a procedure always has a small but real risk of infection or other side effects from minor surgical procedures, which can lead to very serious medical consequences. Development of new algorithms to decrease the chance of inappropriate shock therapy is an active field of research (Morris et al., 1997; Shkurovich et al., 1998).

An implantable ventricular assist device is being tested to aid those individuals waiting for heart transplants (Mussivand et al., 1996). The device might be an important predecessor for a successful artificial heart in that it allows for portability, among other functions. The implantable components include a battery pack, the pump mechanism, and a coil used to receive power and instructions from the external controller. External components are another battery pack, the external portion of the energy and information coil, and controller electronics. These systems are occasionally now just coming online in some cases, as some good candidates for heart transplant await viable hearts.

4.2 Urinary Tract Control

The loss of bladder control is one of the first things to occur in spinal cord injury. Return of bladder control contributes greatly to an individual's feeling better about their quality of life. Because of this, one might surmise that much research occurs in this important area. Unfortunately, this is a false impression. Access to the nervous system is difficult and the various signals controlling the bladder function are not particularly well understood. Some researchers have tested implantable sensors in the bladders of dogs after a spinal cord transection and achieved good results (Sawan et al., 1996). They attached excitatory electrodes both to the bladder and the controlling nerve for bladder function and tested a variety of stimuli to induce muscle fatigue thereby causing urination. These researchers captured the nerve signal originating from the bladder for the timing of bladder release. Between 1989 and 1994, 44 Dutch patients were implanted with a urination control device that stimulated the bladder for three seconds and then did not stimulate it for an interval of time (Creasey and Van Kerrebrock, 1996). Of the 44 patients studied, 41 could empty their bladder completely with this device. Twenty could use this device to assist in defecation, and 62% of the men could achieve erection with this stimulation. Due to anatomical considerations, it is believed that this technique could benefit from a better selection of stimulation pulse shapes and patterns.

4.3 Defecation Control

Along with urinary tract control, control of defecation is one of those high priority items to the SCI patient. The pace of current research does not appear to be making inroads to solve this problem as fast as one might expect given the developments in other areas of prosthetics. The defecation function has unique challenges that do not allow for the information developed in other prosthetics to transfer as easily here as from other research areas. The anal sphincter contains both smooth (involuntary) and voluntary muscle, and the bowels work in a coordinated effort with the many anal sphincters in a manner that is not precisely understood. As smooth (involuntary) and voluntary muscle generate force differently, the transfer of techniques used in other muscular contraction is not trivial. A massive colonic contraction, such as that achieved with FES, would not generate the desired results and might generate more problems than solutions. Implanting hardware that compressed the colon using different geometries has been tried, but this approach caused problems in the bowel walls. A portion of these systems needs to be implantable due to the physiology involved. Much work remains to be done in this area for satisfactory defecation control results.

4.4   Tremor Suppression

For those individuals with hand tremors, picking up an object is sometimes an impossible task. A passive orthotic that would eliminate or minimize the tremors would greatly increase their capability to be self-sufficient. A passive wrist support that acts as a low-pass filter has been developed to minimize tremors that affect the flexion/extension of the wrist (Kotovsky and Rosen, 1998). The device is small and inexpensive.

Another approach is being used to treat high-amplitude debilitating tremors in both Parkinson's Disease and in a condition known as Essential Hand Tremor. Some success with these patients has been achieved with regard to tremor reduction by providing deep brain stimulation to areas of the thalamus with electrodes that are implanted surgically. Generally, the best results of deep brain electrical stimulation are achieved by using frequencies greater than 50 Hz, where the electrical pulse width and amplitude are varied based upon the patient's responses. The power source for the electrical signal is implanted subcutaneously in the upper chest area with wires carrying the stimulating charge up through the neck and into the targeted deep brain region (Figure 25.4). The pulse width and amplitude of the implantable device can be varied by external control signals that the device picks up and interprets from under the skin.

4.5   Respiration

Electrical stimulation has been used for over thirty years to restore breathing to individuals with high quadriplegia causing respiratory paralysis and to individuals with miscellaneous hypoventilation syndromes (Creasey et al., 1996; Glenn, 1980). These systems are comprised of stimulating electrodes implanted on the phrenic nerve, attached to an implanted stimulator, and powered aɪ·d controlled externally. The concept of artificial stimulation to support breathing is conceptually appealing but has had limited application. Initial care for these individuals normally includes mechanical ventilation for its ease of use, but for individuals needing assistance in the long term, mechanical ventilation allows the respiratory muscles to atrophy, which is rarely healthy. Some studies have been completed that look at using stimulation to help the patients be weaned from the mechanical ventilator (Schmit et al., 1998). Implanted long-term respiratory assistance devices use the patient's own muscles for breathing and allows for greater mobility. These devices are currently available commercially.

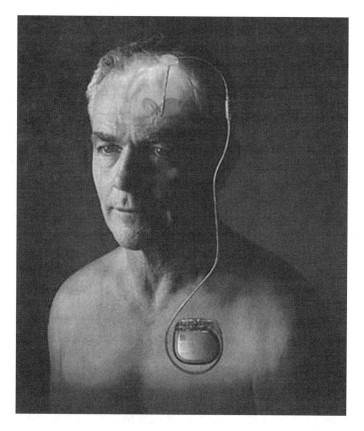

FIG. 25.4. Deep brain stimulation with an implantable chest device (Metronics Web Site).

There has recently been some work with canine models to determine if Functional Magnetic Stimulation (FMS) of respiratory muscle groups to contract, rather than direct, electrode stimulation might be a viable alternative for the SCI population (Lin et al., 1998). The importance of being able to generate a functional cough and reliable breathing for this population cannot be overstated. Pulmonary complications are a very serious danger to the many SCI injured people. It is not clear yet if this approach will ultimately result in using magnetic fields to stimulate respiration in SCI patients; however, canine tests show that there are promising avenues of research for FMS in this area. If FMS turns out to be workable, wearable computers will play an important role for respiratory regulation with feedback control derived from the respiratory status of the patient with a host of sensors that would be transmitting data from the patient that would evaluate acid-base balance, oxygen, and carbon dioxide levels, among several others.

5. LIMITING FACTORS

As with any research and development efforts where interfacing with technology is concerned, there are difficult technical problems to be solved, which might be referred to as limiting factors. While the level of comprehension of neuroanatomy, physiology, and biomechanics is far from complete, efforts are still underway to better understand and design supportive joint interfaces and wearable computers for the body's various systems. Commentary follows with respect to some of the ongoing major challenges.

5.1 Materials/Electrochemistry

The immune system is built to keep aggressors out and as such it creates many challenges for those trying to build implantable devices. Titanium has the characteristics to become osseo-integrated with bone and is accepted as "self" by the immune system. This is not the case with other implanted materials. Typically, devices that are implanted either surgically or injected trigger an immune response, which leads to a fibrous capsule being formed around the implant over time. This has occurred with the well-known case of silicon breast implants. This thickened fibrotic tissue capsule is formed from a variety of reactive cells and contains lymphocytes, macrophages and giant cells (Cameron et al., 1998).

Many studies have shown that a fibrous capsule around an implant requires a change in the implanted electrode current for generating a constant stimulation of surrounding tissue over time; however, a recent study questions that finding and reports a variety of responses from the same stimulus in different muscles. Other studies have shown that attaching a protein layer to the implanted device can reduce the immune reaction but does not eliminate it entirely (Kottke-Marchant et al., 1989; Tang and Eaton, 1993). What is needed is development of a material or coating that does not generate an immune response irrespective of the type of interfacing device with the body that is required.

Other problems with implantable devices for the weight-bearing joints such as knees and hips are that the implants eventually wear out, with replacement needed. While the materials science for these devices is improving with each passing year, often the timing of when a patient will receive such a joint is based not upon functional biomechanical considerations but on how long a patient is expected to live such that several operations to replace worn out prosthetic joints will not be necessary. This is a particular concern as the patient gets older and surgery becomes more of a risk.

5.2 Algorithms

The most commonly used algorithms for evaluating biological signals to generate control signals are those that we know and understand, such as amplitude detection, zero crossings, or frequency detection. Unfortunately, these techniques are not optimized for biological signals and do not provide the insight needed to use a natural signal generated by an individual into a usable, reliable control signal for a prosthetic device. As can be seen in the discussion on recruitment, the CNS utilizes a wide variety of techniques to achieve the desired physiologic result. These techniques are robust, noise insensitive, and tend to be nonlinear. As such, simple linear decomposition algorithms provide limited insight or incomplete guidance for decision making. Algorithms need to be developed that more closely mimic the successful biological algorithms, that is, for FES/FNS, an algorithm needs to be developed that more closely replicates what the CNS generates. Biological algorithms may appear to be very simple, as can be seen in the "all or nothing" concept of motor unit activation, but these algorithms have respectable numbers of processing units (neurons) that work together at multiple levels of analyses to give a "simple" biological outputted result. Until we can more closely mimic these natural biological algorithms, our ability to replace the complexity of biologic function with silicon interfaces for substitutes will be limited.

5.3 Computational Speed

The current speed of computers and chips, when compared to a single neuronal circuit, is impressive. However, this speed advantage is completely lost when compared to neural networked circuits in that there are hundreds and sometimes thousands of neural circuits to perform a single action—and even more to provide feedback responses to the body for that action. Also, many neuronal circuits are built specifically for a single task that has direct connection to the effectors. The chips we design can be designed for a single task, but it is difficult to design chips to adjust to the natural noise inherent in biological systems, and they cannot at present be attached directly to the effectors of the intended neural action. That may be slowly changing as noted at the end of this chapter with the development of the "neurochip," which combines a neural network with a recording electrode interface. The neurochip, however, is only a very preliminary step in the biological/chip interface direction. For the time being, designers must use the primary strength of chips—speed and ever-expanding memory capability—to attempt to remedy these limitations. A good example is that a single neuron in the CNS receives input from one to many different neurons, determines the influence of each input, and then sends out an excitatory or inhibitory output to another neuron(s). Neuronal interactions may change over time (that is, a neuron might have connections to 100 particular neurons at a particular time, and after a month, have connections to 110 neurons, with only 90 of them the same from the previous count). There are an estimated 10 billion neurons in the human brain and with all of the functional neuronal interconnections some have calculated the number of interconnections as being "hyper-astronomical" (Edelman, 1992). This complexity and sheer number of potential neuronal interconnections is what gives the human cortex its amazing intellectual capabilities.

5.4 Control

Control of prostheses, other than passive prosthetics like the cochlear implant, is at present very rudimentary. Historically motor prostheses have been controlled by switches, hands, eyes, breathing, and the like—or with the activation of a specific muscle that has been designated as the "control" muscle for some aspect of the prosthetic's use as noted earlier (e.g., moving the left shoulder forward to close a right artificial hand). While this indirect control method is somewhat functional, it is certainly not optimal. Unfortunately, the technology is not good enough yet to reliably detect the original control signal from the CNS for most prosthetic functions. Detecting and properly decoding neural signals provides much promise to control prosthetic devices but is currently one of the biggest technical challenges that must be surmounted (Crago et al., 1996; Hoffer et al., 1996).

5.5 Power Issues

All neuronal implants, from muscle stimulators to pacemakers, require some sort of power source. Many have lithium or other newer batteries, but batteries with any substantial long-term power output are relatively heavy and may leak, which can be dangerous. Traditional batteries also require regular replacement or changing, which often means some invasive intervention with all of the inherent infection risks. Many prosthetic designs are limited by their need to draw a certain amount of power. One working solution to this problem is to provide power through the skin. This can be done by providing power to the system using a transformer externally and by using an implanted converter. This equipment can provide power and control signals to the implanted electronics and has been in use for many years (Towe, 1986). The FCC has reserved certain frequencies for biomedical applications.

5.6 Electrode Orientation/Waveform Selection for Muscle Contraction

Many different electrode materials and geometries have been tested to determine the best way to generate natural recruitment for muscle contraction. A nerve cuff electrode, developed in 1988, is the style predominantly used for most electrical stimulation (Naples et al., 1988). A wide variety of materials and geometries have been tested for the nerve cuff to simulate natural recruitment. These materials range from platinum and platinum iridium to circular electrodes and electrode arrays for the electrical stimulation geometries (Ison and Walliker, 1987). The electrical current pulse shape and timing significantly affect the neurons, generating many options for the best combination of pulse shape and frequency to achieve the desired results (Fang and Mortimer, 1998).

5.7 Other Limiting Factors

As noted earlier, feedback for normal movement is provided through a complex system of stretch and position sensors. Feedback for prosthetics is minimal at most due to the difficulty in stimulating the correct neurons and is currently limited to visual feedback (seeing the movement). The motoneurons for movement are fairly large and the proprioceptive neurons are very small. Because of their size, these proprioception neurons are very difficult to isolate, stimulate, and study. A single neuron may be all that is needed in some specific cases to signal position, and to detect and isolate this single neuron is an immense challenge. The next biotechinical challenge would be to stimulate it appropriately. Some research is progressing in this area with promising results. Sinkjaer et al. (1997) have measured such muscle afferents to determine if this information can be of some use in later FES systems development.

6. THE PROBLEM OF DIABETES

Diabetes affects nearly 14 million people in the United States alone. This is only approximately 4.6% of the U.S. population, but represents nearly 15% of the direct healthcare costs to the country. This means that diabetes is a very important national healthcare issue. There are two types of diabetes, Type I (insulin dependent) and Type II (non–insulin dependent). Type I diabetes typically has an early onset age and is often referred to as "Juvenile Diabetes." This type of diabetes results from the loss of production of insulin from the beta cells in the pancreas. Researchers believe that an immune system attack upon these beta cells in the pancreas triggered by a wide range of factors is the reason for these cells losing their ability to produce insulin.

Type II diabetes is more strongly linked to lifestyle attributes such as being overweight. In this type of diabetes, the cells in the body essentially lose their sensitivity to insulin such that the glucose in the blood cannot efficiently enter the body's cells and high blood sugars result. High blood sugars from any type of diabetes results in serious health complications such as: retinopathy (which may lead to blindness), kindey failure, peripheral neuropathies, and a variety of heart and vascular diseases. Research has shown that proper diet, exercise, and weight control are the best allies against the development of Type II diabetes. At present there are several approaches to combating Type II diabetes. Exercise and weight control top the list. However, failing that there are many drugs available now that enable the body's cells to better utilize blood glucose.

Type I (insulin-dependent) diabetes presents another problem. Since the pancreas is producing little or no insulin, the biomedical challenge now is to provide it to the body in a near-physiologic way. This is difficult in practice to accomplish. Various short- and long-acting insulin shots (injections) may be taken and are somewhat effective in providing the body with insulin when needed, so long as the patient adheres strictly to dietary and activity constraints. This regimen is required so that the insulin is most effective given the overall fasting metabolic state that the patient finds him- or herself in at the time the type of insulin injected is most effective. Too much insulin in the system at the wrong time produces hypoglycema, while too little insulin produces hyperglycemia. Either extreme physiologic state has

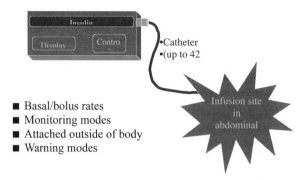

- Basal/bolus rates
- Monitoring modes
- Attached outside of body
- Warning modes

FIG. 25.5. An example of a wearable insulin pump. The various control, display, and monitoring modes are shown.

potentially dangerous short- and long-term consequences. Constant monitoring of blood glucose levels, often by needle stick, is required to monitor and "fine tune" the doses of insulin required.

New technology has now enabled many people to bypass the requirement for injections to deliver insulin into the body. The solution for many is Insulin Pump Therapy. A typical insulin pump is a small wearable microcomputer that has a place for storing a reservoir of insulin. The insulin is delivered from the reservoir by a flexible plastic catheter to an infusion site in the abdominal region. The wearable insulin pump microcomputer has programmable control elements in it for the delivery of insulin at basal and bolus rates into the body (see Figure 25.5). The programmable "basal rate" for insulin input is a continuous flow of insulin at a low level into the body that is required for nonfed state lower energy (glucose) utilization state. A bolus rate is infused when the person eats a meal such that the higher blood glucose levels are able to be handled properly by the body.

Currently, the insulin pump is "open-loop" technology. This means that the person wearing the programmable nsulin pump must handle the determination of the rates of insulin infusion by interfacing directly with the wearable insulin pump. This is done by considering the current blood glucose reading and trends thereof, the amount of food previously consumed (or to be eaten shortly), and the current and anticipated physical activity levels. Maintaining good "tight control" of blood sugar values at the present time with insulin pump therapy is quite possible. To accomplish this feat requires much nutritional science in terms of the types of calories consumed (sugar, protein, or fat) and a lot of guesswork with artful intuition applied based upon previous experiences.

At present time some type of needlestick to draw blood for testing the glucose level is required; however, there is much ongoing research directed toward being able to monitor the level of blood glucose noninvasively. Doing so would enable diabetic patients to know more accurately at a given moment what their blood sugar status and trends are, such that the proper amount of insulin might be infused without the insulin over- and under-shoot problems that are typical of less frequent monitoring.

Future possibilities for "closing the loop" on wearable programmable insulin pump therapy involves not only more nearly continuous monitoring of the body's physiologic state but the ability of the device and person interfacing with it to alter its programmed rates easily and quickly for a change in exercise level or caloric intake factors (Jaremko and Rorstad, 1998). Figure 25.6 highlights some of the key elements needed for closing the loop on insulin pump therapy responses. At any level of analysis the person wearing the pump has to be properly able to assess how well the insulin pump and wearable computer interface is doing its job of achieving blood glucose homeostasis.

Solving the problems of noninvasive blood glucose monitoring with glucose sensors (Gough and Armour, 1995; Sternberg et al., 1995), developing closed-loop control algorithms (Brunetti et al., 1993), and making the device usable under a variety of situations is not a trivial task. It is possibly achievable in the foreseeable future with advancing biomedical

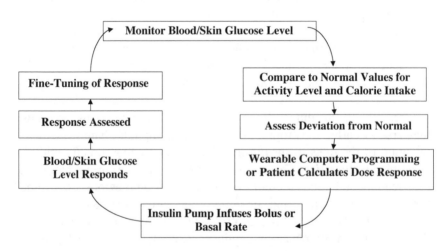

FIG. 25.6. Key points in a flowchart format for blood glucose (sugar) control using Insulin Pump Therapy.

technology for nonextremes in activity states. Technical issues remaining to be addressed, aside from nearly continuous glucose monitoring that would be required, include the potential uncertainty in sensor measurements and insulin absorption rates that may introduce instability in the computer control algorithms. With this is mind, some rather complex mathematical treatments for insulin control have been proposed, including H-infinity theory (Kienitz and Yoneyama, 1993). Other control strategies involve the use and input of clinical judgment in the control system by deciding how much to tolerate moderate hyperglycemia, or to try and retain very "tight" control at the risk of hypoglycemia (Brunetti et al., 1993).

7. MEDICAL SIMULATION

The advent of better simulation technology due to more and faster computer capability has opened a myriad set of possibilities for training physicians and medical research. In recent years numbers of pilots have been using flight simulators to learn basic procedures, test and rehearse dangerous procedures, or to provide a skills refresher for proficiency. Now, doctors and nurses in training, or on staff at some of the larger hospitals, can do the same procedural training with very sophisticated computerized medical dummies. So far there are two manufacturers of medical dummies: Medical Education Technologies of Sarasota, Florida, and MedSim-Eagle Simulation of Binghamton, New York. The "patients" they provide as dummies breathe, have pupils that dilate, airways that close, and hearts sounds that are realistic at a cost of approximately $200,000 per training dummy. A host of crisis situations can be generated from the classic heart attack victim to anaphylatic responses from drug allergy that may be life threatening. Specialized medical simulators for heart sounds and anesthesia team and crises management also exist, and these are gaining more widespread acceptance as the fidelity of the simulations increases (Howard et al., 1992).

Many rapid advances are being made with respect to using virtual/ augmented reality technology and simulation for a wide variety of additional medical training environments including: anatomy training for medical and nursing students, procedural simulators for various types of surgical and endoscopic/laporoscopic procedures, and visualization technologies for enhancing a host of surgical outcomes (Westwood et al., 1999). An expanded discussion of some applications of this technology for neurosurgical cases will be highlighted as an example just a bit later in this chapter.

Other applications of modeling technology are changing the way ortho-pedic medicine will be practiced in the future. There will probably be a slow but perceptible trend toward customizing a wide variety of orthopedic implants by integrating the patient's particular anatomic data from scans with solid-modeling systems from the mechanical engineering domain. For example, an artificial hip joint that is not somewhat customized for a partic-ular patient will be less likely to have a favorable long-term outcome from a variety of anatomical and wear perspectives than a hip joint that has been chosen as more appropriate for that specific individual. Future orthopedics practice will gradually dispense with the concept that "a couple of sizes fit all" and replace such thinking with the notion that there is an optimal size of implant based upon computer scans and specific patient anatomic geometry that is likely a better (albeit more expensive) choice. The implant chosen may be arrived at computationally from among a hundred or so possible design configurations based upon "common" anatomic geometries (Deitz, 1997a).

Orthopedic foot and ankle abnormalities due to structural problems or overuse injuries have traditionally been corrected by educated guesses with orthotic inserts for footwear, which are incrementally adjusted by trial and error over a period of time. Often problems surface years later in other bones or joints because the adverse stresses were simply displaced to other areas of the body. As might be expected, transmission of foot problems to the knee or hip joint are common if the underlying biomechanics are changed unfavorably.

Traditionally the development of good orthotics for a person's foot or ankle problems whether from sports injuries or disease processes such as arthritis has been a mixture of art and science. More recently, better human gait simulation programs have been interfaced with state-of-the-art customized finite-element-analyses software and a patient's specific anatomic information such as age, weight, leg measurements, etc. to pro-duce output from the software that includes animated graphic images of the patient running or walking at various speeds. This presentation may be slowed or magnified for closer gait evaluation. The animation can show the full gait range from heel strike to toe-off and can display limbs as realistic solid models or in skeletal from (Deitz, 1997b). The output of the gait data and the patient's other medical information help the users of this software to make a better orthotics choice and enable modeling the effect of that specific orthotics placement on the patient's gait. Fu-ture improvements to this technology will include incorporating foot force plate data into the simulation equations and using MRI scans to better depict the structure of bone mass as well as ligaments, tendons, and muscle groups.

An excellent general description of the technical challenges and approaches currently used to computationally and graphically model human locomotion may be found in a summary article on "Animating Human Motion" (Hodgins, 1998). A few web sites with human gait animation are referenced in the Hodgins article.

Outcomes from orthopedic surgery intended to improve musculoskeletal function are often unpredictable. This is true to some extent because the development and testing of new surgical procedures relies heavily on clinical trials after the surgery has been accomplished. Reliable tools for quantifying surgical changes in functionality beforehand to predict postoperative results do not yet exist outside of some early laboratory efforts. Dr. Scott Delp at Stanford University believes that the design and analysis of operative procedures can proceed more effectively if musculoskeletal computer models are developed that explain and predict (within bounds) the functional consequences of surgical interventions. His research has been directed in part toward creating a system that can be used in the operating room to implement individualized surgical plans, and for developing virtual environments that can be used to teach surgical techniques (Delp and Loan, 1995; Delp, Stuberg, Davies, Picard, and Leitner, 1998).

There are other interesting uses for medical simulation based upon data taken from an individual patient's anatomic data set. One fruitful area in particular is the study of how a person's carotid artery geometry affects their blood flow dynamics under a variety of pulse pressures, such that assessments can be made as to the likelihood of atherosclerotic plaques forming due to local blood flow patterns in the artery itself. Such plaques result in "clogged" or blocked arteries. These hemodynamic computational results may be eventually extended to other areas of the body where turbulence or flow changes occur that may lead to the formation of atherosclerotic plaque in susceptible individuals. This approach to evaluating hemodynamics can one day be able to help predict those patients with certain medical profiles that are at high risk for artery blockage due to various combinations of blood flow patterns, blood chemistry, and very specific anatomic geometry. Dr. Christopher Zarins at Stanford University has been an active researcher and leader in this area by using fluid dynamics and biomechanics approaches to better define how and why vascular pathologies such as athersclerotic plaque buildup occurs.

8. VIRTUAL REALITY AND NEUROSURGICAL CASES

At a few hospitals and labs, Virtual Reality (VR) technology is finally beginning to realize some of its early promise. As the technology matures and the contributions of several diverse scientific and engineering fields are combined, systems that positively impact patient care are finally being achieved. For example, the technology now exists at the Brigham and Women's Hospital through the Image-Guided Therapy Program to take high-resolution MRI data from a patient's head scan and assign color codes to various brain structures such as gray and white matter, blood vessels, ventricles, and suspected areas of tumor infiltration. This process is obviously not a trivial one and is the result of collaboration between computer scientists, engineers, medical scientists, neurosurgeons, other doctors, and mathematicians (Black et al., 1997).

The process to generate such a database is a complex and tedious problem to solve in practice. Using MRI or Computed Tomography (CT) scans—which are noninvasive techniques doctors use to study different anatomic "slices" of the human body—an accurate anatomic model of the patient in question is generated (Wells, Grimson, Kikinis, and Jolesz, 1996). Then, based upon known anatomic features and geometries, tissue boundaries, and a healthy dose of higher mathematics and signal processing, a picture of the patient's brain may be generated. As an example of such data set manipulations, blood vessels' positions are not usually clearly defined in a traditional MRI. For this reason, information from MR angiograms must be superimposed on the original MRI database and merged into one model for good blood vessel definition and imaging. Using the mathematical concept of "mutual information" and a series of software imaging enhancements programmed by the Artificial Intelligence Laboratory at MIT, the information-rich and information-poor regions of the patient's brain are assigned a metric, which is then translated and converted to a color-coded graphical output. Further volume element (*voxel*) information processing using a variety of fairly sophisticated image processing techniques results in a graphical output that reasonably well defines the different types of tissues in the brain from suspected areas of tumor. This output is interpreted as belonging to a particular part of the anatomy that would be expected given the signal environment from, and around, the voxel of the image in question. Such technological capability as this is redefining the state of the art in healthcare where a high degree of imagery is routinely utilized for treatment decision making.

In addition to MRIs, further studies of the patient's cortex may include "functional mapping" of the brain to assess exactly where parts of the motor cortex reside that initiate certain movements. If surgery then takes place in an area near this region, the patient will not be paralyzed due to pressing the boundaries of the brain areas subserving the critical motor control regions. The functional maps of the cortex may be pictorially overlain on the patient's previously developed brain model and color coded as to how strongly that region affects a particular movement function. This provides the surgeon during the planning stages with additional information by which to assess the best anatomic approach path to the tumor (or other offending site), with the least amount of collateral damage.

Based upon a variety of rather complex assumptions, the tumor's boundaries can be assessed in terms of areas that are clearly tumor and those areas that appear to be "at risk" to be tumor. This aids the physician in assessing whether to go to surgery to begin with, or perhaps to use other therapies such as radiation or medical oncology approaches (Jolesz, 1997). In practice and fact, radiation beam type and targeting to enhance the delivery of the maximum amount of dosage to the appropriate tissue target with the minimum side effects to adjacent tissue is an art and science in and of itself and is by nature very case specific.

After the anatomic model has been constructed and properly "registered" into the computer memory, a new technical challenge appears if the patient goes to the operating room. The brain structures' positions change as the brain is exposed and the neurosurgery proceeds. This creates a novel dilemma that has to be overcome while the patient is on the operating table in terms of updating the shifts of anatomy due to the operation. This refreshing of anatomic information is accomplished by generating new scans of the patient's shifting anatomy that are immediately "registered" with the older ones in real time. This on-site computing under the skin is achieved by having the surgery occur in a newly invented MRI device, as opposed to the more traditional complete cylinder MRI device. In this way, updated anatomic information can be presented to the surgeon in real time to aid in decision making as the surgery proceeds.

An additional benefit of this integrated technology is that of being able to generate a two- or three-dimensional representation of the surgeon's scalpel position superimposed on the computerized graphical output as an aid to more careful surgical navigation. Areas to be avoided may be color coded as "warning zones" that contain key functional attributes based upon known anatomy and prior testing of the patient in question as described earlier. This type of integrative VR technology presentation allows for an intelligent assessment to be made of whether or not the surgeon is resecting all of the tumor possible given the patient's anatomy in real time. Often, it is difficult if not impossible with vision alone to ascertain whether the tumor had been completely removed, or exactly where the tumor margins might be. This allows for the preservation of as much healthy tissues as is possible, with the prospect of as much tumor being removed as is practicable in complex surgical cases.

There are parts of the body where even the powerful MRI cannot properly image tissue. One important tissue where this is the case is with lung tissue. Dr. Tim Beardsley from the University of Virginia Health Sciences Center noted in a short article in *Scientific American* (June, 1999) that specially treated gases such as helium 3 can be hyperpolarized and inhaled to provide much improved imaging of lung tissue. Because hyperpolarized gases such as helium 3 lose their "spin" faster in the presence of oxygen, comparisons between MRI scans made in quick succession should reveal blood flow patterns, since blood carries oxygen to the tissues and changes the helium 3 atom's spin quicker. Surprisingly detailed renderings of hard-to-image lung tissue has been obtained in limited trial studies. Another gas, xenon 129 is also mentioned by Beardsley in his article. This gas binds to certain chemicals in the body, which then are revealed by specially "tuned" MRI's. Luckily, xenon crosses the blood-brain barrier such that it may be used to distinguish different tissue types in the brain, raising as the potential for enhanced neurological imaging. Xenon imaging is not yet at the level of sophistication of helium 3 imaging, but since helium 3 is very rare (except on the Moon) and xenon is more plentiful, the xenon enhanced imaging would appear to deserve further study. Computationally intensive algorithms designed to fine tune the MRI machines to handle such special data sets are currently being investigated and refined. The helium 3 image contrast enhancement approach is expected to enter FDA Phase II clinical trials in 1999, with possible approval of the helium 3 contrast medium for clinical use around 2001.

9. FUTURE DEVELOPMENTS

Recently, attempts have been made to provide communication between the electrodes of a computer chip with the biology of a small group of cultured enurons. Researchers at the California Institute of Technology (Maher et al., 1999) have been able to isolate individual neurons in a one-to-one correspondence with electrodes, obtaining unambiguous access to each well-isolated neuron. This is a potentially dramatic contribution to neuroscience studies (and thus computing under the skin) since in theory it would allow for an exacting description of how neuron groups communicate in a noninvasive, bidirectional manner. Understanding the dynamics of a functioning neural network is thought to be one of the key elements for taking the next steps in being able to describe and predict how certain neuronal groups may respond to a variety of inputs. To accomplish this, simultaneous measurements of all of the neuronal components is necessary. The "neurochip" developed by the above referenced authors provides this capability for a small (16) group of neurons.

This is a first small step in an area that may ultimately become a breakthrough technology for neuroscience. This step is important since previous work on the electrical activity in cultured neurons was traditionally measured using intracellular recordings where a saline-filled glass pipette was used to puncture the cell membrane. Access to the state and activity of multiple cells using such pipette intracellular recording is difficult, but possible (Debanne et al., 1995); however, such simultaneous intracellular recording has several shortcomings for evaluating the neurons' network functions and plasticity: (1) the physical location of the synapses is difficult to determine, (2) penetration of the cell with a pipette alters the cell's properties and slowly kills it—usually within an hour, (3) acceptable recordings are obtained for a small fraction of randomly chosen cells that may in fact be atypical; and (4) simultaneous recording from more than four neurons is virtually impossible. As reported by Maher et al. (1999), other approaches to the study of neuronal interactions have included the use of flat-electrode arrays where metal electrodes are arranged on a surface of glass or silicon substrate, upon which the neurons are cultured (Gross et al., 1997; Jimbo et al., 1993). These systems are capable of recording limited specific aspects of neuronal activity.

Meister et al. (1994) have utilized flat-electrode arrays to describe the manner in which the mammalian retina encodes visual data from ganglion cells to send to the optic nerve. Maher et al. (1999) note that the older flat-electrode arrays have two major drawbacks that make them unsatisfactory for studying network function and plasticity at identified pairs of neurons. First, since the neurons are not confined to the electrodes, electrical stimulation of single neurons is difficult; and secondly, central nervous system neurons in dissociated culture are extremely mobile such that accurately identifying a particular set of neurons may not be possible. All of these difficulties pose some limitations to how much information may be gleaned from functioning neuronal groups under study. This is turn limits how fast the science of neuronal group interactions may develop.

The neurochip overcomes many of these limitations and might be described, as a flat-electrode array with the addition of a small cage (well) around each electrode that contains only a single cell body. This enables a one-to-one correspondence of electrodes to neurons by isolating and immobilizing the cells in a specific well. The neurochip does not compromise the neuronal cellular properties, and the authors report that individual cells may be studied for weeks. This will potentially enable the detailed study of a 16-cel cultured neural network (see Figure 25.7).

Some of the first tangible (albeit tentative) steps toward a direct brain–computer interface have been taken by neuroscience researchers Dr. Roy Bakay and Dr. Phillip Kennedy at Emory University in Atlanta, GA (Figure 25.8). These researchers have developed a neurotropic electrode, which is an electrode implanted with nerve growth factors that have allowed two

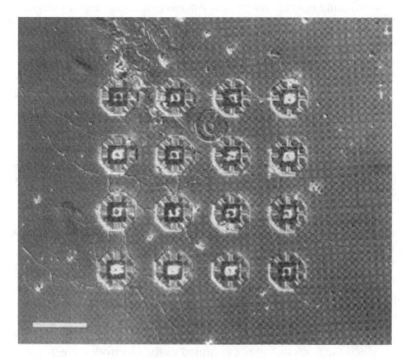

FIG. 25.7. Contrast micrograph of a hippocampal neuron culture after eight days. Several neurites emerge from most of the walls, forming a network. The very small thin "strings" are neuronal processes. (Picture courtesy of Maher and Pine et al., California Institute of Technology).

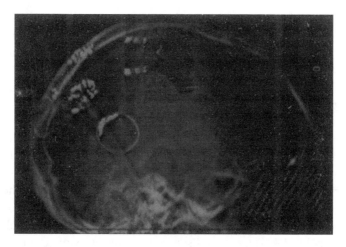

FIG. 25.8. Brain X ray of electrode implanted into the motor cortex. (Picture courtesy of Emory University School of Medicine.)

cognitively intact yet "locked in" patients to communicate in a primitive way via a computer interface. "Locked in" patients are those that are fully aware of their surroundings but cannot communicate with the outside world due to a variety of neurological conditions.

Bakay and Phillip's neurotrophic electrode is implanted into the patient's motor cortex and has a diameter of only 0.1 to 0.4 millimeters by 1.5 millimeters long. After implantation into the motor cortex region for controlling arm motion, the cortical brain cells are induced to grow into the electrode's tip and form contacts since nerve growth factors are placed in the neuroelectrode's tip. Sometimes up to three months may be needed for the quite small amount of motor cortical brain tissue to grow into the electrode and yield stable firing signals. These small neuronal electrical signals are transmitted through the skin to a receiver and amplifier on the outside of the skull. The recording system is powered by an induction coil placed over the scalp so that wires for powering the device do not pass through the skull, which dramatically reduces the chance for any infection occurring. Signal processors help to separate the desired signal from others in the local area. The output signal in then interfaced with a computer cursor that is able to move on the computer screen.

The patient then has to be taught how to "will" the cursor to move across the computer screen to various locations that allow him or her to select icons that are preprogrammed on the computer's user interface. The next step is to try and help the patient communicate to produce speech and word processing, rather than to simply select a phrase to communicate with. If the patient can be taught to communicate very basic commands to a computer, then perhaps even seriously impaired patients would have better control via a computer interface over their environment.

The next step is to try this technology on a large sample of patients. Other applications might include basic brain research, since until this technology was developed, signals been not been recorded directly from a human brain for as long with such reliability. With more neuroscience research and rapidly advancing computer technology, a day may ultimately arrive when the integration of some mammalian neural groups and wearable computer technology will occur in ways that now seem like only the realm of science fiction. However, the idea of directly integrating silicon chips with the central nervous system is not without critics. For example, Dertouzous of MIT (1999) recently discussed the difficulty of communicating abstract concepts like "freedom" through silicon channels to the brain. At present we have only a rudimentary knowledge of how the brain processes information and forms abstract concepts like "freedom," "democracy," or "the theory of relativity." Without this knowledge, how can anyone use a brain implant to download such difficult concepts? Obviously, much research will have to occur in the area of neuroscience before we have the capability to integrate chips with the central nervous system. And, as indicated by Dertouzous, even with this capability, it still may not be a good idea.

Other possible future developments associated with computing under the skin include an electronic nose and tongue. The concept of an electronic nose has been explored by researchers at the California Institute of Technology (Bower, 1995; Freund and Lewis, 1995), Stanford University (Kovacs et al., 1994), the University of Warwick (Gardner et al., 1990), and Oak Ridge National Laboratory (Thundat and Wachter, 1999). Thus far, prototype electron noses can detect a wide range of odors. Such sensors are very small and lightweight and may in the future be integrated with humans. As a word of caution, not much is known about the chemical senses (taste and olfaction) and it would be extremely difficult to integrate such technology with the human's biological systems. However, it may be possible that in the next century may see significant advances in this area. Lastly, researchers at the University of Texas (Lavigne et al., 1998) have developed a prototype electronic tongue that crudely mimics the human tongue. Using a combination of four tastes, sweet, sour, salt, and bitter, it can distinguish between several subtle flavors. And unlike the human tongue, it has the capability to analyze the chemical composition of a substance as well. The added capability to actually analyze the chemical structure of a substance makes the potential integration of an artificial tongue or taste sensor with our biological system a real stimulator of the imagination. However, as with olfaction, little is known about the detailed workings of the chemical sense of taste and it would be extremely difficult to directly integrate such technology with the human sensory system.

10. CONCLUSIONS

This chapter has highlighted many of the state-of-the-art and evolving technologies where computer "under the skin" may be used as prostheses or to improve the delivery of a variety of forms of healthcare to patients. With the convergence of digital technology, medical informatics, neuroscience, noninvasive diagnostics such as MRI, and rapidly evolving sensor technology, there is almost no limit to the gains that can be made with regard to dramatically improving medical care to millions perhaps even billions of people on this planet. The possibility exists for enhancing the sensory capabilities of "normal" healthy individuals as well. As wearable computer processing speeds become faster, and the memory and interfaces with the human body grow more robust, great gains will be realized with applying this technology to a wide spectrum of medical conditions, many of which were discussed in this chapter.

Barfield and Baird (1998) have discussed some of the major themes and issues that must be addressed in the design of wearable computers. Wearable computer design, whether created for medical or other applications, must have as a central tenant the question of the usability of the device from a human-centered engineering standpoint. Simply designing a high-tech device that is not usable, or is not user friendly, results in misuse with a wide assortment of operating errors occurring. In some medical application cases, poor human factors engineering design might very well lead directly to injury or death. Serious consideration with regard to the larger physical and cognitive wearable computer's environment that the device must be operated in should also be carefully thought through early in the design process.

The long-term implications of such technology from a population standpoint are less clear. Indeed, for the first time in human history, the first generation of "cyborgs" are already moving about on the Earth. In the near future we can anticipate that many of the practical applications of wearable computer technology noted in this chapter will continue to evolve and contribute positively to more people's lives as time passes. Taken from a longer term evolutionary perspective, the melding of such rapidly advancing computer technology with many individuals' biology may take Homo Sapiens on an evolutionary trajectory that is unpredictable and fraught with both tremendous opportunity and great peril. Are we perhaps now on a journey that will lead to a part of Homo Sapiens becoming Homo Sapiens-Siliconus faster than we might ever have dreamed of? A final thought is worth pondering: Is Homo Sapiens ready for the journey socially, legally, and spiritually?

Acknowledgments   Dr. Woodrow Barfield would like to thank ONR (NOO0149710388) for an equipment grant that has supported research in the area of wearable computers and augmented and virtual reality.

REFERENCES

Axelson, P. (1996). "Responses of members of the SCI community." Journal of Rehabilitation Research and Development, 33, 187.

Barfield, W., and Baird, K. (1998). "Issues in the design and use of wearable computers." Virtual Reality: Research, Development, and Applications, 3, 157–166.

Barfield, W., Hendrix, C., Bjorneseth, O., Kaczmarak, K., and Lotens, W. (1995). "Comparison of human sensory capabilities with technical specifications for virtual environment equipment." Presence: Teleoperators and Virtual Environments, 4, 329–356.

Beardsley, T. (June, 1999). "Seeing the breath of life." Scientific American, 33–34.

Beebe, D. J., Denton, D. D., Radwin, R. G., and Webster, J. G. (1998). "A silicon-based tactile sensor for finger-mounted aplications." IEEE Transactions on Biomedical Engineering, 45, 151–159.

Black, P. M., Moriart, T., Alexander, E., Stieg, P., Woodward, E. J., Gleason, P. L., Martin, C. H., Kikinis, R., Schwartz, R. B., and Joesz, F. A. (October, 1997). "Development and implementation of intraoperative magnetic resonance imaging and its neurosurgical applications." Neurosurgery, 41, 831–842.

Bower, J. M. (1995). "Reverse engineering the nervous system: an in vito, in vitro, and in computational approach to understanding the mammalian olfactory system." In: An Introduction to Neural and Electronic Networks, Second Edition. S. Zornetzer, J. Davis, and C. Lau, editors. Academic Press.

Brunetti, P., Cobelli, C., Cruciani, P., Fabietti, P. G., Filippucci, F., Santeusanio, F., and Sarti, E. (1993). A simulation study on a self-tuning portable controller of blood glucose. International Journal of Artificial Organs, 16, 51–57.

Cameron, T., Liinamaa, T. L., Loeb, G. E., and Richmond, F. J. R. (1998). "Long-term biocompatibility of a miniature stimulator implanted in feline hind limb muscles." IEEE Transactions on Biomedical Engineering, 45, 1024–1035.

Chase, C. D. (May/June, 1999). "Seeing is believing." MIT Technology Review (44–55). MIT Press.

Chokroverty, S., Deutsch, A., Guha, C., Gonzalalez, A., Kwan, P., Burger, R., et al. (1995). "Thoracic spinal nerve and root conduction: A magnetic stimulation study." Muscle and Nerve, 18, 987–991.

Chow, A. Y., and Chow, V. Y. (1997). "Subretinal electrical stimulation of the rabbit retina." Neuroscience Letters, 225, 13–16.

Crago, P. E., Memberg, W. D., Usey, M. K., Keith, M. W., Kirsch, R. F., Chapman, G. J., Katorgi, M. A., and Perreault, E. J. (1998). "An elbow extension neuroprosthesis for individuals with tetraplegia." IEEE Transactions in Biomedical Engineering, 6, 1–6.

Creasey, G., Elefteriades, J., DiMarco, A., Talonen, P., Bijak, M., Girsch, W., and Kantor, C. (1996). "Electrical stimulation to restore respiration." Journal of Rehabilitation Research and Development, 33, 123–132.

Creasey, G. H., and Van Kerrebroeck, P. E. (1996). "Neuroprosthesis for control of micturation." Journal of Rehabilitation Research and Development, 33, 189–191.

Debanne, D., Guerineau, N. C., Gahwiler, B. H., and Thompson, S. M. (1995). "Physiology and pharmacology of unitary synaptic connections between pairs of cells in areas CA3 and CA1 of rat hippocampal slice cultures." Journal of Neurophysiology, 73, 1282–1294.

Deitz, D. (July, 1997a). "Engineering on the biomedical frontier." Mechanical Engineering (pp. 72–73). The American Society of Mechanical Engineering, New York.

Deitz, D. (July, 1997b). "Optimizing orthotic designs with finite-element-analysis." Mechanical Engineering (pp. 70–71). The American Society of Mechanical Engineering. New York.

Delp, S. L., and Loan, J. P. (1995). "A software system to develop and analyze models of musculoskeletal structures." Computers in Biology and Medicine, 25, 21–34.

Delp, S. L., Stulberg, S. D., Davies, B., Picard, F., and Leitner, (1998). "Computer assisted knee replacement." Clinical Orthopedics and Related Research, 354, 49–58.

Dertouzos, M. (July/August 1999). "Brain implants: A lousy idea", MIT Technology Review, p. 25.

Edelman, G. M. (1992). Bright Air, Brilliant Fire: On the Matter of the Mind. Basic Books, New York.

Fang, Z., and Mortimer, J. T. (1998). "Selective activation of small motor axons by quasitrapezoidal current pulses." IEEE Transactions in Biomedical Engineering, 38, 168–174.

Freund, M. S., and Lewis, N. S. (1995). "A chemically diverse conducting polymer-based electronic nose." Proc. Natl. Acad. Sci. U.S.A., 92, 2652.

Gardner, J. W., Hines, E. L., and Wilkinson, M. (1990). "Application of artificial neural networks to an electronic olfactory system." Meas. Sci. Technol., 1, 446–451.

Glenn, W. L. (1980). "The treatment of respiratory paralysis by diaphragm pacing." Annals of Thoracic Surgery, 30, 106–109.

Gough, D. A., and Armour, J. C. (1995). "Development of the implantable glucose sensor: What are the prospects and why is it taking so long?" Diabetes, 44, 1005–1009.

Graupe, D. (1999). "Independent ambulation by complete paraplegics via FES: Design considerations and walking performance." Proceedings of the IEEE International Workshop on Biosignal Interpretation (pp. 51–54). Chicago.

Grimm, W., Flores, B., and Marlinski, F. (1992). "Electrocardiographically documented unnecessary, spontaneous shocks in 241 patients with implantable cardioverter defibrillators." Pacing and Clinical Electrophysiology, 15, 1667–1673.

Gross, G. W., Rieske, E., Kreutzberg, G. W., and Meyer, A. (1977). "A new fixed-array multielectrode system designed for long-term recording of extracellular single unit activity in vitro." Neuroscience Letters, 6, 101–105.

Hansen, M., Hoffer, J. A., Strange, K. D., and Chen, Y. (1997). "Sensory feedback for control of reaching and grasping using functional electrical stimulation." Web site with abstracts from Fifth Triennial Conference on Neural Prosthesis: Motor Systems V, http://fourier.bme.ualberta.ca/~few/IFESS97/tel_ifess97.html.

Hodgins, J. K. (March, 1998). "Animating human motion." Scientific American, 64–69.

Hoffer, J. A., Stein, R. B., Haugland, M. K., Sinkjaer, T., Durfee, W. K., Schwartz, A. B., Loeb, G. E., and Kantor, C. (1996). "Neural signals for command control and feedback in functional neuromuscular stimulation: A review." Journal of Rehabilitation Research and Development, 33, 145–157.

Hoffer, J. A., Strange, K. D., Christensen, P. R., Chen, Y., and Yoshida, K. (1997). "Multi-channel recordings from peripheral nerves: 1. Properties of multi-contact cuff (MCC) and longitudinal intra-fasicular electrode (LIFE) arrays implanted in cat forelimb nerves." Web site with abstracts from Fifth Triennial Conference on Neural Prosthesis: Motor Systems V, http://fourier.bme.ualberta.ca/~few/IFESS97/tel_ifess97.html.

Ison, K. T., and Walliker, J. R. (1987). "Platinum and platinum/iridium electrode properties when used for extracochlear electrical stimulation of the totally deaf." Medical and Biological Engineering and Computing, 25, 403–413.

Jaffe, D. L. (1995). "Rehabilitation Research and Development Center." Technology & Disability, 4, 149–167.

Jaremko, J., and Rorstad, O. (1998). "Advances toward the implantable artificial pancreas for treatment of diabetes." Diabetes Care, 21, 444–450.

Jimbo, Y., Robinson, H. C., and Kawana, A. (1993). "Simultaneous measurement of intracellular calcium and electrical-activity from patterned neural networks in culture." IEEE Transactions in Biomedical Engineering, 40, 804–810.

Jolesz, F. A. (1997). "Image-guided procedures and the operating room of the future." Radiology, 204, 601–612.

Kagaya, H., Sharma, M., Polando, G., and Marsolais, E. B. (1997). "Closed double helix electrode for an implantable functional electrical stimulation system." Web site with abstracts from Fifth Triennial Conference on Neural Prosthesis: Motor Systems V, http://fourier.bme.ualberta.ca/~few/IFESS97/tel_ifess97.html.

Kientiz, K. H., and Yoneyama, T. (1993). "A robust controller for insulin pumps based upon H- infinity theory." IEEE Transactions in Biomedical Engineering, 40, 1133–1137.

Kotovsky, J., and Rosen, M. J. (1998). "A wearable tremor-suppression orthosis." Journal of Rehabilitation Research and Development, 35, 373–387.

Kottke-Marchant, K., Anderson, J. M., Umemura, Y., and Marchant, R. E. (1989). "Effect of albumin coating on the in vitro blood compatibility of Dacron arterial prosthesis." Biomaterials, 10, 147–155.

Kovacs, G. T. A., Hentz, V. R., and Rosen, J. M. (1994). "Project reports." http://guide.stanford.edu/Publications/dev4.html.

Kurzweil, R. (1990). "The coming merging of mind and machine." In: Your Bionic Future, G. Zorpette and C. Ezzell (eds.), Scientific American, 10(3), 56–61. New York.

Lavigne, J. J., Savoy, S., Clevenger, M. B., Ritchie, J. E., McDoniel, B., Yoo, S. J., Anslyn, E. V., McDevitt, O. T., Shear, J. B., and Neikirk, D. (1998). "Solution-based analysis of multiple analytes by a sensor array: Toward the development of an "Electronic Tongue." JACS, 120, 6429–6430.

Lin, V., Singh, H., Chitkara, R., and Perkash, I. (1998). "Functional magnetic stimulation for restoring cough in patients with Tetraplegia." Archives of Physical and Rehabilitation, 79, 517–522.

Lissens, M. A. (1994). "Motor evoked potentials of the human diaphram elicited through magnetic transcranial brain stimulation." Journal of Neurological Science, 124, 204–207.

Loizou, P. C. (Sept., 1998). "Mimicking the human ear." IEEE Signal Processing Magazine, 101–130.

Lui, W., McGucken, E., Vitchiechom, K., Clements, M., Juan, E. D., and Humayun, M. (1997). "Dual unit visual intraocular prosthesis." Proceedings of the 19th International IEEE/EMBS Conference, 2303–2306.

Lyons, D. R. (March, 1993). "Optical Electronic Multiplexing Reflection Sensor System." U.S. Patent 5,191,458.

Lyons, D. R., and Ndlela, Z. U. (Sept., 1996). "Methods of an Apparatus for Calibrating Precisely Spaced Multiple Transverse Holographic Fibers." U.S. Patent 5,552,882.

Maher, M. P., Pine, J., Wright, J., and Tai, Yu-Chong. (1999). "The neurochip: a new electrode device for stimulating and recording from cultured neurons." Journal of Neuroscience Methods, 87, 45–65.

Meister, M., Pine, J., and Baylor, D. A. (1994). "Multi-neuronal signals from the retina—acquisition and analysis." Journal of Neuroscience Methods, 51, 95–106.

Michael, J. W. (1994). "Prosthetic knee mechanisms." Physical Medicine and Rehabilitation: State of the Art Reviews, 8, 147–164.

Moravec, H. (1999). Robot. Oxford University Press, London.

Morris, M. M., Jenkins, J. M., and Carlo, L. A. (1997). "Band-limited morphometric analysis of the intracardiac signal: Implications for antitachycardia devices." Pacing & Clinical Electrophysiology, 20, 34–42.

Mussivand, T. V., Masters, R. G., Handry, P. J., and Keon, W. J. (1996), "Totally implantable intrathoracic ventricular assist device." Annals of Thoracic Surgery, 61, 444–447.

Naples, G. G., Mortimer, J. T., Scheiner, A., and Sweeney, J. D. (1988). "A spiral nerve cuff electrode for peripheral nerve stimulation." IEEE Transactions in Biomedical Engineering, 35, 905–916.

Parrini, S., Romero, E., Legat, V., and Veraart, C. (1997). "A modeling study to compare tripolar and monopolar cuff electrodes for selective activation of optic nerve fibers." Web site with abstracts from Fifth Triennial Conference on Neural Prosthesis: Motor Systems V, http://fourier.bme.ualberta.ca/~few/IFESS97/tel_ifess97.html.

Paul, G. S., and Cox, E. D. (1996). *Beyond Humanity: CyberEvolution and Future Minds*. Charles River Media, Rockland, MA.

Peyman, G, Chow, A. Y., Liang, C., Chow, V. Y., Perlman, J. I., and Peachy, N. (1998). "Subretinal semiconductor microphotodiode array." Ophthalmic Surgery and Lasers, 29, 234–240.

Sabelman, E. E., Kovacs, G. T. A., Hentz, V. R., Rosen, I. M., Kurjan, C., Merritt, P., Scott, G., Gervin, A., and Teachout, S. (1999). "Tactile transducers to replace lost touch sensation." http://guide.stanford.edu/Publications/dev5.html.

Sawan, M., Hassouna, M. M., Li, J. S., Duval, F., and Elhilali, M. M. (1996). "Stimulator design and subsequent stimulation parameter optimization for controlling micturation and reducing urethral resistance." IEEE Transactions on Rehabilitation Engineering, 4, 39–46.

Schmit, B. D., Stellato, T. A., Miller, M. E., and Mortimer, J. T. (1998). "Laparoscopic placement of electrodes for diaphragm pacing using stimulation to locate the phrenic nerve motor points." IEEE Transactions on Rehabilitation Engineering, 6, 382–390.

Sharma, M., Marsolais, E. B., Polando, G., Triolo, R. J., Davis, J. A., Jr., Bhadra, N., and Uhlir, J. P. (1998). "Implantation of a 16-channel functional electrical stimulation walking system." Clinical Orthopaedics and Related Research, 347, 236–242.

Shkurovish, S., Sahakian, A. V., and Swiryn, S. (1998). "Detection of atrial activity from high-voltage leads of implantable ventricular defibrillators using a cancellation technique." IEEE Transactions on Biomedical Engineering, 45, 229–234.

Similowski, T., Fleury, B., Launois, S., Cathala, H., Bouche, P., and Derenne, J. (1989). "Cervical magnetic stimulation: a new painless method for bilateral phrenic nerve stimulation in conscious humans." Journal of Applied Physiology, 67, 1311–1318.

Sinkjaer, T., Riso, R., and Mosallaie, F. (1997). "Nerve cuff recordings of muscle afferent activity during passive joint motion in a rabbit." Web site with abstracts from Fifth Triennial Conference on Neural Prosthesis.

Sternberg, F., Meyerhoff, C., Mennel, F. J. Bischof, F., and Pfeiffer, E. F. (1995). Subcutaneous glocuse concentration in humans: real estimation and continuous monitoring. Diabetes Care, 18, 1266–1269.

Tang, L., and Eaton, J. W. (1993). "Fibrin (ogen) mediates acute inflammatory response to biomaterials." Journal of Experimental Medicine, 178, 2147–2156.

Thundat, T., and Wachter, E. (1999). "Nose on a chip." http://www.ornl.gov/ORNLReview/meas_tech/shrink.html.

Towe, B. C. (1986). "Passive biotelemetry by frequency keying." IEEE Transactions on Biomedical Engineering, BME 33, 905–909.

Warrick, K. (1999). HYPERLINK, http://www.cyber.rdg.ac.uk/K.Warwick/WWW/tech.html.

Wells, W. M., Grimson, W. L., Kikinis, and Jolesz, F. A. (1996). "Adaptive segmentation of MRI data." IEEE Transactions on Medical Imaging, 15, 429–442.

Westwood, J. D., Hoffman, H. M., Robb, R. A., and Stredney, D. (Eds). (1999). *Medicine Meets Virtual Reality: The Convergence of Physical and Informational Technologies: Options for a New Era in Healthcare*. IOS Press, Washington, DC.

Yoshida, K., Jovanovic, K., and Stein, R. B. (1997). "Neural recordings using longitudinal intrafascicular electrodes." Web site with abstracts from Fifth Triennial Conference on Neural Prosthesis: Motor Systems V, http://fourier.bme.ualberta.ca/~few.IFESS97/tel_ifess97.html.

Index